VOLUME XVIII
MICROWAVE MOLECULAR SPECTRA
Walter Gordy and Robert L. Cook

VOLUME XIX
TECHNIQUES OF MELT CRYSTALLIZATION
Gilbert J. Sloan and Andrew R. McGhie

VOLUME XX
TECHNIQUES FOR THE STUDY OF ION–MOLECULE REACTIONS
Edited by M. Farrar and William H. Saunders, Jr.

VOLUME XXI
SOLUBI[illegible]
ORGANI[illegible]
David J. W. [illegible]

VOLUME XXII
MOLECULAR DESIGN OF ELECTRODE SURFACES
Edited by Royce W. Murray

LIBRARY
ITHACA
COLLEGE
The Caroline Werner Gannett Center

TECHNIQUES OF CHEMISTRY

WILLIAM H. SAUNDERS, JR., *Series Editor*
ARNOLD WEISSBERGER, *Founding Editor*

VOLUME XXII

MOLECULAR DESIGN OF ELECTRODE SURFACES

TECHNIQUES OF CHEMISTRY

VOLUME XXII

MOLECULAR DESIGN OF ELECTRODE SURFACES

Edited by

ROYCE W. MURRAY

Kenan Laboratories of Chemistry
University of North Carolina
Chapel Hill, North Carolina

A WILEY-INTERSCIENCE PUBLICATION

JOHN WILEY & SONS, INC.

New York • Chichester • Brisbane • Toronto • Singapore

Library of Congress Cataloging in Publication Data:

Murray, Royce W.
Molecular design of electrode surfaces / Royce Murray.
p. cm.—(Techniques of chemistry; v. 22)
"A Wiley-Interscience publication."
Includes bibliographical references and index.
ISBN 0-471-55773-0
1. Electrodes—Design and construction. 2. Surface chemistry.
I. Title. II. Series.
QD61.T4 vol. 22 1992
[QD571]
542 s—dc20
[541.3′724] 91-25499
CIP

Printed in the United States of America
10 9 8 7 6 5 4 3 2 1
Printed and bound by Malloy Lithographing, Inc.

CONTRIBUTORS

C. P. ANDRIEUX, *Laboratorie D'Electrochimie, Universite de Paris 7,2,Place Jussieu, 75251 Paris Cedex 05, France*

ALLEN J. BARD, *Department of Chemistry, University of Texas, Austin, Texas*

JOHN S. FACCI, *Xerox Webster Research Center, 114/39D, Webster, New York*

ARTHUR T. HUBBARD, *Department of Chemistry, University of Cincinnati, Cincinnati, Ohio*

CHARLES R. LEIDNER, *Department of Chemistry, Purdue University, West Lafayette, Indiana*

MARCIN MAJDA, *Department of Chemistry, University of California, Berkeley, California*

THOMAS MALLOUK, *Department of Chemistry, University of Texas, Austin, Texas*

CHARLES R. MARTIN, *Department of Chemistry, Colorado State University, Fort Collins, Colorado*

ROYCE W. MURRAY, *Kenan Laboratories of Chemistry, University of North Carolina, Chapel Hill, North Carolina*

TAKEO OHSAKA, *Department of Electronic Chemistry, Graduate School of Nagatsuta, Tokyo Institute of Technology, 4259 Nagatsuta, Midori-ku, Yokohama 227, Japan*

NOBORU OYAMA, *Department of Applied Chemistry, Faculty of Technology, Tokyo University of Agriculture and Technology, 2-24-16 Naka-machi, Koganei, Tokyo 184, Japan*

J.-M. SAVÉANT, *Laboratorie D'Electrochimie, University de Paris 7,2,Place Jussieu, 75251 Paris Cedex, 05 France*

GHALEB N. SALAITA, *Department of Chemistry, University of Cincinnati, Cincinnati, Ohio*

LEON S. VAN DYKE, *Department of Chemistry, Colorado State University, Fort Collins, Colorado*

INTRODUCTION TO THE SERIES

Techniques of Chemistry is the successor to Technique of Organic Chemistry and its companion, Technique of Inorganic Chemistry. The newer series reflects the fact that many modern techniques are applicable over a wide area of chemical science. All of these series were originated by Arnold Weissberger and edited by him for many years.

Following in Dr. Weissberger's footsteps is no easy task, but every effort will be made to uphold the high standards he set. The aim remains the same: the comprehensive presentation of important techniques. At the same time, authors will be encouraged to illustrate what can be done with a technique rather than cataloging all known applications. It is hoped in this way to keep individual volumes to a reasonable size. Readers can help with advice and comments. Suggestions of topics for new volumes will be particularly welcome.

WILLIAM H. SAUNDERS, JR.

Department of Chemistry
University of Rochester
Rochester, New York

PREFACE

The notion of synthetically attaching electron transfer and catalytically active molecules to electrode surfaces is now well into its second decade of research and exploration. Interest in chemically modified electrodes shows no signs of abating and indeed novel off-shoots of the original ideas continue to appear each year. There is hardly a facet of electrochemical science that has been untouched by the now-accepted ability to design and control the molecular character of electrode interfaces. The literature of the subject has become sufficiently massive that single review articles in practical terms have great difficulty in explaining all of the important ideas of electrode modification. It thus seems timely to incorporate these ideas into a larger and more detailed volume, which we offer here to interested readers.

I was fortunately able to persuade many of the leading contributors to the development and understanding of molecularly designed electrodes to contribute to this volume. I gave them substantial freedom in their choice of topic. This produced inevitably some overlap between chapters, and a few gaps, but also it produced chapters that reflect the special flavor of insight that writing by leading research scholars entails. I was additionally delighted to find many original insights embedded in the chapters, and insights that clarify current literature issues; I hope that readers both novice and expert will enjoy discovering them.

I prepared Chapter I to set the stage with some of the motivations for molecular electrode design (i.e., why does this subject exist?), an introduction to some concepts of electroreactivity of molecules attached to electrode surfaces, and a general review of chemical tactics for surface attachments and the types of thus attached molecules. Next follows a chapter by Salaita and Hubbard (Chapter II) on using adsorption to immobilize organic species on well-defined electrode surfaces. Professor Hubbard carried out seminal experiments on using chemisorption to modify electrode surfaces. This chapter presents a summary and progress report of his subsequent experiments quantifying and analytically characterizing adsorption in the light of the atomic order of the metal electrode surface and the chemical reactivities that adsorption reflects. Chapter III, prepared by Facci, explains a different type of monomolecular adsorption, which is controlled substantially by hydrophobic interactions. This subject includes so-called self-assembled films and those prepared by spreading and compression of monolayers in Langmuir troughs. This area is a recently blossomed one with a particularly active research profile.

Chapter IV, by Majda, begins descriptions of electrodes coated with multimolecular layers of electron-transfer active sites, so-called redox polymers. These molecular films entail fascinating issues of how electrons hop from donor to acceptor sites and what forces and chemical effects govern the rate at which such electron self-exchange reactions occur. This chapter is followed by Chapter V by Andrieux and Savéant, which presents theoretical aspects and experimental examples of redox and chemical catalysis at redox polymer coated electrodes. The electrocatalytic oxidation–reduction of a substrate species by an electroactive molecular film potentially depends on quite a variety of different kinetic events: electron hopping, permeation, boundary crossing, as well as the catalytic reaction itself. Understanding how these interplay is important in considering applications of modified electrodes to catalysis. Chapter VI, by Bard and Mallouk, describes a relatively recent development in electrode surface design, the use of clays and zeolites as modifying layers. These layers add to the chemical reactivity features of designing electrodes, the element of steric selectivity with which reactants can penetrate the ordered lattices of clays and zeolites.

Chapter VII, by Leidner, describes a particular type of electron-transfer mediation reaction using redox polymer films, designed to reveal their free-energy rate connections. Oyama and Ohsaka, in Chapter VIII, expand on Majda's introduction to charge transport in redox polymers, in describing additional methodology to charge-transport studies and a useful survey of many of the results that have been obtained. The closing chapter, Chapter IX, by Martin and Van Dyke describes another element of molecular design of electrodes, as embodied in electronically conducting polymers. Martin particularly focuses on important aspects of proper measurement of the kinetics of charge transport in these important materials.

I am exceedingly grateful to these scholars for preparation of these chapters on the molecular design of electrode surfaces. I am also grateful to Mrs. Kathy Justice for her help in keeping this project and my office organized, and to my student Hong-hua Zhang for her expert assistance with artwork.

ROYCE W. MURRAY

Chapel Hill, North Carolina
January 1992

CONTENTS

TECHNIQUES OF CHEMISTRY

WILLIAM H. SAUNDERS, JR., *Series Editor*
ARNOLD WEISSBERGER, *Founding Editor*

VOLUME XXII

MOLECULAR DESIGN OF ELECTRODE SURFACES

Chapter I

INTRODUCTION TO THE CHEMISTRY OF MOLECULARLY DESIGNED ELECTRODE SURFACES

Royce W. Murray
Kenan Laboratories of Chemistry,
University of North Carolina, Chapel Hill, North Carolina

1.1 SOME MOTIVES FOR MOLECULARLY DESIGNED ELECTRODE SURFACES

The beauty of electrochemical oxidation and reduction reactions is that one can employ the electrode potential as a source or sink of pure, uncomplicated electrons of flexibly chosen free energy. In a thermodynamic sense, such control permits predictive inciting of oxidation–reduction reactions according to their

Molecular Design of Electrode Surfaces,
Edited by Royce W. Murray. Techniques of Chemistry Series, Vol. XXII.
ISBN 0-471-55773-0

free-energy characteristics. This beauty is indeed realized for so-called reversible, outer-sphere, one-electron-transfer reactions. For example, changing the potential of a Pt electrode in an equimolar, aprotic solution of *p*-benzoquinone and its radical anion to a value slightly more negative than the equilibrium potential, causes a reduction current to flow. First the electrode–solution interphase, and eventually the entire cell solution, become enriched in the reduced form $Q^{\dot{-}}$ to an extent predictable by the applied potential and the Nernst equation. Resetting the electrode potential to the original, equilibrium value, results in an oxidation current that restores the electrode–solution interphase and eventually the entire cell solution to the original equimolar $Q/Q^{\dot{-}}$ composition. The ability to recover the original composition defines this reaction as *chemically reversible*. If the rate of the electron-transfer reaction is faster than any other step in the process (such as diffusion to the electrode), the electrode reaction is also referred to as an *electrochemically reversible* reaction, at least in the practical sense (1).

Then again, the beauty of reversible electrochemistry often gives way to the reality of finite chemical and electrochemical reaction kinetics. Many electrochemical oxidation–reduction half-reactions exhibit slow heterogeneous reaction kinetics and are termed *electrochemically irreversible* (i.e., *not* reversible). Furthermore, the kinetics of a particular electrochemical reaction, and sometimes even the eventual electrode reaction product, can also depend on the composition of the electrode. In that case, the electrode serves as a chemical reagent. For example, the electrochemical reduction of hydrogen ion occurs rapidly and reversibly at a (clean) Pt electrode surface, but its reaction kinetics are very slow at a Hg surface. There is a chemical reason for the difference; an H-atom bound surface intermediate is formed with favorable energetics and kinetics on Pt but not on Hg. On Hg, the electron free energy must be greatly elevated (i.e., more negative potentials applied) to produce a rapid H^+/H_2 reaction; the extra potential is called an *over-potential*. The H^+/H_2 reaction illustrates the role of the electrode surface in electrochemical catalysis, or *electrocatalysis*.

Seeking understanding and control of electrochemical reaction kinetics and catalysis has motivated electrochemical research for decades. Much research in the 1950s and 1960s was directed at choosing different metal alloy electrode surfaces to somehow increase the rates of desired reactions. Many fundamental aspects of electrode reaction kinetics were elucidated during these studies (2, 3). The relation of metal H-atom binding to H_2 reaction kinetics (4) was only one example of how surface binding (*adsorption*) exerts a strong influence on the electrode kinetics of numerous organic and inorganic electrode reactions. A constraint on classical approaches to electrocatalysis of slow reactions was the requirement for an electronically conducting electrode, that is, a metal or carbon. This constraint severely limited the repertoire of surface chemistry available for electrocatalytic purposes.

A new approach to electrocatalysis emerged in the mid-1970s from studies in which electrochemically reactive molecular materials were attached to electrode surfaces (5–13) using systematic synthetic chemistry. Access to molecular electrode surfaces generated, in effect, an enormous diversification in the nature of electrode surfaces. Choice of the applied potential determines whether the attached molecules are in the electron donor or acceptor forms. The electrode reactions of the target *substrate* species now occur on a *molecular* electrode surface, whose reactivity towards the substrate is now a hopefully more *predictable* subject. This is a crucial point; by choice and design of the attached molecules, we have new ways by which to seek appropriate forms of substrate binding, intermediate states, steric effects, and so on, that lead to a rapid electrode reaction of the substrate. With such electrodes, which were dubbed *chemically modified electrodes* (14–16), the problem was transformed from one of surface metal atom chemistry into a problem that could be addressed using rational choices of chemical reactivities of known non-immobilized catalyst molecules that are capable of binding substrates and of delivering electrons. By and large, reactivities of non-immobilized electron-transfer molecules are preserved when immobilized on an electrode. The chemical diversity of electrochemical science was greatly enlarged by this transformation of the problem.

Electrocatalysis of the reduction of molecular dioxygen (O_2) is an important illustration of the benefits of choice and design of molecules attached to electrode surfaces. The synthetic efforts of Collman, Anson, and their co-workers (17–22) produced a dimeric cobalt porphyrin, termed Co_2FTF4 (Fig. 1.1, lower structure) designed, in its reduced state, to be a multiple electron donor and to exhibit an affinity for binding of O_2. When this molecule was immobilized (by chemisorption) on a carbon electrode surface, the resulting modified electrode was shown to effect a four-electron reduction of O_2 to water. Rotated ring–disk voltammetry in Fig. 1.1*b* illustrates this process. The potential of the electrocatalytic O_2 reduction wave reflects that required to produce the oxygen-binding catalytically active porphyrin state (22). The research indicated (20) that the Co(III)Co(II)(Por) (Por = porphyrin) form of the molecule strongly binds O_2. It is possible then that formation of this adsorbed species is the leading step in a subsequent series of proton and electron-transfer reactions that culminate in the formation of water. A crucial point was minimal escape of intermediate states of reduction (viz., H_2O_2) from the porphyrin cavity (wherein the reaction ostensibly occurred). In the absence of the porphyrin layer, with monomeric porphyrins, or, with those of nonoptimum binding characteristics (Fig. 1.1*a*), O_2 reduction on the carbon electrode is both slower and at comparable potentials yields H_2O_2 as the primary (two-electron) reaction product. Other examples of analogously designed four-electron chemistry have since appeared (22, 23).

Electrocatalysis has been a continuing, strong theme of chemically modified electrode research. Its principles and experimental progress are discussed by Andrieux and Savéant in Chapter V. Modified electrode research has taken much more diverse, and still-widening directions, however, both practical and fundamental (5–13). Electrode modification involves chemical synthesis and reactivity studies (24, 25) on thin molecular films, including monomolecular layer films; this topic has been a frontier in other areas as well, such as

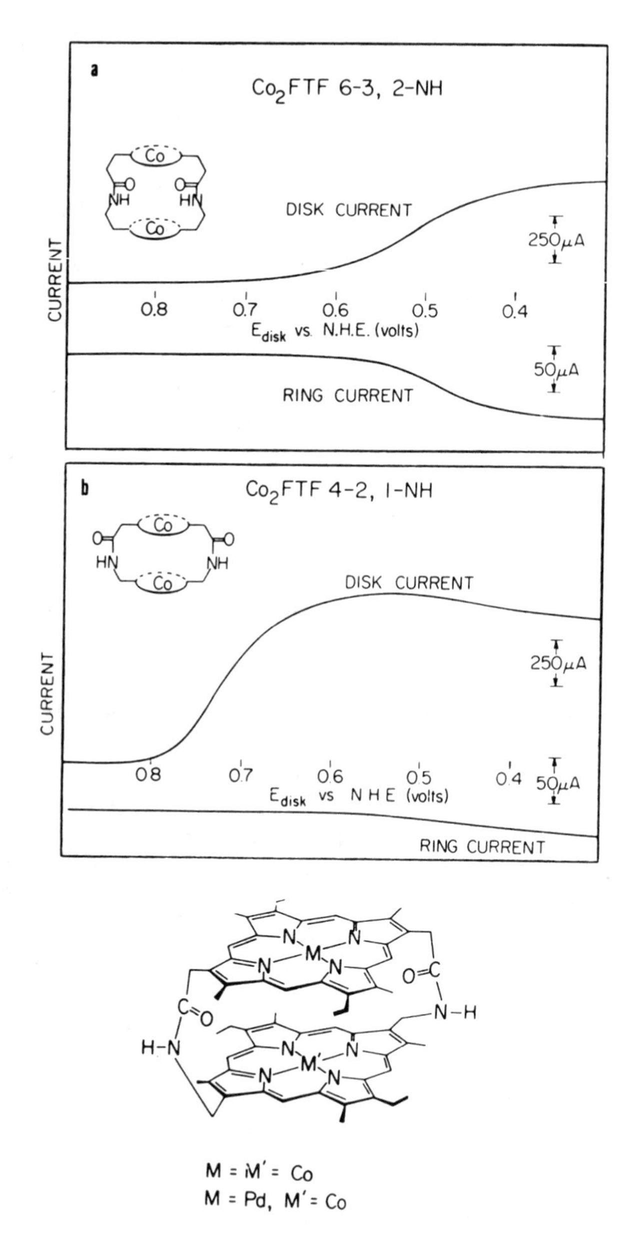

Figure 1.1. Electrocatalytic rotated disk electrode currents versus applied potential for Co_2FTF6 (upper) and Co_2FTF4 (lower) porphyrin (Por) dimers (six atom and four atom, β linked) coated on the rotated graphite disk, in O_2-saturated 0.5*M* CF_3CO_2H, and currents at rotated Pt ring showing that H_2O_2 is generated at the more positive potentials of the disk electrocatalysis in the case of Co_2FTF6 but not in the case of Co_2FTF4. [Reprinted with permission from J. P. Collman, P. Denisevich, Y. Konai, M. Marrocco, C. Koval, and F. C. Anson, *J. Am. Chem. Soc.*, **1980**, *102*, 6027. Copyright © (1980) American Chemical Society.]

biomembranes and the surfaces of polymeric materials. Modified electrode surfaces have been both models and targets for modern and novel approaches to surface analysis and structure (26–40). The introduction (41–46) of electroactive polymer films on electrodes provoked considerable research on synthetic and electropolymerization assembly of polymeric metal complexes (12, 13) and on mass and electron transport inside mixed-valent polymer films (47). It is possible to fruitfully design electrode surfaces with, on the one hand, electroanalysis (48–50) in mind and on the other aiming at photocorrosive protection of semiconductor electrodes (51–53). Controlling the physical spacing between the electrode atoms and an attached electron donor or acceptor (54–58) is an important entré into the basics of long-range electron transfer and of the relation of the "double layer" to electrode reactions. Other forms of spatial control of electrode-attached chemicals lead to rectifying molecular junctions, molecular transistors, and other features of molecular electronics interest (59–60). There has been great interest in electrochemical formation of films of so-called conducting polymers and of their electrochemical properties (61–65).

The following chapters expand upon many aspects of modified electrodes. The purpose of the remainder of this introductory chapter (Chapter I) is to give the reader an outline of the chemistry of electrode surface modifications and their relevance to the various areas of study of modified surfaces. Suffice it to say that the topic of molecular design of electrode surfaces has expanded tremendously beyond its beginnings and is a now strongly established component of electrochemical science.

1.2 ATTACHING AND USING MOLECULES ON ELECTRODE SURFACES

1.2.1 What "Immobilization" Means

Molecular surface design necessarily involves some form of molecular immobilization on the electrode surface. In the modified electrode context, immobilization actually refers to several interwoven goals, not all of which may be successfully attained, or even desired. First, one generally seeks a *physically and chemically stable immobilization* of the target molecule. Characterization of this stability is an important component of researching the surface preparation. Typical criteria are that the "immobilized" species remain on the electrode surface when contacted or washed by an electrolyte solution, and when the potential of the electrode is varied between values leading to useful electroactivity. Important characteristics of immobilized electroactive molecules are their stability upon repeated oxidation and reduction, and good chemical and electrochemical reversibility. An equivalent criterion refers to use in an electrocatalytic reaction; the surface molecule should stably survive many reaction turnovers in the catalytic scheme. Some applications may, on the other hand, actually aim to electrochemically provoke a controlled *detachment* of the immobilized molecules, as in using the electrode surface as a microscopic reagent source (66).

Second, the term "immobilization" is usually employed in a macroscopic sense and does not (necessarily) denote absence of molecular-scale mobility of the surface molecule. For example, there are many cases (5–7) of monolayers of electroactive electron donors or acceptors attached to the electrode surface by insulating but flexible molecular chains. While such donor–acceptor sites may have a noncontacting equilibrium spacing from the underlying electrode, they apparently still possess thermal fluctuations for sufficiently close approach to yield facile electron transfers with the electrode (5–7, 67). These molecules are accordingly immobilized at, but not with respect to, the electrode. In other, and more recent, monolayer cases (54–58), in which the molecular connectors are both insulating *and* nonflexible, the electron donor–acceptor couple may be sufficiently immobilized with respect to the underlying metal electrode surface, that electron transfers are constrained to larger than contact distances. When electrodes are coated with electroactive polymer films, multiple monolayers of the target immobilized molecules are present, some adjacent to but most others remote from the electrode. When the target molecules are counterions of an ion-exchange polymer film, they are immobilized by the film, but may nonetheless diffuse within it (46, 68, 69) to the electrode surface. Thus, "immobilization" is more an operational than a molecularly descriptive term, and there can be many interesting microscopic elements of transport remaining for the researcher of a newly immobilized molecular design to uncover.

Electrode-immobilized molecular species fall into three broad categories: *monomolecular layers*, *multimolecular layers*, and *spatially defined, molecularly heterogeneous layers*. A further categorization can be made with respect to the electronic character of the electron donor–acceptor sites. Polymeric, multilayer species with *delocalized electronic states*, such as poly(pyrrole) (63–65), are usually referred to as *conducting polymers* (61, 62). Molecular layers in which the donor–acceptor sites are electronically well defined and localized as molecular states, such as ferrocene or the dicobalt porphyrin in Fig. 1 are referred to as *redox monolayers* or *redox polymers* (5–13). Most of the material in this volume will refer to electronically localized, immobilized redox species, although conducting polymers are also often used in molecular electrode design.

The manner of preparation and the uses of monomolecular layers, multimolecular layers, and spatially defined, molecularly heterogeneous layers are generally distinctive to each category, as will be indicated in the sections that follow. In our discussion of immobilization chemistry and applications, the intent is to liberally provide appropriate illustrations, but not comprehensively catalog the extensive literature on the subject. We apologize in advance to authors whose work is not included.

1.2.2 Cyclic Voltammetry of Immobilized Redox Molecules

Before proceeding further, it will be useful to outline the cyclic voltammetric behavior of immobilized, electroactive molecules on electrodes. Cyclic voltammetry is the method most commonly employed to characterize the electron-transfer activity of electroactive films.

Figure 1.2 is a commonly used schematic of electron transfers in surface films of electronically discrete redox molecules (e.g., an immobilized ferrocene species). Figure 1.2*B* refers to monolayer films; Fig. 1.2*C* and *D* represent multimolecular layers, and Fig. 1.2*A* shows a cyclic voltammogram of a reversibly reacting redox (Ox and Red) layer (monolayer or multilayer) with

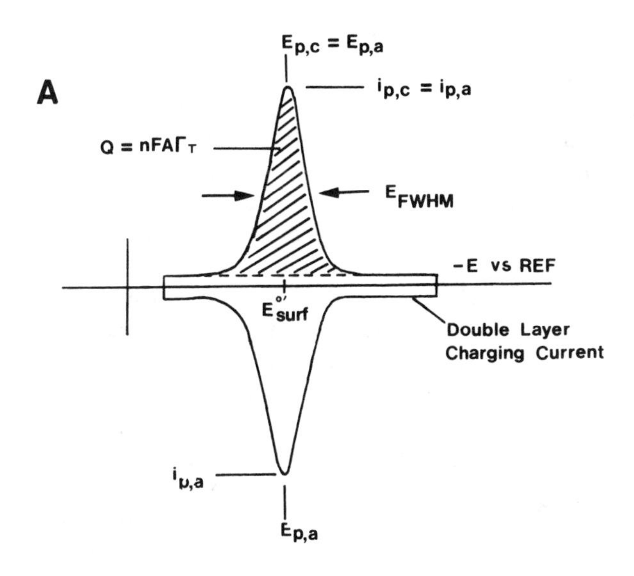

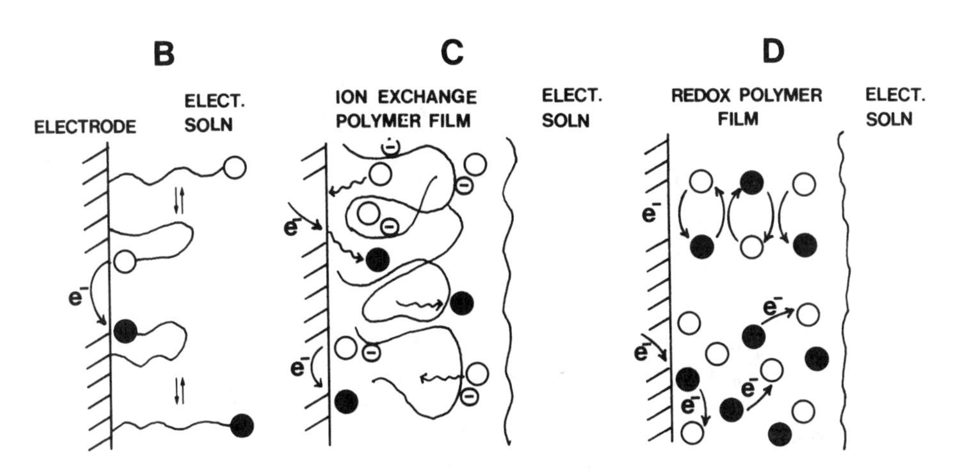

Figure 1.2. Curve *A*, ideal reversible cyclic voltammetric wave of an immobilized redox layer, whether monolayer ($\Gamma_T \lesssim$ ca. 10^{-10} mol cm^{-2}) or multilayer ($\Gamma_T \gg 10^{-10}$ mol cm^{-2}); Curve *B*, schematic reduction of monolayers attached by insulating yet flexible molecular chain; Curves *C* and *D*, schematic reduction of multimonolayers of ion exchange and redox polymers.

formal potential E°_{surf}. Both monomolecular and multimolecular layer reactions have (at least) two parts: transport of the redox molecule within the immobilization matrix to the site of the electron-transfer reaction (whether contact collision with the electrode or some longer range electron exchange), and the electron-transfer step itself. For the monolayer case, transport is a microscopic issue related to the flexibility of the molecular immobilization apparatus as discussed above. For the multilayer case, transport is a more complicated issue since the multiple monolayers of Ox and Red are held within a macroscopically thick film on the electrode. Ox and Red may be confined within the film but free to diffuse within it to the electrode–film interface (Fig. 1.2*C*). This is the case for ion-exchanged electroactive films (Section 1.4.2). Alternatively, if Ox and Red are also attached to the molecular framework (Sections 1.4.1 and 1.4.3) then their electron exchange with the electrode is indirect and depends on a succession of *electron self-exchanges* as shown in Fig. 1.2*D*.

In all of the cases in Fig. 1.2, understanding the electroactivity of the redox molecular layer involves (a) the thermodynamic characteristics of the donor–acceptor couple (i.e., what is E°_{surf} and is it similar to E°_{soln} of an analogous, nonimmobilized molecule) (16), (b) the kinetics of the electron-transfer step, and (c) the kinetics of transport of electrochemical charge (i.e., transport as in Fig. 1.2*C* or *D*) within the immobilized molecular layer.

The simplest immobilized redox layer undergoes a reversible reaction, that is, interlayer transport and electron transfers are fast on the experimental timescale (i.e., in cyclic voltammetry the potential sweep timescale, v volts/sec). In this case, the Ox/Red composition of the redox layer maintains equilibrium with the applied potential and is described by the Nernst equation at each $E_{applied}$. The observed current peak is symmetrical and the same whether observed in a positive or negative-going potential sweep, and for a reduction is given by (1, 5)

$$i = \frac{-4i_p \exp[(n\mathscr{F}/RT)(E - E^{\circ}_{surf})]}{\langle 1 + \exp[(n\mathscr{F}/RT)(E - E^{\circ}_{surf})]\rangle^2} \tag{1.1}$$

where the current peak occurs at E°_{surf} and is

$$i_p = \frac{(n\mathscr{F})^2 A\Gamma_T v}{4RT} \tag{1.2}$$

and Γ_T is total *electroactive coverage* (e.g., the sum of the mol cm^{-2} quantities of Ox and Red, $\Gamma_{Ox} + \Gamma_{Red}$), A is the electrode area, n is the number of electrons, and v is potential scan rate. The significant experimental observables are the proportionality between i_p and potential sweep rate v, equal peak potentials ($\Delta E_p = 0$, see Fig. 1.2*A*), identical waveshapes for the cathodic and anodic current–potential curves, which are called *surface waves*, and the peak width $E_{fwhm} = 90.6/n$ mV (where fwhm = full width at half-maximum).

Examining the proportionality between i_p and v is straightforward and, when obtained, generally means exhaustive oxidation–reduction of the molecular layer's redox sites during the potential sweep. If transport processes are slow, then the i_p versus v proportionality is most closely approached at small rather than large v. The quantity of electroactive material, Γ_T (mol/cm^{-2}), is obtained from the charge under the surface wave (Fig. 1.2*A*). In any real experiment, the surface wave rides above a background current (double-layer charging plus any extraneous faradaic process), which must be subtracted. This is usually done by extrapolating (Fig. 1.2*A*) the background current under the surface wave, which implicitly assumes that the double-layer capacitance undergoes no dispersion with the changing redox state of the film.

The current–potential curve shapes observed for monolayer and for multilayer redox layers often differ from that prescribed by Eq. 1.1. The curve shapes may be symmetrical, but exhibit an E_{fwhm} more narrow or broader than the ideal value. The curve shapes may also be nonsymmetrical, and/or have $\Delta E_p > 0$, and sometimes even exhibit multiple features. Selected examples of modified electrode voltammograms (46, 56, 57, 70–76) are shown in Fig. 1.3. The preparations of these various surfaces are discussed later.

Broadening or narrowing (Fig. 1.3*B*, *F*, and *H*) of the surface redox wave can signal deviations from the assumptions made in Eq. 1.1, that all of the redox sites have the same effective $E^{\circ}_{\mathrm{surf}}$ and that their activities can be approximated with the surface quantities Γ_{Ox} and Γ_{Red}. Which of these two assumptions fails in a specific instance is in fact difficult to ascertain.

There have been several theoretical analyses (77–83) of how surface waveshapes are changed by activity effects. These variously incorporate a quantity-dependent activity, a parameterized interaction between the Ox and Red sites, or rely on other thermodynamic models. In the context of parameterized interactions (80), occupied (reduced) sites with "positive" or "repulsive" interactions tend to avoid one another; in the limit this promotes alternating site occupancy, which is the equivalent of compound formation. The voltammetric peaks in this case are broader than Eq. 1.1. When the interactions are "negative" or "attractive," the occupied sites tend to group together, which leads to phase separation and voltammetric peaks that are sharper than Eq. 1.1. In Fig. 1.3*A*, the ferrocene sites in a monolayer film are diluted (55, 56) by nonelectroactive sites, which minimizes their interactions and leads to near-ideal peak widths and symmetry. Figure 1.3*C*, *F*, and *H* are examples of very narrow and broad waves obtained with multilayer films (71, 74, 76). Their peak widths can vary with solvent (i.e., the solvation of the redox polymer) (74) and with the counterion (Fig. 1.3*H*) that becomes intercalated into the film to maintain electrical charge neutrality during film electrolysis (76). These results show that much information about the thermodynamics of mixed-valent redox systems is available by observations of voltammetric waveshapes.

Figure 1.3*I* is an example where the surface reaction exhibits multiple overlapping peaks (75); this is most readily rationalized as actual differences in $E^{\circ}_{\mathrm{surf}}$ among the redox layer's sites. The example shown, a poly(vinylferrocene) multimolecular layer film, was analyzed (75) in terms of different kinds of slowly

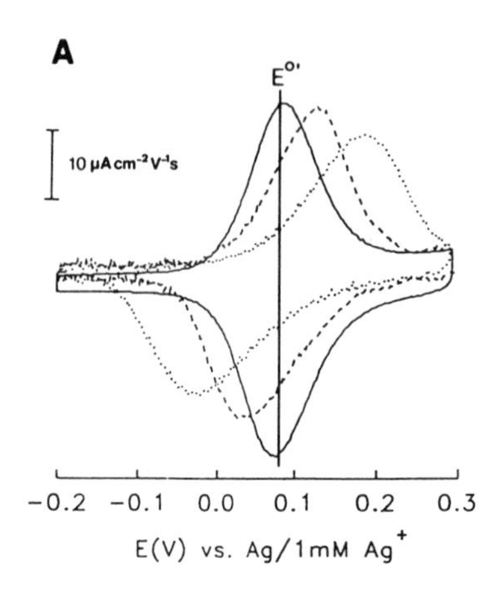
A
E°'
10 μA cm⁻² V⁻¹s
-0.2 -0.1 0.0 0.1 0.2 0.3
E(V) vs. Ag/1mM Ag+

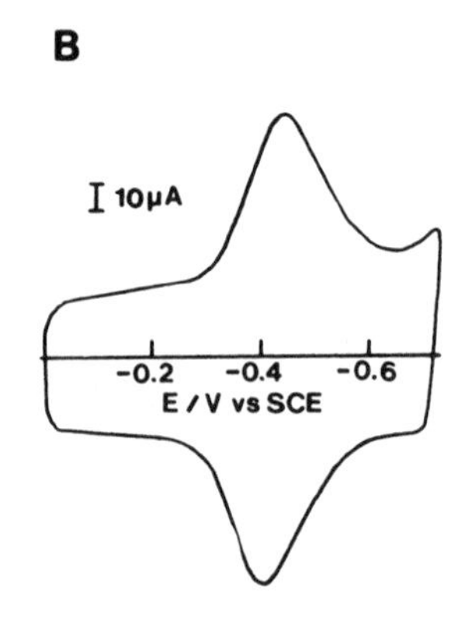
B
10μA
-0.2 -0.4 -0.6
E / V vs SCE

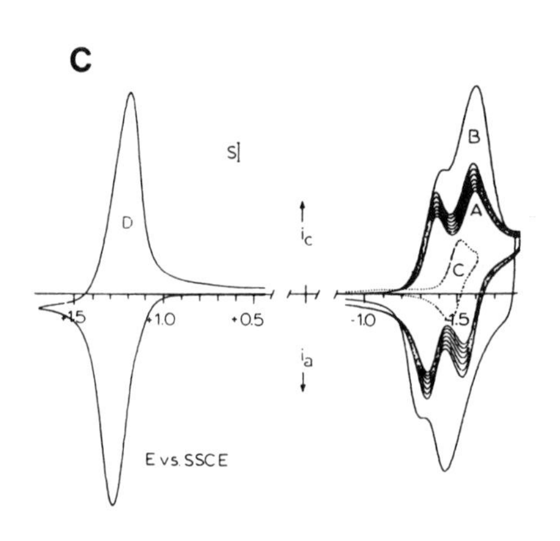
C
S
D
B
A
C
ic
ia
+1.5 +1.0 +0.5
-1.0 -1.5
E vs. SSCE

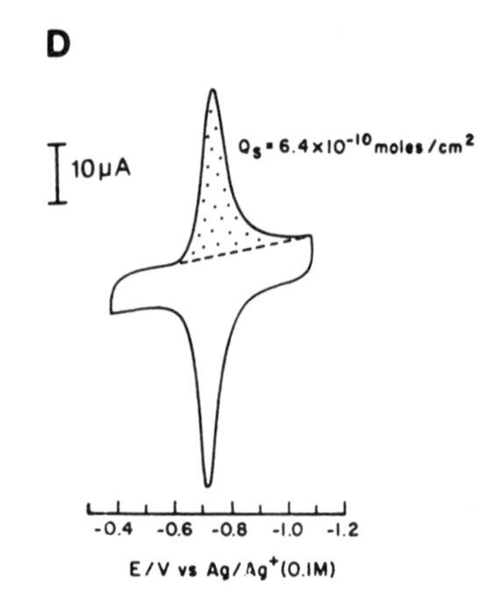
D
10μA
Qs = 6.4×10⁻¹⁰ moles/cm²
-0.4 -0.6 -0.8 -1.0 -1.2
E/V vs Ag/Ag+(0.1M)

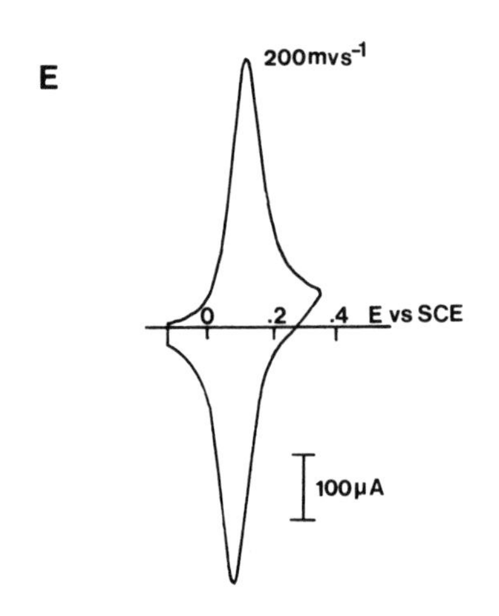
E
200mvs⁻¹
0 .2 .4 E vs SCE
100μA

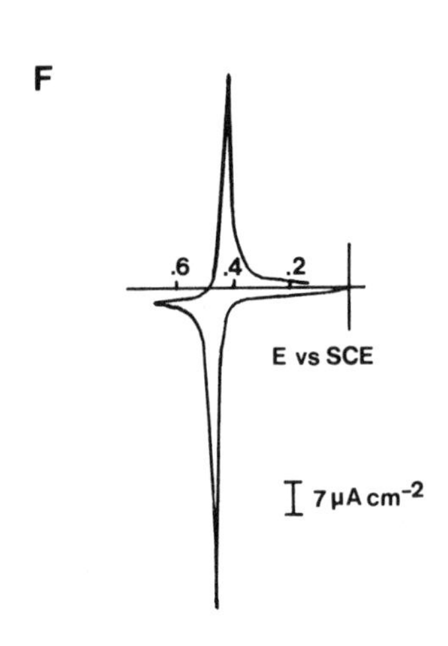
F
.6 .4 .2
E vs SCE
7 μA cm⁻²

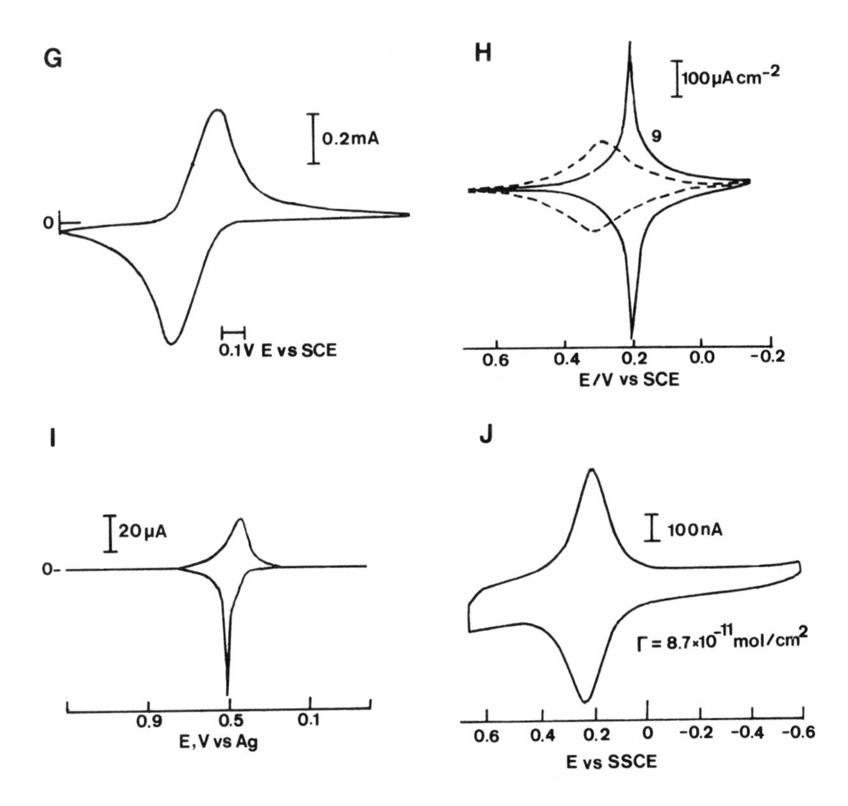

Figure 1.3. Illustrative cyclic voltammograms of electrode surface immobilized redox molecules: *A.* A mixed monolayer of $(\eta^5C_5H_5)Fe(\eta^5C_5H_4)CO_2(CH_2)_{16}SH$ and $CH_3(CH_2)_{15}SH$ self-assembled on Au, in $1M$ $HClO_4$, potential scan rate $10\,mV\,s^{-1}$ (solid), $120\,mV\,s^{-1}$ (dashed), $500\,mV\,s^{-1}$ (dotted), an electron-transfer distance effect; (adapted from Ref. 50); *B.* The LB monolayer film of $[CH_3(CH_2)_{15}—^{+}N\langle\text{pyridine}\rangle]_2$ transferred to conducting oxide (ITO) electrode, $0.1M$ KCl (adapted from Ref. 69); *C.* A series of voltammograms illustrating the electrochemical polymerization and film growth of $[Ru(bpy)_2(vinyl\text{-}py)_2]^{2+}$ (where bpy = 2,2′-bipyridine and py = pyridine) in $0.1M$ Et_4NClO_4/CH_3CN by scanning through the bpy reduction waves [Reprinted with permission from H. D. Abruña, P. Denisevick, M. Umaña, T. J. Meyer, and R. W. Murray, *J. Am. Chem. Soc.*, **1981**, *103*, 1. Copyright © (1981) American Chemical Society.]; *D.* $6 \times 10^{-10}\,mol\,cm^{-2}$ film formed on Au from $(CH_3O)_3Si(CH_2)_2—C_6H_4—CH_2—^{+}N\langle\text{pyridine}\rangle—\langle\text{pyridine}\rangle N^{+}—CH_3$, in $0.1M$ Bu_4NClO_4/CH_3CN, $100\,mV\,s^{-1}$ (adapted from Ref. 72); *E. Multilayer film* ($5 \times 10^{-9}\,mol\,cm^{-2}$) formed on Pt by hydrolysis of *N,N,N′,N′*-tetrakis(trimethoxysilyl-3-propyl)-1,4-benzenediamine (Structure V) in $0.1M$ Bu_4NClO_4/CH_3CN (adapted from Ref. 73); *F. multilayer* ($\approx 10^{-7}\,mol\,cm^{-2}$) film formed on Pt by coupling of aminophenylferrocene to spin coated, hydrolyzed layer of $(CH_3O)_3Si(CH_2)_3NH(CH_2)_2NH_2$, in $0.1M$ Et_4NClO_4/CH_3CN, $2\,mV\,s^{-1}$ (adapted from Ref. 74); *G.* Multilayer film of $[Fe(CN)_6]^{3-}$ ion exchanged into poly(vinylpyridine) film adsorbed on pyrolytic graphite electrode, $0.2M$ CF_3CO_2Na (adapted from Ref. 46); *H.* Multilayer film of Prussian Blue, on Au, in $1.0M$ KCl (solid) and $1M$ RbCl (dashed), $5\,mV\,s^{-1}$ (adapted from Ref. 76); *I.* Multilayer ($8 \times 10^{-8}\,mol\,cm^{-2}$) film of poly(vinylferrocene), MW50,000, on Pt, in $0.1M$ Et_4N(tosylate)/CH_3CN, $10\,mV\,s^{-1}$ (adapted from Ref. 75); *J.* Self-assembled monolayer of $[Os(bpy)_2 \times (dipy)Cl]^{1+}$, where dipy = 4,4′-trimethylenedipyridine, on Pt, in $0.1M$ Bu_4NClO_4/CH_2Cl_2, $100\,mV\,s^{-1}$ (adapted from Ref. 54).

interconverting ferrocene sites in the polymer matrix. It is likely that other examples of dispersion in E°_{surf} exist that are less severe and produce broadening but not splitting of current–potential curves. These are difficult to distinguish from activity effects (see above), especially if the E°_{surf} dispersion is roughly symmetrical about the peak's center. The dispersion may, on the other hand, be nonsymmetrical, such (71, 72) as in Fig. 1.3*C* and *D*, where the steepness of the curves differs on the positive and negative potential sides. Such effects could be caused by a second, but minor, population of redox sites, having a different E°_{surf} because of variations in local structure or environment.

It is also common that ΔE_p of a surface redox wave exceeds zero by a few or by many millivolts. If ΔE_p is observed to increase as the experimental timescale is shortened (faster potential sweep rate), this effect is some kind of kinetic limitation. The kinetic possibilities are slow immobilized molecule–electrode electron transfers (84–88), slow transport (Fig. 1.2*C* and *D*) of the electrochemical charge within the immobilizing layer to the electrode interface (5–13, 89–93), slow transport of charge compensating counterions, and a resulting uncompensated potential drop within the layer (94, 95), or slow changes in solvation or structural organization attendant to a change in redox state (75, 80). Understanding these effects and their chemical origins is a rich part of modified electrode research that appears again in later chapters.

A number of other electrochemical methodologies can also be used to study chemically modified electrodes. Many have been reviewed earlier (5). Some are more and some less sensitive to the waveshape nonidealities mentioned above, and some are more suited for detailed quantitative study than cyclic voltammetry (96). Alternating current (ac) voltammetry is, for example, very sensitive to surface kinetic phenomena over a wide timescale and with the added advantage of phase-selective observations (26, 80, 97–100). Potential-step chronoamperometry (PSCA) and potential-step chronocoulometry (PSCC) are more sensitive to transport effects than to electron-transfer rates or uncompensated resistances at the electrode–film interface, and has been extensively employed to measure charge diffusion in redox polymer films (5, 8, 12, 13, 94, 101, 102). Multiple electrode contacts to redox layers allow steady state current observations, which is especially valuable for charge-transport study when counterion diffusivity effects are to be avoided (13, 47, 59, 103–108). Rotated disks are good to study permeability of the film (as a transport barrier) to other electroactive molecules (109–111).

1.3 IMMOBILIZATION OF MONOMOLECULAR LAYERS ON ELECTRODES

Monomolecular layers can be attached to electrode surfaces through strong, irreversible adsorption (*chemisorption*) or by design and synthesis of covalent linkages to the molecular layer. This is a convenient but artificial distinction since chemisorption certainly involves strong bonding formation. Hydrophobic ordering and adsorption effects can also be invoked to form monolayers.

1.3.1 Chemisorption

Chemisorption is an old topic in electrochemical phenomena. Coating of carbon and other electrodes with metal macrocycles was, for example, extensively researched in the 1960s for the reduction of O_2 in fuel cells, in a search for suitable spacecraft energy technology. A useful set of references to this older work has been given by Yeager and his co-workers (112).

Modern electrode modification with chemisorption attachment chemistry began in 1973 with the pioneering work of Lane and Hubbard (84, 113), who exploited the tendency of alkenes to irreversibly adsorb on Pt surfaces. Three major effects were demonstrated. By attaching ionic moieties to the alkene, thus tethering these ions in the electrode double layer, surface charge effects could be observed in the electrode reactions of several Pt complexes dissolved in the contacting solution. Second, a Pt electrode exposed to 3-allylsalicylate was

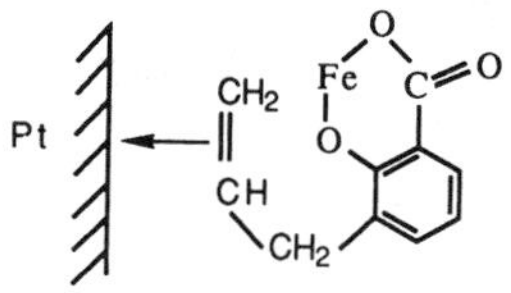

Structure I

shown (by passage of an additional amount of electrochemical charge) to be capable of coordinating and accumulating Fe(II) at the electrode surface, presumably as the complex shown. Third, evidence was given that electrodes exposed to allylamine were capable of assembling an electroactive bromo–Pt

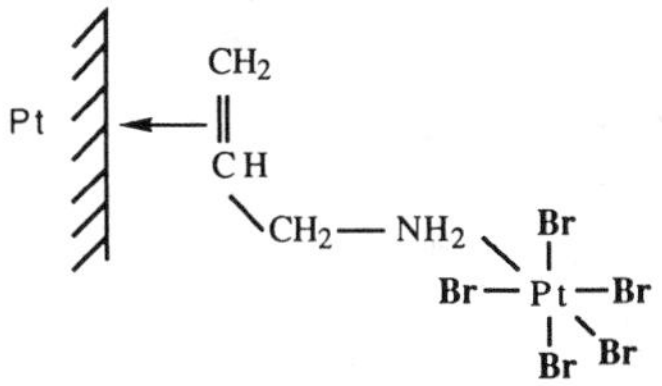

Structure II

complex on the electrode surface, via complexation to the chemisorbed amine ligand as shown.

This theme of chemisorbable moieties linked to interesting target molecules has been extended by Hubbard and his co-workers over a wide range of species and particularly at well-defined Pt electrode surfaces. This work is described in Chapter II.

Chemisorption on carbon is another important route to monolayer immobilization. Molecules with aromatic π systems show a particular proclivity to chemisorb on carbon. Anson and his co-workers provided several examples (88, 114–116) of chemisorption-modified carbon electrodes, including an especially clean extension of the original Lane–Hubbard (84) chemistry showing separate functions of different parts of the chemisorbed molecule. In the case

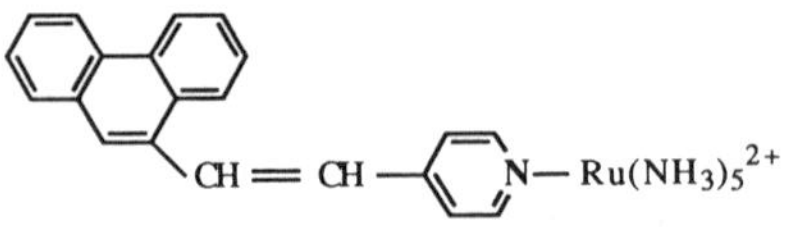

CH = CH — N — $Ru(NH_3)_5^{2+}$

Structure III

shown (116), the phenanthroline moiety provides the carbon surface chemisorption, whereas the Ru complex was bound by its pendant pyridine. Importantly, comparison to appropriate control molecules demonstrated the similarity of the Ru(III/II) formal potentials of structurally analogous surface-immobilized and of dissolved ruthenium complexes.

Most aromatic systems chemisorbed on carbon involve metal macrocycles: porphyrins and phthalocyanines, such as the above-cited example (Fig. 1.1) of chemisorbed Co_2FTF4 dioxygen reduction catalyst. For "flat" macrocycles, carbon surface chemisorption seems to be a rather reliable strategy. For porphyrins with noncoplanar substituents, such as tetraphenylporphyrins, the chemisorption is less strong (more reversible), especially in good solvents for the porphyrin.

Chemisorption experiments are operationally fairly simple. The electrode surface is first cleaned or pretreated in some manner. For Pt, this can involve an ultrahigh vacuum environment and procedures and a single-crystal face of the metal. For carbon, surface structure is less easy to control, and the preferred pretreatment is often established empirically. The electrode surface is exposed to a solution of the adsorbate, and rinsed, after which, if the chemisorption was stable, the cyclic voltammetric redox transformations (or some other property) of the chemisorbed target molecule are observed in fresh electrolyte solution.

In contrast to a relatively good understanding of how alkenes interact with Pt, the binding details of aromatic species to carbon remain somewhat obscure. The literature contains speculations on interactions between the aromatic and carbon basal plane π systems (18) and of oxygenated edge-plane carbon surface functionalities as axial ligands–metal macrocycles (18, 117, 118), both of which imply coplanarity between the adsorbed π system and the carbon surface. On the other hand, Yeager and his co-workers (119, 120) questioned such pictures based on spectroscopic studies of chemisorbed macrocycles. Carbon surface chemisorption work is hampered by the complexity (and diversity) of carbon surface chemistry; perhaps current novel approaches (121) to pretreating the carbon surface will lead to further understanding of its chemisorption chemistry.

Chemisorption has been involved in several studies of redox protein behavior at electrode surfaces. Proteins are especially prone to adsorb on electrodes, which often complicates bioelectrochemical investigations. Sometimes direct electrochemical measurements are possible (122–126). In studies aimed at promoting favorable modes of redox protein adsorption, the groups of Hawkridge (127, 128), Hill (129–131), and later Taniguchi (132, 133) observed that adsorption of nitrogen heterocycles on Au and Cr complexes (134) on carbon electrode surfaces promoted the electroactivity of the redox proteins. These modified surfaces were not electron-transfer mediator-based but rather surface-binding controllers, and provided interesting avenues to study the surface interactions of proteins.

Another important extension of the dual-function Lane–Hubbard chemistry was made by Li and Weaver (54), who chemisorbed thiolated Co(III) complexes onto Au electrodes and observed the rate of the (chemically irreversible) Co(III→II) reaction. The electron-transfer reaction rate in the apparently very

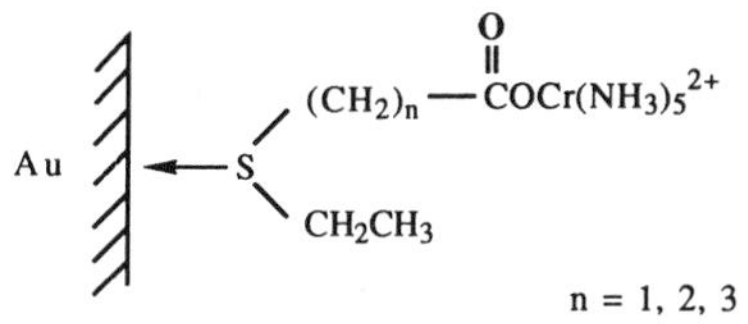

$$\text{Au} \leftarrow S \begin{cases} (CH_2)_n - \overset{O}{\overset{\|}{C}}OCr(NH_3)_5^{2+} \\ CH_2CH_3 \end{cases}$$

n = 1, 2, 3

Structure IV

tightly packed layer ($\Gamma_T = 1.5–2 \times 10^{-10}\,\text{mol cm}^{-2}$) was established to decrease as the number of connecting bonds between Au and Co(III) increased, evidence for successful manipulation of the distance over which electron-transfer occurred. All previously reported monolayer immobilizations on electrodes, by whatever means (5–13), revealed no clean restrictions by the immobilization structure on electron transfer, even though the importance of doing so was early recognized (67). Indeed, the reversible occurrence of electron transfers when using insulating but flexible molecular connectors was called (67) the "floppy model." The Li–Weaver result (54) was the first example of a "stiff" model, as we will term it. Further examples have appeared in reports by Abruña and his co-worker (57) and Majda and his co-worker (58).

1.3.2 Hydrophobic Layers

The "stiff" model theme has become important in recent work in which hydrophobic chain interactions have been invoked to produce structurally organized monolayers on electrodes. As discussed in Chapter III, these monolayers are of two broad types, those relying for molecular organization on hydrophobic effects plus compression as Langmuir–Blodgett (LB) films (7, 58, 70, 135–151) and those based on self-organization (*self-assembly*) of hydrophobic chains with a chemisorbable terminus (55–57, 144–170). In the LB experiment, a long-chain hydrophobic target molecule, which may have an electron donor or acceptor group at one terminus, is spread and compressed as a monolayer at the air–water interface in a Langmuir trough, and transferred to the electrode. Figure 1.3*B* is an example of voltammetry of such a monolayer (70). Langmuir–Blodgett films are ordinarily transferred to electrodes by passing the electrode vertically through the plane of the compressed monolayer, but it has been recently shown that films can also be transferred to a horizontally oriented electrode surface (148–151). Less-well-defined layers can be adsorbed (58) onto the electrode from a nearly saturated solution of the target molecule in a non- or weakly-surface-adsorbing solvent.

Self-assembly is also a powerful method for electrode modification. It has the virtue of operational simplicity, but lacks the molecular layer compression variable (Langmuir trough surface pressure) of the LB experiment. In self-assembled films, chemisorption occurs onto the electrode from solutions of functionalized hydrophobic molecules, such as fatty thiols, sulfides, disulfides, silanes onto Au surfaces, nitriles on Pt, and phosphonates onto metal phosphonates. Figure 1.3*A*, *D*, and *J* show voltammetry of examples of this (56, 57, 72). Structurally organized films prepared by self-assembly onto electrodes can be traced to the work of Sagiv (171–173), who studied films self-assembled from fatty silanes onto silica surfaces.

Some illustrative examples of hydrophobic redox active molecules are shown in Fig. 1.4.

While the LB and self-assembled films are generally deposited and studied while on the electrode, Majda and his co-workers (145–147, 152–154) introduced an important tactic with physical arrangements that place the plane of the oriented molecular layers parallel to the electrode surface. In one scheme, a

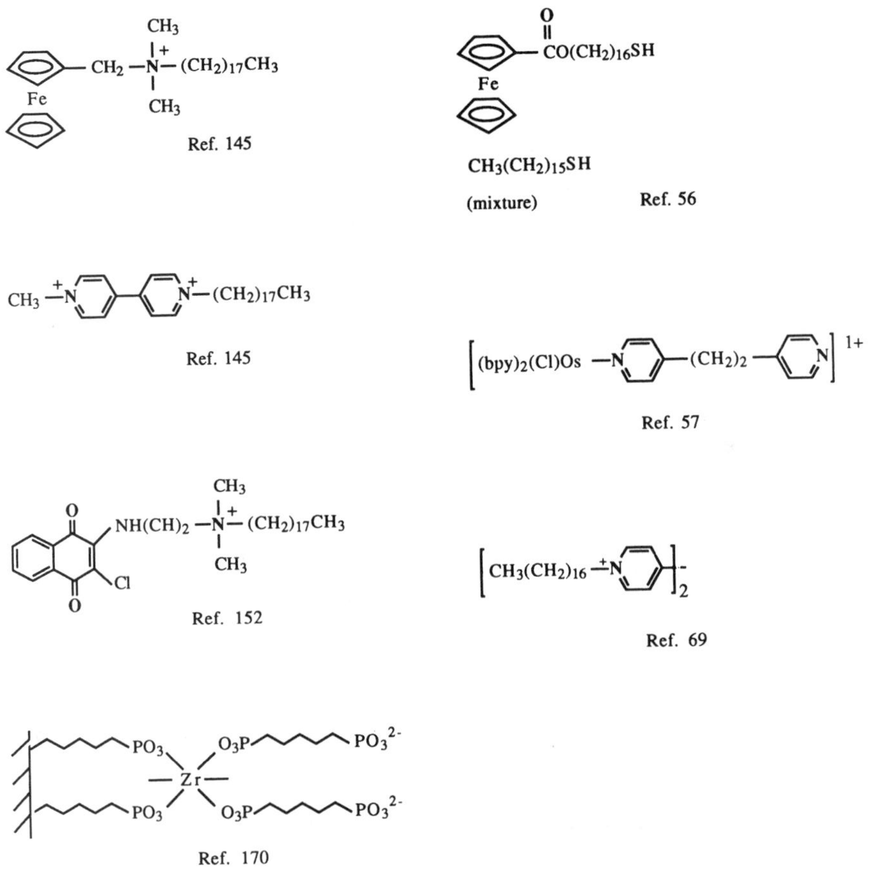

Fe — CH_2 — $N^+(CH_3)_2$ — $(CH_2)_{17}CH_3$

Ref. 145

Fe — $CO(CH_2)_{16}SH$

$CH_3(CH_2)_{15}SH$

(mixture) Ref. 56

CH_3-N^+ (bipyridinium) $N^+-(CH_2)_{17}CH_3$

Ref. 145

$[(bpy)_2(Cl)Os-N \ldots -(CH_2)_2- \ldots N]^{1+}$

Ref. 57

NH$(CH)_2$ — $N^+(CH_3)_2$ — $(CH_2)_{17}CH_3$ (chloronaphthoquinone)

Ref. 152

$[CH_3(CH_2)_{16}-N^+ \ldots]_2^-$

Ref. 69

PO_3 — Zr — O_3P ... PO_3^{2-}

Ref. 170

Figure 1.4. Examples of redox molecules using hydrophobic chains for ordering in monolayer films.

Al_2O_3 film with roughly cylindrical pores was placed over the electrode, and hydrophobic films were self-assembled on the inner surfaces of the Al_2O_3 pores (152, 153). In a second scheme (154), an electrode was contacted to the edge of a hydrophobic film compressed in a trough at the air–water interface. In both experiments, *lateral* diffusion of redox species within the molecular film was studied, and was found to be rather rapid.

Another interesting twist in ordered layer formation (Fig. 1.4) was introduced by Mallouk and his co-workers (170), based on the formation of alkyl phosphonate complexes with Zr(IV). First, an alkyl sulfide that was terminated with phosphonic acid groups was chemisorbed onto a Au surface. (An alkyl silanol similarly terminated was also used, on Si/SiO_2.) The exposed "phosphonic acid surface" was next complexed with Zr(IV), and the residual Zr coordination sites on this surface then reacted with another phosphonic acid or a bis(phosphonic acid). This succession of reaction steps produced a plane of Zr alkyl phosphonate coordination that could be detected and that apparently enforced further order in the film. Thus, specific chemical reactivity can be used as a tactic adjunct to hydrophobic effects, in ordered film formation.

Langmuir–Blodgett and self-assembled redox species have a substantial potential in fundamental studies of electron-transfer environment and distance effects at electrode surfaces. The fatty ferrocene, at the top of Fig. 1.4, has been recently shown (56) to cochemisorb and self-assemble with $CH_3(CH_2)_{15}SH$ onto Au surfaces, producing the voltammetry in Fig. 1.3*A*. At slow potential scan rates, the current–potential behavior is nearly reversible, but at higher scan rates a peak potential splitting appears that reflects an electron-transfer rate limitation. These electron transfer, and other capacitance effects (163), are consistent with achieving the "stiff" model form of immobilization, that is, the ferrocene sites are relatively immobile with respect to the electrode surface (55, 56). The electron transfer is interpreted as a tunneling event, and density of state effects are accounted for (56). Abruña and his co-worker (57) described another example of apparently distance-modulated electron transfer in a self-assembled osmium bipyridine complex film (Fig. 1.4). The study of "stiff" hydrophobic structures promises to be quite fruitful.

Hydrophobic interactions (actually the disruption thereof) can also be exploited (164) as a method for depositing organic films on electrodes. Micelles based on ferrocene surfactants have been used to solubilize a variety of substances in water, which are released as films on electrodes when the ferrocene surfactant is oxidized and the micelle thereby disrupted.

1.3.3 Covalent Bonding to Electrode Surface Functionalities

The useful chemistry of covalent bonding between a molecular monolayer and an electrode is determined by the natural or inducible functional groups available on the electrode surface. On (metal and metal oxide) electrodes with oxide surfaces, metal hydroxyl is a natural terminator of the oxide phase. For carbon, the edges of basal plane sheets tend to be terminated with oxidized sites including carboxylic acid groups. These two kinds of surface functionalities have

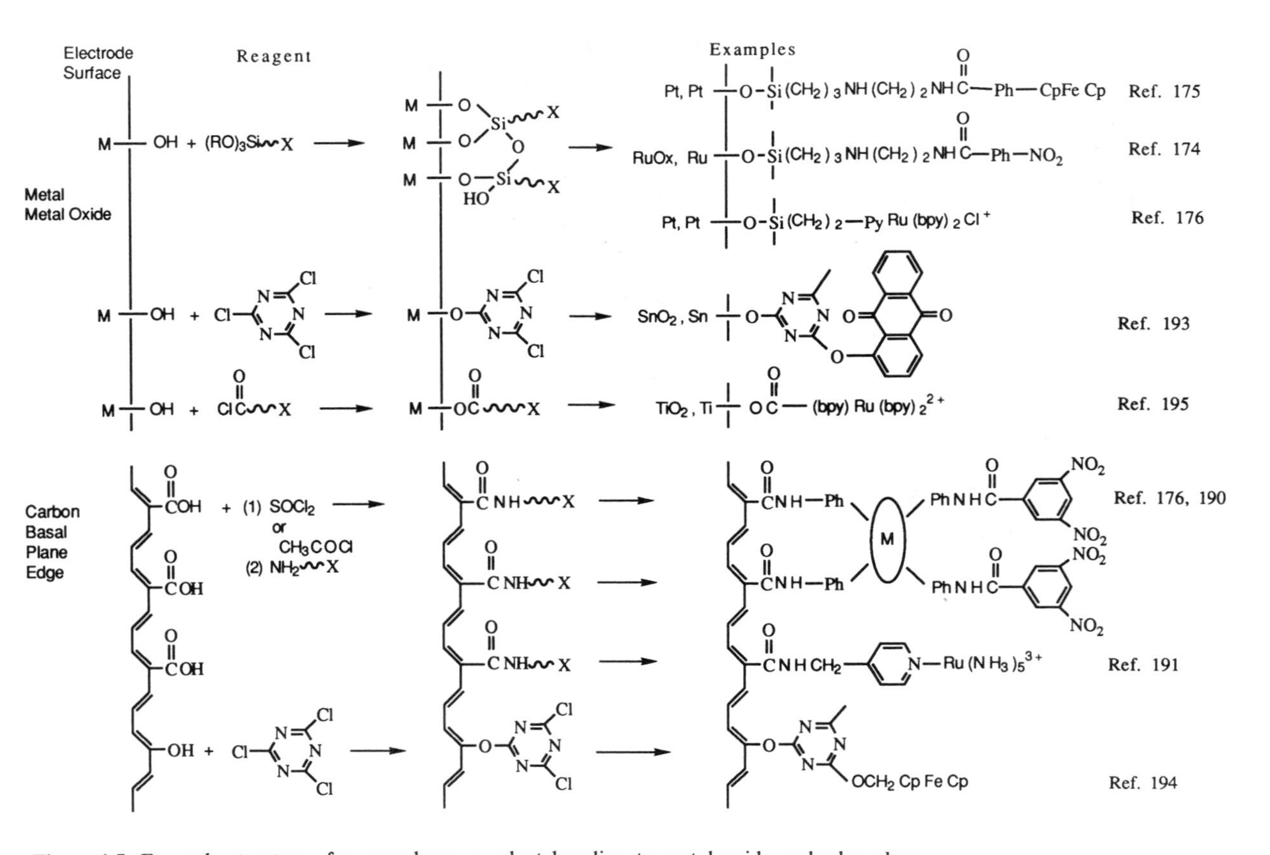

Figure 1.5. Example structures for monolayer covalent bonding to metal oxide and edge plane pyrolytic graphite electrodes.

led to a fairly versatile monolayer surface bonding chemistry (5–11), examples of which are given in Fig. 1.5.

Most modifications of metal oxide surfaces have relied (5) on reactive organosilane reagents (*silanization*). Chlorosilane and alkoxysilane reagents react readily with the hydroxyls present on the surfaces of conducting bulk metal oxide electrodes (14, 67, 174) like RuO_2, doped SnO_2, doped TiO_2, and the superficial (few monolayer) oxide layer on Pt metal electrodes (175, 176). The organosilane reagents typically have multiple chloro- or alkoxysilyl groups, which can result in a microheterogeniety in the surface bonds as shown schematically in Fig. 1.5 (top). When such bonding heterogeneity occurs, the surfaces lack precise control of the molecular ordering, but at the same time the multiple surface bonds enhance the generally good stability exhibited in aprotic media by silane-bonded monolayer films. (Stability is, however, limited in aqueous media.) Multiple reactive Si—OR or Si—Cl groups on the reagent can also lead to premature oligomerization of the reactive silane reagent, prior to its encounter with and bonding to the surface. This can produce multiple monolayers on the electrode. Premature oligomerization reactions are minimized by using strictly aprotic solvents and well-controlled reagent purity (177), but sometimes the reagent is just so reactive that multilayers are routinely obtained nonetheless (43, 177–181).

To produce an interesting electrode surface, the reactive silane should bear an electrochemically or chemically active functionality ("X" in Fig. 1.5, top). Among choices of commercially available silane reagents, "X" can be the nominally electroinactive amine, nitrile, vinyl, halide, pyridine, or acyl halide. Silanization of metal hydroxyl surfaces with such reagents produces, for example, an "amine surface," that can be subsequently amidized with carboxylic acid derivatives of redox species. The many possibilities of this particular silanization reaction have been explored (5) and examples of surface structures produced are shown in Fig. 1.5 (top). Since multiple reaction steps are required to build such silane-based monolayers, and monolayer synthesis is uncommon, special attention was given to characterization of these layers. For example, the reaction yields (24, 182) were investigated, in addition to counting the surface electroactive site population with cyclic voltammetry (see above). X-ray photoelectron spectroscopy was heavily used in these investigations (24, 177, 181, 183, 184), being sensitive enough to detect monolayers with ease and selective as to element and oxidation state.

Alternatively, "X" in Fig. 1.5 can be a synthetically incorporated redox moiety, such as those designed (Structure V) by Wrighton and his co-workers (43, 73, 178–181, 185–188). These reagents provided a second versatile approach to covalent surface modification. The tetraalkylphenylenediamine structure shown produced the voltammetry in Fig. 1.3*E*, this being also an example of how reagent oligomerization yields multilayer coverage by the electroactive species. Such multilayer coverage has of course, positive aspects, since these reagents lead to an important class of redox-polymer modified electrodes.

Carbon electrodes, whether fractured or polished in air, invariably bear some

$Si(OCH_2CH_3)_3$

Fe

$[(CH_3O)_3Si(CH_2)_3]_2N-C_6H_4-N-[(CH_2)_3Si(OCH_3)_3]_2$

$(CH_3O)_3Si(CH_2)_3-{}^{+}N \quad N^{+}-(CH_2)_3Si(OCH_3)_3$

Structure V

reasonable population of carboxylic acid groups on their surfaces. On edge plane pyrolytic graphite surfaces, which are generally rather rough, carboxylic acid sites are probably the dominant functionality. The behavior of glassy carbon indicates that the carboxylate surface population is also considerable there, since monolayer level immobilization of redox reagents is observable. Nonetheless, for use in electrode immobilization, carbon surface carboxylic acid sites must be activated in some manner, which can be done with thionyl chloride or acyl halide (Fig. 1.5, bottom). The activated sites are then reactive towards amine or hydroxyl-functionalized target reagents, provided in the synthetic protocol they are used immediately so as to avoid hydrolytic decay.

Carbon electrodes were first modified in this manner by Miller and his co-workers (189), using amide bonds and chiral attached reagents. A variety of different redox reagents (5–8, 11, 16, 190–192) have been attached via amide bonds to carbon (both glassy and edge plane graphite), starting with the work of Lennox and Murray (190) and as illustrated by the structures in Fig. 1.5 (bottom). In the porphyrin example shown, it was demonstrated that multiple amide bonds were formed between the carbon surface and the tetra(amino)-phenylporphyrin redox moiety (26). The number of bonds formed with the carbon surface is probably influenced by the steric register of amine sites on the redox molecule with the acid chloride sites on the carbon surface (5). For the case shown, this problem was investigated (26) by reacting (and thus labeling) unused porphyrin amine sites with a nitro group containing acid chloride reagent, which produced the tagged structure shown. Counting the different kinds of N 1*s* peaks on the carbon surface (amide + porphyrin vs. nitro) with X-ray photoelectron spectroscopy established that only two of the four porphyrin amine substituents (on the average) formed bonds with the surface.

Carbon surfaces also bear groups (apparently hydroxyl) that stably couple with cyanuric chloride (Fig. 1.5). Metal oxide surfaces also undergo this coupling. The residual aryl halide functions can subsequently be coupled to hydroxylated redox species, such as anthraquinone (193) and hydroxymethylferrocene (194), as shown in the figure. This chemistry has not been widely explored, but given the diversity of reactions that have been studied with this reagent (in dye chemistry), there appear to be additional directions available with cyanuric chloride for surface modifications. Molecular surface binding sites (such as the shape-selective cyclodextrin) have been immobilized using ester linkages to carbon (195, 196). Ester coupling to metal oxides (Fig. 1.5) has also been explored (197) but such bonds are not hydrolytically stable (193).

Following the suggestions and experiments of Mazur et al. (198) on the reactivity of thermally cleaned (deoxygenated?) edge plane carbon electrodes, two other tactics have been explored for producing so-called "oxide-free" carbon. In one case, "oxide" (carboxylic, hydroxylic, etc.) groups were removed by Ar plasma bombardment, followed by exposure of the activated surface to ammonia (199) or to amine reagent gases (200). The most successful reagent gas was diethylenetriamine, which produced (200) a surface with an "amine" reactivity that could be coupled to test redox probes. In the other study, the carbon was mechanically abraded in the presence of a vinyl reagent potentially capable of coupling to exposed carbon radicals at the fracture sites. Such coupling was achieved with vinylferrocene and vinylpyridine (201). While these two "oxide-free" procedures were reasonably effective in a practical sense, the molecular details of their carbon surface attachment chemistry are not known. For fundamental purposes these chemistries are consequently not as appealing as others involving more well-defined reactivities.

1.4 MULTIMOLECULAR LAYER FILMS ON ELECTRODES

The introduction of electrochemically reactive polymer materials was an important development in molecularly designed electrode surfaces. This work started with reports from Merz and Bard (41) and Miller and Van de Mark (42) on the electrochemistry of thin films of poly(vinylferrocene) and poly(4-nitrostyrene), respectively, deposited on electrodes. Publications from other groups (43–46, 177, 202–204) quickly followed, and research on polymeric and other multimolecular layer films rapidly grew and diversified over the following decade.

Electroactive polymers offer several significant advantages, and some less certain ones, relative to monomolecular layers on electrodes:

1. The electrochemical signals from charging of films containing multiple monolayers of redox sites are larger, and consequently easier to detect and study. Recall the relation between current and electroactive coverage Γ_T in Eq. 1.2; it applies equally well to monolayers and multilayers, provided its requirements are satisfied. Since many monolayer electrochemical responses exhibit (Fig. 1.3) meager signal-to-background (charging) current ratios, this is a significant advantage.

2. Until recently (54–58), no heterogeneous electron-transfer rate constants between immobilized molecular monolayers and the adjacent electrode had been successfully measured, since they were too fast (5, 6, 88). In electroactive polymeric films composed of electronically discrete redox sites (as opposed to delocalized ones as in "conducting polymers"), electrons are transported through the polymer by hopping between electron donor and acceptor sites, as illustrated in Fig. 1.2*C* and by the reaction,

$$\text{DONOR AT } X_1 + \text{ACCEPTOR AT } X_2 \Rightarrow \text{ACCEPTOR AT } X_1 + \text{DONOR AT } X_2 \qquad (1.3)$$

Electron hopping from position X_1 to X_2 (generating an electrode-detected current flow) is axtually an *electron self-exchange reaction.* When the nature of this electron-transport process was realized (89–93, 205, 206), the way was opened to study the electron dynamics of immobilized molecules that had (at least for monolayers thereof) been previously unfruitful. The study of electron hopping in electroactive films on electrodes has since been an intensively pursued topic, the current state of which is discussed in Chapter IV.

3. The stability of a molecular mediator film is a crucial practical issue with respect to its potential use in electrocatalysis. It is widely perceived that multilayer films survive larger numbers of oxidation state turnovers than do monolayer films, implying that redox sites in multilayer films have an enhanced stability. This may often in fact not be a real stability effect, since the characteristics of electrocatalytic mechanisms at modified electrodes (Chapter V) make it difficult to distinguish between truly enhanced molecular site stability in multilayers, and the necessity to lose (by decay) a much larger fraction of a multilayer film than a monolayer film, before significant loss in activity is witnessed. Nonetheless, the larger currents from multilayer films do make electrocatalysis measurements easier.

4. When used for electrocatalysis, the presence on the electrode surface of a large electron-transfer mediator molecule population should in a first-order analysis yield a large reaction flux between mediator and the reaction substrate. This does occur, but only if the rate of the mediator–substrate reaction is slow relative to (a) the rate of permeation of the substrate throughout (so as to react with) the entire polymer film, and (b) the rate of the electron-hopping delivery of electrons between electrode and mediator sites. When neither of these conditions is met, it turns out that the rate of the net mediator–substrate reaction may be actually no faster than that generated by a monomolecular layer of mediator. The overall rate of an electrocatalytic reaction is, in other words, subject in very important ways to competing transport steps. These have been the subject of extensive research and are fully discussed in Chapter V.

5. Another useful aspect of polymer films is their role in assembly of spatially defined microstructures on electrode surfaces, described in Section 1.5. Microstructures are of molecular electronic interest and are also useful measurement tools for study of the polymers themselves.

The use of multilayer films on electrodes raises some issues not apparent (or less obvious) with monolayer films, that experience continuing investigation. First, the electrochemical charging of a redox polymer film (i.e., oxidation–reduction of multiple monolayers of redox sites) requires for electroneutrality an accompanying influx or egress of charge compensating counterions from the adjoining solution. The transport rate of these counterions can influence the rate of charging. If their diffusivity is less than that of the electron [its effective diffusivity via Reaction 1.3 is called the electron diffusion coefficient D_e], then during a transient charging, an internal electrical field in the polymer is created that enhances the Reaction 1.3 electron hopping rate (94, 95). This coupling of electron and counterion transport process can, if present, complicate interpretation of the transient charging currents. Our laboratory has investigated the electrical field dependency of the electron hopping (207, 208). Second, polymer films are solvated and swelled by the solvent according to their specific chemical nature, any cross-linking present, and importantly, their oxidation state. Accordingly, there can be substantial (but not readily anticipated) solvent effects on the electrochemical charging, ion and electron transport, and accordingly, electrocatalytic behavior of the polymer film. Third, the electroactive polymer film does not have to be chemically attached to the electrode surface, merely adherent to it. Little is known about the actual details of the polymer–electrode interface. Electron transfers seem to occur rapidly there, but whether polymer chain folding and mobility, swelling, and charged redox and/or counterion population differ between the electrode–polymer interface and bulk are experimentally unknown.

We now turn to an illustrative survey of electroactive multimolecular layer materials that have been applied to electrodes. Several other reviews of this topic are available (12, 13, 209).

1.4.1 Redox Polymers

Redox polymers consist of electronically localized electron donor and acceptor sites that are bonded to a polymer chain, or are linked together to form a polymeric chain. They include both strictly organic systems and polymers containing metal complex or organometallic sites. Redox polymers were known (210) as a class of materials long before research on modified electrodes began. Modified electrode research has, however, produced substantial diversification of the available redox polymer materials and a better understanding of their electron-transfer reactivity.

Redox polymers can be preassembled and then applied as a film to the electrode surface, or they can be assembled from monomers directly as a film on the electrode. Both approaches have been widely researched and each offers certain advantages. Both offer substantial synthetic versatility. Generally, larger amounts of material are available through direct synthesis of the redox polymer, which allows a relatively better analytical and structural characterization. On the other hand, fabrication of very thin, uniform films from preassembled redox polymers can be difficult, since they are often multiply charged and reluctantly soluble. In situ assembly of redox polymer films by hydrolytic or electrochemical polymerization of monomers can yield superb thin-film forming characteristics, but usually at the expense of a less thorough analytical characterization since only ultrathin film is available.

1.4.1A Preassembled

The book by Cassidy and Kum (210) summarizes a number of tactics for attaching target molecules to polymer chains. Such chemistry is also used for heterogenizing of homogeneous catalysts (211). The common pathways are polymerization of vinyl-substituted redox monomers, pendant attachment of redox monomer to a suitably functionalized linear polymer, and condensation polymers involving redox monomers. Illustrative examples of redox polymers made in this way (89, 212–222) are found in Fig. 1.6.

Only a few vinyl-substituted redox monomers have been preassembled into polymers. Poly(vinylferrocene) and poly(4-nitrostyrene) were used in the early studies of redox polymer films on electrodes (41, 42). Poly(nitrostyrene) is highly reactive and not very stable in its radical anion state. Poly(vinylferrocene) (Fig. 1.6) gives reasonably stable voltammetry and its transport characteristics

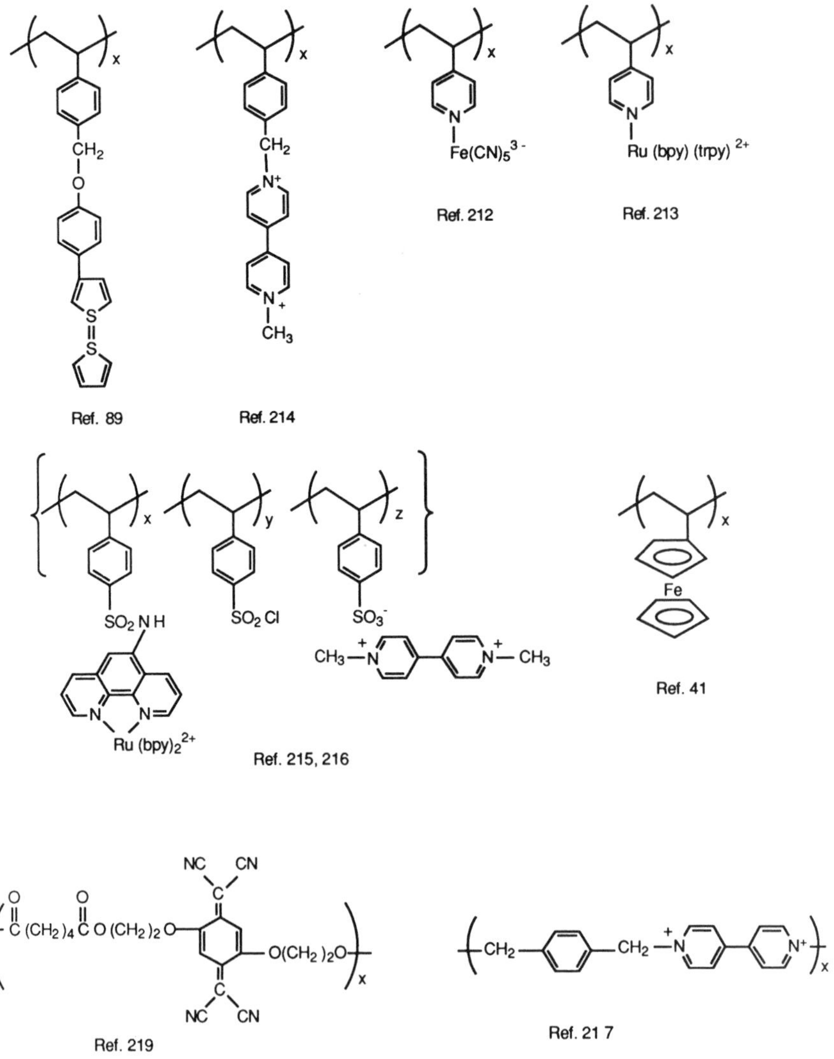

Figure 1.6. Example structures for pendant and condensation redox polymers.

have been widely investigated, including electron transport (75) and counterion transport, and solvent swelling and transport (223). Metal poly(pyridine) complexes with vinyl or acrylate (224, 225) substituents can be thermally polymerized but in the case of the vinyl-substituted complexes thin films are more effectively made via electrochemical polymerization (12, 226) see below).

Redox polymers made by condensation reactions (13, 217, 219, 227–230) are illustrated in Fig. 1.6 (bottom). Poly(xylylviologen) is formed (217, 228–230) from the reaction between α,α'-dibromoxylene and 4,4′-bipyridine, and the TCNQ (where TCNQ = tetracyanoquinodimethane) hydroxy derivative shown via esterification with a bis(acyl halide) (218–222, 231).

Pendant group attachment is a very flexible synthetic strategy, wherein well-characterized (and sometimes conveniently available) functionalized, linear polymers are reacted with suitably substituted redox molecules. Examples are shown (89, 212–216) in Fig. 1.6; the parent polymer reactivities were based on chloromethyl, pyridine, and sulfuryl chloride substituents. Pendant group coupling can also be based on acyl halide chemistry. The redox polymers that have been prepared in this way comprise a large body of materials. Besides those redox moieties shown in the figure, pyrazoline (232), anthraquinone (233), *o*-orthoquinone (234), and a variety of Ru complexes (25, 213, 215, 215, 235–244) have also been affixed as pendant groups to linear functionalized polymers. Mixtures of redox moieties can be incorporated.

It now seems reasonable to assert that, via direct synthesis, any suitably functionalized redox molecule should yield to incorporation into a polymer chain and should exhibit electroactivity as a film on an electrode. In the film's electrochemistry, however, the electron-transfer dynamics of the chosen redox site will determine whether *multi*layer electrochemical reactivity is actually observed. If electron self-exchange between redox sites is very slow, or is coupled to slow chemical steps (i.e., proton or ligand transfers), then although a multilayer film is present, potentially only a monolayer or two of the redox sites may electrochemically react on a reasonable timescale (245). That is, the elements of increased electrochemical currents, electrocatalytic reactivity, electroanalytical capacity, and so on, potentially attendant to multilayer films are attained *only* by fast electron-transfer dynamics of the chosen redox donor–acceptor couple.

Electron-hopping rates additionally depend on the concentration of the redox sites in the polymer, since electron self-exchange is a bimolecular reaction. Site dilution (i.e., spatial separation) additionally can degrade the donor–acceptor electronic coupling. Redox polymers can differ substantially with respect to redox site concentration and whether dilution of sites by nonactive parts of the polymer occurs in a controlled or uncontrolled fashion. In pendant group polymers, reactions of pendant sites with redox molecule are often rarely quantitative, so the pendant redox sites are diluted (hopefully at random) by unreacted pendant sites or in some cases their hydrolytic products. The redox sites are in contrast not seriously diluted in polymerized vinyl monomers or in condensation polymers, unless nonelectroactive co-monomers have been deliberately incorporated in the polymerization reaction medium This is also the case with hydrolytically or electrochemically polymerized redox monomers, again provided nonelectroactive co-monomers are not added. This characteristic has to be balanced against the synthetic flexibility offered by direct synthesis of pendant group redox polymers.

1.4.1b Assembled in Situ

There are four general kinds of in situ deposition of polymer films onto electrodes from solutions of their monomers, two lead to redox polymer films, and two to electronically conducting and insulating films.

Redox polymer films can be made in situ by the hydrolytic polymerization and by the electrochemical polymerization of redox group containing monomers and subsequent (or concurrent) binding (or precipitation) to the electrode surface. The chemistries of making these films were introduced by the Wrighton (51–53, 60, 73, 178–181, 185–188, 246–257), and Murray–Meyer (71, 103, 104, 107, 110, 258–277) groups, respectively, and many significant contributions have been made by others (226, 278–301).

The hydrolytically reactive monomers introduced by Wrighton and his co-workers are based on attaching reactive organosilane functionalities to redox couples such as ferrocenes and viologens (see above, Structure V). As discussed above, these reagents can be employed to prepare bonded monolayers on metal oxide electrodes, but their hydrolytic reactivity is so great that many electrode preparations yield multilayer amounts of the redox groups. The hydrolytically produced multilayers generally exhibit good electron-transfer activity; Fig. 1.3*E* displays that for a 1,4-diaminobenzene derivative (73).

Redox monomers used in electrochemical polymerization contain some polymerizing moiety that is activated by a change in the monomer's oxidation state. Figure 1.7 shows some examples (71, 107, 260, 263, 266, 275, 276, 271, 294, 300). Oxidation of the tetra(aminophenyl)porphyrin shown can be viewed as oxidation of an aniline derivative with a porphyrin substituent. Reduction of a metal poly(pyridine) complex to the (formal) metal(I) or metal(0) state is actually reduction of the metal-bound pyridine or bipyridine ligand, whose "vinyl-substituted radical anion" then undergoes polymer-forming coupling reactions. Such electropolymerization chemistry is reasonably reliable, and its applicability is governed by the extent to which functionalities like those in Fig. 1.7 can be synthetically attached to the desired redox moiety.

The metal(III/II) voltammetry of electropolymerized metal poly(pyridine) complexes is particularly robust, and films of these complexes and of the polymerized tetraphenylporphyrins have seen extensive study of electron transport and electrocatalytic behavior as detailed in later Chapters IV, V, VII, and VIII. Less investigated has been the polymerization reaction itself. Electropolymerization is actually a rather complex series of events whose relative kinetics are important. The process involves diffusion of monomer to the electrode, activating reduction or oxidation, and then competing diffusion away

$M(vbpy)_3{}^{2+}$ (M = Ru, Os, Fe, Zn) Ref. 260

$M(bpy)_2(vpy)_2{}^{2+}$ (M = Ru, Os) Ref. 71

$Re(vbpy)(CO)_3Cl$ Ref. 276

$M(bpy)_2(p\text{-}cinn)_2{}^{2+}$ (M = Ru, Os) Ref. 263

$[Ru(5\text{-}NH_2\text{-}phen)_3]^{2+}$ Ref. 266

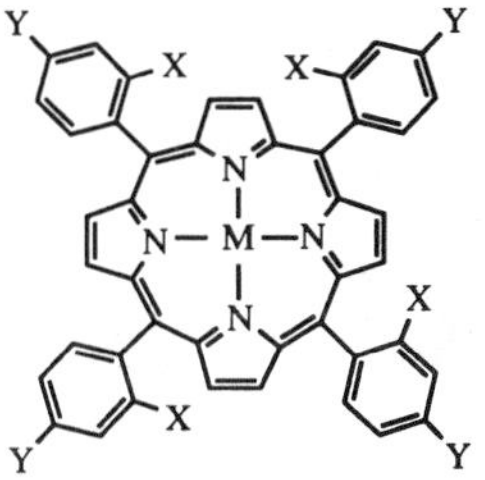

(M = Fe, Co, Mn, Zn, Cu, H_2, Ni, Ru)

(X or Y(not both) = $-NH_2$, -OH, —N(pyrrole))

Ref. 107, 271, 275

$[Ru(bpy)_2(py\text{-}CH_2\text{-}pyrrole)_2]^{2+}$ Ref. 300

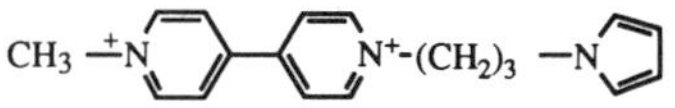

$CH_3\text{-}^{+}N$(bipyridinium)$N^{+}\text{-}(CH_2)_3$ —N(pyrrole) Ref. 294

Figure 1.7. Example structures of electrochemically polymerizable redox monomers, where vbpy = 4-vinyl-4′-methyl-2,2′-bipyridine, vpy = 4-vinylpyridine, p-cinn = *N*-(4-pyridyl)cinnamamide, 5-NH_2-phen = 5-amino-1,10-phenanthroline, py-CH_2-pyrrole = 3-(pyrrol-1-ylmethyl) pyridine.

from the electrode versus oligomerization and precipitation onto the electrode (270). The efficiency of deposition (of converting activated monomer to deposited polymer) can vary widely depending on the diffusion–reaction competition, but the details of this process have not been quantitated. Also generally unknown is the degree of cross-linking of the multifunctional monomers, the typical chain lengths, and the incidence of microstructure or clustering. It was earlier thought (267) that vinyl couplings of the metal poly(pyridine) complexes, lead mainly to terminated dimers, but Elliott et al. (226) presented recent evidence that longer chain lengths may be involved. An impediment to the characterization of these polymeric materials is the thinness and general insolubility of their films.

Another significant feature of electropolymerized films is their relative freedom from defects that are much larger than molecular size (i.e., pinholes). Pinholes are, apparently, efficiently filled during electropolymerization, because current fluxes and thus populations of activated monomer are enhanced at the pinhole sites. Because the monomers that make up the films are rather bulky, they continue to exhibit permeability to small molecule and counterion permeants that pass through the interstices between monomer sites. Bulky permeants can, however, be strongly excluded by electropolymerized films. This means that the films act as molecular sieves, admitting permeants according to their molecular size, as has been shown for metal poly(pyridine) (109, 111) and tetra(aminophenyl)porphyrin films (110). This permeability characteristic can also be used to force the locus of electrocatalytic reactions with bulky substrates out to the polymer–solution interface (259, 261, 265, 302, 303), as discussed in Chapter VIII.

Electronically conducting polymer films made by electrochemical polymerization of monomers like pyrrole, thiophene, and aniline comprise an important class of materials. The oxidative electropolymerization of pyrrole reported by Diaz et al (63–65), for example, yields black polymer films exhibiting ohmic conductivities that are generally much larger than those obtainable with redox polymers (although important exceptions exist) (225, 304–308). The important feature of poly(pyrrole) is the electronic charge delocalization that exists along the chain of α,α'-linked, fractionally oxidized pyrrole units. While activated electron-hopping events are still required for interchain electron exchange and at defects (chain ends, cross-linked sites, and counterion coulomb wells) the overall electron-transport process in a "conducting polymer" is more facile than by electron hopping alone between, for example, a mixture of ferrocene and ferricenium groups in poly(vinylferrocene). Conducting polymers have been extensively studied and reviewed (61, 62) and are further discussed in Chapter IX.

Electrochemical polymerization of many monomers, an example being phenol (111, 309–324), produce adherent polymer films (often quite efficiently) that are nonconducting in *both* the electron-hopping and delocalization sense [e.g., poly(phenylene oxide), an insulator]. Such electropolymerization reactions tend to be *passifying*, meaning that the electrochemically driven film growth slows and eventually stops. Passification occurs when, (a) owing to the absence of suitable electronic states, there is no electron transport across the film to the fresh monomer pool driving at the film–solution interface, and (b) when permeation of monomer through the film to the electrode–film interface is slow owing to low permeability of the polymer. In the case of poly(phenylene oxide), film passivation is strongly dependent on electropolymerization conditions, and film growth may be just slowed (318–323) or it may be stopped altogether (111, 324). Such characteristics make passifying films of interest as protective coatings and as selective transport barriers.

1.4.2 Ion-Exchange Polymers

Coating an ion-exchange polymer onto an electrode surface, and incorporating ion redox species into it as its counterions, is an elegantly simple route to electrode modification. This strategy was introduced by Anson and Oyama and their co-workers (46, 68, 69, 93, 325–339), who in their initial reports on films of poly(vinylpyridine) adsorbed on carbon electrodes, observed that, in aqueous acid–dilute $[Fe(CN)_6]^{4-}$ solution, currents for $[Fe(CN)_6]^{4-/3-}$ voltammetry gradually increased as the electrode potential was repeatedly cycled. The (protonated) cationic polymer, acting as an anion exchanger, gradually incorporates the anionic metal complex at the expense of the less highly charged electrolyte anion. Placing the electrode in a fresh electrolyte solution causes gradual out-partitioning, which can be slowed or halted by adding a minute concentration of $[Fe(CN)_6]^{4-}$ to the solution. Making electroactive films on electrodes in this manner has been since extensively explored (239, 340–373).

Forming an electroactive, ion-exchanged film requires (a) a method for casting or coating an adherent film of the ion-exchange polymer onto the electrode and (b) the use of a solvent–electrolyte system that adequately swells the ion-exchange phase and in which redox counterion partitioning is favored on either kinetic or equilibrium grounds. The optimum solvent is generally water, so most of the literature deals with aqueous media. The ion-exchanging procedure offers the advantage that, once a ion-exchanger film-forming method is devised, that film can be employed with a wide variety of target redox ions, derivatization of which is not required but a multivalent charge usually is. It is important to rember that the stability of in-partitioning depends on the concentration and type of supporting electrolyte employed in subsequent voltammetry. High electrolyte concentration and multiply charged electrolyte ions are usually undesirable.

Examples of ion-exchange polymers that have been employed on electrodes are shown in Fig. 1.8. Of these, poly(vinylpyridine) and Nafion have seen the most attention. Poly(vinylpyridine) can be adsorbed onto electrodes from solutions of high molecular weight material (46, 68, 69), cast from solution droplets, and cross-linked in place for added stability with reagents such as 1,2-diiodoethane. Poly(vinylpyridine) can be quaternized before or after film forming, or simply used in acidic medium. Other quaternized nitrogeneous base

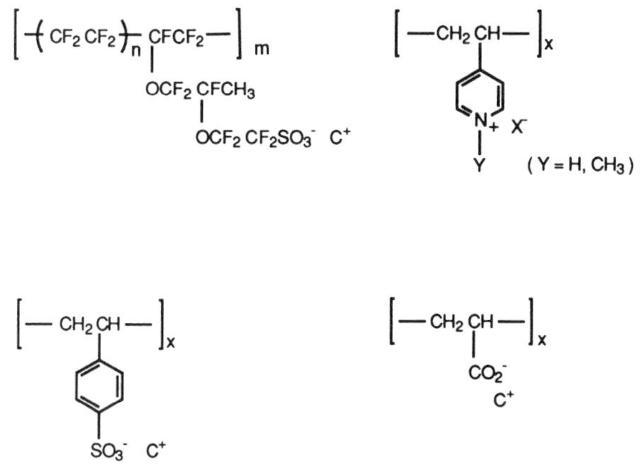

Figure 1.8. Example structures of ion-exchange polymers into which redox counterions can be exchanged.

polymers, including viologens, can be handled in similar fashion (326, 330). Hydrogels like poly(lysine) can also be employed in ion-exchanging films in aqueous media (327, 369).

The sulfonated perfluorinated poly(ether), Nafion, shown in Fig. 1.8 was developed by DuPont as a highly stable cationic ion conductor membrane for synthetic electrochemical applications. Analogous structures have been pursued by other organizations (367), and cationic analogues are also known (333). Nafion is available in membrane form, which can be too thick for electrode coating purposes, but which can be dissolved and recast by a procedure reported by Martin and his co-workers (365–367). Nafion is also available in a low molecular weight (1100) form, as a polymer solution. Proper annealing of films cast from these solutions is important for film stability. Nafion is a particularly good illustration of the importance of a swelling solvent (i.e., water); dehydration of the Nafion phase yields a hard, brittle phase with very poor transport properties.

The electrochemical behavior of the ion-exchange-based electroactive polymers is generally similar to that of redox polymers (see above). Important differences between these classes of electroactive polymers arise, however, from the existence of physical diffusivity of the redox counterions within the ion-exchange phase. The transport of electrochemical charge (i.e., donor or acceptor states) occurs by both electron self-exchange between donor and acceptor states and by physical diffusion of the redox ion. Research directed at understanding this interesting situation is discussed in Chapter IV. In an important paper, Buttry and Anson (335) showed that transport of $[Co(bpy)_3]^{2+}$ during the $[Co(bpy)_3]^{2+/1+}$ reaction is aided by Co(II/I) self-exchange, whereas its transport during the $[Co(bpy)_3]^{2+/3+}$ oxidation is by physical diffusion alone. The difference in behavior is attributable to the much larger self-electron exchange rate constant for the former reaction.

A second consequence of redox ion physical diffusivity in ion-exchange polymer films is that their electroactivity is observable at much lower concentrations than is typically possible with (nondiffusive) donor–acceptor sites bound to redox polymers. That is, the demands of electron self-exchange bimolecularity and electronic coupling in redox polymers are relaxed by the physical mobility of the redox site.

A third important difference is in the electrochemical response stability. Decay of a few monolayers of redox sites at a redox polymer–electrode interface can choke off the electroactivity of the more remote sites, whereas in the ion exchangers, the redox site diffusivity would ordinarily minimize the consequences of loss of a few redox sites.

A special characteristic of Nafion and related perfluoro polymers is their internal phase segregation. This polymer has a relatively high density of charged hydrophilic sites, has long segments of hydrophobic chain, and is not rigidified by cross-linking. The charged sulfonate sites consequently segregate themselves from the hydrophobic chains into water-filled pockets and channels, that are surrounded by coiled hydrophobic perfluoro-poly(ether) chains. The microscopic phase segregation greatly modifies Nafion's ion-exchange character in that in-partitioning is governed by *both* charge and hydrophobicity of the redox ion (349). Transport properties are also affected, since a hydrophobic redox cation can during its diffusion (and/or electron self-exchanges), partition *internally*, between water-filled anion sites and the perfluoro phase. The phase segregation of Nafion thus adds both an interpretive complexity and a transport flexibility in its applications to electroactive films.

Finally, there have been a number of studies using polymer–polymer complexes (i.e., redox polymers with polymeric counterions) (217, 370–372), and of polymeric electrolytes in experiments in the electroactive polymer films (305–308). Oyama and his co-workers (370, 371) attempted to manipulate the ion-exchange property (towards a monomeric redox ion) of an ionic redox polymer (i.e., viologen) film by forming a redox polymer–poly(styrene sulfonate) ionomer complex film on an electrode. Elliott and his co-workers (305–308) and Pressprich (373) used soluble ionomers as electrolyte ions in schemes aimed at controlling entry–egress of charge-compensating counterions into redox polymer films during their electrolysis. These tactics seem to be promising probes of how ionomer counterions interact and of how this affects their transport characteristics.

1.4.3 Inorganic Polymers

Electroactive films can also be prepared on electrodes from inorganic lattice-forming reactions. These films are mainly represented by the metallocyanates, the first example of which were Prussian Blue films reported by Neff (374) and Itaya et al. (375). Electrochemical transformations of the intensely colored Prussian Blue ferric ferrocyanide to Prussian White (reduction) and to Prussian Yellow (oxidation), and of related films, have been studied in some detail and have been reviewed (376). Charge transport and further compositional investigations have also been made (76, 377–379). Films of other metallocyanates, such as the particularly well-studied (29, 36, 380–383) nickel hexacyanoferrate by Bocarsly and his co-workers (384), can be made by oxidation of a metal film (e.g., Ni) in the presence of a ferricyanide solution. This versatile approach can be extended to other metals as well. The metallocyanates differ from the redox and ion-exchange polymers (see above) in several important respects: (a) The potentials of the redox sites (e.g., ferricyanide) are generally strongly perturbed by their incorporation into the lattice compound, in contrast to the potentials of redox polymer sites that are generally little altered from their dissolved analogues. (b) Their redox potentials are also strongly influenced by guest, charge compensating, counterions that are intercalated into the lattice during electrochemical charging (76, 380, 381). (c) The inorganic polymer lattice also exhibits strong size–charge selectivity among different charge-compensating counterions. This feature leads to electroanalytical applications in detection of the preferred counterion species.

Tungsten oxides and oxometallates provide additional examples of multilayer films based on inorganic structures. Faulkner and his co-workers (385–

388) showed that electroreduction of a solution containing tungsten oxide suspension and chloroplatinate produces an electroactive tungsten oxide–Pt microparticle film with interesting electrocatalytic properties, particularly for hydrogen evolution. Nadjo and his co-workers (389–391) entrapped oxometallates such as $[SiW_{12}O_{40}]^{4-}$ in polymeric ion-exchange films and report remarkable electrochemical stability for these electroactive films. These are interesting new directions for modified electrode electrocatalysis.

1.5 HETEROGENEOUS AND SPATIALLY DEFINED LAYERS AND MICROSTRUCTURES

A wide assortment of electrode coatings have been devised based on electroactive films that are heterogeneous in some manner (nonuniform in composition or structure), or are based on *multiple layers* of uniform composition or *multiple contacting electrodes*, or both. These include films with particles deliberately added during film preparation (such as zeolites and clays as discussed in Chapter VI), or formed in situ within a polymer film (such as metal particles dispersed in polymer film) (186, 187, 250, 251, 254–256, 392, 393), multiple layers of redox polymers (called bilayer electrodes) (59, 71, 104, 258, 261, 262, 394, 395), redox polymers sandwiched between electrodes (sandwich electrodes) (59, 103–105, 107), or resting in the interelectrode gaps of interdigitated array electrodes (59, 60, 106, 108, 144), combinations of lithographically defined electrodes and redox polymers that act as molecular transistors (60, 396–406), redox polymers or surfactants coated on porous electrodes as a membrane between two electrolyte solutions (ion gate electrodes) (59, 62, 407–410), porous aluminum oxide films on electrodes the pores of which can be filled with redox polymers or coated with self-assembled films (145–147, 152–154), (Chapter III) films of porous polycarbonate (Nuclepore membranes) combined in a variety of ways with thin-film electrodes and electroactive polymer films (Chapter IX), electrodes coated with thin films of porous (sol gel-formed) glass (411), and carbon (particle) paste electrodes containing redox species dissolved in the paste or adsorbed on the carbon (412–415). We briefly discuss some of these arrangements and the reasons underlying their design.

Zeolite and clay particles can be coated on electrodes through casting from colloidal dispersions, electrophoretic deposition, adsorptive effects, and using various polymeric materials as binders. The first reports on clay particle-coated electrodes were by Ghosh and Bard (416) and on zeolites by Rolison and co-workers (417) have been joined by papers from a number of other workers on clays (418–429) and zeolites (430–439), and this is an active research topic. Zeolites had been used previously in other ways in electrochemical cells (440). The key features of these inorganic lattice electrode coating materials, as discussed in Chapter IV, are their shape and size selective cavities. Clays exhibit sheet or layerlike structures, and zeolites have well-defined pores and channels. In both cases, one gains the prospect of molecular recognition effects in the design of electrocatalysts for specific substrates and of analytical devices for specific analytes. Most of the studies to date have been concerned with the mechanism(s) of incorporation of redox species or reaction substrates into these particle-based films on electrodes and the manner in which they interact with their environment including charge transport.

The incorporation of metal particles into polymer films on electrodes aims at gaining the catalytic activity of such particles in certain electrochemical reactions. The tactic was introduced by Wrighton et al. (186, 187) who formed the metal particles by ion exchanging a metal complex into a redox polymer (e.g., a viologen polymer), then used the redox polymer to reduce the complex to the metal form by reducing the redox polymer. Metal particles in films on p-type Si semiconductor electrodes were exploited for the photoelectrocatalytic reduction of hydrogen ion (186, 187, 250, 251, 254) and in films on conductor electrodes for the electrocatalytic reduction of bicarbonate (255, 256, 392, 393).

Bilayer electrodes are prepared by coating the electrode first with a layer of redox polymer and then with a second layer of a different redox polymer. There is no contact between the electrode and the second polymer layer, except by way of the electrons that are transported to it by the first, innermost layer. In contact with an electrolyte solution, the redox polymer–polymer interface acts as a rectifying junction because it consists of redox species with different formal potentials $E^{\circ\prime}$ on opposing sides of the interface. Figure 1.9 illustrates this scheme based on an inner Ru polymer and an outer ferrocene polymer (71). Figure 1.10 (top) shows the voltammetry of the film components as single films, and Fig. 1.10 (bottom) their voltammetry in the Fig. 1.9 bilayer format. In Fig. 1.10 (lower) the ferrocene oxidation occurs as a prepeak on the rising part of the Ru(III/II) wave potential, which is its only electron-transport communication with the electrode. Following its oxidation, the ferrocene film is trapped in the oxidized form since reduction of ferricenium by Ru(II) is thermodynamically disfavored. The ferrocene layer as shown in Fig. 1.10 (lower) can be reset to its reduced form by using electrode potentials that reduce the underlying Ru polymer film.

Bilayer electrodes are useful as models of chemical diodes and as electrodes that can trap and store electrochemical charge (or associated film color). Another use has been to measure the rate at the interface between the two polymers (265, 303), which being a reaction between contacting monolayers of metal complexes, is a chemically novel reaction. Wrighton (441) has shown that solution pH can also serve as a charge-trapping signal for bifunctional (quinone–viologen not layered) redox polymers.

Contact of an electroactive polymer film by a second electrode can be accomplished by evaporating a thin, semiporous metal layer into the outer surface of a redox polymer film; this is called a sandwich electrode (59). An equivalent arrangement places the electroactive polymer in the interelectrode gap of an interdigitated array electrode, termed an IDA electrode (59, 60). The sandwich and IDA microstructured electrodes have been employed by the Murray and Wrighton groups for study of electron transport through the polymer films. They offer the important advantage in such studies of measuring

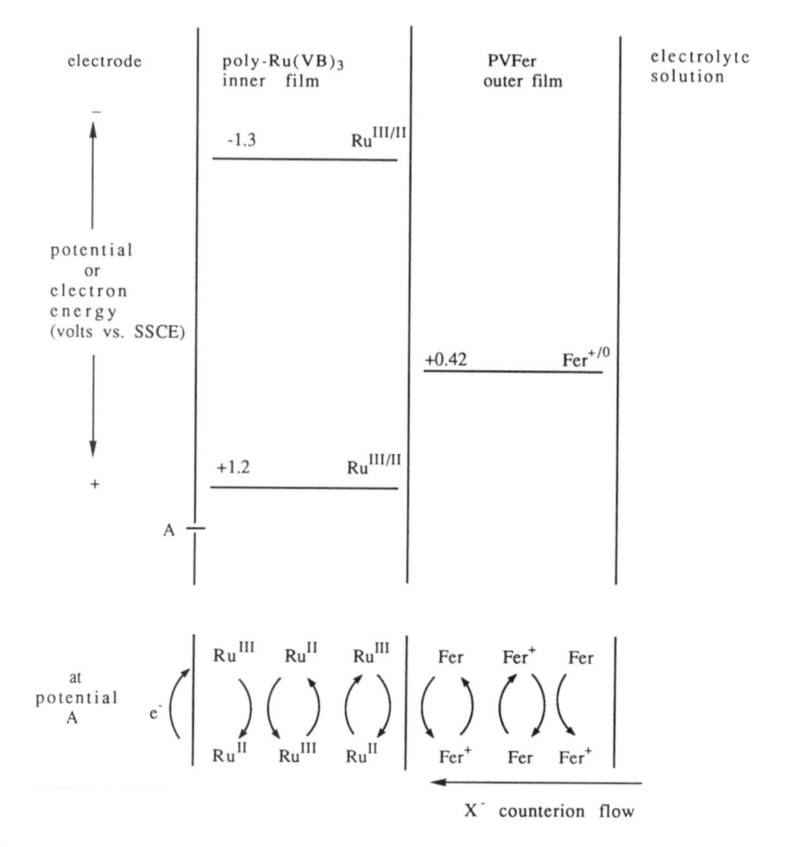

Figure 1.9. Schematic representation of electron energy levels for a Pt/poly-$[Ru(vbpy)_3]^{2+}$ /poly(vinylferrocene) bilayer electrode (adapted from Ref. 71).

electron transport under *steady state* conditions, thus avoiding macroscopic counterion diffusivity effects. A second use, mainly exploited in the IDA format and by the Wrighton group (60) is to use the sensitivity of the electron conductance of the redox polymer to its oxidation state (it is appreciably conductive only in the mixed-valent state) to amplify the electrolytic effects of changing the average film potential and its oxidation state. The sensitivity of the current across the IDA gap to the applied potential, makes this device act as a "molecule-based" transistor. This idea can be used in fluid electrolyte solutions, and in a "solid state" mode with a polymer electrolyte solution (406) as illustrated in Fig. 1.11. In the experiment shown, the conductance of poly(3-methyl-thiophene) film (the drain current) changes sharply as the potential of the film relative to a reference electrode (the gate voltage) is changed.

Finally, a new chapter in electrode microstructures was recently opened by Meyer and his co-workers (442). By a combination of photochemical ligand loss

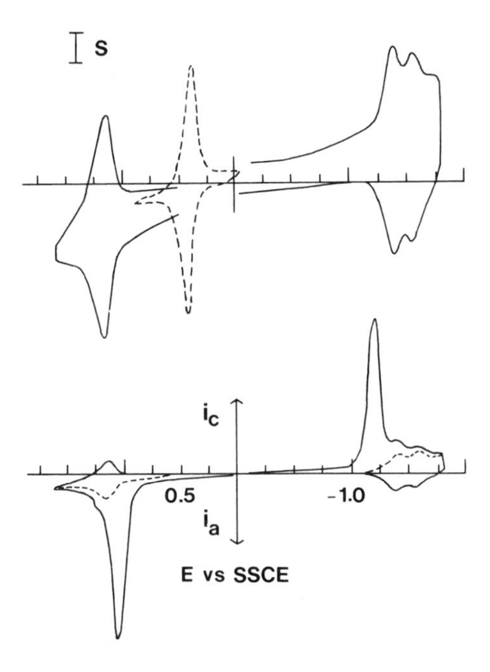

Figure 1.10. Top: Cyclic voltammetry of two polymer coated Pt electrodes 4×10^{-9} mol cm^{-2} poly-$[Ru(vbpy)_3]^{2+}$ film (solid, S = 46 μA cm^{-2}) and of 4×10^{-9} mol cm^{-2} poly(vinylferrocene) film (dashed, S = 91 μA cm^{-2}); bottom: cyclic voltammogram of bilayer electrode, 1.2×10^{-8} mol cm^{-2} of poly(vinylferrocene) on top of 2.4×10^{-9} mol cm^{-2} of poly-$[Ru(vbpy)_3]^{2+}$ on Pt, S = 227 μA cm^{-2}, solid line is $0 \rightarrow +1.6 \rightarrow -1.6 \rightarrow$ 0-V excursion, the dashed line results if scan reversed at 0.6 V (following a positive $0 \rightarrow +1.6 \rightarrow +0.6$-V scan) or at 0 V (following a negative $0 \rightarrow -1.6 \rightarrow 0$ V scan). All in 0.1*M* Et_4NClO_4/CH_3CN. [Reprinted with permission from H. D. Abruña, T. J. Meyer, and R. W. Murray, *J. Am. Chem. Soc.*, **1981**, *103*, 1. Copyright © (1981) American Chemical Society.]

in electrochemically polymerized $[Ru(bpy)_2(vpy)_2]^{2+}$ films and electrochemical polymerization of $[Os(vbpy)_3]^{2+}$, they were able to form an image of the latter polymerized complex corresponding to the initial image of light that had caused ligand loss and depolymerization of the former. This technique provides a literally three-dimensional control of microstructuring of films that will be a fruitful topic of future study.

ACKNOWLEDGMENTS

I gratefully acknowledge research support from the Office of Naval Research and the National Science Foundation that has led to some of the research described in this Chapter. I also express congratulations to my many student/colleagues for the excellence of their contributions to the chemical literature.

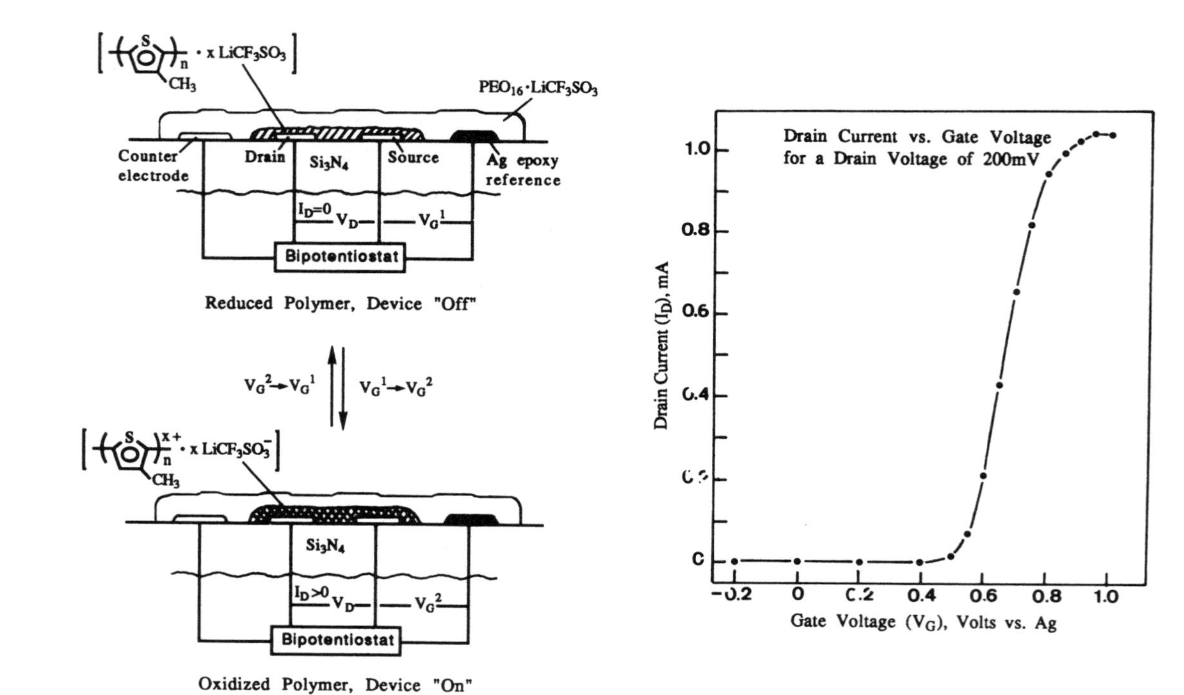

Figure 1.11. Characteristics of a poly(3-methyl thiophene) based solid state electrochemical transistor, at 95°C in N_2. The drain current is steady state. [Reprinted with permission from S. Chao and M. S. Wrighton, *J. Am. Chem. Soc.*, **1987**, *109*, 2197. Copyright © (1987) American Chemical Society.]

REFERENCES

1. A. J. Bard and L. R. Faulkner, *Electrochemical Methods*, Wiley, New York, **1980**.
2. J. O'M. Bockris and A. K. N. Reddy, *Modern Electrochemistry*, Vol. 1, Plenum, New York, **1970**.
3. J. O'M. Bockris and A. K. N. Reddy, *Modern Electrochemistry*, Vol. 2, Plenum, New York, **1973**.
4. R. Parsons, *Surf. Sci.*, **1964**, *2*, 418.
5. R. W. Murray, *Electroanalytical Chemistry*, Vol. 13, A. J. Bard (Ed.), Marcel Dekker, New York, **1984**, p. 191.
6. R. W. Murray, *Acc. Chem. Res.*, **1980**, *13*, 135.
7. L. R. Faulkner, *Chem. Eng. News*, **1982**, *Feb. 27*, p. 28.
8. M. Fujihira, *Topics in Organic Electrochemistry*, A. J. Fry and W. R. Britton (Eds.), Plenum, New York, **1986**, p. 255.
9. J. S. Miller (Ed.), *Chemically Modified Surfaces in Catalysis and Electrocatalysis*, ACS Symposium Series No. 192, American Chemical Society, Washington DC, **1982**.
10. W. J. Albery and A. R. Hillman, *Ann. Rev. C. R. Soc. Chem. London*, **1981**, p. 377.
11. K. D. Snell and A. G. Keenan, *Chem. Soc. Rev.*, **1979**, *8*, 259.
12. R. W. Murray, *An. Rev. Mater. Sci.*, **1984**, *14*, 145.
13. H. D. Abruña, "Electrode Modification with Polymeric Reagents", in Electroresponsive Molecular and Polymeric Systems, T. Skotheim (Ed.), Marcel Dekker, New York, **1988**.
14. P. R. Moses, L. M. Wier, and R. W. Murray, *Anal. Chem.*, **1975**, *78*, 1882.
15. J. R. Lenhard and R. W. Murray, *J. Electroanal. Chem.*, **1977**, *78*, 195.
16. J. R. Lenhard, R. Rocklin, H. Abruña, K. Willman, K. Kuo, R. Nowak, and R. W. Murray, *J. Am. Chem. Soc.*, **1978**, *100*, 5213.
17. J. P. Collman, M. Marocco, P. Denisevich, C. Koval, and F. C. Anson, *J. Electroanal. Chem.*, **1979**, *101*, 117.
18. J. P. Collman, P. Denisevich, Y. Konai, M. Marrocco, C. Koval, and F. C. Anson, *J. Am. Chem. Soc.*, **1980**, *102*, 6027.
19. J. P. Collman, F. C. Anson, C. S. Bencosme, R. R. Durand, Jr., and R. P. Kreh, *J. Am. Chem. Soc.*, **1983**, *105*, 2699.
20. Y. Le Mest and M. L'Her, *J. Am. Chem. Soc.*, **1986**, *108*, 533.
21. K. Kim, J. P. Collman, and J. A. Ibers, *J. Am. Chem. Soc.*, **1988**, *110*, 4242.
22. J. P. Collman and K. Kim, *J. Am. Chem. Soc.*, **1986**, *108*, 7847.
23. C. K. Chang, H. Y. Liu, and I. Abdalmuhdi, *J. Am. Chem. Soc.*, **1984**, *106*, 2725.
24. P. R. Moses, L. M. Wier, J. C. Lennox, H. O. Finklea, J. R. Lenhard and R. W. Murray, *Anal. Chem.*, **1978**, *50*, 576.
25. N. Oyama and F. C. Anson, *J. Am. Chem. Soc.*, **1978**, *100*, 4248.
26. J. C. Lennox and R. W. Murray, *J. Am. Chem. Soc.*, **1978**, *100*, 3710.
27. C. M. Carlin, L. J. Kepley, and A. J. Bard, *J. Electrochem. Soc.*, **1985**, *132*, 353.
28. S. R. Snyder, H. S. White, S. Lopez, and H. D. Abruña, *J. Am. Chem. Soc.*, **1990**, *112*, 1333.
29. A. Hamnett, S. Higgins, R. S. Mortimer, and D. R. Rosseinsky, *J. Electroanal. Chem.*, **1988**, *255*, 315.
30. L. J. Amos, M. H. Schmidt, S. Sinha, and A. B. Bocarsly, *Langmuir*, **1986**, *2*, 561.
31. S. Bruckenstein, C. P. Wilde, M. Shay, A. R. Hillman, and D. C. Loveday, *J. Electroanal. Chem.*, **1989**, *258*, 457.
32. M. J. Albarelli, J. H. White, G. M. Bommarito, M. McMillan, and H. D. Abruña, *ACS Symposium Series* M. P. Soriaga (Ed.), *378*, ACS, Washington, DC, p. 216, **1988**.

33. A. E. Kaifer and A. J. Bard, *J. Phys. Chem.*, **1986**, *90*, 868.
34. J. G. Gaudiello, P. K. Ghosh, and A. J. Bard, *J. Am. Chem. Soc.*, **1985**, *107*, 3027.
35. C. Lee and A. J. Bard, *Anal. Chem.*, **1990**, *62*, 1906.
36. B. D. Humphrey, S. Sinha, and A. B. Bocarsly, *J. Phys. Chem.*, **1984**, *88*, 736.
37. B. B. Kaul, R. E. Holt, V. L. Schlegel, and T. M. Cotton, *Anal. Chem.*, **1988**, *60*, 1580.
38. Y.-M. Tsou, H.-Y. Liu, and A. J. Bard, *J. Electrochem. Soc.*, **1988**, *135*, 1669.
39. D. R. Rosseinsky, J. D. Slocombe, A. M. Soutar, P. M. S. Monk, and A. Glidle, *J. Electroanal. Chem.*, **1989**, *258*, 233.
40. M. J. Albarelli, J. M. White, M. McMillan, and H. D. Abruña, *J. Electroanal. Chem.*, **1988**, *248*, 77.
41. A. Merz and A. J. Bard, *J. Am. Chem. Soc.*, **1978**, *100*, 3222.
42. L. L. Miller and M. R. van de Mark, *J. Electroanal. Chem.*, **1978**, *100*, 3223.
43. M. S. Wrighton, R. G. Austin, A. B. Bocarsly, J. M. Bolts, and O. Hass, K. D. Legg, L. Nadjo, and M. C. Palazzotto, *J. Electroanal. Chem.*, **1978**, *87*, 429.
44. P. Daum, J. R. Lenhard, D. R. Rolison, and R. W. Murray, *J. Am. Chem. Soc.*, **1980**, *102*, 4649.
45. R. J. Nowak, F. A. Schultz, M. Umaña, R. Lam, and R. W. Murray, *Anal. Chem.*, **1980**, *52*, 315.
46. N. Oyama and F. C. Anson, *J. Electrochem. Soc.*, **1980**, *127*, 247.
47. N. Surridge, J. C. Jernigan, F. Dalton, R. P. Buck, M. Watanabe, T. T. Wooster, H. Zhang, M. Pinkerton, M. L. Longmire, J. S. Facci, and R. W. Murray, *Discus. Faraday Soc.*, **1990**, *88*, 1.
48. A. R. Guadalupe and H. D. Abruña, *Anal. Chem.*, **1985**, *57*, 142.
49. L. M. Wier, A. R. Guadalupe, and H. D. Abruña, *Anal. Chem.*, **1985**, *57*, 2009.
50. M. J. Gehron and A. Brajter-Toth, *Anal. Chem.*, **1986**, *58*, 1488.
51. M. S. Wrighton, R. G. Austin, A. B. Bocarsly, J. M. Bolts, O. Haas, K. D. Legg, L. Nadjo, and M. C. Palazzotto, *J. Electroanal. Chem.*, **1978**, *87*, 429.
52. M. S. Wrighton, M. C. Palazzotto, A. B. Bocarsly, J. M. Bolts, A. Fischer, and L. Nadjo, *J. Am. Chem. Soc.*, **1978**, *100*, 7264.
53. M. S. Wrighton, *Acc. Chem. Res.*, **1979**, *12*, 303.
54. T. T. Li and M. J. Weaver, *J. Am. Chem. Soc.*, **1984**, *106*, 6107.
55. C. E. D. Chidsey, C. R. Bertozzi, T. M. Pritvinski, and A. M. Mujsce, *J. Am. Chem. Soc.*, **1990**, *112*, 4301.
56. C. E. D. Chidsey, *Science*, **1991**, *251*, 919.
57. D. Acevedo and H. D. Abruña, *5. Phys. Chem.*, in press.
58. D. A. Van Galen and M. Majda, *Anal. Chem.*, **1988**, *60*, 1549.
59. C. E. D. Chidsey and R. W. Murray, *Science*, **1986**, *231*, 25.
60. M. S. Wrighton, *Science*, **1986**, *231*. 32.
61. *Handbook of Conducting Polymers*, Vol. 1, T. A. Skotheim (Ed.), Marcel Dekker, New York, **1986**.
62. *Handbook of Conducting Polymers*, Vol. 2, T. A. Skotheim (Ed.), Marcel Dekker, New York, **1986**.
63. A. F. Diaz, K. K. Kanazawa, and G. P. Gardini, *J. Chem. Soc. Chem. Commun.*, **1979**, 635.
64. A. F. Diaz, et al. *J. Chem. Soc. Chem. Commun.*, **1979**, 854.
65. A. F. Diaz, J. Castillo, K. K. Kanazawa, and J. A. Logan, *J. Electroanal. Chem.*, **1982**, *133*, 233.
66. L. L. Miller and B. Zinger, *J. Am. Chem. Soc.*, **1984**, *106*, 6861.
67. P. R. Moses and R. W. Murray, *J. Am. Chem. Soc.*, **1976**, *98*, 7435.
68. N. Oyama and F. C. Anson, *Anal. Chem.*, **1980**, *52*, 1192.
69. N. Oyama, T. Shinomura, K. Shigehara, F. C. Anson, *J. Electroanal. Chem.*, **1980**, *112*, 271.

70. Y. S. Obeng, A. Founta, and A. J. Bard, *New J. Chem.*, in press, **1991**.
71. H. D. Abruña, P. Denisevich, M. Umaña, T. J. Meyer, and R. W. Murray, *J. Am. Chem. Soc.*, **1981**, *103*, 1.
72. C. A. Widrig and M. Majda, *Anal. Chem.*, **1987**, *59*, 754.
73. R. M. Buchanan, G. S. Calabrese, T. J. Sobieralski, and M. S. Wrighton, *J. Electroanal. Chem.*, **1983**, *153*, 129.
74. K. W. Willman, R. D. Rocklin, R. Nowak, K. Kuo, F. A. Schultz, and R. W. Murray, *J. Am. Chem. Soc.*, **1980**, *102*, 7629.
75. P. J. Peerce and A. J. Bard, *J. Electroanal. Chem.*, **1980**, *114*, 89.
76. J. W. McCarger and V. D. Neff, *J. Phys. Chem.*, **1988**, *92*, 3598.
77. A. P. Brown and F. C. Anson, *Anal. Chem.*, **1977**, *49*, 1589.
78. H. Angerstein-Kozlowska, J. Klinger, and B. E. Conway, *J. Electroanal. Chem.*, **1977**, *75*, 45.
79. T. Ikeda, C. R. Leidner, and R. W. Murray, *J. Electroanal. Chem.*, **1982**, *138*, 343.
80. C. E. D. Chidsey and R. W. Murray, *J. Phys. Chem.*, **1986**, *90*, 1479.
81. D. Ellis, M. Eckhoff, and V. D. Neff, *J. Phys. Chem.*, **1981**, *85*, 1225.
82. S. T. Coleman, W. R. McKinnon, and J. R. Dahn, *Phys. Rev. B*, **1984**, *29*, 4147.
83. E. Laviron, *J. Electroanal. Chem.*, **1981**, *122*, 37.
84. R. F. Lane and A. T. Hubbard, *J. Phys. Chem.*, **1973**, *77*, 1401.
85. E. Laviron, *J. Electroanal. Chem.*, **1979**, *100*, 263.
86. H. Angerstein-Kozlowska, B. E. Conway, and J. Klinger, *J. Electroanal. Chem.*, **1978**, *87*, 321.
87. M. Sharp and M. Petersson, *J. Electroanal. Chem.*, **1981**, *122*, 409.
88. A. P. Brown and F. C. Anson, *J. Electroanal. Chem.*, **1978**, *92*, 133.
89. F. B. Kaufman, A. H. Schroeder, E. M. Engler, S. R. Kramer, and J. Q. Chambers, *J. Am. Chem. Soc.*, **1980**, *102*, 483.
90. P. Daum and R. W. Murray, *J. Electroanal. Chem.*, **1979**, *103*, 289.
91. P. Daum, J. R. Lenhard, D. R. Rolison, and R. W. Murray, *J. Am. Chem. Soc.*, **1980**, *102*, 4649.
92. R. J. Nowak, R. A. Schulz, M. Umaña, R. Lam, and R. W. Murray, *Anal. Chem.*, **1980**, *52*, 315.
93. N. Oyama and F. C. Anson, *J. Electrochem. Soc.*, **1980**, *127*, 640.
94. C. P. Andrieux and J.-M. Savéant, *J. Phys. Chem.*, **1988**, *92*, 6761.
95. J.-M. Savéant, *J. Phys. Chem.*, **1988**, *92*, 1011.
96. J. Leddy and A. J. Bard, *J. Electroanal. Chem. Interf. Electrochem.*, **1985**, *189*, 203.
97. M. Senda and P. Delahay, *J. Phys. Chem.*, **1961**, *65*, 1580.
98. T. Ikeda, K. Toriyama, and M. Senda, *Bull. Chem. Soc. Jpn.*, **1979**, *52*, 1937.
99. T. Ohsaka, T. Ushirogouchi, and N. Oyama, *Bull. Chem. Soc. Jpn.*, **1985**, *58*, 3252.
100. R. D. Armstrong, B. Lindholm, and M. Sharpe, *J. Electroanal. Chem. Interfacial Electrochem.*, **1986**, *202*, 69.
101. C. J. Miller, C. A. Widrig, D. H. Charych, and M. Majda, *J. Phys. Chem.*, **1988**, *92*, 1928.
102. C. A. Goss, C. J. Miller, and M. Majda, *J. Phys. Chem.*, **1988**, *92*, 1937.
103. P. Pickup and R. W. Murray, *J. Am. Chem. Soc.*, **1983**, *105*, 4510.
104. P. Pickup, W. Kutner, C. R. Leidner, and R. W. Murray, *J. Am. Chem. Soc.*, **1984**, *106*, 1991.
105. J. C. Jernigan, C. E. D. Chidsey, and R. W. Murray, *J. Am. Chem. Soc.*, **1985**, *107*, 2824.
106. B. J. Feldman and R. W. Murray, *Anal. Chem.*, **1986**, *58*, 2844.
107. B. A. White and R. W. Murray, *J. Am. Chem. Soc.*, **1987**, *109*, 2576.
108. B. J. Feldman and R. W. Murray, *Inorg. Chem.*, **1987**, *26*, 1702.
109. T. Ikeda, R. Schmehl, P. Denisevich, K. Willman, and R. W. Murray, *J. Am. Chem. Soc.*, **1982**, *104*, 2683.

110. K. Pressprich, S. Maybury, R. Thomas, R. W. Linton, E. A. Irene, and R. W. Murray, *J. Phys. Chem.*, **1989**, *93*, 5568.
111. R. L. McCarley, E. A. Irene, and R. W. Murray, *J. Phys. Chem.*, **1991**, *95*, 2492.
112. J. Zagal, R. K. Sen, and E. Yeager, *J. Electroanal. Chem.*, **1977**, *83*, 207.
113. R. F. Lane and A. T. Hubbard, *J. Phys. Chem.*, **1973**, *77*, 1411.
114. A. P. Brown and F. C. Anson, *Anal. Chem.*, **1977**, *49*, 1589.
115. A. P. Brown, C. Koval, and F. C. Anson, *J. Electroanal. Chem.*, **1976**, *72*, 379.
116. A. P. Brown and F. C. Anson, *J. Electroanal. Chem.*, **1977**, *83*, 203.
117. C. P. Jester, R. D. Rocklin, and R. W. Murray, *J. Electrochem. Soc.*, **1980**, *127*, 1979.
118. A. Bettelheim, R. J. H. Chan, and T. Kuwana, *J. Electroanal. Chem.*, **1979**, *99*, 391.
119. B. Z. Kikolic, R. R. Adzic, and E. B. Yeager, *J. Electroanal. Chem.*, **1979**, *103*, 281.
120. R. Kotz and E. Yeager, *J. Electroanal. Chem.*, **1980**, *113*, 113.
121. R. J. Bowling, R. T. Packard, and R. L. McCreery, *J. Am. Chem. Soc.*, **1989**, *111*, 1217.
122. E. F. Bowden, F. M. Hawkridge, J. F. Chlebowski, E. E. Bancroft, C. Thorpe, and H. N. Blount, *J. Am. Chem. Soc.*, **1982**, *104*, 7641.
123. K. B. Koller and F. M. Hawkridge, *J. Am. Chem. Soc.*, **1985**, *107*, 7412.
124. D. E. Reed and F. M. Hawkridge, *Anal. Chem.*, **1987**, *59*, 2334.
125. K. B. Koller, F. M. Hawkridge, G. Fauque, and J. LeGall, *Biochem. Biophys. Res. Commun.*, **1987**, *145*, 619.
126. F. A. Armstrong, H. A. O. Hill, N. J. Walton, and B. N. Oliver, *J. Am. Chem. Soc.*, **1984**, *106*, 921.
127. H. L. Landrum, R. T. Salmon, and F. M. Hawkridge, *J. Am. Chem. Soc.*, **1977**, *99*, 3154.
128. J. F. Stargardt, F. M. Hawkridge, and H. L. Landrum, *Anal. Chem.*, **1978**, *50*, 930.
129. M. J. Eddowes and H. A. O. Hill, *J. Am. Chem. Soc.*, **1979**, *101*, 4461.
130. M. J. Eddowes and H. A. O. Hill, *J. Chem. Soc. Chem. Commun.*, **1977**, 771.
131. W. J. Albery, M. J. Eddowes, H. A. O. Hill, and A. R. Hillman, *J. Am. Chem. Soc.*, **1981**, *103*, 3904.
132. I. Taniguchi, M. Iseki, T. Eto, K. Toyosawa, H. Yamaguchi, and K. Yasukouchi, *Bioelectrochem. Bioenerg.*, **1984**, *13*, 373.
133. I. Taniguchi, T. Funatsu, M. Iseki, H. Yamaguchi, and K. Yasukouchi, *J. Electroanal. Chem. Interfacial Electrochem.*, **193**, 295.
134. F. A. Armstrong, P. A. Cox, H. A. Hill, B. N. Oliver, and A. A. Williams, *J. Chem. Soc. Chem. Commun.*, **1985**, 1236.
135. J. S. Facci, P. A. Falcigno, and J. M. Gold, *Langmuir*, **1986**, *2*, 732.
136. J. S. Facci, *Langmuir*, **1987**, *3*, 525.
137. X. Zhang and A. J. Bard, *J. Am. Chem. Soc.*, **1989**, *111*, 8098.
138. X. Zhang and A. J. Bard, *J. Phys. Chem.*, **1988**, *92*, 5566.
139. M. Fujihira and S. Poosittisak, *J. Electroanal. Chem.*, **1986**, *199*, 481.
140. M. Fujihira and T. Araki, *Bull. Chem. Soc. Jpn.*, **1986**, *59*, 2375.
141. K. Aoki, K. Tokuda, and H. Matsuda, *J. Electroanal. Chem.*, **1986**, *199*, 69.
142. H. Matsuda, J. Aoki, and K. Tokuda, *J. Electroanal. Chem.*, **1987**, *217*, 15.
143. C.-W. Lee and A. J. Bard, *Chem. Phys. Lett.*, **1990**, *170*, 57.
144. J. J. Hickman, C. Zou, D. Ofer, P. D. Harvey, M. S. Wrighton, P. E. Laibinis, C. D. Bain, and G. M. Whitesides, *J. Am. Chem. Soc.*, **1989**, *111*, 7271.
145. C. J. Miller, C. A. Widrig, D. H. Charych, and M. Majda, *J. Phys. Chem.*, **1988**, *92*, 1928.
146. C. J. Miller and M. Majda, *Anal. Chem.*, **1988**, *60*, 1168.

147. C. A. Widrig and M. Majda, *Langmuir*, **1989**, *5*, 689.
148. J. Ouyang and A. B. P. Lever, *J. Phys. Chem.*, **1991**, *95*, 2101.
149. C. J. Miller and A. J. Bard, *Anal. Chem.*, submitted, **1991**.
150. X. Zhang and A. J. Bard, *J. Am. Chem. Soc.*, **1989**, *111*, 8089.
151. M. Fujihara and T. Araki, *Chem. Lett.*, **1988**, 921.
152. C. A. Goss, C. J. Miller, and M. Majda, *J. Phys. Chem.*, **1988**, *92*, 1937.
153. C. J. Miller and M. Majda, *J. Am. Chem. Soc.*, **1986**, *108*, 3118.
154. C. A. Widrig, C. J. Miller, and M. Majda, *J. Am. Chem. Soc.*, **1988**, *110*, 2009.
155. H. C. De Long and D. A. Buttry, *Langmuir*, **1990**, *6*, 1319.
156. H. O. Finklea, D. A. Snider, and J. Fedyk, *Langmuir*, **1990**, *6*, 371.
157. I. Rubinstein, S. Steinberg, Y. Tor, A. Shanzer, and J. Sagiv, *Nature (London)*, **1988**, *332*, 426.
158. O. J. Garcia, P. A. Quintela, and A. E. Kaifer, *Anal. Chem.*, **1989**, *61*, 979.
159. K. A. Bunding-Lee, *Langmuir*, **1990**, *6*, 709.
160. C. E. D. Chidsey and D. N. Loiacono, *Langmuir*, **1990**, *6*, 682.
161. R. G. Nuzzo, L. H. Dubois, and D. L. Allara, *J. Am. Chem. Soc.*, **1990**, *112*, 558.
162. L. H. Dubois, B. R. Zegarski, and R. G. Nuzzo, *J. Am. Chem. Soc.*, **1990**, *112*. 570.
163. M. D. Porter, T. B. Bright, D. L. Allara, and C. E. D. Chidsey, *J. Am. Chem. Soc.*, **1987**, *109*, 3559.
164. T. Safi, K. Hoshino, Y. Ishii, and M. Goto, *J. Am. Chem. Soc.*, **1991**, *113*, 450.
165. N. Tillman, A. Ulman, J. S. Schildkraut, and T. L. Penner, *J. Am. Chem. Soc.*, **1988**, *110*, 6136.
166. N. Tillman, A. Ulman, and J. F. Elman, *Langmuir*, **1989**, *5*, 1020.
167. B. J. Barner and R. M. Corn, *Langmuir*, **1990**, *6*, 1023.
168. Y. Okahata, M. Yokobori, Y. Ebara, H. Ebato, and K. Ariga, *Langmuir*, **1990**, *6*, 1148.
169. T. M. Putvinski, M. L. Schilling, H. E. Katz, C. E. D. Chidsey, A. M. Mujsce, and A. B. Emerson, *Langmuir*, **1990**, *6*, 1567.
170. H. Lee, L. J. Kepley, H.-G. Hong, S. Akhter, and T. E. Mallouk, *J. Phys. Chem.*, **1988**, *92*, 2597.
171. J. Sagiv, *J. Am. Chem. Soc.*, **1980**, *102*, 92.
172. S. R. Cohen, R. Naaman, and J. Sagiv, *J. Phys. Chem.*, **1986**, *90*, 3054.
173. E. Sabatani, I. Rubinstein, M. Maoz, and J. Sagiv, *J. Electroanal. Chem.*, **1987**, *219*, 365.
174. P. R. Moses and R. W. Murray, *J. Electroanal. Chem.*, **1977**, *77*, 393.
175. J. R. Lenhard and R. W. Murray, *J. Electroanal. Chem.*, **1977**, *78*, 195.
176. H. Abruña, T. J. Meyer, and R. W. Murray, *Inorg. Chem.*, **1979**, *11*, 3233.
177. J. R. Lenhard and R. W. Murray, *J. Am. Chem. Soc.*, **1978**, *100*, 7878.
178. M. S. Wrighton, M. C. Palazzotto, A. B. Bocarsly, J. M. Bolts, A. B. Fischer, and L. Nadjo, *J. Am. Chem. Soc.*, **1978**, *100*, 7629.
179. J. M. Bolts and M. S. Wrighton, *J. Am. Chem. Soc.*, **1978**, *100*, 5257.
180. D. C. Bookbinder, N. S. Lewis, M. G. Bradley, A. B. Bocarsly, and M. S. Wrighton, *J. Am. Soc.*, **1979**, *101*, 7721.
181. A. B. Fischer, M. S. Wrighton, M. Umaña, and R. W. Murray, *J. Am. Chem. Soc.*, **1979**, *101*, 3442.
182. H. S. White and R. W. Murray, *Anal. Chem.*, **1979**, *51*, 236.
183. K. W. Willman, E. Greer, and R. W. Murray, *Nouv. J. Chem.*, **1979**, *3*, 455.
184. K. W. Willman, R. D. Rocklin, R. Nowak, K. Kuo, F. A. Schultz, and R. W. Murray, *J. Am. Chem. Soc.*, **1980**, *102*, 7629.
185. J. M. Bolts, A. B. Bocarsly, M. C. Palazzotto, E. G. Walton, N. S. Lewis, and M. S. Wrighton, *J. Am. Chem. Soc.*, **1979**, *101*, 1378.

186. D. C. Bookbinder and M. S. Wrighton, *J. Am. Chem. Soc.*, **1980**, *102*, 5123.
187. D. C. Bookbinder, J. A. Bruce, R. N. Dominey, N. S. Lewis, and M. S. Wrighton, *Proc. Natl. Acad. Sci. USA*, **1980**, *77*, 6280.
188. A. B. Fischer, J. B. Kinney, R. H. Staley, and M. S. Wrighton, *J. Am. Chem. Soc.*, **1979**, *101*, 7863.
189. B. F. Watkins, J. R. Behling, E. Kariv, and L. L. Miller, *J. Am. Chem. Soc.*, **1975**, *97*, 3549.
190. J. C. Lennox and R. W. Murray, *J. Electroanal. Chem.*, **1977**, *78*, 395.
191. C. A. Koval and F. C. Anson, *Anal. Chem.*, **1978**, *50*, 223.
192. R. D. Rocklin and R. W. Murray, *J. Electroanal. Chem.*, **1979**, *100*, 271.
193. M. A. Fox, F. J. Nabs, and T. A. Voynick, *J. Am. Chem. Soc.*, **1980**, *102*, 4029.
194. A. M. Yacynych and T. Kuwana, *Anal. Chem.*, **1978**, *50*, 640.
195. T. Matsue, M. Fujihira, and T. Osa, *J. Electrochem. Soc.*, **1981**, *128*, 1473.
196. T. Matsue, U. Akiba, and T. Osa, *Anal. Chem.*, **1986**, *58*, 2096.
197. S. Anderson, E. C. Constable, M. P. Dare-Edwards, J. B. Goodenough, A. Hamnett, K. R. Seddon, and R. D. Wright, *Nature (London)*, **1979**, *280*, 571.
198. S. Mazur, T. Matusinovic, and K. Camman, *J. Am. Chem. Soc.*, **1977**, *99*, 3888.
199. J. F. Evans and T. Kuwana, *Anal. Chem.*, **1977**, *49*, 1632.
200. N. Oyama, A. P. Brown, and F. C. Anson, *J. Electroanal. Chem.*, **1978**, *87*, 435.
201. R. Nowak, F. A. Schultz, M. Umaña, H. Abruña, and R. W. Murray, *J. Electroanal. Chem.*, **1978**, *94*, 219.
202. N. Oyama and F. C. Anson, *J. Am. Chem. Soc.*, **1979**, *101*, 739.
203. N. Oyama and F. C. Anson, *J. Am. Chem. Soc.*, **1979**, *101*, 3450.
204. O. Haas and J. G. Vos, *J. Electroanal. Chem.*, **1980**, *113*, 139.
205. E. Laviron, *J. Electroanal. Chem.*, **1980**, *112*, 1.
206. C. P. Andrieux and J. M. Savéant, *J. Electroanal. Chem.*, **1980**, *111*, 377.
207. J. C. Jernigan, N. Surridge, M. E. Zvanut, M. Silver, and R. W. Murray, *J. Phys. Chem.*, **1989**, *93*, 4620.
208. N. A. Surridge, M. E. Zvanut, F. Richard Keene, M. Silver, and R. W. Murray, *J. Phys. Chem.*, in press.
209. H. Abruña, "Electrode Modification with Polymeric Reagents," in *Electroresponsive Molecular and Polymeric Systems*, Vol. 1, T. A. Skotheim (Ed.), Marcel Dekker, New York, **1988**.
210. H. G. Cassidy and K. A. Kun, *Oxidation Reduction Polymers (Redox Polymers)*, Wiley-Interscience, New York, **1965**.
211. A. Akelah and D. C. Sherrington, *Chem. Rev.*, **1981**, **1981**, *81*, 557.
212. K. Shigehara, N. Oyama, and F. C. Anson, *J. Am. Chem. Soc.*, **1981**, *103*, 2552.
213. J. M. Calvert and T. J. Meyer, *Inorg. Chem.*, **1981**, *20*, 27.
214. P. Burgmayer and R. W. Murray, *J. Electroanal. Chem.*, **1982**, *135*, 335.
215. N. A. Surridge, S. F. McClanahan, J. T. Hupp, E. Danielson, S. Gould, and T. J. Meyer, *J. Phys. Chem.*, **1989**, *93*, 294.
216. N. A. Surridge, J. T. Hupp, S. F. McClanahan, S. Gould, and T. J. Meyer, *J. Phys. Chem.*, **1989**, *93*, 304.
217. N. Oyama, N. Oki, H. Ohno, Y. Ohnuki, H. Matsuda, and E. Tsuchida, *J. Phys. Chem.*, **1983**, *87*, 3642.
218. G. Inzelt, J. Q. Chambers, J. F. Kinstle, and R. W. Day, *J. Phys. Chem.*, **1984**, *88*, 3906.
219. R. W. Day, G. Inzelt, J. F. Kinstle, and J. Q. Chambers, *J. Am. Chem. Soc.*, **1982**, *104*, 6804.
220. G. Inzelt, R. W. Day, J. F. Kinstle, and J. Q. Chambers, *J. Phys. Chem.*, **1983**, *87*, 4592.

221. R. W. Day, H. Karimi, C. V. Francis, J. F. Kinstle, and J. Q. Chambers, *J. Polym. Sci. Chem. Edit.*, **1986**, *24*, 645.
222. G. Inzelt, J. Bacskai, J. Q. Chambers, and R. W. Day, *J. Electroanal. Chem.*, **1986**, *201*, 301.
223. P. Varineau and D. A. Buttry, *J. Phys. Chem.*, **1987**, *91*, 1295.
224. C. M. Elliott and E. J. Hersenhart, *J. Am. Chem. Soc.*, **1982**, *104*, 7219.
225. C. M. Elliott, J. G. Redepenning, and E. M. Balk, *J. Am. Chem. Soc.*, **1985**, *107*, 8302.
226. C. M. Elliott, C. J. Baldy, L. Nuwaysir, and C. L. Wilkins, *Inorg. Chem.*, **1990**, *29*, 389.
227. R. M. Kellett and T. G. Spiro, *Inorg. Chem.*, **1985**, *24*, 2378.
228. P. Martigny and F. C. Anson, *J. Electroanal. Chem.*, **1982**, *139*, 383.
229. R. J. Mortimer and F. C. Anson, *J. Electroanal. Chem.*, **1982**, *138*, 325.
230. H. D. Abruña and A. J. Bard, *Anal. Chem.*, **1981**, *103*, 6898.
231. H. Karimi and J. Q. Chambers, *J. Electroanal. Chem.*, **1987**, *217*, 313.
232. F. B. Kaufman, A. H. Schroeder, V. V. Patel, and K. H. Nichols, *J. Electroanal. Chem.*, **1982**, *132*, 151.
233. A. Kitani and L. L. Miller, *J. Am. Chem. Soc.*, **1981**, *103*, 3595.
234. M. Fukui, C. DeGrand, and L. L. Miller, *J. Am. Chem. Soc.*, **1982**, *104*, 28.
235. K. Shigehara and F. C. Anson, *J. Electroanal. Chem.*, **1982**, *132*, 107.
236. T. D. Westmoreland, J. M. Calvert, R. W. Murray, and T. J. Meyer, *J. Chem. Soc. Chem. Commun.*, **1983**, 65.
237. G. J. Samuels and T. J. Meyer, *J. Am. Chem. Soc.*, **1981**, *103*, 307.
238. O. Haas and J. G. Vos, *J. Electroanal. Chem.*, **1980**, *113*, 139.
239. O. Haas, M. Kriens, and J. G. Vos, *J. Am. Chem. Soc.*, **1981**, *103*, 1318.
240. N. Oyama and F. C. Anson, *J. Am. Chem. Soc.*, **1979**, *101*, 739.
241. C. D. Ellis and T. J. Meyer, *Inorg. Chem.*, **1984**, *23*, 1748.
242. J. T. Hupp, J. P. Otruba, S. J. Paurs, and T. J. Meyer, *J. Electroanal. Chem.*, **1985**, *190*, 287.
243. L. D. Margerum, R. W. Murray, and T. J. Meyer, *J. Phys. Chem.*, **1986**, *90*, 728.
244. O. Haas, H. R. Zumbrunnen, and J. G. Vos, *Electrochim. Acta*, **1985**, *30*, 1551.
245. C. DeGrand, L. Roullier, L. L. Miller, and B. Zinger, *J. Electroanal. Chem.*, **1984**, *178*, 101.
246. J. M. Bolts and M. S. Wrighton, *J. Am. Chem. Soc.*, **1979**, *101*, 6179.
247. A. B. Bocarsly, E. G. Walton, and M. S. Wrighton, *J. Am. Chem. Soc.*, **1980**, *100*, 3390.
248. N. S. Lewis, A. B. Bocarsly, and M. S. Wrighton, *J. Phys. Chem.*, **1980**, *83*, 2033.
249. J. A. Bruce and M. S. Wrighton, *J. Am. Chem. Soc.*, **1982**, *104*, 74.
250. R. N. Dominey, N. S. Lewis, J. A. Bruce, D. C. Bookbinder, and M. W. Wrighton, *J. Am. Chem. Soc.*, **1982**, *104*, 467.
251. R. N. Dominey, T. J. Lewis, and M. S. Wrighton, *J. Phys. Chem.*, **1983**, *87*, 5345.
252. R. A. Simon, A. J. Ricco, and M. S. Wrighton, *J. Am. Chem. Soc.*, **1982**, *104*, 2031.
253. G. S. Calabrese, R. M. Buchanan, and M. S. Wrighton, *J. Am. Chem. Soc.*, **1983**, *105*, 5594.
254. D. J. Harrison and M. S. Wrighton, *J. Am. Chem. Soc.*, **1984**, *106*, 3932.
255. C. J. Stadler, S. Chao, and M. S. Wrighton, *J. Am. Chem. Soc.*, **1984**, *106*, 3673.
256. J.-F. Andree and M. S. Wrighton, *Inorg. Chem.*, **1985**, *24*, 4288.
257. T. E. Mallouk, V. Cammarata, J. A. Crayston, and M. S. Wrighton, *J. Phys. Chem.*, **1986**, *90*, 2150.
258. K. W. Willman and R. W. Murray, *J. Electroanal. Chem.*, **1982**, *133*, 211.
259. T. Ikeda, C. R. Leidner, and R. W. Murray, *J. Am. Chem. Soc.*, **1981**, *103*, 7422.
260. P. Denisevich, H. D. Abruña, C. R. Leidner, T. J. Meyer, and R. W. Murray, *J. Am. Chem. Soc.*, **1982**, *21*, 2153.

261. T. Ikeda, C. R. Leidner, and R. W. Murray, *J. Electroanal. Chem.*, **1982**, *138*, 343.
262. P. G. Pickup, C. R. Leidner, P. Denisevich, and R. W. Murray, *J. Electroanal. Chem.*, **1984**, *164*, 39.
263. J. S. Facci, R. H. Schmehl, and R. W. Murray, *J. Am. Chem. Soc.*, **1982**, *104*, 4959.
264. R. H. Schmehl and R. W. Murray, *J. Electroanal. Chem.*, **1983**, *152*, 97.
265. C. R. Leidner and R. W. Murray, *J. Am. Chem. Soc.*, **1985**, *107*, 551.
266. C. D. Ellis, L. D. Margerum, R. W. Murray, and T. J. Meyer, *Inorg. Chem.*, **1983**, *22*, 2152.
267. J. M. Calvert, R. H. Schmehl, B. P. Sullivan, J. S. Facci, T. J. Meyer, and R. W. Murray, *Inorg. Chem.*, **1983**, *22*, 2151.
268. W. J. Vinning, N. A. Surridge, and T. J. Meyer, *J. Phys. Chem.*, **1986**, *90*, 2281.
269. A. Bettelheim, D. Ozer, and R. Harth and R. W. Murray, *J. Electroanal. Chem.*, **1988**, *246*, 139.
270. R. L. McCarley, R. E. Thomas, E. A. Irene, and R. W. Murray, *J. Electrochem. Soc.*, **1990**, *137*, 1485.
271. A. Bettelheim, D. Ozer, R. Harth, and R. W. Murray, *J. Electroanal. Chem.*, **1989**, *266*, 93.
272. T. R. O'Toole, B. P. Sullivan, M. R.-M. Bruce, L. D. Margerum, R. W. Murray, and T. J. Meyer, *J. Electroanal. Chem.*, **1989**, *259*, 217.
273. C. P. Horwitz and R. W. Murray, *Mol. Cryst. Liq. Cryst.*, **1988**, *160*, 389.
274. A. Bettelheim, B. A. White, and R. W. Murray, *J. Electroanal. Chem.*, **1987**, *217*, 271.
275. A. Bettelheim, B. A. White, S. A. Raybuck, and R. W. Murray, *Inorg. Chem.*, **1987**, *26*, 1009.
276. T. R. O'Toole, L. D. Margerum, T. D. Westmoreland, W. J. Vining, R. W. Murray, and T. J. Meyer, *J. Chem. Soc. Chem. Commun.*, **1985**, 1416.
277. N. A. Surridge, F. Richard Keene, B. A. White, J. S. Facci, M. Silver, and R. W. Murray, *Inorg. Chem.*, **1990**, *29*, 4950.
278. P. K. Ghosh and T. G. Spiro, *J. Electrochem. Soc.*, **1981**, *121*, 1281.
279. W. J. Albery, M. G. Boutelle, P. Colby, and A. R. Hillman, *J. Electroanal. Chem.*, **1982**, *133*, 135.
280. L. R. Faulkner and M. Majda, *J. Electroanal. Chem.*, **1982**, *137*, 149.
281. K. A. Macor and T. G. Spiro, *J. Am. Chem. Soc.*, **1983**, *105*, 5601.
282. J. M. Calvert, D. L. Peebles, and R. J. Nowak, *Inorg. Chem.*, **1985**, *24*, 3111.
283. K. T. Potts, D. Usifer, A. Guadalupe, and H. D. Abruña, *J. Am. Chem. Soc.*, in press.
284. P. G. Pickup and R. A. Osteryoung, *J. Electrochem. Soc.*, **1983**, *130*, 1965.
285. T. F. Guarr and F. C. Anson, *J. Phys. Chem.*, **1987**, *91*, 4037.
286. K. A. Macor and T. G. Spiro, *J. Am. Chem. Soc.*, **1983**, *105*, 5601.
287. R. M. Kellet and T. G. Spiro, *Inorg. Chem.*, **1985**, *24*, 2378.
288. S. Cosneir, A. Deronzier, and J.-C. Moutet, *J. Electroanal. Chem.*, **1985**, *89*, 193.
289. S. Cosneir, A. Deronzier, and J.-C. Moutet, *J. Phys. Chem.*, **1985**, *89*, 4895.
290. S. Cosneir, A. Deronzier, and J.-C. Moutet, *J. Electroanal. Chem.*, **1986**, *207*, 315.
291. J. G. Eaves, H. S. Munro, and D. Parker, *J. Chem. Soc. Chem. Commun.*, **1985**, 684.
292. G. Schiavon, G. Zotti, and G. Bontempelli, *J. Electroanal. Chem.*, **1984**, *161*, 323.
293. J. Asseraf, F. Bedioui, O. Reges, Y. Robin, F. Devynck, and C. Bed-Charreton, *J. Electroanal. Chem.*, **1983**, *170*, 255.
294. G. Bidan, A. Deronzier, and J.-C. Moutet, *J. Chem. Soc. Chem. Commun.*, **1984**, 1185.
295. H. C. Hurrell and H. D. Abruña, *Inorg. Chem.*, **1990**, *29*, 736.
296. A. R. Guadalupe, D. A. Usifer, K. T. Potts, H. C. Hurrell, A. E. Mogstad, and H. D. Abruña, *J. Am. Chem. Soc.*, **1988**, *110*, 3462.
297. V. Le Berre, L. Angely, N. Gueguen-Simonet, and J. Simonet, *J. Electroanal. Chem. Int. Electrochem.*, **1986**, *206*, 115.

298. J. Berthelot-Rault and J. Simonet, *J. Electroanal. Chem. Int. Electrochem.*, **1985**, *182*, 187.
299. P. C. Lacaze, J. E. Dubois, A. Monvernay-Desbene, P. L. Desbene, J. J. Basselier, and D. Richard, *J. Electroanal. Chem. Interf. Electrochem.*, **1983**, *147*, 107.
300. G. Bidan, A. Deronzier, and J. C. Moutet, *Nouv. J. Chim.*, **1984**, *8*, 501.
301. V. K. Gater, M. D. Liu, M. D. Love, and C. R. Leidner, *J. Electroanal. Chem.*, **1988**, *257*, 133.
302. C. R. Leidner and R. W. Murray, *J. Am. Chem. Soc.*, **1984**, *106*, 1606.
303. J. C. Jernigan and R. W. Murray, *J. Am. Chem. Soc.*, **1990**, *112*, 1034.
304. C. M. Elliott, J. G. Redepenning, and R. M. Balk, *J. Electroanal. Chem.*, **1986**, *213*, 203.
305. C. M. Elliott, S. J. Schmittle, J. G. Redepenning, and E. M. Galk, *J. Macromol. Sci. Chem.*, **1988**, 1215.
306. C. M. Elliott, J. G. Redepenning, S. J. Schmittle, and E. M. Balk, *ACS Symp. Ser. 360*, M. Zeldin, K. J. Wynne, and J. R. Allcock (Eds.), Washington, DC, pp. 420–429, **1988**.
307. C. J. Baldy, C. M. Elliott, and S. W. Feldberg, *J. Electroanal. Chem.*, **1990**, *283*, 53.
308. C. M. Elliott, A. B. Kopelove, W. J. Albery, and Z. Chen, *J. Phys. Chem.*, **1991**, *95*, 1743.
309. G. Mengoli and M. M. Usiani, *J. Electrochem. Soc.*, **1987**, *134*, 643C.
310. F. Bruno, M. C. Pham, and J. E. Doubois, *Electrochim. Acta*, **1977**, *22*, 451.
311. N. Oyama, T. Ohsaka, Y. Ohnuki, and T. Suzuki, *J. Electrochem. Soc.*, **1987**, *131*, 3068.
312. G. Mengoli, *Adv. Polym. Sci.*, **1978**, *33*, 26.
313. R. V. Subramanian, *Adv. Polym. Sci.*, **1979**, *33*, 43.
314. M. C. Pham, P. C. Lacaze, and J. E. Dubois, *J. Electroanal. Chem.*, **1978**, *86*, 147.
315. J. E. Dubois, P. C. Lacaze, and M. C. Pham, *J. Electroanal. Chem.*, **1981**, *117*, 233.
316. M. C. Pham, J. E. Dubois, and P. C. Lacaze, *J. Electroanal. Chem.*, **1979**, *99*, 331.
317. C. Iwakura, M. Tsunga, and H. Tamura, *Electrochim. Acta*, **1972**, *17*, 1391.
318. G. Mengoli, S. Daolio, U. Giulio, and C. Folonari, *J. App. Electrochem.*, **1979**, *9*, 483.
319. G. Mengoli, S. Daolio, and M. M. Musiani, *J. App. Electrochem.*, **1980**, *10*, 459.
320. G. Mengoli, M. M. Musiani, B. Pelli, M. Fleischmann, and I. R. Hill, *Electrochim. Acta*, **1983**, *28*, 1733.
321. G. Mengoli and M. M. Musiani, *Electrochim. Acta*, **1986**, *31*, 201.
322. G. Mengoli, P. Bianco, S. Daolio, and M. T. Munari, *J. Electrochem. Soc.*, **1981**, *128*, 2276.
323. M. M. Musiani, C. Pagura, and G. Mengoli, *Electrochim. Acta*, **1985**, *30*, 501.
324. R. L. McCarley, R. E. Thomas, E. A. Irene, and R. W. Murray, *J. Electroanal. Chem.*, **1990**, *290*, 79.
325. N. Oyama, T. Shimomura, K. Shigehara, and F. C. Anson, *J. Electroanal. Chem.*, **1980**, *112*, 271.
326. R. J. Mortimer and F. C. Anson, *J. Electroanal. Chem.*, **1982**, *138*, 325.
327. F. C. Anson, J.-M. Savéant, and K. Shigehara, *J. Am. Chem. Soc.*, **1983**, *105*, 1096.
328. N. Oyama and F. C. Anson, *Anal. Chem.*, **1980**, *52*, 1192.
329. N. Oyama, T. Ohsaka, and T. Ushirogouchi, *J. Phys. Chem.*, **1984**, *88*, 5274.
330. N. Oyama, K. Sato, and H. Matsuda, *J. Electroanal. Chem.*, **1980**, *115*, 149.
331. F. C. Anson, T. Ohsaka, and J.-M. Savéant, *J. Am. Chem. Soc.*, **1983**, *105*, 4883.
332. T. Ohsaka, T. Okajima, and N. Oyama, *J. Electroanal. Chem.*, **1986**, *215*, 191.
333. N. Oyama, T. Ohsaka, and T. Okajima, *Anal. Chem.*, **1986**, *58*, 979.
334. D. A. Buttry and F. C. Anson, *J. Am. Chem. Soc.*, **1982**, *104*, 4824.
335. D. A. Buttry and F. C. Anson, *J. Am. Chem. Soc.*, **1983**, *105*, 685.
336. D. A. Buttry, J.-M. Savéant, and F. C. Anson, *J. Phys. Chem.*, **1984**, *88*, 3086.
337. K. Shigehara, E. Tsuchida, and F. C. Anson, *J. Electroanal. Chem.*, **1984**, *175*, 291.

338. F. C. Anson, J.-M. Savéant, and Y.-M. Tdou, *J. Electroanal. Chem.*, **1984**, *178*, 113.
339. F. C. Anson, C.-L. Ni, and J.-M. Savéant, *J. Am. Chem. Soc.*, **1985**, *107*, 3442.
340. J. Redepenning and F. C. Anson, *J. Phys. Chem.*, **1987**, *91*, 4549.
341. J. Redepenning and F. C. Anson, *J. Phys. Chem.*, **1986**, *90*, 6227.
342. C. F. Shu and F. C. Anson, *J. Phys. Chem.*, **1990**, *94*, 8345.
343. J. S. Facci and R. W. Murray, *J. Phys. Chem.*, **1981**, *85*, 2870.
344. J. S. Facci and R. W. Murray, *J. Electroanal. Chem.*, **1981**, *124*, 339.
345. K. Doblhofer, H. Braun, and R. Lange, *J. Electroanal. Chem.*, **1986**, *206*, 93.
346. K. Niwa and K. Doblhofer, *Electrochim. Acta*, **1986**, *31*, 553.
347. E. S. De Castro, E. W. Huber, D. Villaroel, C. Galiatsatos, J. E. Mark, W. E. Heineman, and P. T. Murray, *Anal. Chem.*, **1987**, *59*, 134.
348. W. J. Vining and T. J. Meyer, *Inorg. Chem.*, **1986**, *25*, 2023.
349. M. N. Szentirmay and C. R. Martin, *Anal. Chem.*, **1984**, *56*, 1898.
350. M. Majda and L. R. Faulkner, *J. Electroanal. Chem.*, **1984**, *169*, 97.
351. I. Rubinstein and A. J. Bard, *J. Am. Chem. Soc.*, **1980**, *102*, 6641.
352. H. S. White, J. Leddy, and A. J. Bard, *J. Am. Chem. Soc.*, **1982**, *104*, 4811.
353. C. R. Martin, I. Rubinstein, and A. J. Bard, *J. Am. Chem. Soc.*, **1982**, *104*, 4817.
354. T. P. Henning, H. S. White, and A. J. Bard, *J. Am. Chem. Soc.*, **1983**, *103*, 3937.
355. A. J. Bard, T. P. Henning, and H. S. White, *J. Am. Chem. Soc.*, **1982**, *104*, 5362.
356. A. E. Kaifer and A. J. Bard, *J. Phys. Chem.*, **1986**, *90*, 868.
357. J. Leddy and A. J. Bard, *J. Electroanal. Chem.*, **1985**, *189*, 203.
358. G. Nagy, G. A. Gerhardt, A. F. Oke, M. E. Rice, R. N. Adams, R. B. Moore, III, M. N. Szentirmay, and C. R. Martin, *J. Electroanal. Chem.*, **1985**, *188*, 85.
359. C. R. Martin, *J. Chem. Soc. Faraday Trans. 1*, **1986**, *82*, 1051.
360. R. M. Penner and C. R. Martin, *J. Electrochem. Soc.*, **1985**, *132*, 514.
361. C. M. Lieber, M. H. Schmidt, and N. S. Lewis, *J. Am. Chem. Soc.*, **1986**, *108*, 6103.
362. C. M. Lieber, M. H. Schmidt, and N. S. Lewis, *J. Phys. Chem.*, **1986**, *90*, 1002.
363. N. K. Cenas, A. K. Pocius, and J. J. Kulys, *Bioelectrochem. Bioenerg.*, **1984**, *12*, 583.
364. M. E. Rice and C. Nicholson, *Anal. Chem.*, **1989**, *61*, 1805.
365. C. R. Martin, *Anal. Chem.*, **1982**, *54*, 1639.
366. R. B. Moore III and C. R. Martin, *Anal. Chem.*, **1986**, *58*, 2569.
367. L. D. Whiteley and C. R. Martin, *J. Phys. Chem.*, **1989**, *93*, 4650.
368. C.-F. Shu and F. C. Anson, *J. Am. Chem. Soc.*, **1990**, *112*, 9227.
369. F. C. Anson, T. Ohsaka, and J.-M. Savéant, *J. Phys. Chem.*, **1983**, *87*, 640.
370. M. Kato, N. Oki, H. Ohno, E. Tsuchida, and N. Oyama, *Polymer*, **1983**, *24*, 846.
371. N. Oyama, T. Ohsaka, K. Sato, and H. Yamamoto, *Anal. Chem.*, **1983**, *55*, 1429.
372. P. Ugo and F. C. Anson, *Anal. Chem.*, **1989**, *61*, 1799.
373. K. Pressprich, Ph.D. Thesis, University of North Carolina, **1989**.
374. V. D. Neff, *J. Electrochem. Soc.*, **1978**, *125*, 886.
375. K. Itaya, A. Tasuaki, and S. Toshima, *J. Am. Chem. Soc.*, **1982**, *104*, 4767.
376. K. Itaya, I. Uchida, and V. D. Neff, *Acc. Chem. Res.*, **1986**, *19*, 162.
377. C. Lundgren and R. W. Murray, *Inorg. Chem.*, **1988**, *27*, 933.
378. B. J. Feldman and R. W. Murray, *Anal. Chem.*, **1986**, *58*, 2844.
379. B. J. Feldman and R. W. Murray, *Inorg. Chem.*, **1987**, *26*, 1702.
380. A. B. Bocarsly and S. Sinha, *J. Electroanal. Chem.*, **1982**, *137*, 157.

381. A. B. Bocarsly and S. Sinha, *J. Electroanal. Chem.*, **1982**, *140*, 167.
382. A. B. Bocarsly, S. A. Galvin, and S. Sinha, *J. Electrochem. Soc.*, **1983**, *130*, 1319.
383. S. Sinha, B. D. Humphrey, and A. B. Bocarsly, *Inorg. Chem.*, **1984**, *23*, 203.
384. C. Hidalgo-Luangdilok and A. B. Bocarsly, *Inorg. Chem.*, **1990**, *29*, 2894.
385. P. J. Kulesza and L. R. Faulkner, *J. Electroanal. Chem.*, **1988**, *248*, 305.
386. P. J. Kulesza and L. R. Faulkner, *J. Am. Chem. Soc.*, **1988**, *110*, 4905.
387. P. J. Kulesza and L. R. Faulkner, *J. Electroanal. Chem.*, **1989**, *259*, 81.
388. P. J. Kulesza and L. R. Faulkner, *J. Electrochem. Soc.*, **1989**, *136*, 707.
389. B. Keita, K. Essaadi, and L. Nadjo, *J. Electroanal. Chem.*, **1989**, *259*, 127.
390. B. Keita and L. Nadjo, *J. Electroanal. Chem.*, **1988**, *243*, 87.
391. B. Keita, L. Nadjo, and J.-M. Savéant, *J. Electroanal. Chem.*, **1988**, *243*, 105.
392. K. A. Daube, D. J. Harrison, T. E. Mallouk, A. J. Ricco, S. Chao, and M. S. Wrighton, *J. Photochem.*, **1985**, *29*, 71.
393. K. M. Kost, D. Bartak, B. Kazee, and T. Kuwana, *Anal. Chem.*, **1990**, *62*, 151.
394. J. R. Schneider and R. W. Murray, *Anal. Chem.*, **1972**, *54*, 1508.
395. H. D. Abruña, P. Denisevich, M. Umaña, T. J. Meyer, and R. W. Murray, *J. Am. Chem. Soc.*, **1981**, *103*, 1.
396. G. P. Kittlesen, H. S. White, and M. S. Wrighton, *J. Am. Chem. Soc.*, **1984**, *106*, 7389.
397. G. P. Kittlesen, H. S. White, and M. S. Wrighton, *J. Am. Chem. Soc.*, **1985**, *107*, 7373.
398. E. W. Paul, A. J. Ricco, and M. S. Wrighton, *J. Phys. Chem.*, **1985**, *89*, 1441.
399. J. W. Thackeray, H. S. White, and M. S. Wrighton, *J. Phys. Chem.*, **1985**, *89*, 5133.
400. M. J. Natan, T. E. Mallouk, and M. S. Wrighton, *J. Phys. Chem.*, **1987**, *91*, 648.
401. D. Ofer, D. M. Crooks, and M. S. Wrighton, *J. Am. Chem. Soc.*, **1990**, *112*, 7689.
402. C.-F. Shu and M. S. Wrighton, *J. Phys. Chem.*, **1988**, *92*, 5221.
403. E. T. T. Jones, O. M. Chyan, and M. S. Wrighton, *J. Am. Chem. Soc.*, **1987**, *109*, 5526.
404. D. Belanger and M. S. Wrighton, *J. Am. Chem. Soc.* ???
405. H. S. White, G. P. Kittelsen, and M. S. Wrighton, *J. Am. Chem. Soc.*, **1984**, *106*, 5375.
406. S. Chao and M. S. Wrighton, *J. Am. Chem. Soc.*, **1987**, *109*, 2197.
407. P. Burgmayer and R. W. Murray, *J. Am. Chem. Soc.*, **1982**, *104*, 6139.
408. P. Burgmayer and R. W. Murray, *J. Electroanal. Chem.*, **1983**, *147*, 339.
409. P. Burgmayer and R. W. Murray, *J. Phys. Chem.*, **1984**, *88*, 2515.
410. Y. Okahata, S. Hachiya, K. Ariga, and T. Seki, *J. Am. Chem. Soc.*, **1986**, *108*, 2863.
411. C. A. Lundgren and R. W. Murray, *J. Electroanal. Chem.*, **1987**, *227*, 287.
412. E. S. Takeuchi and R. W. Murray, *J. Electroanal. Chem.*, **1985**, *107*, 872.
413. M. K. Halbert and R. P. Baldwin, *J. Chromatogr.*, **1985**, *345*, 43.
414. M. K. Halbert and R. P. Baldwin, *Anal. Chem.*, **1985**, *57*, 591.
415. J. Wang, B. Greene, and C. Morgan, *Anal. Chim. Acta*, **1984**, *158*, 15.
416. P. K. Ghosh and A. J. Bard, *J. Am. Chem. Soc.*, **1983**, *105*, 5691.
417. C. G. Murray, R. J. Nowak, and D. R. Rolison, *J. Electroanal. Chem.*, **1984**, *164*, 205.
418. P. V. Kamat, *J. Electroanal. Chem.*, **1984**, *163*, 389.
419. A. Yamagishi and A. J. Aramata, *J. Chem. Soc., Chem. Commun.*, **1984**, 452.
420. P. K. Ghosh, A. W-H. Mau, and A. J. Bard, *J. Electroanal. Chem.*, **1984**, *169*, 315.
421. D. Ege, P. K. Ghosh, J. R. White, J. F. Equey, and A. J. Bard, *J. Am. Chem. Soc.*, **1985**, *107*, 5644.
422. H. Y. Liu and F. C. Anson, *J. Electroanal.* Chem., **1985**, *184*, 411.
423. N. Oyama and F. C. Anson, *J. Electroanal. Chem.*, **1986**, *199*, 467.

423. J. R. White and A. J. Bard, *J. Electroanal. Chem.*, **1986**, *197*, 233.
424. M. T. Carter and A. J. Bard, *J. Electroanal. Chem.*, **1987**, *229*, 191.
425. C. M. Castro-Acuna, F.-R. F. Fan, and A. J. Bard, *J. Electroanal. Chem.*, **1987**, *236*, 43.
426. R. D. King, D. G. Nocera, and T. J. Pinnavaia, *J. Electroanal. Chem.*, **1987**, *236*, 43.
427. H. Inoue, S. Haga, C. Iwakura, and H. Yoneyama, *J. Electroanal. Chem.*, **1988**, *249*, 133.
428. A. Fitch, A. Lavy-Feder, S. A. Lee, and M. T. Kirsch, *J. Phys. Chem.*, **1988**, *92*, 6665.
429. W. E Rudzinski and A. J. Bard, *J. Electroanal. Chem.*, **1986**, *199*, 323.
430. B. de Vismes, F. Bedioui, J. Devynck, and C. Bied-Charreton, *J. Electroanal. Chem.*, **1985**, *187*, 197.
431. H. A. Gemborys and B. R. Shaw, *J. Electroanal. Chem.*, **1986**, *208*, 95.
432. B. R. Shaw, K. E. Creasy, C. J. Lanczycki, J. A. Sargeant, and M. Tirhado, *J. Electrochem. Soc.*, **1988**, *135*, 869.
433. Z. Li and T. E. Mallouk, *J. Phys. Chem.*, **1987**, *91*, 643.
434. Z. Li, C. M. Wang, L. Persaud, and T. E. Mallouk, *J. Phys. Chem.*, **1988**, *92*, 2592.
435. Z. Li, C. Lai, and T. E. Mallouk, *Inorg. Chem.*, **1989**, *28*, 178.
436. D. R. Rolison, R. J. Nowak, S. Pns, J. Ghoroghchian, and M. Fleischmann, in *Molecular Electronic Devices III*, F. L. Carter, R. E. Siatkowski, and H. Wohltjen (Eds.), Elsevier: Amsterdam, **1988**.
437. D. R. Rolison, E. A. Hayes, and W. E. Rudzinski, *J. Phys. Chem.*, in press.
438. B. de Vismes, F. Bedioui, J. Devynck, C. Bied-Charreton, and M. Perree-Fauvet, *Nouv. J. Chim.*, **1986**, *10*, 81.
439. K. E. Creasy and B. R. Shaw, *Electrochim. Acta*, **1988**, *33*, 551.
440. G. A. Ozin, A. Kuperman, and A. Stein, *Angew. Chem. Intl. Ed.*, in press.
441. D. K. Smith, L. M. Tender, G. A. Lane, S. Licht, and M. S. Wrighton, *J. Am. Chem. Soc.*, **1989**, *111*, 1099.
442. S. Gould, T. R. O'Toole, and T. J. Meyer, *J. Am. Chem. Soc.*, **1990**, *112*, 9490.

Chapter **II**

ADSORBED ORGANIC MOLECULES AT WELL-DEFINED ELECTRODE SURFACES

Ghaleb N. Salaita and Arthur T. Hubbard
Department of Chemistry, University of Cincinnati,
Cincinnati, Ohio

2.1 INTRODUCTION

In this chapter we discuss the chemisorption of aromatic molecules from aqueous solution at well-defined electrode surfaces prepared and characterized under ultrahigh vacuum (UHV) conditions. "Well-defined surfaces" are atomically clean, ordered single-crystal planes (1–3). Single crystals used in this work were oriented parallel to low-index planes, such as Pt(111) and Pt(100). Each plane of course provides different types of binding sites, different densities, and different coordination of the surface atoms, Fig. 2.1. Low-index surfaces are widely used in surface investigations because of their low surface free-energy, high symmetry, and stability (3, 4). Platinum electrodes combine the above characteristics with interesting chemical and catalytic properties. Platinum surfaces adsorb a wide variety of ions, atoms, and molecular functional groups, often accompanied by oxidation–reduction or dissociation. Rapid progress has

Molecular Design of Electrode Surfaces,
Edited by Royce W. Murray. Techniques of Chemistry Series, Vol. XXII.
ISBN 0-471-55773-0

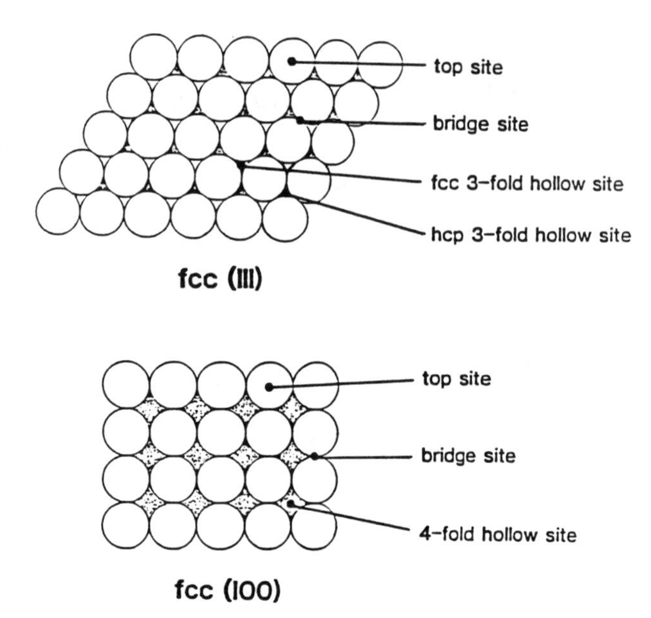

Figure 2.1. Top views of the atomic arrangement in the (111) and (100) planes of face centered cubic (fcc) metals. Surface atoms are unshaded while second layer atoms are shaded. High symmetry binding sites on the surfaces are indicated.

been achieved recently in studying the surface chemistry of organic molecules at platinum–liquid interfaces and significant data have begun to accumulate (5–13). The objective of the work is to establish a firm foundation for understanding solid–liquid interfaces.

At solid–liquid interfaces, the surface properties of adsorbed species are influenced by such variables as pH, concentration, electrode potential, and metal surface crystallographic orientation (14–22). In particular, these variables influence adsorbate composition, orientation, molecular symmetry, electronic structure, conformation, mode of attachment, electrochemical reactivity, and chemical reaction pathways. These dependencies are associated with important technologies and industrial applications. Examples are adhesives, lubricants, batteries, fuel cells, electrocatalysts, agents for inhibition of corrosion, semiconductor devices, electroplating processes, polymer films, and surface-related biochemical technology.

With the advent of surface-sensitive electron spectroscopic methods and UHV technology coupled with cyclic voltammetry and chronocoulometry, it is now possible to investigate at the molecular level the processes occurring at the electrode–solution interfaces. In our laboratory, for example, such measurements are conducted in a specially designed and constructed instrument with a background pressure of better than 1×10^{-10} torr. The best method of cleaning electrode surfaces is usually argon ion bombardment, but in any event the surface must be carefully characterized after cleaning and annealing. Surface structure is observed by low-energy electron diffraction (LEED). Surface elemental composition and cleanliness are monitored by Auger electron spectroscopy (AES). Residual gas composition in the main vacuum chamber and the kinetics of adsorbate desorption from the electrode surface are probed by thermal desorption mass spectroscopy (TDMS). Vibrational bands are established by high resolution electron energy loss spectroscopy (EELS) and by infrared reflection–adsorption spectroscopy (IRRAS).

The molecular orientations of a large number of organic compounds adsorbed at polycrystalline Pt surfaces were studied recently by use of thin-layer electrodes (TLE) (23, 24). Aromatic compounds in the absence of competing surfactants display a series of packing density ($nmol\,cm^{-2}$) plateaus as a function of adsorbate concentration (at constant pH and electrode potential). Comparison of plateau values with molecular models based on covalent and van der Waals radii (25), suggests that each plateau corresponds to a specific molecular orientation. A drawback of TLE is that the method is limited to electroactive compounds. On the other hand, it has an important advantage: It gives reliable and quantitative packing density information. Hence, it can be used as a first method for testing the feasibility of the formation of an adsorbed layer of any new electroactive compound at an electrode surface (11). Also, a comparison between data obtained from polycrystalline platinum TLE (23, 26–31) with data obtained at well-defined surfaces by means of surface spectroscopy has now become possible. Well-defined electrode surfaces are of interest at the very least as ideal model systems and the data obtained could in many cases be adopted as standards (11, 32).

2.2 KINETICS OF ADSORPTION

Comparisons of the chemical changes that take place at the gas–solid (G–S) and liquid–solid (L–S) interfaces are of utmost importance in heterogeneous catalysis. The succession of possible events in these two types of heterogeneous processes are intrinsically similar:

1. Trapping of the adsorbate in a shallow physisorption state that acts as a precursor of chemisorption. Formation of this state can be virtually reversible and nonactivated.
2. Formation of a chemisorbed state, typically as an irreversible process at low temperatures or more nearly reversible at higher temperatures. This adsorption is often described by Lennard–Jones potential-energy diagrams (33, 34). However, the molecular structures and detailed chemical reaction pathways involved in G–S and L–S interfacial processes may differ as a result of:
   a. Differing molecular fluxes.
   b. Solvent effects.
   c. Influences of interfacial electrode potential at the L–S interface, Fig. 2.2 (35).

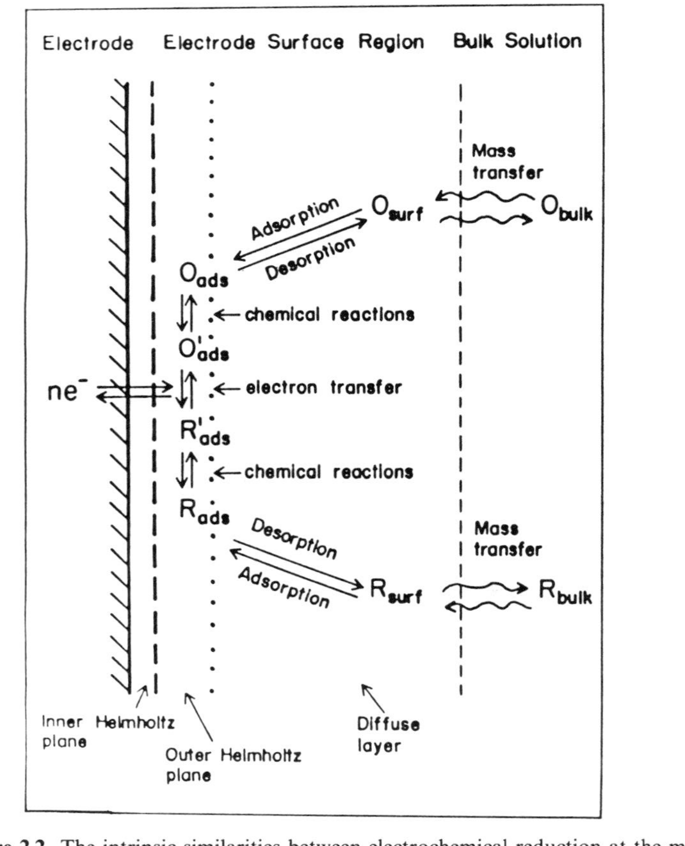

Figure 2.2. The intrinsic similarities between electrochemical reduction at the metal electrode in solution (*A*) and reaction of gas-phase molecules at a metal surface (*B*). The presence of solvent molecules and the effect of the electrode potential in electroreductions are the main differences between these two processes. [Reprinted with permission from A. Wieckowski, S. D. Rosasco, G. N. Salaita, A. T. Hubbard, B. E. Bent, F. Zaera, D. Godkey, and G. A. Somarjai, *J. Am. Chem. Soc.*, *107*, 5910 (1985). Copyright © (1985) American Chemical Society.]

For instance, a comparison has been made between reduction of ethylene at L–S and G–S interfaces under otherwise identical conditions (35). Ethylene was chosen as the adsorbate because it has been the center of considerable attention at G–S (36–42) and L–S interfaces (43–45). The final hydrogenation product of ethylene is ethane in G–S and L–S processes. In these studies several surface science and electrochemical techniques such as LEED, Auger spectroscopy, temperature programmed desorption (TPD), EELS, cyclic voltammetry, and chronocoulometry were employed in UHV to characterize the adsorbed species before, during, and after the reduction process. The results show that L–S hydrogenation occurs directly on the Pt surface and involves adsorbed hydrogen atoms, whereas in G–S hydrogenation, the primary pathway is thought to be a transfer of hydrogen atoms from the Pt surface to the ethylene that is adsorbed on top of a layer of irreversibly adsorbed hydrocarbon fragments (35). The G–S hydrogenation behavior approaches L–S behavior most closely when the Pt surface is presaturated with H_2.

Low-energy electron diffraction shows a (2 × 2) pattern for adsorbed ethylene under G–S conditions at Pt(111). From LEED (46, 47), UPS (48), and TPD (49–51) studies it was found that the (2 × 2) LEED pattern for adsorbed ethylene at the Pt(111) surface match those for ethylidyne at the same surface (46, 49). From this it was concluded that ethylene adsorbs dissociatively on the clean Pt(111) surface to from ethylidyne (CCH_3, Fig. 2.3). The conversion of

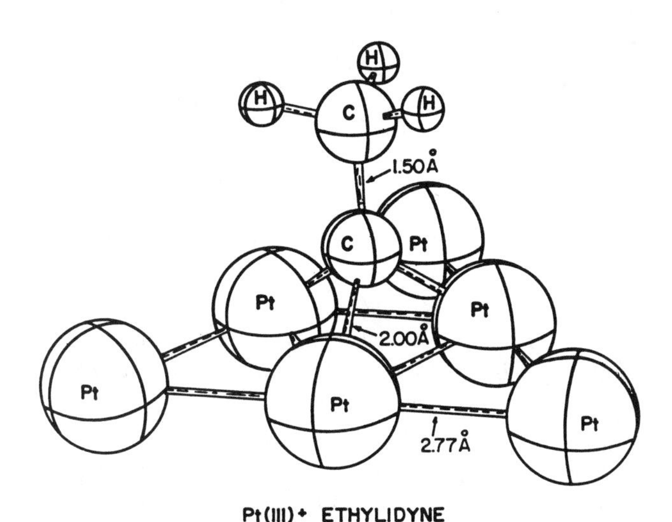

Figure 2.3. The structure of ethylidyne, produced by ethylene chemisorption from gas phase on a Pt(111) single-crystal surface in the temperature range 300–450 K. The bond lengths of this ethylidyne species were determined from LEED analysis. [Reprinted with permission from A. Wieckowski, S. D. Rosasco, G. N. Solaita, A. T. Hubbard, B. E. Bent, F. Zaera, D. Godkey, and G. A. Somarjai, *J. Am. Chem. Soc.*, *107*, 5910 (1985). Copyright © (1985) American Chemical Society.]

ethylene to ethylidyne involves transfer of a hydrogen atom from ethylene to the surface, followed by recombination with another hydrogen atom and desorption as H_2. The remaining C_2H_3 fragment undergoes a hydrogen atom shift to form a methyl group, which sits above the dehydrogenated carbon in a threefold hollow site in the Pt(111) surface. Ethylene can only be physisorbed on a Pt(111) surface covered with ethylidyne. On the other hand, chemisorbed ethylene at Pt(111) under L–S conditions in UHV gives a (1×1) LEED pattern, which indicates densely packed, undissociated ethylene (35). A useful compilation of data related to surface structures of adsorbed gases on a large number of single-crystal surfaces has been published by Castner and Somorjai (42, 52).

2.3 EXPERIMENTAL ASPECTS OF LIQUID–SOLID ADSORPTION

In typical experiments, Pt(111) and Pt(100) single-crystal planes are polished (53) and oriented (54) so that all faces of the crystal are crystallographically equivalent to the plane of interest. Prior to each experiment, all six faces of the crystal are cleaned simultaneously by Ar^+ bombardment (500 eV, 4×10^{-6} A cm^{-2}) and annealed by resistance heating in UHV. Then the surface is examined for cleanliness by Auger spectroscopy and tested by LEED for ordered structure. The clean crystal is then transferred and isolated in an argon filled antechamber, immersed into the adsorbate solution at controlled potential for approximately 3 min, and then rinsed thoroughly with pure supporting electrolyte. The covered Pt surface is then transferred back into vacuum and recharacterized by the various sensitive electron spectroscopies.

A schematic diagram of a surface-electrochemistry instrument utilized in L–S research is displayed in Fig. 2.4. The instrument consists of three stainless steel vacuum chambers interconnected by gate valves:

1. The sample chamber houses LEED, AES, EELS, TDMS, ion bombardment gun, leak valve-doser tube assembly, and a viewport through which to photograph LEED patterns.
2. A separate antechamber that encloses the electrochemical (or high pressure) cell and the IR reflection–adsorption components. A motorized translation stage is employed to transfer the sample between the UHV chamber and the antechamber.
3. The pumping chamber consists of two sorption pumps (Zeolite beads cooled to liquid nitrogen temperature) to lower the pressure from atmosphere to 10^{-3} torr; a cryogenic pump (activated charcoal at 14 K) to lower the pressure from 10^{-3} to 10^{-7} torr, an ionization pump to lower the pressure from 10^{-7} to below 10^{-10} torr, and a titanium sublimation pump (surrounded by a liquid nitrogen cooled cryopanel to lower the level of residual gases such as CO during ion bombardment of the sample surface with Ar^+ ions (11, 55).

Adsorption of organic molecules from aqueous solution at a well-

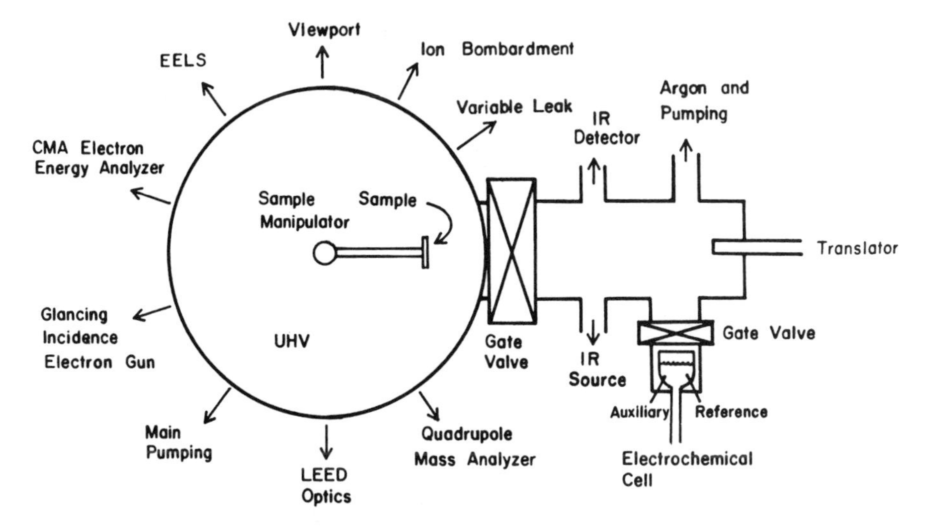

Figure 2.4. The UHV–electrochemistry system used in the L–S interfacial work. Crystal cleaning and adsorbate characterization is accomplished with the surface science techniques in the UHV chamber. The crystal was transferred through a gate valve to the antechamber for electrochemical measurements. [Reprinted with permission from (*Electrochim. Acta, 34*, N. Batina et al., "Oriented adsorption at Well-Defined Electrode Surfaces Studied by Auger, LEED, and EELS Spectroscopy") Copyright © (1989) Pergamon Press, New York.

defined electrode surface Pt(111) and Pt(100) can lead to at least three different types of surface species:

a. A chemisorbed molecular layer of one atomic layer thickness, such as horizontally oriented hydroquinone.
b. A chemisorbed layer of multiatomic layer thickness, such as vertically oriented pyridine or benzene.
c. On an oriented layer of mixed thickness characteristics (e.g., both horizontal and vertical orientation).

The orientation of the adsorbate can be deduced from accurate measurements of the packing densities, *F* (moles of adsorbed atoms or molecules per area), based upon two independent Auger methods (5–13).

1. The second derivative Auger signal (I_X), due to each element X is measured, from which the packing densities, F_X, are calculated. When applying this method to chemisorbed species of one atomic layer thickness as in case (a) the initial calculation neglects scattering of Auger electrons within the adsorbed layer, Eq. 2.1:

$$\Gamma_X = \frac{I_X/I^{\circ}_{Pt}}{B_X} \tag{2.1}$$

where I_X is the Auger signal for element X (X = C, O, N, S, and any other non-hydrogen atoms) and I_{Pt}° is the Auger signal for clean Pt substrates (normalization peak). The term I_{Pt}° was measured at 161 eV for non-sulfur containing compounds and at 235 eV for sulfur containing compounds to avoid overlap of the sulfur signal with Pt at 161 eV, as shown in Fig. 2.5. The term B_X is obtained by means of calibration experiments using an ordered atomic layer containing the appropriate element.

2. The second Auger method for determination of the packing density, Γ (nmol cm^{-2}) is based on attenuation of the Pt substrate Auger signal by the adsorbed layer, Eq. 2.2,

$$\frac{I_{Pt}}{I_{Pt}^{\circ}} = (1 - J_1 K_X \Gamma) \tag{2.2}$$

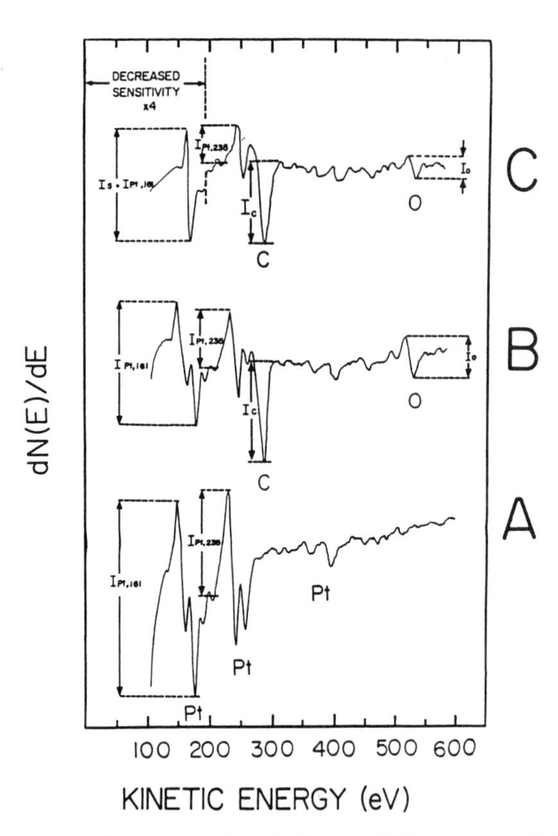

Figure 2.5. Measurements of Auger signals: (*A*) Clean Pt(111) spectrum. (*B*) Spectrum of HQ adsorbed on Pt(111). (*C*) Spectrum of DMBM adsorbed on Pt(111). Experimental conditions: incident beam, 10^{-7} A at 2000 eV, normal to the surface; modulation amplitude, 5 V peak to peak; electrolyte, 10 m*M* KF adjusted to pH = 4 with HF; electrode potential, 0.20 V versus Ag/AgCl reference.

where I°_{Pt} is the Pt Auger signal for clean Pt substrate before adsorption, I_{Pt} is the Pt Auger signal with the adsorbed layer present, J_1 is the number of non-hydrogen atoms in the molecule, and K_X is a calibration constant ($cm^2\,nmol^{-1}$). The term K_X at 161 eV is found to be equal to $0.16\,cm^2\,nmol^{-1}$ for light elements such as C, N, and O; for heavier elements such as S, K_S is found to be equal to $0.229\,cm^2\,nmol^{-1}$. However, when the Auger signal was measured at 235 eV, K_S is found to be equal to $0.153\,cm^2\,nmol^{-1}$. For example, for horizontally oriented adsorbed hydroquinone (HQ) the packing densities, Γ_C ($nmol\,cm^{-2}$) and Γ_O ($nmol\,cm^{-2}$), are found from the carbon and oxygen Auger signals, respectively, Eqs. 2.3 and 2.4:

$$\Gamma_C = (I_C/I^{\circ}_{Pt})/B_C \tag{2.3}$$

$$\Gamma_O = (I_O/I^{\circ}_{Pt})/B_O \tag{2.4}$$

where $B_C = 0.377\,cm^2\,nmol^{-1}$ and $B_O = 0.476\,cm^2\,nmol^{-1}$. The packing density, Γ($nmol\,cm^{-2}$), of horizontally oriented HQ based on the attenuation of Pt substrate Auger signal, I_{Pt}/I°_{Pt}, is

$$\frac{I_{Pt}}{I^{\circ}_{Pt}} = (1 - 8\,K_X\Gamma) \tag{2.5}$$

Adsorbed HQ displays a Pt(111)(3 × 3) diffraction pattern, Fig. 2.6*A*. Based upon this (3 × 3) symmetry and the presence of one HQ molecule per unit cell, the packing density amounts to one HQ molecule per nine surface Pt atoms, $\Gamma = 0.277\,nmol\,cm^{-2}$. This packing density value is adopted as the calibration point for measurements of packing density. A diagram of the (3 × 3) pattern appears in Fig. 2.6*B*. It is useful to note that the theoretical packing density of horizontally adsorbed HQ based on covalent and van der Waals radii (Ref. 24 and Fig. 2.7) is $0.293\,nmol\,cm^{-2}$, compared with the observed horizontal plateau value of $0.29–0.30\,nmol\,cm^{-2}$ based upon Auger spectroscopy.

For adsorbed layers of multiatomic layer thickness, the scattering of Auger electrons within the layers is considered. The above equations can be generalized as follows:

$$\frac{I_X}{I^{\circ}_{Pt}} = B_X\Gamma_X(L_1a_1 + \cdots + L_ia_i + \cdots + L_nA_n) \tag{2.6}$$

where L_i is the fraction of atoms of type X located in level i ($i = 1$ is adjacent to the solid surface and n is the outermost layer). The scattering factor of the ith carbon atom in the layer is

$$a_i = f_C^{M_i} \tag{2.7}$$

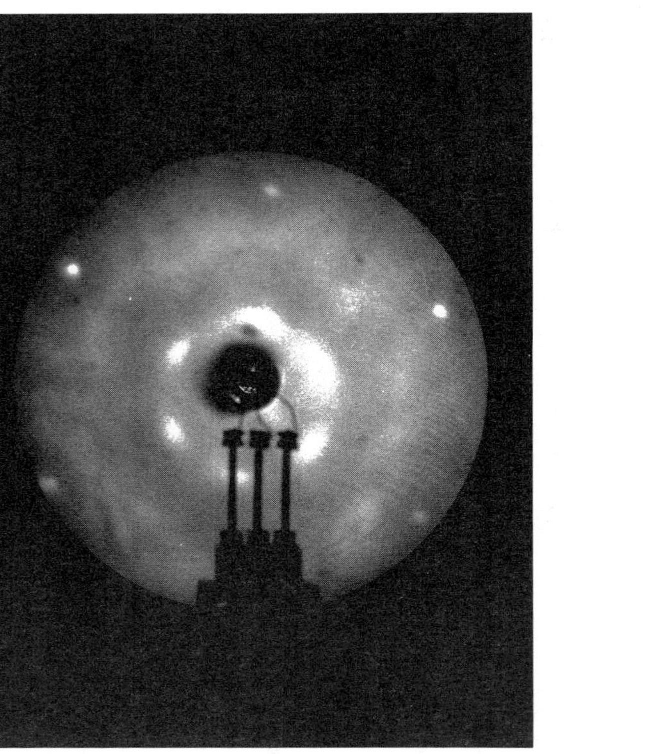

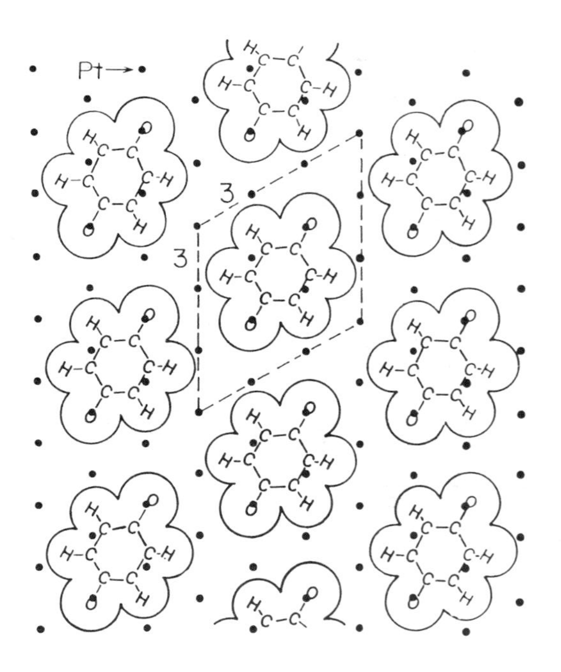

Figure 2.6. LEED pattern for HQ at Pt(111). (*A*) HQ. (*B*) Pt(111)(3 × 3) – HQ, $\Theta = \frac{1}{9}(\Gamma = 0.277\ \text{nmol cm}^{-2})$. [Reprinted with permission from F. Lee, et al., *Langmuir*, *4*, 637 (1988). Copyright © American Chemical Society.]

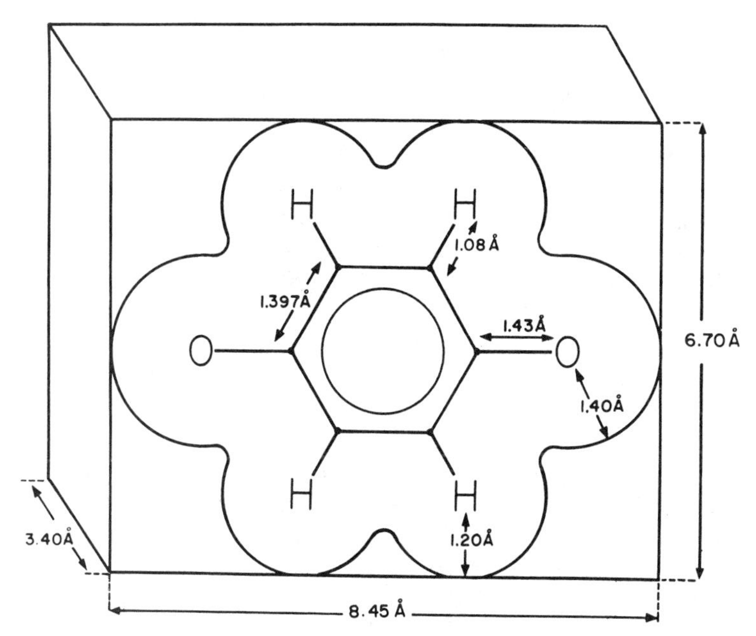

Figure 2.7. Model structure used for calculating the theoretical packing densities of horizontally and vertically oriented HQ. [Reprinted with permission from (*Electrochem, Acta, 34,* N. Batina et al. "Oriented Adsorption at Well-Defined Electrode Surfaces Studied by Auger, LEED, and EELS Spectroscopy", Copyright © (1989) Pergamon Press, New York.

where M_i is the number of atoms located on the path from the emitting carbon atom to the acceptance angle (42° ± 3°) of the cylindrical mirror analyze (CMA) detector. The term $f_C = f_O = 0.70$ is the attenuation of the Pt Auger signal at 235 eV based upon data for horizontally oriented adsorbed HQ on Pt(111).

Based upon Pt Auger signal attenuation

$$\frac{I_{Pt}}{I^{\circ}_{Pt}} = (1 - J_1 K_X \Gamma) \cdots (1 - J_n K_X \Gamma) \tag{2.8}$$

where J_i is the number of non-hydrogen atoms per molecule located at the ith level of the adsorbed layer, and n is the outermost layer. It could be mentioned here that Γ (mol cm^{-2}) can be converted to Θ (molecule per surface Pt atom) using Eq. 2.9.

$$\Theta = \frac{\Gamma}{\Gamma_{Pt}} \tag{2.9}$$

where $\Gamma_{Pt} = 1.5 \times 10^{15}$ atoms cm^{-2} for Pt(111). An example of this large class of multiatomic layer adsorbates is 2,5-dihydroxy-4-methylbenzylmercaptan

(DMBM) in which the aromatic ring is oriented perpendicularly to the surface as shown in Fig. 2.8. From the figure it is evident that the Auger electrons emitted from the sulfur atom and from the carbon and oxygen atoms situated near the Pt substrate surface are influenced as a result of the scattering by atoms situated enroute to the detector.

Referring to Fig. 2.8, the Auger electrons emitted by C_3 and C_8 are not attenuated (no atoms lie above them) while the Auger electrons from C_2, C_4, C_1, and C_7 are attenuated in about one-half of the possible directions of travel to the spectrometer, giving rise to the term in Eq. 2.6. Auger electrons from C_5 are attenuated once, while C_6 is attenuated twice in one-half of its possible directions and once in the other half. Combining terms leads to Eq. 2.10.

$$\Gamma_C = (I_C/I^\circ_{Pt})/[B_C(\tfrac{1}{2} + \tfrac{7}{16}f_C + \tfrac{1}{16}f^2_C)] \tag{2.10}$$

where $B_C = 0.848\,\mathrm{cm}^2\,\mathrm{nmol}^{-1}$ and $f_C = (I_{Pt}/I^\circ_{Pt})$ at 235 eV, HQ = 0.70. Similarly, the Auger signal from oxygen on C_2 is unattenuated while oxygen on C_5 is attenuated by one atom over half of its possible direction and by three atoms over the other half. Combining terms:

$$\tfrac{1}{2} + \tfrac{1}{4}f_O + \tfrac{1}{4}f^3_O$$

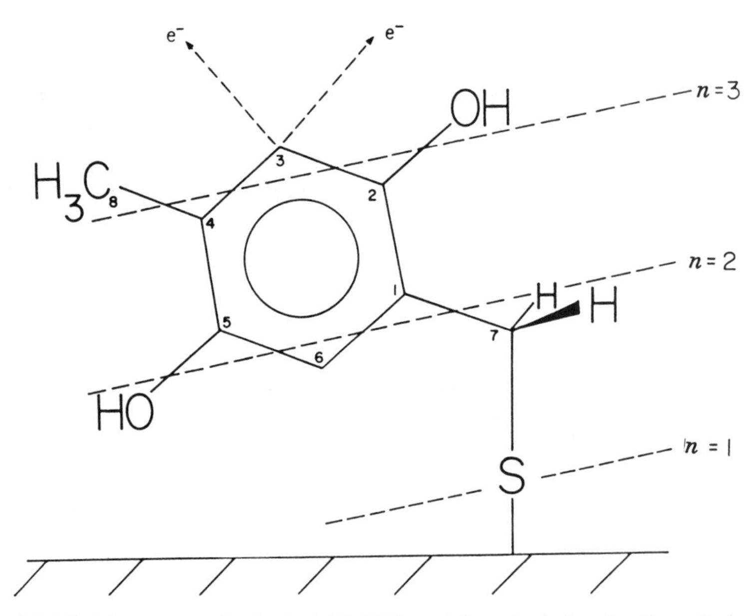

Figure 2.8. Model structure of adsorbed DMBM used for calculating the theoretical packing density ($\Gamma_{\text{theoretical}} = 0.399\,\mathrm{nmol\,cm}^{-2}$).

Similarly, the packing density based on the oxygen Auger signal is given by Eq. 2.11.

$$\Gamma_O = (I_O/I^{\circ}_{Pt})/[B_O(\tfrac{1}{2} + \tfrac{1}{4}f_O + \tfrac{1}{4}f_O^3)] \tag{2.11}$$

where $B_O = 1.27\,\mathrm{cm^2\,nmol^{-1}}$, and $f_O = (I^{\circ}/I^{\circ}_{Pt})$ at 235 eV, HQ = 0.70. The Auger electrons from the sulfur atom are attenuated by one atomic layer one-half of the time and by two monolayers the other one-half: $f_S/2 + f_S^2/2$. Hence, the packing density based on the sulfur Auger signal is

$$\Gamma_S = (I_S/I^{\circ}_{Pt})/[B_S(\tfrac{1}{2}f_S + \tfrac{1}{2}f_S^2)] \tag{2.12}$$

f_S is the sulfur attenuation coefficient obtained from the attenuation of the Pt Auger signal at 161 eV because of horizontally oriented HQ.

$$f_S = (I_{Pt}/I^{\circ}_{Pt}) \text{ at } 161\,\text{eV} \qquad \text{HQ} = 0.62 \tag{2.13}$$

B_S was obtained from data for a monolayer of sulfur formed by immersion of a Pt(111) surface into 0.1-m*M* aqueous Na_2S.

$$B_S = 9.69\,\mathrm{cm^2\,nmol^{-1}}$$

In Fig. 2.8 are shown the three "atomic layers" that constitute the adsorbed DMBM molecular layer at Pt(111). The first layer consists only of a sulfur atom from which we get the term: $(1 - K_S\Gamma)$. The second layer ($n = 2$) and third layer ($n = 3$) each contain five atoms that give rise to the term:

$1 - 5K_C\Gamma)(1 - 5K_C\Gamma)$. Combining terms gives an equation for Γ based upon Pt Auger signal attenuation.

$$\frac{I_{Pt}}{I^{\circ}_{Pt}} = (1 - K_S\Gamma)(1 - 5K_C\Gamma) \tag{2.14}$$

Examples of adsorbates that display orientational transitions, case (c), are HQ and 2,2′,5,5′-tetrahydroxybiphenyl (THBP). Hydroquinone at low concentrations adopts a horizontal orientation (ring parallel to the surface) while at high concentrations it forms a vertical or mixed vertical–horizontal orientation (5, 8). Up to the limiting plateau value ($\Gamma_{hp} = 0.293\,\mathrm{nmol\,cm^{-2}}$) Eqs. 2.3–2.5 are employed. However, beyond the horizontal plateau (Fig. 2.9) the adsorbed layer involves both the horizontal ($\Gamma_{X,h}$) and vertical ($\Gamma_{X,v}$) orientation components:

$$\Gamma_X = \Gamma_{X,h} + \Gamma_{X,v} \tag{2.15}$$

where X is either carbon or oxygen.

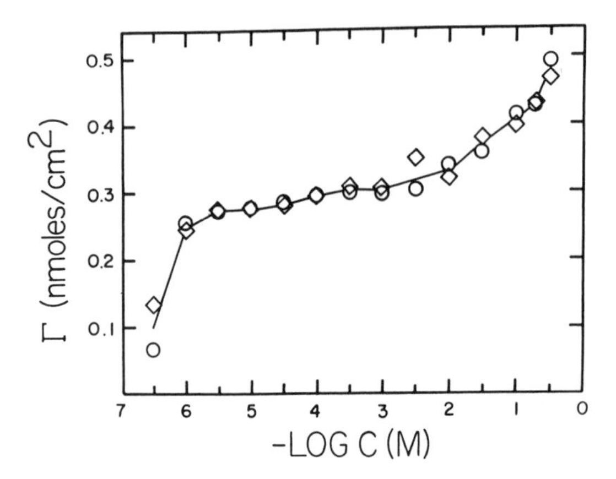

Figure 2.9. Packing densities of HQ/BQ at Pt(111): (○) based upon I_C/I°_{Pt}; (◇) based upon I_{Pt}/I°_{Pt}. Experimental conditions: The incident beam was 10^{-7} A at 2000 eV, normal to the surface; modulation amplitude was 5 V peak to peak; 10 m*M* KF adjusted to pH = 4 with HF; electrode potential 0.20 V versus Ag/AgCl (1*M* KCl reference).

The packing density of the vertically oriented component is

$$\Gamma_{X,v} = \frac{\left(\left\{\dfrac{I_X}{I^\circ_{Pt}}\right\} - B_X \Gamma_{X,hp}\right)}{B_X(a_X - \Gamma_{hp}/\Gamma_{X,vp})} \tag{2.16}$$

where $a_C = 0.85$ for carbon and $a_O \times 1.00$ for oxygen. The packing density of the horizontally oriented component ($\Gamma_{X,h}$) is

$$\Gamma_{X,h} = (1 - \Gamma_{X,v}/\Gamma_{X,vp})\Gamma_{X,hp} \tag{2.17}$$

where $\Gamma_{C,hp}$ and $\Gamma_{O,hp}$ are the theoretical horizontal packing densities of carbon and oxygen, respectively, $\Gamma_{C,hp}/6 = \Gamma_{hp} = 0.293\,\text{nmol}\,\text{cm}^{-2}$, and $\Gamma_{O,hp}/2 = 0.293\,\text{nmol}\,\text{cm}^{-2}$. Similarly, $\Gamma_{C,vp}$ and $\Gamma_{O,vp}$ are the theoretical vertical packing densities of carbon and oxygen:

$$\Gamma_{C,vp}/6 = \Gamma_{vp} = 0.578\,\text{nmol}\,\text{cm}^{-2} \quad \text{and} \quad \Gamma_{O,vp}/2 = 0.578\,\text{nmol}\,\text{cm}^{-2}$$

Packing density of adsorbed HQ based on the attenuation of the Pt substrate Auger signal is found from:

$$\text{v}\Gamma = \Gamma_{vp}\left(a_h - \frac{I_{Pt}}{I^\circ_{Pt}}\right)\Big/(a_h - a_v) \tag{2.18}$$

$$\Gamma_h = \Gamma_{hp}(1 - \Gamma_v/\Gamma_{vp}) \tag{2.19}$$

where $a_h = 1 - 8K\Gamma_{hp} = 0.625$ and $a_v = (1 - 2K\Gamma_{vp})^2(1 - 4K\Gamma_{vp}) = 0.418$.

Another compound that displays mixed orientational transitions is THBP. A fraction of the adsorbed THBP layer exhibits reversible electroactivity, Fig. 2.10. The solid curve in Fig. 2.10*A* is the voltammogram obtained for adsorbed THBP immediately after adsorption from solution. The dotted curve in Fig. 2.10*A* is the voltammogram obtained after evacuation of the adsorbed layer for about 1 h. The similarity of the two curves in Fig. 2.10*A* is an indication of the stability of the adsorbed layer. However, cyclic voltammetric oxidation and reduction of the adsorbed layer as depicted in Fig. 2.10*B* leads to removal of some of the adsorbed material; following this electrolysis treatment, the Auger signals for C and O are decreased and peaks of the EELS spectrum are attenuated in agreement with the voltammetric data.

Because of the mixed orientations of THBP, measurement of its packing density requires use of voltammetric or coulometric data to distinguish the contributions of the two principal orientations to the total packing density. The electroactive fraction of THBP can be found from the Faraday law for electroactive adsorbate:

$$\Gamma_{el} = \frac{Q - Q_{bg}}{n\mathscr{F}A} \tag{2.20}$$

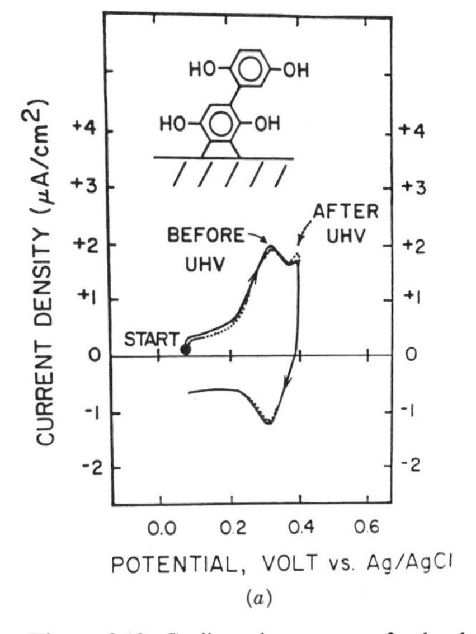

(*a*)

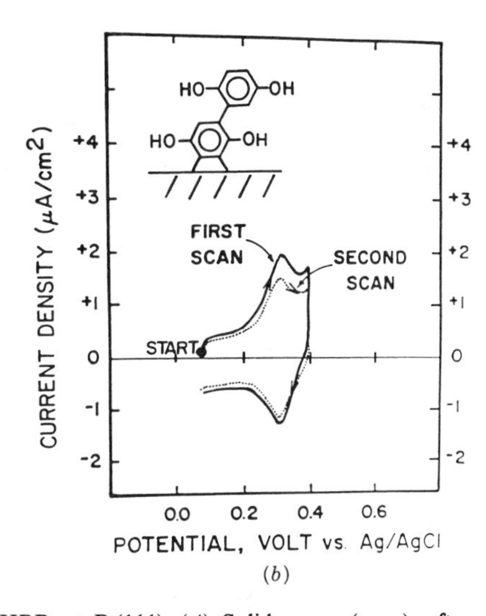

(*b*)

Figure 2.10. Cyclic voltammetry of adsorbed THBP at Pt(111). (*A*) Solid curve (——): after immersion into 0.2 m*M* THBP (0.2 V versus A Ag/AgCl reference) followed by rinsing with 10 m*M* TFA (Trifluoroacetyl). Dotted curve (····): as above, followed by 1 h in vacuum prior to voltammetry. (*B*) Solid curve (——): after immersion into 2.0 m*M* THBP (0.2 V) followed by rinsing with 10 m*M* TFA, first scan. Dotted curve (····): second scan. Scan rate: 5 mV s^{-1}. (Reprinted with permission from Ref. 7.)

where Q denotes the coulometric charge for the oxidation of the adsorbed species, Q_{bg} is the background charge, $n = 2$, $\mathscr{F}$ is the Faraday constant, and A is the electrode area. The total packing density can be obtained by treating THBP as though it were horizontally oriented, as the contribution of Γ_{el} is small at smooth surfaces.

$$\Gamma = (I_C/I^{\circ}_{Pt})/(12B_C) \tag{2.21}$$

Alternatively, the total packing density can be obtained by use of Pt Auger signal attenuation data.

$$\Gamma = (1 - I_{Pt}/I^{\circ}_{Pt})/(16K_C) \tag{2.22}$$

Let us now turn our attention to some recent findings concerning the nature of the species formed at single-crystal surfaces as a result of adsorption from aqueous solutions. Solutions employed for such measurements ordinarily contain a comparatively weakly adsorbing electrolyte, such as 10 mM KF adjusted to pH = 4 with HF, to provide buffering capacity and sufficient conductivity for potentiostatic control of electrode potential. Adsorbate solutions should be prepared using pyrolytically distilled water to avoid introducing impurities that might act as competing adsorbates (43). Prior to each experiment, the surface is cleaned by argon ion bombardment, followed by annealing in UHV. After examination by Auger spectroscopy for cleanliness and testing by LEED for ordered structure, the surface is transferred to an adjoining antechamber that is then backfilled with argon gas. The electrode is immersed into the adsorbate solution at controlled electrode potential, usually for about 3 min, then rinsed several times with dilute ($\sim$0.1 mM) eledtrolyte, and brought back to the UHV chamber for characterization by LEED, Auger spectroscopy, and EELS.

2.3.1 Hydroquinone

Adsorption of HQ at annealed polycrystalline Pt electrodes from aqueous solution has been the subject of detailed, systematic study although numerous aspects remain unexplored (14–16, 23, 25–31). From these studies it is concluded that at low concentrations, HQ is adsorbed in a horizontal orientation (the ring parallel to the surface), while at high concentration HQ is adsorbed in an edgewise orientation.

Intriguing studies of aromatic compounds adsorbed from the gas phase have been reported. For instance, benzene received particular attention at transition metal surfaces such as Pt (56–68) and Ni (57, 67, 69–72) under UHV conditions utilizing several surface sensitive techniques. Naturally, the choice of benzene for study is prompted to some extent by its symmetric structure, and the possible involvement of the π orbitals in surface bonding. Studies based on EELS, LEED, AES, TDMS, Ultraviolet photoelectron spectroscopy (UPS), and measurement of work function concluded that benzene is adsorbed with its ring parallel to the surface (46–49, 56, 57, 62, 67, 68). The alkylbenzenes behave similarly, but are adsorbed less strongly due to steric hindrance between the alkyl group and the surface (64, 73–75).

Studies of adsorption from aqueous solution have emphasized water-soluble benzene derivatives such as HQ. Hydroquinone has been studied over a wide concentration range, at controlled potential utilizing EELS to study adsorption from solution for the first time (5–13). Low energy electrondiffraction, AES, TDMS, and cyclic voltammetry were also employed in those initial studies. Figure 2.5*B* shows the Auger spectrum of chemisorbed HQ at 0.2 V (vs. Ag/AgCl/1*M* KCl reference). Based on carbon and platinum signals of the Auger spectra, the packing density can be calculated separately by means of Eqs. 2.3 and 2.4 (Table 2.1). The data are plotted as a function of concentration in Fig. 2.9. Both equations lead to similar results, indicating the reliability, reproducibility, and self-consistency of the experimental data that can be obtained, at least with strongly adsorbed, chemically stable adsorbate such as HQ. The isotherms in Fig. 2.9 are suggestive of the transition from horizontal to vertical orientation postulated in earlier work, as described above. However, the annealed, atomically smooth Pt(111) surface favors the horizontal orientation relative to the vertical orientation, in comparison with electrochemically cycled polycrystalline Pt surfaces.

Electron energy loss spectroscopy of 1.0 m*M* HQ/BQ (BQ = benzoquinone) adsorbate, recorded with detection at the specular reflection angle (5), is shown in Figs. 2.1*A* and *B*. Also shown in Fig. 2.11 is the mid-IR spectrum of solid HQ. From Fig. 2.11 it can be seen that all the bands present in the EELS spectrum of HQ are also present in the IR spectrum except for the absence of the O—H stretching frequency at 3260 cm^{-1}. Vibrational assignments of the EELS spectra can be made in accordance with the accepted assignments of the IR bands of HQ (D_{2h} symmetry) (5, 76, 77). Such assignments are given in Table 2.2. Based on the above EELS observations we suggest that the phenolic hydrogen is removed during adsorption, in agreement with the thin-layer electrochemistry findings (15, 16, 23, 24, 26–31), as shown in Eqs. 2.21 and 2.22. However, the absence of an O—H stretching band could also be due to a low sensitivity of EELS to that particular vibration. The EELS spectra are essentially identical regardless of whether the starting adsorbate is HQ or BQ.

A further indication of an orientational transition from horizontal (at concentrations below 1 m*M*) to vertical orientation (at concentrations above 1 m*M*) is provided by the EELS spectrum for adsorbed HQ at high concentration (0.5*M*), Fig. 2.11*C*. That is, the spectrum shows an increase in the ring bending modes and O—H bending modes indicative of decreased ring interaction with the surface.

Electrochemical oxidation of chemisorbed HQ at Pt(111) was studied by means of cyclic voltammograms such as that shown in Fig. 2.12*A*. A second cycle of reduction and oxidation was carried out to determine the background charge. The coulometric charges, Q_{OX} and Q'_{bg}, respectively, were recorded after the charge–time curves became parallel, Fig. 2.12*B*. Electrocatalytic oxidation

Table 2.1. Packing Densities and Oxidation Factors at Pt(111)

	Packing Density					*Oxidation Factor*	
	From Elemental Auger Signals					*From Pt Auger Signal Attenuation*[a]	
−log C	Γ_C (nmol cm^{-2})	Γ_O (nmol cm^{-2})	Γ_F (nmol cm^{-2})	Γ_S (nmol cm^{-2})	Γ (nmol cm^{-2})	Γ (nmol cm^{-2})	$n_{Ox} = (Q_{Ox} - Q_{bg})\mathscr{F}A\Gamma$ (e^- $molecule^{-1}$)
				HYDROQUINONE (HQ)			
6.52	0.395	0.139			0.066	0.132	24.2
6.00	1.518	0.353			0.253	0.246	
5.52	1.619	0.538			0.270	0.275	24.9
5.00	1.662	0.553			0.277	0.277	
4.52	1.698	0.595			0.283	0.280	23.9
4.00	1.759	0.586			0.294	0.295	24.8
3.52	1.780	0.584			0.299	0.307	24.8
3.00	1.762	0.599			0.295	0.304	
2.52	1.810	0.655			0.302	0.348	23.7
2.00	2.039	0.685			0.340	0.320	
1.52	2.141	0.613			0.357	0.380	21.4
1.00	2.479	0.712			0.413	0.397	
0.70	2.531	0.769			0.422	0.430	16.8
0.52	2.948	0.908			0.491	0.464	16.3
				BENZOQUINONE (BQ)			
3.16	*1.374*	*0.462*			*0.229*	*0.239*	

[a] Q_{bg} = background charge.

Table 2.2. Assignments of EELS Bands for Adsorbed Diphenols

Compound[a]	*Concentration* (m*M*)	*Peak Frequency* (cm^{-1})	*Symmetry Species*	*Description*
HQ/BQ	0.1	2979	$D_{2h}: B_{2u}, B_{3u}$	O—H stretch
		1611	B_{3u}	CC stretch
		1496	B_{1g}	CC stretch
		1247	B_{2u}	O—O stretch
		905	B_{1u}	O—H bend
		785	B_{2u}	Ring bend
		506	B_{1u}	Ring bend
DMBM	0.1	3568	$C_1: A$	O—H stretch (meta)
		3374		O—H stretch (ortho)
		2981		C—H, O—H stretches
		1600		CC stretch (O)
		1411		CC stretch
		1194		O—O stretch
		1021		O—H stretch
		865		C—H bend
		710		Ring bend
		646		C—S stretch
		425		Ring bend

[a]HQ = hydroquinone, BQ = benzoquinone, DMBM = 2,5-dihydroxy-4-methylbenzylmercaptan.

can be characterized by n_{Ox}, the number of electrons transferred during oxidative desorption of a chemisorbed molecule (31, 55). The term n_{Ox} provides a measure of the stoichiometry of the electrocatalytic oxidation reaction as in Eq. 2.23.

$$\text{Adsorbed layer} \rightarrow \text{products} + n_{Ox}e^- \tag{2.23}$$

The value of n_{Ox} depends on the initial molecular orientation, being smaller for edgewise orientation than for horizontal orientation (16, 17, 31). Combining the data obtained from Fig. 2.12*B*, $(Q_{Ox} - Q'_{bg})$, with the packing densities based upon Auger data, $\Gamma(\text{nmol cm}^{-2})$, gives n_{Ox}, Eq. 2.24:

$$n_{Ox} = (Q_{Ox} - Q'_{bg})/\mathcal{F}/A\Gamma \tag{2.24}$$

The results are graphed as a function of HQ concentration in Fig. 2.13. It is noteworthy that n_{Ox} begins to decrease at the same HQ concentration where the

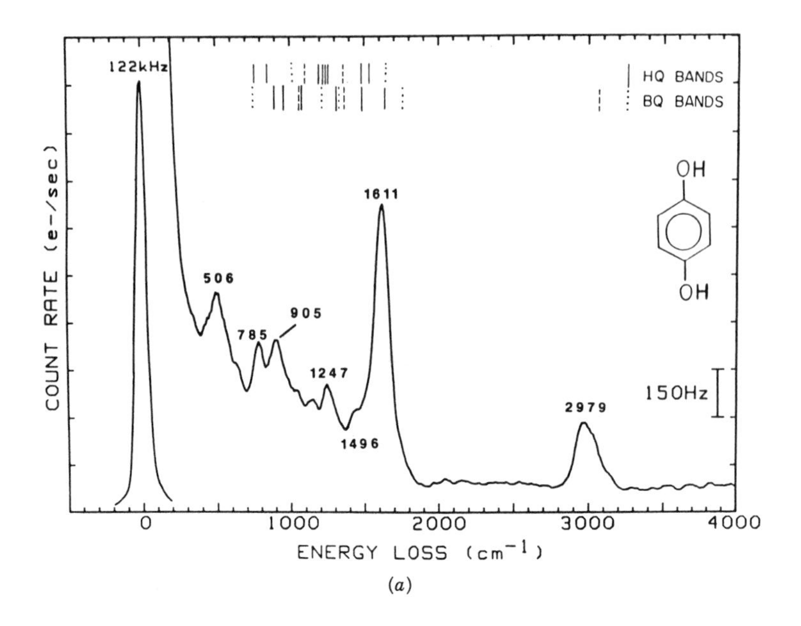

(*a*)

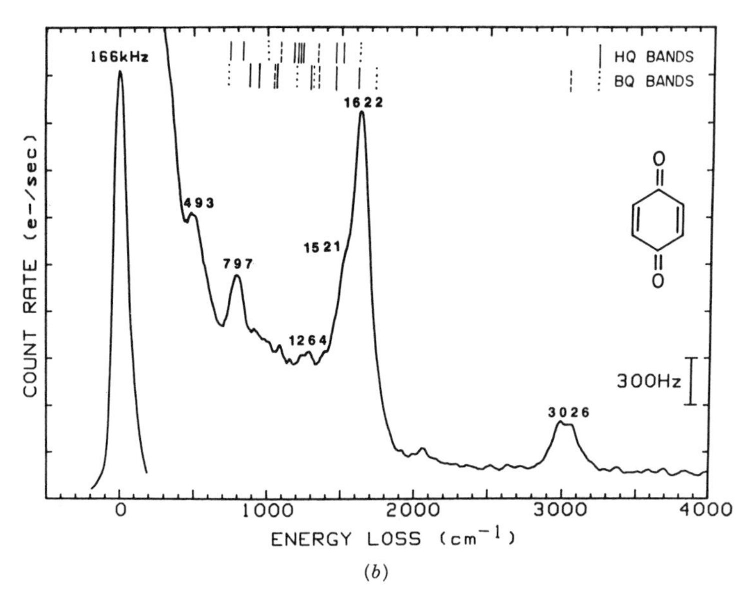

(*b*)

Figure 2.11. Electron energy loss spectra at Pt(111). (*A*) HQ (1.0 m*M*). (*B*) BQ (1.0 m*M*). (*C*) HQ (0.50 *M*). Experimental conditions: supporting electrolyte, 10 m*M* KF adjusted to pH = 4 with HF; temperature 23 ± 1°C; beam energy, 4 eV; beam current, 0.15 nA; incidence and detection angles, 62° from surface normal. (Reprinted with permission from Ref. 7.)

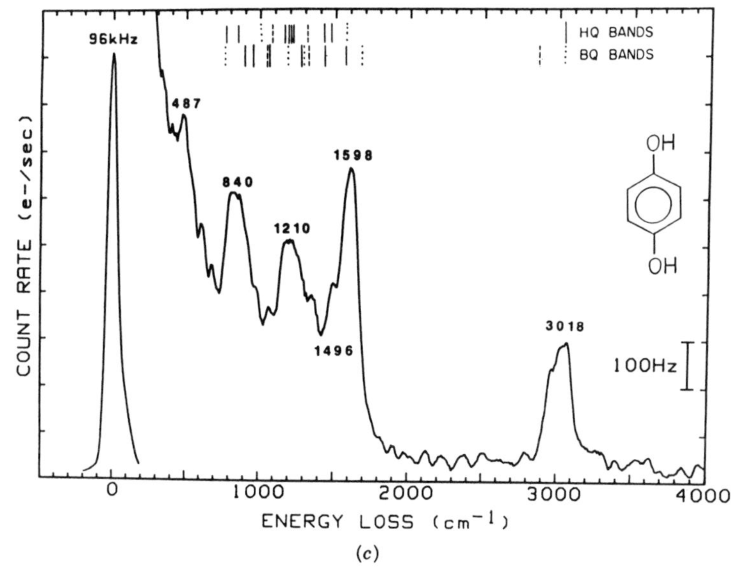

(*c*)

Figure 2.11. Continued.

packing density begins to increase (1 mM). When measured at the horizontal orientation packing density plateau, n_{Ox} is found to be $24 \pm 0.5\,e^-$ molecule^{-1}, indicating complete oxidation of adsorbed HQ to CO_2.

$$+\ 10H_2O \rightarrow 6CO_2 + 24H^+ + 24e^- \qquad (2.25)$$

However, as the packing density increased above the plateau value, the n_{Ox} value decreased. This suggests that the vertically adsorbed HQ component leads to other products; maleic acid is a likely possibility.

$$HO{-}C_6H_4{-}OH \xrightarrow{6H_2O} O{=}C(OH){-}CH{=}CH{-}C(OH){=}O + 2CO_2 + 12H^+ + 12e^- \qquad (2.26)$$

At an HQ concentration of 0.2M the adsorbed HQ layer consists of 61% vertical species (based on Γ_C) and exhibits an n_{Ox} value of 16.7; n_{Ox} observed for the vertical orientation at polycrystalline Pt is 17.6 (5).

Examination of phenol (P_hOH) adsorption is of interest as its structure is intermediate between that of HQ and benzene. Adsorption of PL at Pt(111) from aqueous solution resembles that of HQ: The hydroxyl hydrogen is apparently lost (or the OH stretching band for EELS is unexpectedly weak),

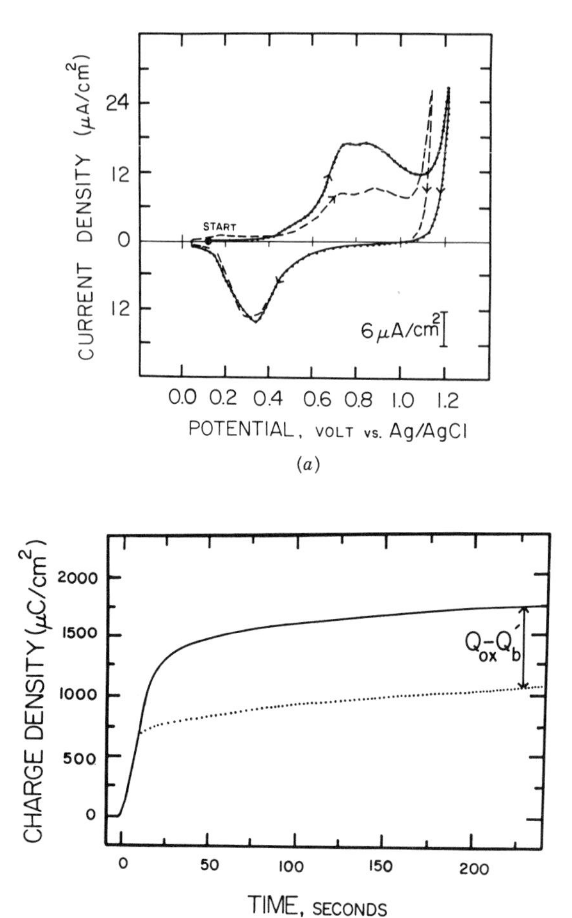

Figure 2.12. Cyclic voltammetry and chronocoulometry of adsorbed HQ/BQ at Pt(111). (*A*) Cyclic voltammetry (scan rate $5\,\mathrm{mV\,s^{-1}}$). Solid curve (——): first oxidation/reduction cycle. Dashed curve (– – –): second oxidation/reduction cycle. Dotted curve (· · · ·): first oxidation/reduction cycle, afrer 1 h in UHV. (*B*) Chronocoulometry (electrode potential, 1.00 V). Solid curve (——): first oxidation step. Dotted curve (· · · ·): second oxidation step. Experimental conditions: electrolyte was 10 m*M* TFA; HQ concentration, $3 \times 10^{-5} M$; surface was rinsed with pure water before evacuation; scan rate $5\,\mathrm{mV\,s^{-1}}$; temperature, 23 ± 1°C; oxidation at 1.00 V; reduction at -0.20 V. [Reprinted with permission from F. Lu et al., *Langmuir*, *4*, 637 (1988). Copyright © (1988) American Chemical Society.]

while aromaticity of the ring is retained, based upon the aromatic CC stretches at 1620 and $1480\,\mathrm{cm^{-1}}$.

$$C_6H_5OH_{(aq)} \underset{}{\overset{\text{Pt surface}}{\rightleftharpoons}} C_6H_5\text{—O (adsorbed, flat on Pt surface)} + H^+ + e^- \qquad (2.27)$$

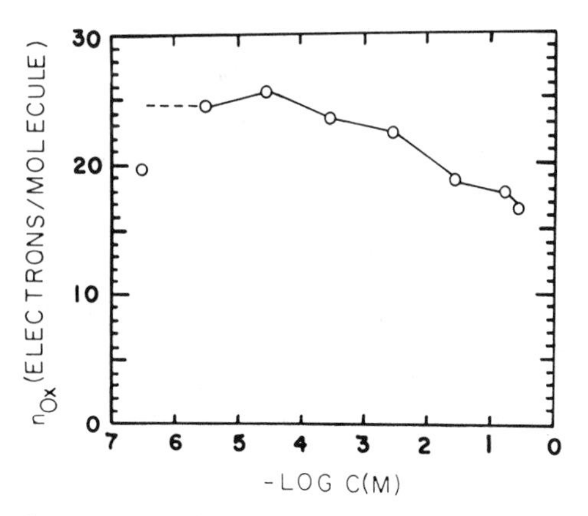

Figure 2.13. Number of electrons, n_{Ox}, for oxidative desorption of HQ/BQ versus solute concentration at which adsorption was carried out. Supporting electrolyte was 10 m*M* TFA. [Reprinted with permission from F. Lu et al., *Langmuir*, *4*, 637 (1988). Copyright © (1988) American Chemical Society.]

If the hydroxyl hydrogen is in fact lost during adsorption, then presumably the odd electron is involved in bonding to the metal. The observed n_{Ox} value of 26.4 e^- molecule^{-1}, suggests that the adsorbate is horizontally oriented and partially dehydrogenated, Eq. 2.28, such that oxidative desorption proceeds to CO_2.

$$C_6H_5O\text{ (adsorbed)} + 11H_2O \rightarrow 6CO_2 + 27H^+ + 27e^- \tag{2.28}$$

Adsorbed PhOH displays a (3 × 3) LEED pattern similar to those for HQ and BQ (5).

A large number of investigations were reported concerning organic adsorbates at platinum and related surfaces. Particular emphasis has been given to phenols and derivatives (78–83), aromatic amines (78, 81), benzoic acids and their esters (84–90), and naphthalene derivatives (78–81, 91, 92). Previous work dealt mainly with the transformations of these compounds in the presence of supported and unsupported catalysts (Pt, Pd, Ni, Rh, Ru, Ir, and bimetals) and the stereochemistry at various temperatures and pressures (52, and references cited therein). The stereochemical course varies among adsorbates and catalysts. For example, phenol on Pt catalysts hydrogenates easily and with very high selectivity via a cyclohexanone intermediate (52), while aromatic amines are hydrogenated with difficulty both on supported and unsupported catalysts, requiring high temperature and pressure. Almost none of the previous studies involved surfaces characterized in UHV.

2.3.2 2,2′,5,5′-Tetrahydroxybiphenyl

We have chosen to study this compound because of its reversible electroactivity in solution and in the chemisorbed state. Being soluble and stable in water, it has been studied previously by thin-layer electrochemistry (31, 93). We are not aware of any previous studies of THBP involving AES or EELS. The goals of these studies were to quantitate the adsorbed layer, to identify the adsorbed species, and to characterize their electrochemical reactivity. The compound THBP was synthesized according to a published procedure (94).

Figure 2.10*A* shows the cyclic current–potential curve for THBP in the chemisorbed state at a clean Pt(111) surface. The figure shows that a reversibly electroactive adsorbed material is present in the chemisorbed layer (solid curve). The chemisorbed material remains attached to the surface during evacuation (dotted curve). The voltammetric curves are similar prior to and following evacuation indicating that the adsorbed layer is very stable in vacuum. On the other hand, evidence for damage to the adsorbed layer by cyclic voltammetry is shown in Fig. 2.10*B*, as is obvious from the difference between the first (solid curve) and the second (dotted curve) voltammetric scans. Based upon quantitative coulometric and Auger data, electrolysis of adsorbed THBP proceeds according to Eq. 2.29.

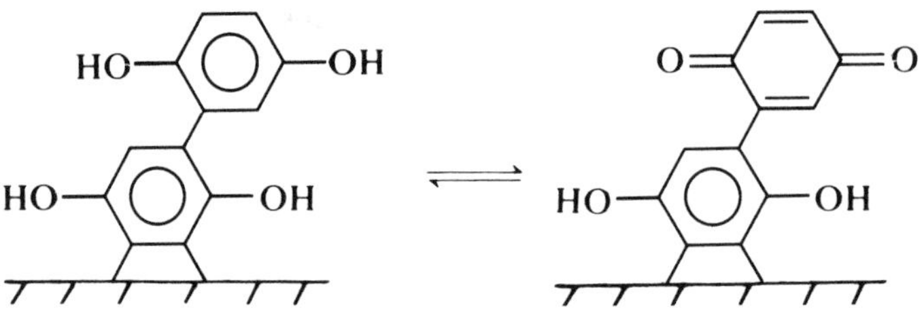

HO—⟨○⟩—OH / HO—⟨○⟩—OH (on surface) ⇌ O=⟨ ⟩=O / HO—⟨○⟩—OH (on surface) $+ 2H^+ + 2e^-$ (2.29)

Earlier work based upon thin-layer coulometry (14, 23) has shown that direct attachment of the HQ or catechol moiety to the Pt surface destroys its reversible electroactivity. Evidently, in some of the adsorbed molecules, one of the two rings of THBP is not directly attached to the Pt surface, thus allowing reversible redox reactions to occur in the pendant ring. Quantitative determination of the THBP packing density from Auger data, Fig. 2.14*B*, was carried out by combining electrochemical methods with Auger spectroscopic methods, Eq. 2.30:

$$\Gamma = \Gamma_{el} + \Gamma_{h} \tag{2.30}$$

where Γ_{el} is obtained from Eq. 2.20 and Γ is obtained from the carbon Auger signal based on Eq. 2.21 and can be written in the form of Eq. 2.31 and 2.32:

$$I_C/I^{\circ}_{Pt} = B_C\Gamma_{C,h}a_{C,h} + 12B_C\Gamma_{el}a_{C,v} \tag{2.31}$$

$$\frac{I_O}{I^{\circ}_{Pt}} = B_O\Gamma_{O,h}a_{O,h} + 4B_O\Gamma_{el}a_{O,v}, \tag{2.32}$$

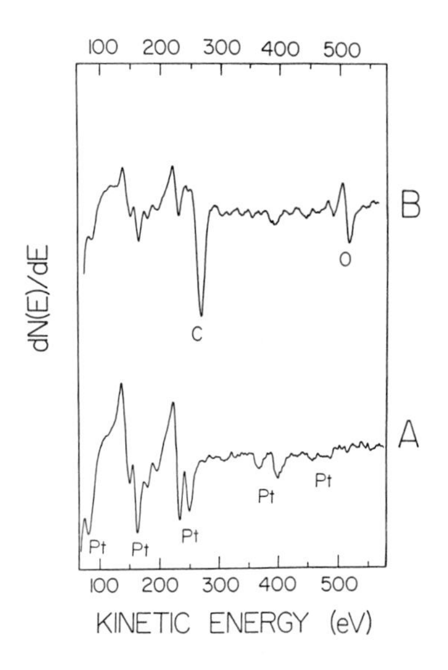

Figure 2.14. Auger spectra. (*A*) Clean Pt(111). (*B*) THBP (20 m*M*) adsorbed at Pt(111). Experimental conditions: Electron beam at normal incidence, 10^{-7} A, 2000 eV. [Reprinted with permission from F. Lee, *Langmuir*, *4*, 637 (1988). Copyright © (1988) American Chemical Society.]

I°_{Pt} is the substrate Auger signal at 161 eV, where

$$a_{\mathrm{C,v}} = \left(\frac{1}{3} + \frac{9}{24f_{\mathrm{C}}} + \frac{5}{24f_{\mathrm{C}}^2} + \frac{1}{12f_{\mathrm{C}}^3}\right) = 0.727$$

and

$$a_{\mathrm{C,h}} = 1 - 8(1 - f_{\mathrm{C}})(6 \times 10^{-2}\Gamma_{\mathrm{el}})^{1/2}$$

are the attenuation factors for carbon and Auger electrons from vertically and horizontally oriented THBP. The attenuation factors for oxygen are similar.

$$a_{\mathrm{O,v}} = (\tfrac{5}{8} + \tfrac{1}{8}f_{\mathrm{O}} + \tfrac{1}{4}f_{\mathrm{O}}) = 0.835 \quad \text{and} \quad a_{\mathrm{O,h}} = a_{\mathrm{C,h}}$$

The molecular packing density is related to the elemental packing density by Eq. 2.33.

$$\Gamma = \frac{\Gamma_{\mathrm{C}}}{12} + \frac{\Gamma_{\mathrm{O}}}{4} \tag{2.33}$$

Similarly, Γ_{h} was obtained from the Pt Auger signal attenuation at 161 eV as given by Eq. 2.22. The resulting packing densities of THBP are plotted as a function of concentration as shown in Fig. 2.15 and Table 2.3. The two Auger methods exhibit similar results as shown in Fig. 2.15. There is a gradual increase in the packing density of electroactive material (Γ_{el}) with increasing adsorbate concentration. Near the solubility limit the molecular packing density ($\Gamma_{\mathrm{THBP}} = 0.200$ nmol cm^{-2}) was only slightly larger than the theoretical limit for

Table 2.3. Packing Densities of THBP as a Function of Concentration

	Packing Densities						
	From Elemental Auger Signals				*From Pt Auger Signal Attenuation*	*Reversibly Electroactive Packing Density*	
−log C	Γ_C (nmol cm^{-2})	Γ_O (nmol cm^{-2})	Γ_S (nmol cm^{-2})	Γ (nmol cm^{-2})	Γ (nmol cm^{-2})	$\Gamma_{el} = (Q - Q_{bg})/(2\mathscr{F}A)$ (nmol cm^{-2})	*Oxidation Factor*[a] $n_{Ox} = (Q_{Ox} - Q_{bg})\mathscr{F}A\Gamma$
				2,5,2′,5′-Tetrahydroxybiphenyl (THBP)			
6.00	0.451	0.227		0.038	0.048		
5.52	1.462	0.538		0.122	0.118		
5.00	1.538	0.853		0.137	0.137	0.0122	41.3
4.52	1.670	0.817		0.154	0.158	0.0332	44.4
4.00	1.817	0.827		0.170	0.165	0.0431	45.1
3.52	1.928	0.782		0.181	0.164	0.0502	39.8
3.00	1.943	0.711		0.183	0.182	0.0574	35.8
2.52	1.971	0.688		0.187	0.179	0.0644	37.6
2.00	2.050	0.617		0.195	0.200	0.0715	36.6

[a]Experimental conditions: n_{Ox} (oxidation factor) was calculated by use of Eq. 2.15 in which the packing density (Γ) was the average of the values obtained by means of the Pt signal attenuation method and the carbon elemental Auger signal method; coulometric charges are given in Table 2.1; other conditions as in Table 2.1.

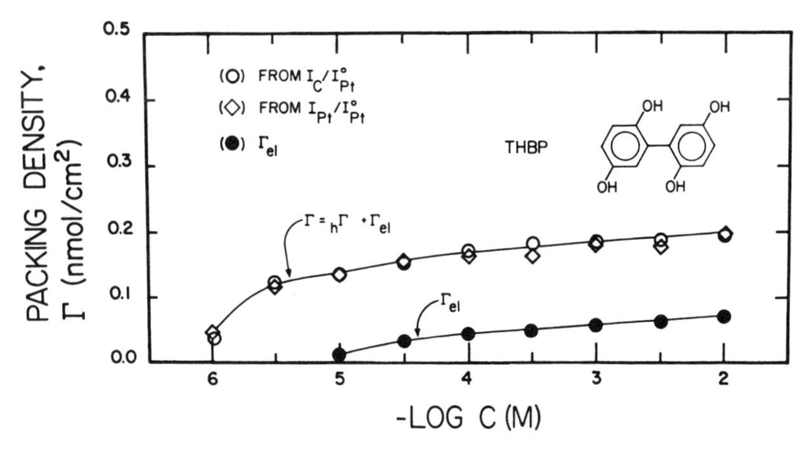

Figure 2.15. Packing density of THBP at Pt(111) based upon Auger data and coulometry: (O) Γ from elemental Auger signals; ( ) Γ from Pt Auger signal attenuation; ( )...Γ_{el} from Eq. 2.18. Experimental conditions: 10 mM KF–HF electrolyte (pH = 4); temperature 23 ± 1°C; electrode potential, 0.2 V versus Ag/AgCl (1M) reference.

horizontally oriented THBP ($\Gamma_{hp} = 0.187\,\text{nmol cm}^{-2}$), and markedly lower than the theoretical limit of vertically oriented THBP ($\Gamma_{vp} = 0.460\,\text{nmol cm}^{-2}$). This suggests that a significant fraction of the adsorbed THBP is present in a horizontal orientation, along with a small fraction of a vertical orientation. The fraction that is vertically oriented increases as the concentration of the chemisorbed THBP increases. This indicates a transition in orientation from horizontal at low concentration to edge–pendant at high concentration. A similar conclusion as to the orientation of adsorbed THBP at Pt(111) surface has been reached on the basis of electrocatalytic oxidation of THBP (n_{Ox} measurements) (5–7, 22, 24, 31, 55). The results are summarized in Table 2.3 and graphed as a function of concentration in Fig. 2.16. The term n_{Ox} is large and nearly constant at Pt(111) over a wide concentration range (10^{-5}–10^{-2} M). Only a small decrease in n_{Ox} value is observed with increasing THBP concentration. The highest measured value of n_{Ox} (45.1) is reasonable for horizontally oriented THBP, since the theoretical n_{Ox} for horizontal orientation is 46.

$$n^{12} - C_{12}H_6O_4 + 20H_2O \rightarrow 12CO_2 + 46H^+ + 46e^- \quad (2.34)$$

The decrease in n_{Ox} value upon increase in adsorbate concentration (see Table 2.3, Fig. 2.15) is further evidence that the adsorbed THBP layer at high concentration consists of horizontally oriented molecules, mixed with a small fraction of vertically oriented molecules (7).

Electron energy loss spectra of chemisorbed THBP at Pt(111) from 0.1 and 5.0-mM solutions are shown in Fig. 2.17A and B. Also shown in Fig. 2.17C are the IR bands of solid THBP (KBr pellet). Assignment of the EELS bands of

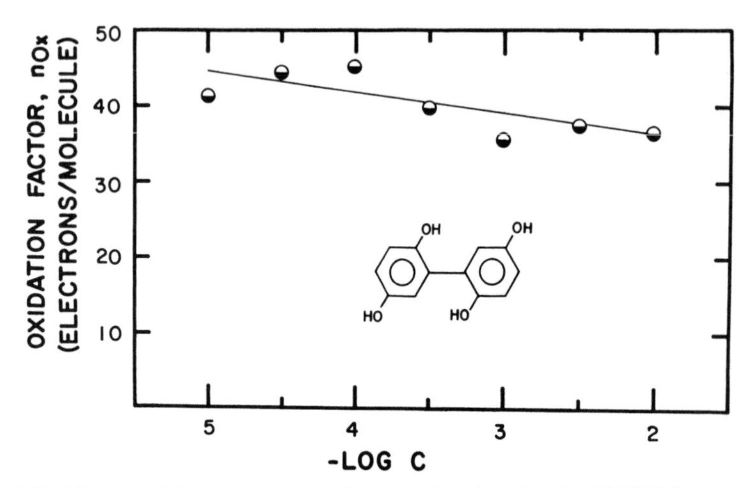

Figure 2.16. Number of electrons, n_{Ox}, to oxidatively desorb molecule of THBP from the Pt(111) surface versus the THBP concentrations at which adsorption was carried out. Experimental conditions: 10 m*M* TFA electrolyte; electrode potentials, 0.20 V (adsorption) and 1.00 V (oxidation); temperature 23 ± 1°C. (Reprinted with permission from Ref. 7.)

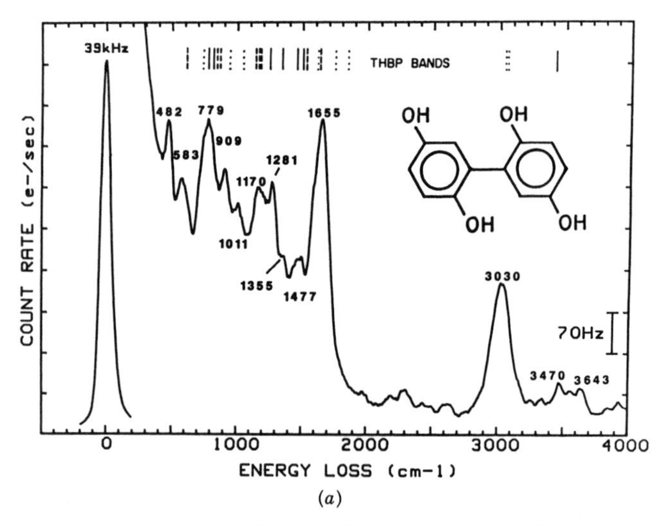

(*a*)

Figure 2.17. Vibrational Spectra of THBP: (*A*) EELS spectrum of Pt(111) after immersion into 0.1 m*M* THBP. (*B*) EELS spectrum of Pt(111) after immersion into 5 m*M* THBP. Experimental conditions: EELS spectra: incidence and detection angles, 62° from surface normal; beam energy, 4 eV; beam current, 1.5×10^{-10} A; resolution 10 meV (80 cm^{-1}) fwhm; electrolyte, 10 m*M* KF/HF (pH = 4). (*C*) Infrared spectra: resolution 4 cm^{-1}; 10% THBP in KBr. (Reprinted with permission from Ref. 7.)

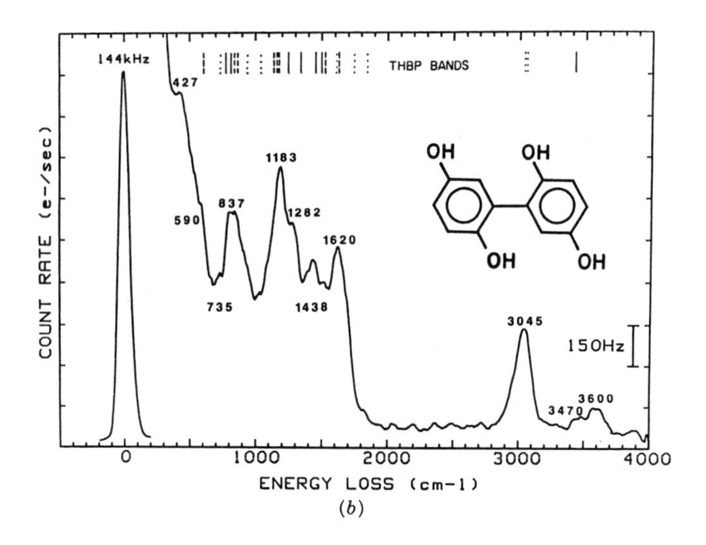

(*b*)

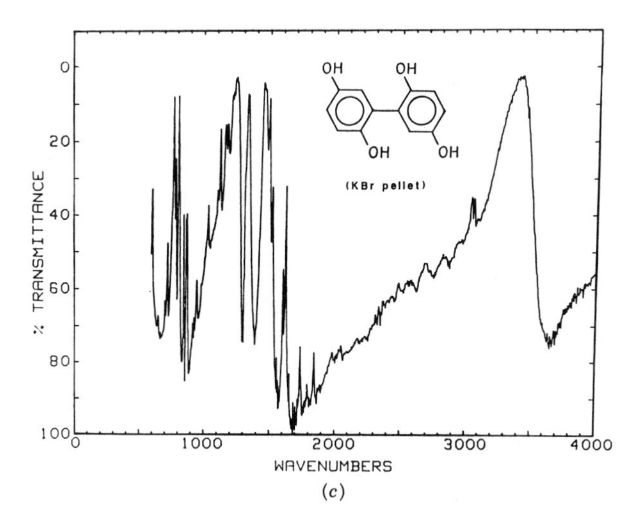

(*c*)

Figure 2.17. Continued.

THBP in line with the accepted IR assignments of THBP, HQ, and related compounds (76, 77) are given in Table 2.4. The heights of the O—H stretching bands (3643 and 3470 cm^{-1}) and the phenolic C—O stretch band (1281 cm^{-1}) increase with increasing THBP concentrations. This indicates that at concentrations below 0.1 mM, THBP loses its phenolic hydrogens during the adsorption process and forms horizontally oriented quinone:

$$\text{(HO)}_2\text{C}_6\text{H}_3\text{—C}_6\text{H}_3\text{(OH)}_2\text{(aq)} \longrightarrow (\text{Surface—O})_2\text{C}_6\text{H}_3\text{—C}_6\text{H}_3(\text{O—Surface})_2 + 4\text{H}^+ + 4\text{e}^- \qquad (2.35)$$

Table 2.4. Assignments of EELS Bands of Adsorbed Phenols

Compound	*Concentration* (*mM*)	*Peak Frequency* (cm^{-1})	*Symmetry Species*	*Description*
THBP	5.0	3622	$C_s: A'$	OH stretch
		3411	A'	OH stretch
		2944	A'	CH stretch
		1653	A'	CC stretch
		1416	A'	CC stretch
		1278	A'	CO stretch, CH bend
		1190	A'	CO stretch
		839	A'	CH bend
		738	A'	CH bend, ring bend
		610	A''	CC stretch (O—O)
		464	A''	Ring bend, C═O bend
		391	A''	Ring bend, OH bend
THBP	0.1	3643	$C_s: A'$	OH stretch
		3470	A'	OH stretch
		3030	A'	OH stretch
		1655	A'	C═O stretch, CC stretch
		1477	A'	CC stretch
		1355	A'	OH bend
		1281	A'	CO stretch, CH bend

Table 2.4. (Continued)

Compound	Concentration (mM)	Peak Frequency (cm^{-1})	Symmetry Species	Description
		1170	A'	CO stretch
		1011	A''	CH bend
		909	A'', A'	CH bend, CC stretch
		779	A''	CH bend, ring bend
		583	A''	C—C stretch (O—O)
		482	A''	Ring bend, C═O bend
DMBM	0.1	3568	$C_1:A$	OH stretch (meta)
		3374		OH stretch (ortho)
		2981		CH stretch
		1600		CC stretch
		1411		CC stretch
		1194		CO stretch
		1021		CH bend
		865		CH bend
		710		Ring bend
		646		CS stretch
		425		Ring bend
DHTP	0.1	3508	$C_1:A'$	OH stretches
		3048	A'	CH stretch
		1574	A'	CC stretch
		1463	A'	CC stretch
		1171	A'	CO stretch
		821	A''	CH bend
		821	A'	Ring bend
		409	A''	Ring bend
		290	A'	PtS stretch
HQ/BQ	0.1	2979	C_{2h}(HQ) B_u(HQ), B_{2u}(BQ)	CH stretch (HQ, BQ)
		1611	D_{2h}(BQ) B_2, B_3(BQ)	C═O stretch, C═C stretch
		1486	B_u(HQ)	CC stretch (HQ)
		1247	B_u(HQ), B_{2u}(BQ)	CO stretch (HQ), CH bend
		905	B_u(HQ)	CH bend (HQ)
		785	B_u(HQ), B_{3u}(BQ)	Ring bend
		506	B_u(HQ), B_u(BQ)	Ring bend (HQ), C═O bend

[a]THBP = 2,5,2′,5′-tetrahydroxybiphenyl, DHTP = 2,5-dihydroxythiophenol, DMBM = 2,5-dihydroxy-4-methylbenzylmercaptan, HQ = hydroquinone, BQ = benzoquinone.

while above 0.1 m*M* the EELS spectra indicate at least partial retention of the O—H groups. This is as expected if a fraction of the adsorbed THBP molecules are oriented in an edge–pendant configuration:

OH OH

HO HO (aq) $\longrightarrow$ HO—OH HO—OH $+ 2H^+ + 2e^-$

(2.36)

Comparison of the EELS spectrum of HQ (Fig. 2.11*A*) with the EELS spectrum of horizontally oriented THBP in Fig. 2.17*A* reveals a number of striking similarities: The C—H stretching band of HQ at 2979 cm^{-1} and of THBP at 3030 cm^{-1}; the aromatic C—C stretching bands of THBP at 1655 and 1477 cm^{-1} compared to 1611 and 1496 cm^{-1} for HQ; the C—O stretching bands of THBP at 1281 cm^{-1} and of HQ at 1247 cm^{-1}; the C—H bending band of THBP at 909 cm^{-1} and of HQ at 787 and 506 cm^{-1}. The EELS spectrum of THBP, however, contains additional peaks at 1355, 1170, 1011, and 583 cm^{-1}, which are attributable to the lower symmetry of THBP (C_s), rather than HQ (D_{2h}). The absence of an O—H stretching band in the EELS spectrum of adsorbed THBP in Fig. 2.17*A* is a clear indication that the phenolic hydroxyl hydrogens are lost upon adsorption. This is evidence that at concentrations below 0.1 m*M* THBP adsorbs with both rings essentially parallel to the surface. At higher concentrations, a fraction of the THBP adsorbs with one ring attached edge-wise to the surface and the other ring in pendant. Similar results were obtained with HQ. At concentrations below 0.1 m*M* HQ adsorbed parallel to the surface, while at higher concentrations a fraction adsorbs edge-wise (see Ref. 7).

2.3.3 Thiophenol and Related Compounds

2.3.3A 2,5-Dihydroxy-4-methylbenzylmercaptan

Immersion of the clean Pt(111) crystal into a 0.5-m*M* DMBM solution at 0.0 V (vs. Ag/AgCl reference) forms an adsorbed layer that is not rinsed away by supporting electrolyte (10 m*M* KF, pH = 4). Shown in Fig. 2.18*A* are voltammograms of adsorbed DMBM in 10 m*M* of KF obtained immediately following adsorption (solid curve) or after evacuation into UHV for about 1 h (dotted curve). The two voltammograms are essentially identical indicating that the adsorbed layer is stable in vacuum. However, the adsorbed layer of DMBM is

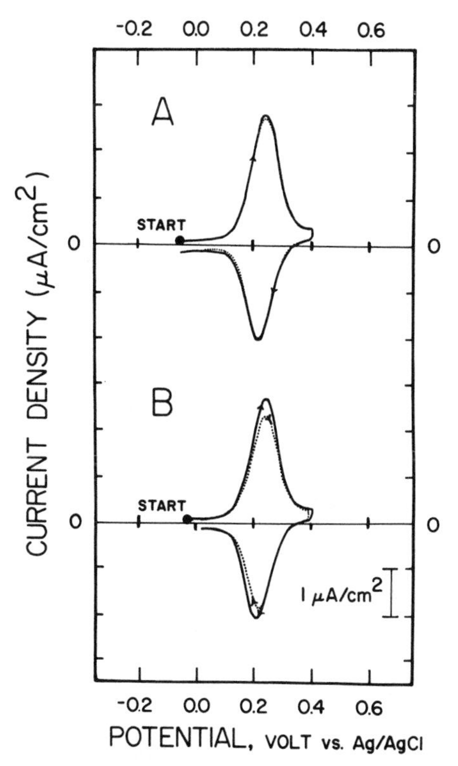

Figure 2.18. Cyclic voltammetry of adsorbed DMBM at Pt(111). (*A*) Solid curve (——): immersion into 7×10^{-4} DMBM followed by rinsing with $10^{-2}M$ TFA. Dotted curve (....): as above followed by one hour in vacuum prior to voltammetry. (*B*) Solid curve (——): first scan. Dotted curve (....): second scan. Scan rate: $5\,\mathrm{mV\,s^{-1}}$. [Reprinted with permission from D. A. Stern, E. Wellner, G. N. Salaita, L. Laguren–Davidson, F. Lu, N. Batina, D. G. Frank, D. C. Zapien, N. Walton, and A. T. Hubbard, *J. Am. Chem. Soc.*, *110*, 4885 (1988). Copyright © (1988) American Chemical Society.]

damaged by cyclic voltammetry, as can be seen from the difference between the first (solid curve) and second (dotted curve) voltammetric scans in Fig. 2.18*B*. Adsorbed DMBM undergoes reversible electrode reaction, Eq. 37:

$$\mathrm{HO\text{-}C_6H_2(CH_3)(OH)\text{-}CH_2\text{-}S\text{-}[Pt]} \rightleftharpoons \mathrm{O{=}C_6H_2(CH_3){=}O\text{-}CH_2\text{-}S\text{-}[Pt]} + 2H^+ + 2e^- \qquad (2.37)$$

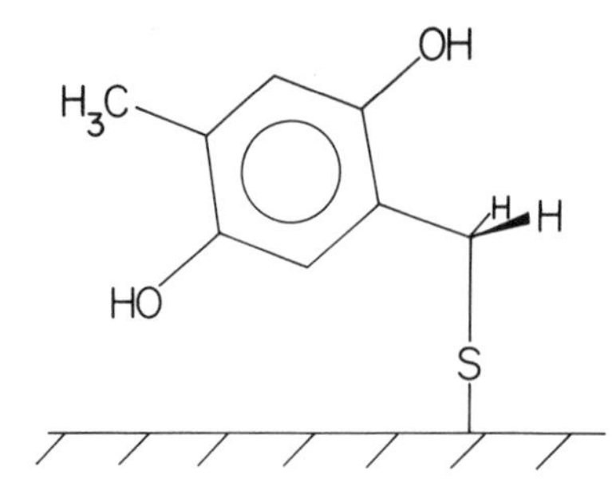

Figure 2.19. Structural model of DMBM adsorbed at Pt(111) at saturation coverage. [Reprinted with permission from D. A. Stern, E. Willman, G. N. Solaita, L. Oaguren–Davidson, F. Lee, N. Batina, D. G. Frank, D. C. Zapien, N. Walton, and A. T. Hubbard, *J. Am. Chem. Soc.*, *110*, 4885 (1988). Copyright © (1988) American Chemical Society.]

The Auger spectrum for adsorbed DMBM on Pt(111) is shown in Fig. 2.5*C*. Packing densities were calculated from C, O, and S Auger signals based on Eqs. 2.9–2.11 in which I°_{Pt} was measured at the positive lobe of the Pt signal at 235 eV. The limiting molecular packing density (Γ) from Γ_C is $\Gamma_C/8 = 0.373\ \mathrm{nmol\ cm^{-2}}$. Molecular packing density (Γ) of DMBM can also be calculated from the attenuation of the Pt signal at 235 eV due to the adsorbed layer by means of Eq. 2.12. The molecular packing density (Γ) based on Eq. 2.12, is $\Gamma = 0.386\ \mathrm{nmol\ cm^{-2}}$. The theoretical limiting molecular packing density based on covalent and van der Waals radii (25) in the model structure shown in Fig. 2.19 is $\Gamma = 0.399\ \mathrm{nmol\ cm^{-2}}$ (41.7*A* molecule^{-1}) (6, 11). The molecular packing density was also determined from the coulometric charge based on Eq. 2.18, $\Gamma_{\mathrm{el}} = 0.381\ \mathrm{nmol\ cm^{-2}}$. The similarities of the packing density values based on the two Auger spectroscopic methods when compared with the value obtained from coulometry suggest that all of the chemisorbed DMBM is electroactive. Packing densities obtained from all these methods are graphed versus concentration (Fig. 2.20). Results based upon Auger spectra and coulometry of DMBM are essentially identical.

Figure 2.21*A* shows the EELS of adsorbed DMBM at Pt(111). Figure 2.21*B* shows the mid-IR (600–4000 cm^{-1}) spectrum of solid DMBM (KBr pellet). As can be seen, these two spectra are similar except that the S—H stretching band at 2500 cm^{-1} is absent from the EELS spectrum indicating that the mercaptan hydrogen of DMBM is lost upon adsorption.

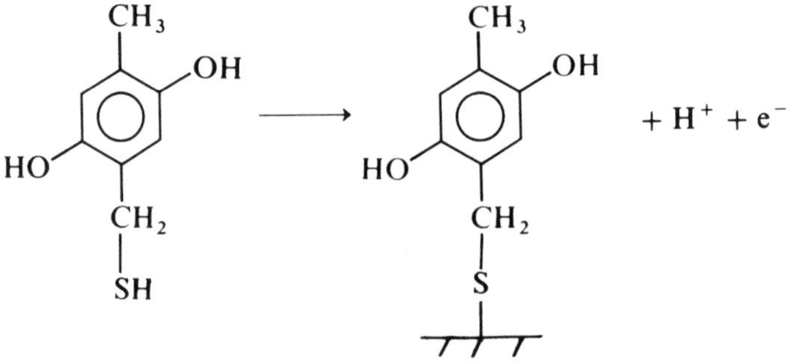

$$\text{(CH}_3\text{)(OH)(HO)C}_6\text{H}_2\text{–CH}_2\text{–SH} \longrightarrow \text{(CH}_3\text{)(OH)(HO)C}_6\text{H}_2\text{–CH}_2\text{–S–Pt} + \mathrm{H^+} + e^- \qquad (2.38)$$

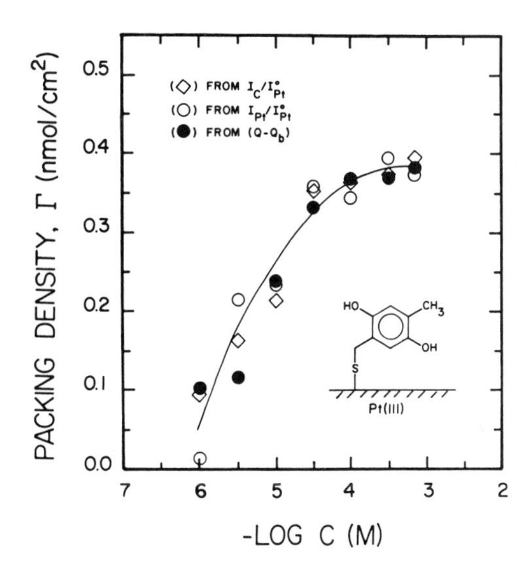

Figure 2.20. Packing density of DMBM at Pt(111) based upon I_{Pt}/I_{Pt} (◇), $I_C/I^\circ 1^{Pt}$ (○), and cyclic voltammetry (●). [Reprinted with permission from D. A. Stern, E. Wallner , G. N. Salaita, L. Laguren–Davidson, F. Lu, N. Batina, D. G. Frank, D. C. Zapien, N. Walton, and A. T. Hubbard, *J. Am. Chem. Soc.*, *110*, (1988). Copyright © American Chemical Society.]

The LEED patterns for adsorbed DMBM at Pt(111) surface displayed beams characteristic of $(2\sqrt{3} \times 2\sqrt{3})R\text{-}30^\circ$ symmetry as depicted in Fig. 2.22*A*. A surface structure model based on the $(2\sqrt{3} \times 2\sqrt{3})$ symmetry was proposed as shown in Fig. 2.22*B*. The theoretical packing density of DMBM in the $2\sqrt{3}$ structure, 0.208 nmol cm^{-2}, resembles that obtained by Auger spectroscopy, Eq. 2.12 at 10^{-5} *M*, 0.207 nmol cm^{-2}. The packing density of DMBM in this ordered stucture is lower than the maximum observed packing density, $\Gamma = 0.4$ nmol cm^{-2}, suggesting that the DMBM molecule is configured so as to span the $2\sqrt{3}$ unit cell. As shown in Fig. 2.22*B*, the benzyl moiety can rotate freely around the C—S bond allowing the benzyl pendant to sweep through the whole area of the $2\sqrt{3}$ cell (6, 11).

2.3.3B 2,5-Dihydroxythiophenyl

In Fig. 2.23*A* is shown the cyclic voltammogram of adsorbed 2,5-dihydro-opythiophenyl (DHTP) at Pt(111). The DHTP layer is not removed by rinsing and displays reversible electroactivity (Fig. 2.23*A*, solid curve). After adsorption of DHTP followed by rinsing with the pure electrolyte, the crystal was brought to vacuum (LEED, Auger, and EELS spectra were recorded), then dipped in pure electrolyte. The cyclic voltammogram was then recorded (dotted curve). As can be seen, the two voltammetric scans are similar, demonstrating the stability of chemisorbed DHTP on Pt(111) in vacuum (6). On the other hand, cyclic voltammetry caused damage to the adsorbed DHTP layer. Each successive

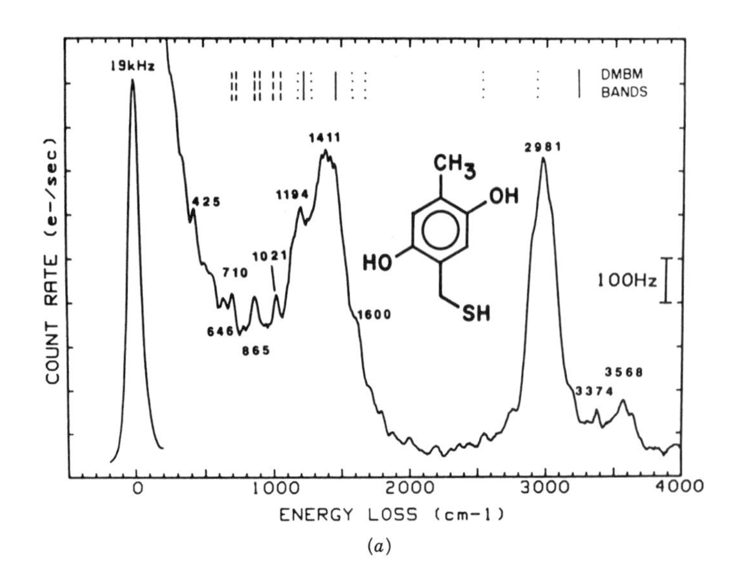

(*a*)

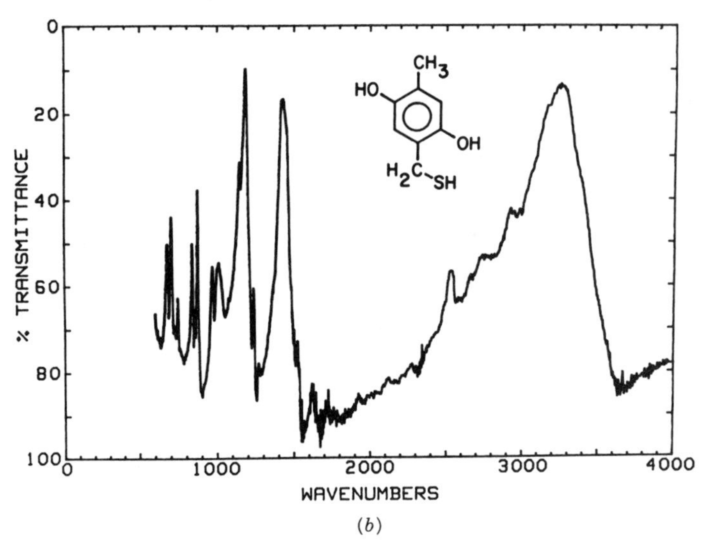

(*b*)

Figure 2.21. Vibrational spectra of DMBM. (*A*) EELS spectrum of adsorbed DMBM at Pt(111). The locations of the mid-IR adsorption bands are also shown (⁚) = weak, (|) = strong, and (|) very strong. (*B*) IR spectrum of solid DMBM in KBr. Experimental conditions: DMBM solution concentration, 0.1 m*M* in 10 m*M* KF/HF electrolyte (pH 4); EELS incidence and detection angles, 62° from surface normal; beam energy, 4 eV; beam current, 0.15 nA; EELS resolution, 10 meV (80 cm^{-1}) fwhm; IR resolution, 4 cm^{-1} fwhm. [Reprinted with permission from D. A. Stern, E. Willner, G. N. Salaita, L. Laguren-Davidson, F. Lu, N. Batina, D. G. Frank, D. C. Zapier, N. Walton, and A. T. Hubbard, *J. Am. Chem. Soc.*, *110*, (1988). Copyright © American Chemical Society.]

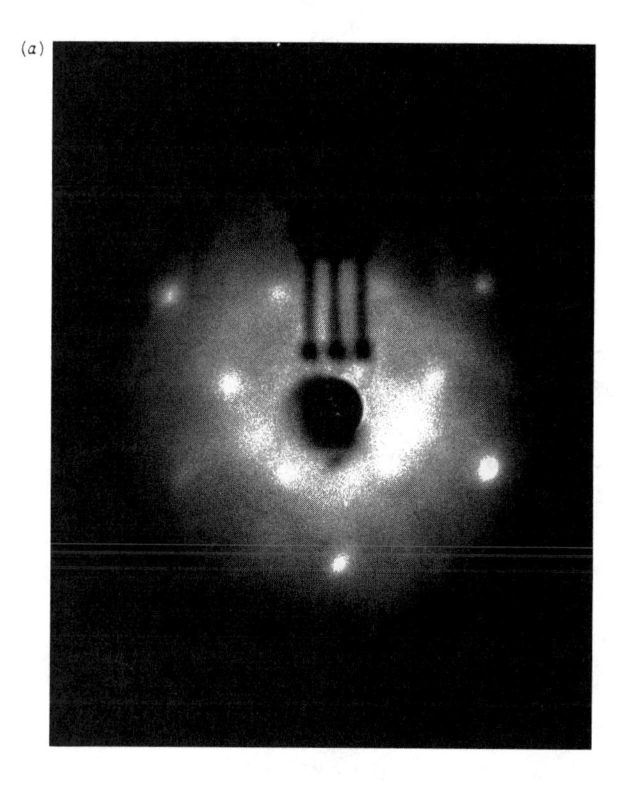

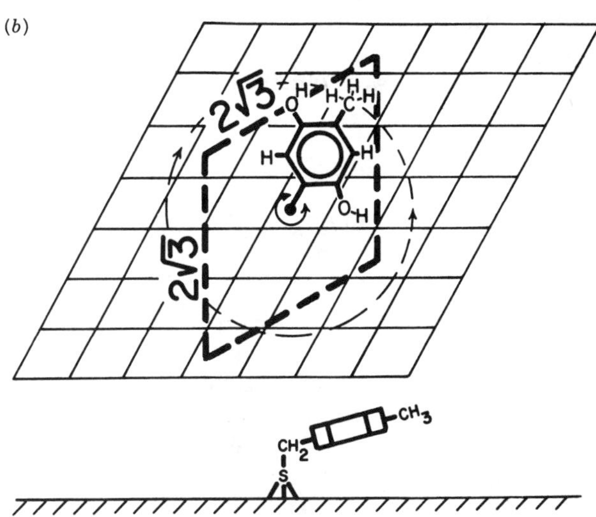

Figure 2.22. LEED pattern and structural models of species adsorbed at Pt(111). (*A*) Pt(111)($2\sqrt{3} \times 2\sqrt{3}$)R30° LEED pattern of adsorbed DMBM. (*B*) DMBM structural model. Experimental conditions: LEED beam energy, 70 eV; electrolyte, 10 m*M* KF/HF (pH = 4); electrode potential, 0.2 versus Ag/AgCl (1*M*) reference; DMBM concentration, 0.01 m*M*. [Reprinted with permission from D. A. Stern, E. Willner, G. N. Salaita, L. Laguren-Davidson, F. Lu, N. Batina, D. G. Frank, D. C. Zapier, N. Walton, and A. T. Hubbard, *J. Am. Chem. Soc.*, *110*, (1988). Copyright © American Chemical Society.]

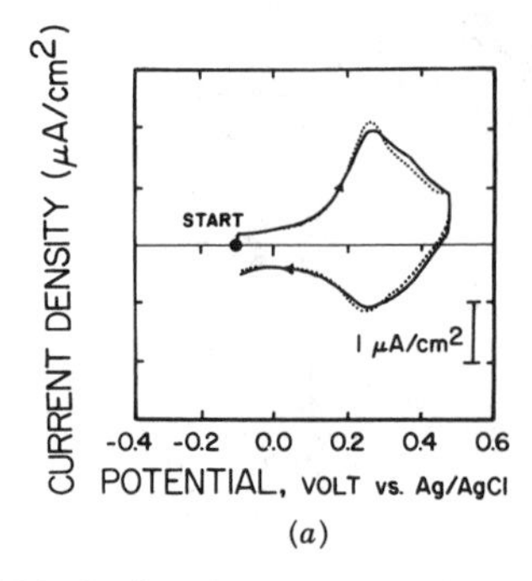

(*a*)

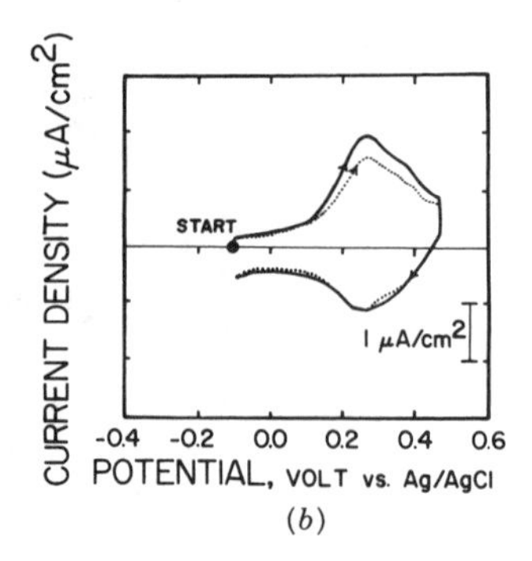

(*b*)

Figure 2.23. Cyclic voltammetry of adsorbed DHTP at Pt(111). (*A*) Solid curve (——): immersion into 0.7 m*M* DHTP followed by rinsing with 10 mf*M* TFA. Dotted curve (....): as above, but with 1 h in vacuum prior to voltammetry. (*B*) Solid curve (——): first scan; dotted curve (....), second scan. Scan rate $5\,\mathrm{mV\,s^{-1}}$. [Reprinted with permission from D. A. Stern, E. Willner, G. N. Salaita, L. Laguren-Davidson, F. Lu, N. Batina, D. G. Frank, D. C. Zapier, N. Walton, and A. T. Hubbard, *J. Am. Chem. Soc.*, *110* (1988). Copyright © American Chemical Society.]

voltammetric scan leads to a smaller peak, Fig. 2.23*B*. The redox peak widths for adsorbed DHTP are broader than those for chemisorbed DMBM, Figs. 2.18 and 2.23.

Soriaga and co-workers (23, 31d, 95) attribute this broadening in peak widths for adsorbed DHTP and DMBM to adsorbate–adsorbate interactions mediated through the metal electrode.

Packing densities of DHTP were obtained from Auger data by use of Eqs. 39–42.

$$I_{\mathrm{Pt}}/I_{\mathrm{Pt}}^{\circ} = (1 - K_x\Gamma)(1 - 4K_{\mathrm{C}}\Gamma)^2 \tag{2.39}$$

$$\Gamma_{\mathrm{C}} = (I_{\mathrm{C}}/I_{\mathrm{Pt}}^{\circ})/[B_{\mathrm{C}}(\tfrac{1}{4} + \tfrac{7}{12}f_{\mathrm{C}}/ + \tfrac{1}{6}f_{\mathrm{C}}^2)] \tag{2.40}$$

$$\Gamma_{\mathrm{O}} = (I_{\mathrm{O}}/I_{\mathrm{Pt}}^{\circ})/[B_{\mathrm{O}}(\tfrac{3}{4} + \tfrac{1}{4}f_{\mathrm{O}}^3)] \tag{2.41}$$

$$\Gamma_{\mathrm{S}} = (I_{\mathrm{S}}/I_{\mathrm{Pt}}^{\circ})/[B_{\mathrm{S}}(\tfrac{1}{2}f_{\mathrm{S}} + \tfrac{1}{2}f_{\mathrm{S}}^2)] \tag{2.42}$$

The molecular packing density obtained from Eq. 2.39, $\Gamma = 0.259\,\mathrm{nmol\,cm^{-2}}$, is small compared to the theoretical packing density, $\Gamma = 0.572\,\mathrm{nmol\,cm^{-2}}$, $29.01\,\text{Å}^2\,\mathrm{molecule^{-1}}$. The packing density Γ_{el} based on Eq. 2.18 was found to be equal to $0.261\,\mathrm{nmol\,cm^{-2}}$, which is close to $\Gamma = 0.259$ obtained from Eq. 2.39.

HO, OH, SH (2,5-dihydroxythiophenol) ⟶ OH, OH (hydroquinone) + S (2.43)

This suggests loss of the diphenol moiety from about one-half of the adsorbed DHTP molecules.
This conclusion is in agreement with previous results (31) obtained by means of capillary gas chromatography combined with thin-layer electrochemistry.

Figure 2.24*A* shows the EELS for adsorbed DHTP at the Pt(111) surface. Shown in Fig. 2.24*B* is the mid-IR spectrum of solid DHTP (KBr pellet). The two spectra are similar except for the absence of the S—H stretching frequency at 2500 cm^{-1} and the S—H bending frequency at 1020 cm^{-1} from the EELS spectrum. This is indicative of the loss of mercaptan hydrogen upon adsorption as shown in Eq. 2.44.

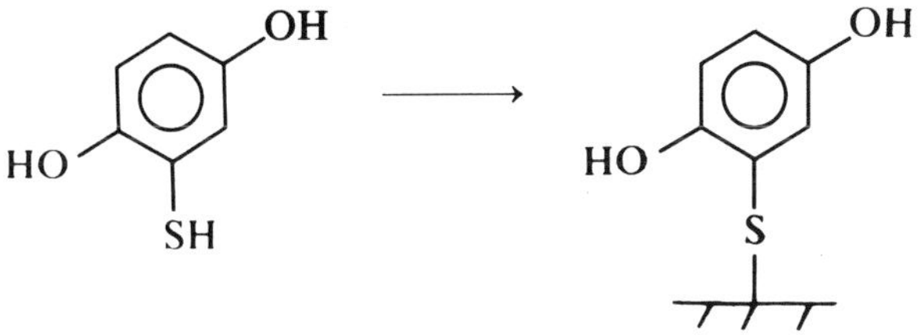

OH, HO, SH → OH, HO, S $+ H^+ + e^-$ (2.44)

2.3.4 Amino Acids

L-DOPA (LD): Shown in Fig. 2.25*B* is the Auger spectrum for adsorbed LD at the Pt(100) surface. Similar results were obtained for Pt(111). The packing density for adsorbed LD at Pt(100) was calculated based on the two independent Auger methods described earlier. The elemental packing densities of Γ_C, Γ_O, and Γ_N were calculated from the elemental Auger signals I_C, I_O, and I_N according to Eqs. 2.45–2.47.

$$I_C = (I_C/I^{\circ}_{Pt})/[B_C(\tfrac{5}{9} + \tfrac{4}{9}f_C)] \qquad (2.45)$$

where $f_C = f_N = f_O = 0.70$ and $B_C = 0.377\ cm^2\ nmol^{-1}$;

$$\Gamma_O = (I_O/I^{\circ}_{Pt})/B_O \qquad (2.46)$$

where $B_O = 0.476\ cm^2\ nmol^{-1}$;

$$\Gamma_N = (I_N/I^{\circ}_{Pt})/B_N \qquad (2.47)$$

where $B_N = 1.176\ cm^2\ nmol^{-1}$.

The molecular packing density (Γ) was obtained from the elemental packing density (Γ_C) as given in Eq. 2.48.

$$\Gamma = \tfrac{1}{9}\Gamma_C \qquad (2.48)$$

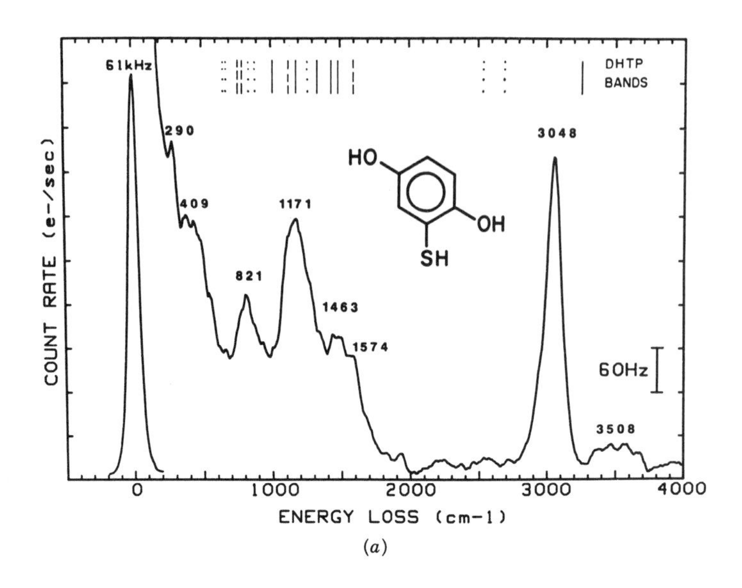

(*a*)

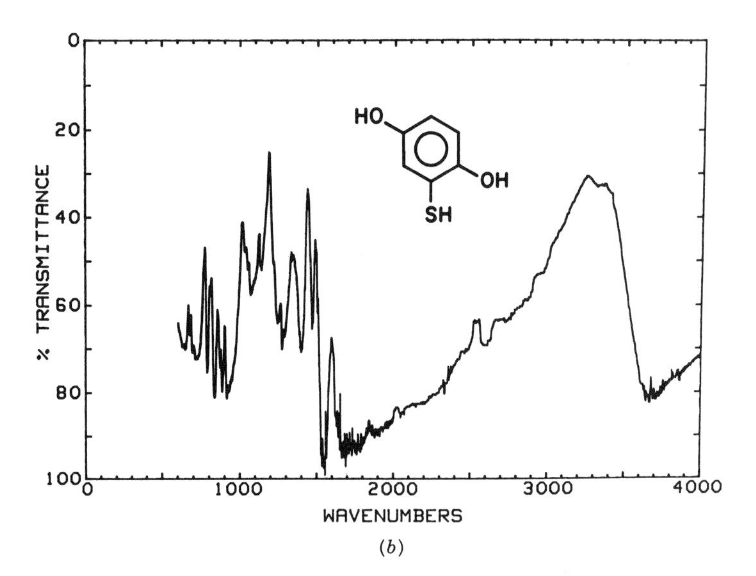

(*b*)

Figure 2.24. Vibrational spectra of DHTP. (*A*) EELS spectrum of adsorbed DHTP at Pt(111). (*B*) IR spectrum of solid DHTP (KBr pellet). Experimental conditions: DHTP solution concentration, 0.5 m*M*; other conditions as in Fig. 2.21. [Reprinted with permission from D. A. Stern, E. Willner, G. N. Salaita, L. Laguren-Davidson, F. Lu, N. Batina, D. G. Frank, D. C. Zapier, N. Walton, and A. T. Hubbard, *J. Am. Chem. Soc.*, *110*, (1988). Copyright © American Chemical Society.]

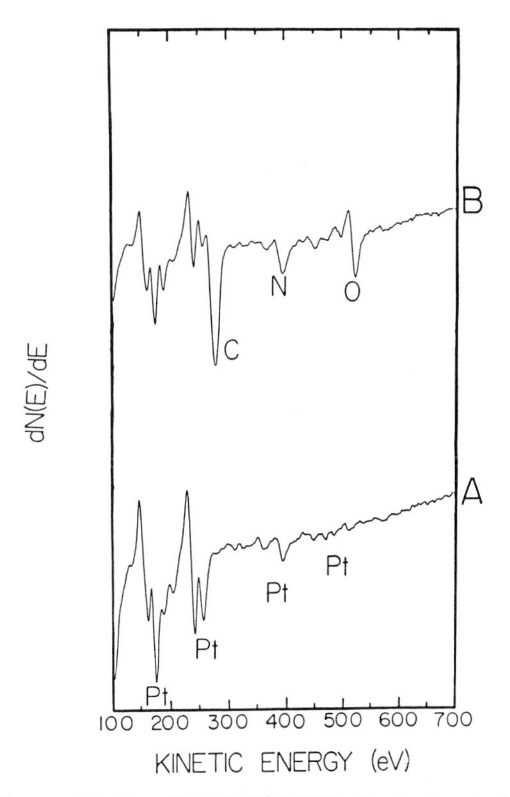

Figure 2.25. Auger Spectra. (*A*) Clean Pt(100). (*B*) Pt(100) after treatment with L-DOPA (3 m*M*). Experimental conditions: incident beam, 10^{-7} at 2000 eV, normal to the surface; modulation amplitude, 5 V peak to peak; electrolyte was 10 m*M* KF, adjusted to pH = 4 with HF; electrode potential was 0.20 V versus Ag/AgCl (1*M* KCl).

Similarly, the molecular packing density was calculated independently by the attenuation of the Pt Auger signal of the clean surface I°_{Pt} at 161 eV as given in Eq. 2.49.

$$I_{\mathrm{Pt}}/I^{\circ}_{\mathrm{Pt}} = (1 - 9K_{\mathrm{C}}\Gamma)(1 - 5K_{\mathrm{C}}\Gamma) \tag{2.49}$$

where $K_{\mathrm{C}} = 0.160\,\mathrm{cm^2\,nmol^{-1}}$. The measured packing densities based upon the above mentioned Auger methods are graphed as a function of concentrations for both Pt(100) and Pt(111) as shown in Fig. 2.26*A* and *B*, respectively. The plateau packing density (above 1 m*M* LD) based upon Eq. 2.48 for Pt(100) is $\Gamma = 0.23\,\mathrm{nmol\,cm^{-2}}$ and for Pt(111) is $\Gamma = 0.19\,\mathrm{nmol\,cm^{-2}}$. Based upon Eq. 2.49 for Pt(100), $\Gamma = 0.22\,\mathrm{nmol\,cm^{-2}}$ and for Pt(111) $\Gamma = 0.17\,\mathrm{nmol\,cm^{-2}}$. The packing density of adsorbed LD at Pt(111) is lower than at Pt(100) by about 20% (13) under identical conditions. These values are in agreement with the ideal

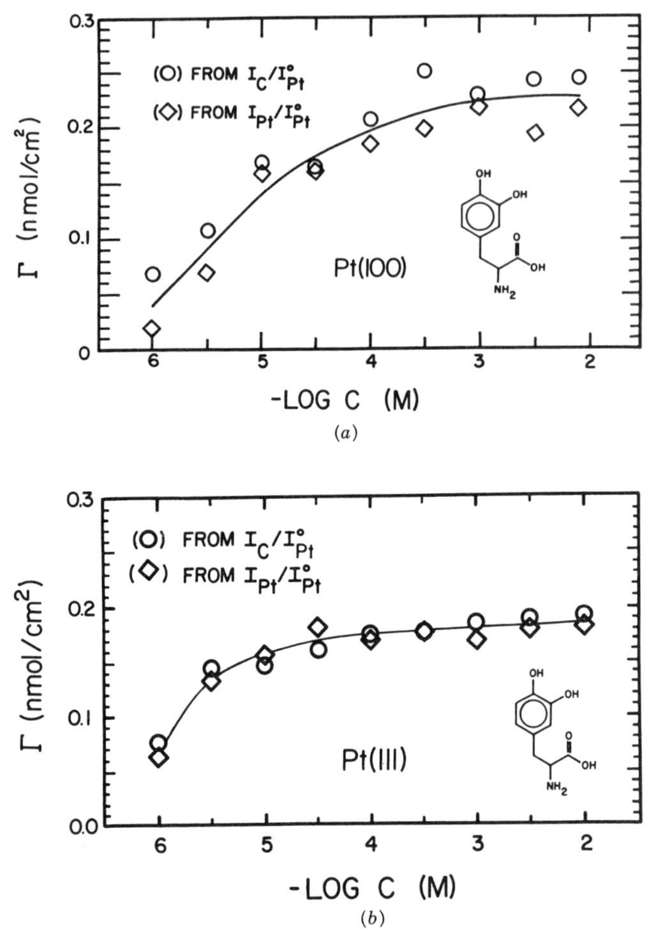

Figure 2.26. Packing density of L-DOPA versus concentration. (*A*) Pt(100). (*B*) Pt(111). (○) = packing density from carbon Auger signal; (◇) = packing density from Pt Auger signal attenuation. [Reprinted with permission from D. A. Stern et al., *Langmuir*, *4*, 711 (1988). Copyright © (1988) American Chemical Society.]

theoretical packing densities of adsorbed LD at Pt(100) as shown in the surface structural model (Fig. 2.27) $\Gamma = 0.213\,\text{nmol}\,\text{cm}^{-2}$, $78.0\,\text{Å}^2$ molecule^{-1}.

From the results, the most likely conformation for adsorbed LD at Pt(100) is the orientation displayed in Fig. 2.27 in which the catechol ring is parallel to the surface. On the other hand, the packing density of adsorbed LD at Pt(111) is in good agreement with the calculated packing density for the surface structural model shown in Fig. 2.27. The amino acid moiety and catechol ring of adsorbed LD at Pt(111) are both in contact with the surface, Fig. 2.27.

Shown in Fig. 2.28 is the EELS spectrum for adsorbed LD at Pt(111) surface. Shown in the same figure are the IR bands of solid LD (KBr pellet). The two

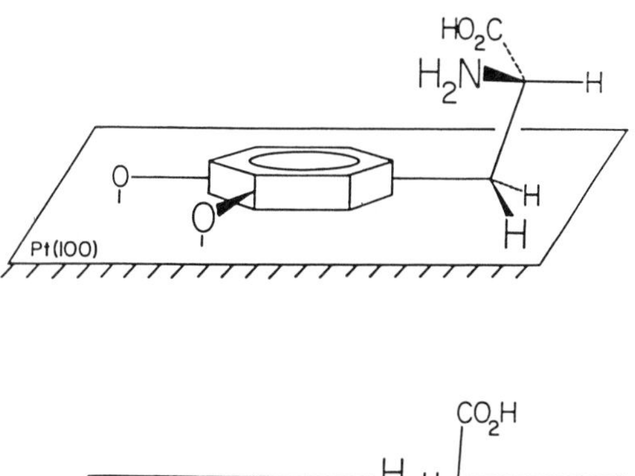

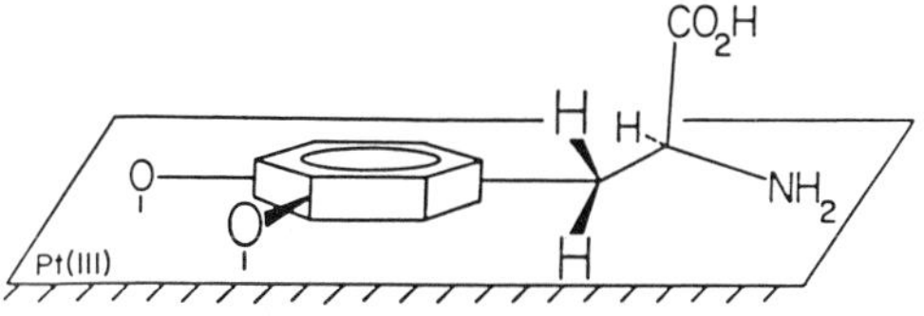

Figure 2.27. Structural models of adsorbed molecules at Pt(100) and Pt(111) surfaces, L-DOPA. [Reprinted with permission from D. A. Stern et al., *Langmuir*, *4*, 711 (1988). Copyright © (1988) American Chemical Society.]

spectra are similar except that the phenolic O—H stretch is absent from the EELS that is expected. The carboxylic O—H stretching region of the EELS spectrum shows a band at 2558 cm^{-1}, which is evidence of a pendant amino acid moiety (9, 13). The bands that appeared at 3338 and 3158 cm^{-1} are a result of N—H stretching. The C—H stretching region shows a band at 3070 cm^{-1}, which is due to aromatic C—H stretching, in contrast to the aliphatic C—H bands that appeared at 2980 and 2923 cm^{-1}. This suggests that the aromatic character of adsorbed LD is retained upon adsorption. The peak at 1758 cm^{-1} is attributable to carboxylic C═O stretching.

The EELS spectrum of adsorbed catechol is shown in Fig. 2.28*B* for comparison with the accepted assignment of the IR spectra of solid catechol (9, 13). The phenolic O—H stretching bands in the IR spectrum of solid catechol are absent from the EELS spectrum of adsorbed catechol, indicating the loss of phenolic hydrogen upon adsorption as shown in Eq. 2.50.

OH, OH (catechol) ⟶ O, O (adsorbed) $+ 2H^+ + 2e^-$ (2.50)

It has been shown that the hydroxyl hydrogen atoms of HQ and phenol are lost

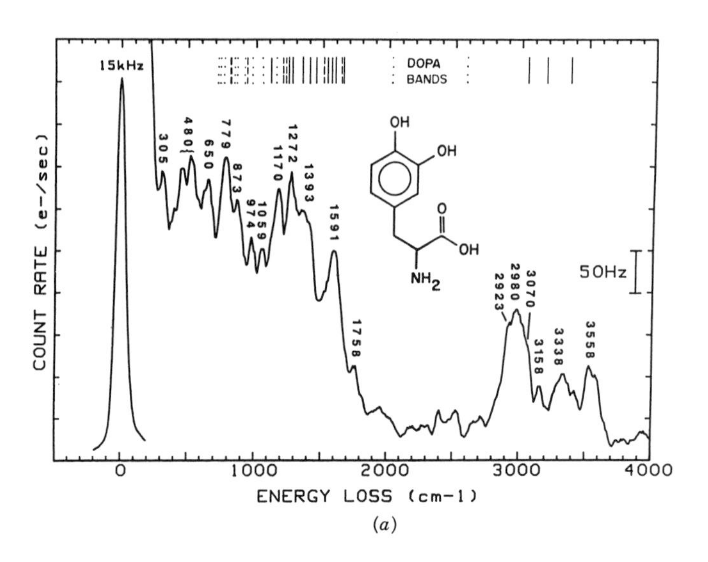

(a)

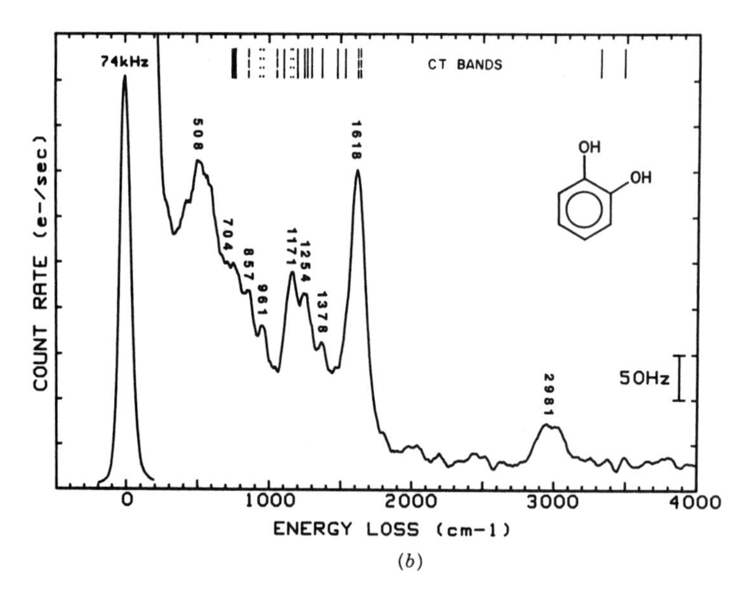

(b)

Figure 2.28. EELS spectra. (*A*) L-DOPA adsorbed at Pt(111). (*B*) CT adsorbed at Pt(111). Experimental conditions: electrolyte was 10 m*M* KF/HF (pH = 4); L-DOPA concentration 5 m*M*; EELS incidence and detection angles 62° from surface normal; beam energy 4 eV; beam current 0.15 nA; 10 meV (80^{-1}) fwhm resolution; adsorption potentials were +0.2, +0.1, and −0.1 V versus Ag/AgCl (1*M*), respectively. Also shown for reference are the mid-IR bands of the parent compound in KBr: (|) = strong bands, (|) = medium strong; (:) = weak bands. [Reprinted with permission from D. A. Stern et al., *Langmuir*, *4*, 711 (1988). Copyright © (1988) American Chemical Society.]

upon adsorption (5). Hence, it is very possible that phenolic hydrogens of LD are lost upon adsorption as shown in Eq. 2.51.

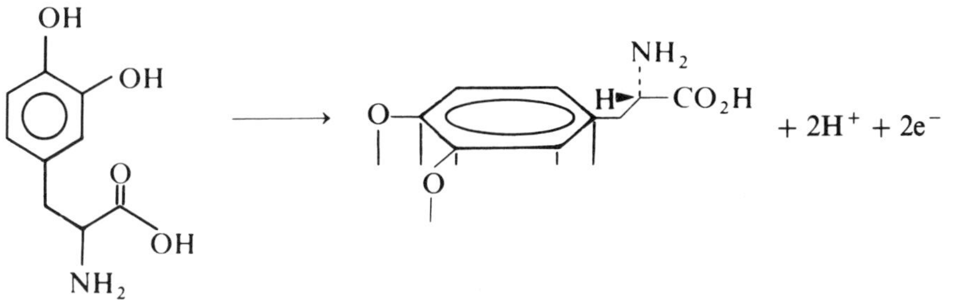

$$\longrightarrow \quad + 2H^+ + 2e^- \qquad (2.51)$$

n_{Ox} measurements were made at Pt(100) and Pt(111) as shown in Fig. 2.29*A* and *B*. After adsorbing LD (0.00 V vs. Ag/AgCl) on Pt(100) and Pt(111) the surface and adsorbed layer were oxidized [for Pt(100) 0.83 V vs. Ag/AgCl, 1*M* KCl reference] followed by reduction at −0.12 V [for Pt(111), oxidation at 0.91 V vs. Ag/AgCl, 1*M* KCl reference] and reduction at −0.12 V. The background charge, Q_{bg}, was measured in the absence of the adsorbed layer, Fig. 2.29*C*.

Combining the $Q_{Ox} - Q'_{bg}$ values with the packing density Γ based on I_C/I°_{Pt} allowed calculation of n_{Ox}.

$$n_{Ox} = \frac{Q_{Ox} - Q'_{bg}}{\mathcal{F} A \Gamma} \qquad (2.52)$$

n_{Ox} for adsorbed LD at Pt(111) was equal to 33.0 while for adsorbed LD at Pt(100) it was 17.8. This suggests that the oxidation of LD at Pt(100) leads to different products than at Pt(111). Complete oxidation of adsorbed LD to CO_2 and NO_2 would require 41 e^-. Several other amino acids have been studied; details of that work are given in Refs. 5a and 13.

2.3.5A Pyridine

The adsorption of pyridine (pyr) from the gas phase under UHV conditions was extensively studied by EELS, LEED, UPS, IR, near-edge X-ray absorption fine structure spectroscopy (NEXAFS), work function, and isotopic methods at Pt(111) and Pt(100) (63, 96–99). At room temperature and high coverage it was reported that pyridine adsorbs through the nitrogen atom with the ring perpendicular to the surface (64, 97). However, at higher temperature and lower coverage there is participation of the electrons of the ring upon adsorption and the molecule is adsorbed in a tilted orientation. Recently, we obtained such information for pyridine adsorption from solution.

Figure 2.30*B* shows the Auger spectrum of adsorbed pyr at a Pt(111) surface.

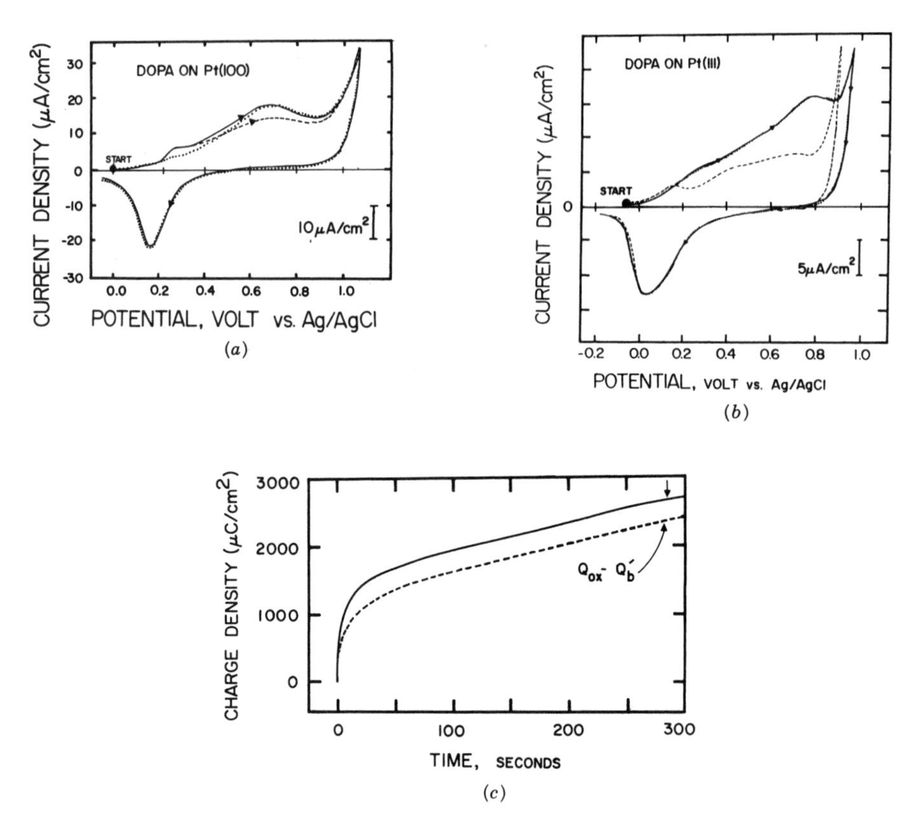

Figure 2.29. Cyclic voltammetry and coulometry. (*A*) Cyclic voltammetry of adsorbed L-DOPA at Pt(100). (*B*) Cyclic voltammetry of adsorbed L-DOPA at Pt(111). (*C*) Chronocoulometry of adsorbed L-DOPA at Pt(111). Solid curve (——): first oxidation–reduction cycle. Dotted curve (....): first oxidation–reduction cycle, after 1 h in vacuum. Dashed curve (---): second oxidation–reduction cycle. Experimental conditions: electrolyte was 10 m*M* KF adjusted to pH 4 with HF; L-DOPA concentration was 1.0 m*M*; scan rate, $5\,\mathrm{mV\,s^{-1}}$; temperature, 23 ± 1°C; electrode potentials, 0.83 V (Ox), −0.12 V (Red) versus Ag/AgCl (1*M*). [Reprinted with permission from D. A. Stern et al., *Langmuir*, *4*, 711 (1988). Copyright © (1988) American Chemical Society.]

Packing densities were calculated from the Auger data (Table 2.5) based on Eqs. 2.53–2.56.

$$\Gamma_C = (I_C/I^\circ_{Pt})[B_C(\tfrac{2}{5} + \tfrac{3}{5}f)] \tag{2.53}$$

$$\Gamma = \tfrac{1}{5}\Gamma_C \tag{2.54}$$

$$\Gamma_N = (I_N/I^\circ_{Pt})/(B_N f^2) \tag{2.55}$$

$$I_{Pt}/I^\circ_{Pt} = 1 - 6K\Gamma \tag{2.56}$$

The saturation molecular packing density (Table 2.6) based upon carbon Auger signal (Eqs. 2.53 and 2.54) is $0.45\,\mathrm{nmol\,cm^{-2}}$, while that based upon calculated

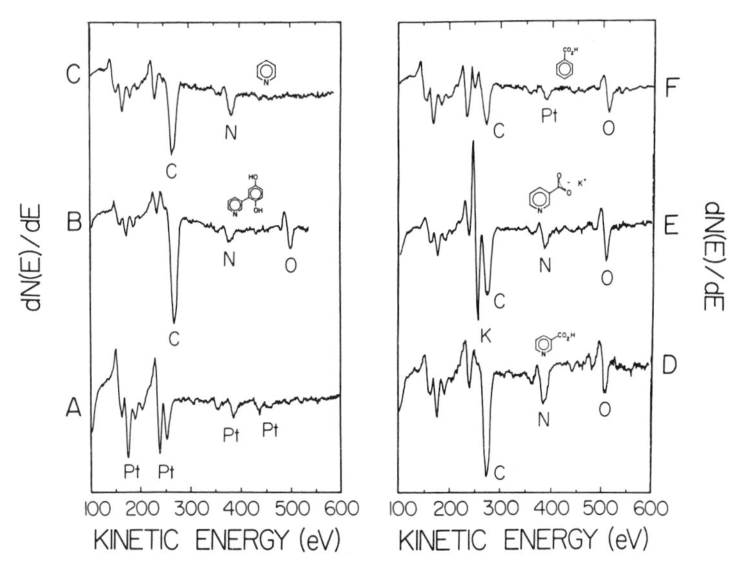

Figure 2.30. Auger spectra of adsorbed layers at Pt(111). Curve A, Clean Pt(111). Curve B, 3PHQ (0.5 mM, pH = 4; rinsed at pH = 4; −0.1 V). Curve C, PYR (1 mM, pH = 3; rinsed at pH = 3; −0.1 V). Curve D, NA (1 mM, pH = 3; rinsed at pH = 3; −0.15 V). Curve E, KNA (1 mM, pH = 3; rinsed at pH = 10; −0.3 V). Curve F, BA (1 mM, pH = 3; rinsed at pH = 3; −0.3 V). Experimental conditions: adsorption from 10 mM KF electrolyte, following rinsing with 0.1 mM KOH (pH = 10); 0.1 mM KF (pH = 4) or 2 mM HF (pH = 3); electrode potentials versus Ag/AgCl (1 mM KCl); electron beam at normal incidence, 100 nA, 2000 eV. [Reprinted with permission from D. A. Stern, L. Laguren-Davidson, D. G. Frank, J.Y-P. Gul, C-H. Lin, F. Lu, G. N. Salaita, N. Walton, D. C. Zapien, and A. T. Hubbard, *J. Am. Chem. Soc.*, *111*, 877 (1989). Copyright © (1989) American Chemical Society.]

Pt Auger attenuation (Eq. 2.56) is 0.46 nmol cm^{-2}. The theoretical packing density based on covalent and van der Waals radii (25) is 0.450 nmol cm^{-2} (36.91 Å^2 molecule^{-1}) for the surface model structure shown in Fig. 2.31 in which the angle between the plane of the ring and the Pt surface is 71°.

$$\text{Molecular area} \qquad \text{Å}^2 = a(b \cos \phi + 3.45\phi) \qquad (2.57)$$

where $a = 6.71$ and $b = 6.99$. A graph of packing densities versus concentration is shown in Fig. 2.32. Constancy of pyridine density over a wide range of concentrations (from 10^{-6} to neat pyr) is an indication of the remarkable strength of adsorption of pyr and of the uniformity of the pyr layer (Fig. 2.31). An angle of 74 ± 10° was reported for pyr adsorbed at Pt(111) from the gas phase utilizing the NEXAFS method (99). An angle of 55° was also reported for pyr deposited at low temperature on Ag(111) from the gas-phase using EELS as a surface probe (100). On the other hand, vertically oriented pyr would have resulted in a packing density of 0.728 nmol cm^{-2} (22.8 Å^2 molecule^{-1}), much larger than the observed value of 0.450 nmol cm^{-2}, while the horizontal

Table 2.5. Auger and Electrochemical Data for Molecules Adsorbed at Pt(111)

Compound	$-\log C$	*Electrode Potential* (V)	*Adsorbed* pH	*Rinse* pH	*Normalized Auger Intensity*				
					I_C/I°_{Pt}	I_O/I°_{Pt}	I_N/I°_{Pt}	I_{Pt}/I°_{Pt}	$I_K/I^\circ_{Pt}(*)$
pyr	6.00	−0.1	7	7.0	0.483		0.161	0.644	
pyr	5.00	−0.1	7	7.0	0.680		0.177	0.539	
pyr	4.00	−0.1	7	7.0	0.678		0.192	0.499	
pyr	3.00	−0.1	7	7.0	0.710		0.159	0.515	
pyr	2.00	−0.1	7	7.0	0.697		0.146	0.512	
pyr	1.00	−0.1	7	7.0	0.687		0.157	0.491	
pyr	0.00	−0.1	7	7.0	0.698		0.158	0.495	
pyr	Neat	Open			0.713		0.190	0.521	
NA	6.00	−0.2	7	3.3	0.370	0.276	0.075	0.704	
NA	5.00	−0.2	7	3.3	0.593	0.377	0.142	0.551	
NA	4.00	−0.2	7	3.3	0.619	0.461	0.162	0.493	
NA	3.00	−0.2	7	3.3	0.647	0.370	0.156	0.474	
NA	2.00	−0.2	7	3.3	0.677	0.405	0.151	0.458	
NA	1.00	−0.2	7	3.3	0.679	0.416	0.150	0.451	

NA	3.00	−0.3	7	7.0	0.634	0.455	0.144	0.386	0.83
NA	3.00	0.3	3	7.0	0.677	0.489	0.151	0.406	0.05
NA	6.00	0.3	7	3.3	0.227	0.169	0.086	0.834	
NA	5.00	0.3	7	3.3	0.404	0.312	0.150	0.651	
NA	4.00	0.3	7	3.3	0.479	0.332	0.155	0.648	
NA	3.00	0.3	7	3.3	0.574	0.441	0.169	0.533	
NA	2.00	0.3	7	3.3	0.624	0.423	0.174	0.482	
NA	1.00	0.3	7	3.3	0.642	0.446	0.176	0.470	
NA	3.00	−0.3	3	10.0	0.728	0.474	0.171	0.376	0.768
NA	3.00	−0.2	3	10.0	0.685	0.492	0.184	0.360	0.779
NA	3.00	−0.1	3	10.0	0.677	0.476	0.210	0.386	0.888
NA	3.00	0.0	3	10.0	0.688	0.478	0.150	0.367	0.944
NA	3.00	0.1	3	10.0	0.653	0.463	0.156	0.406	0.938
NA	3.00	0.1	3	10.0	0.653	0.463	0.156	0.406	0.938
NA	3.00	0.2	3	10.0	0.647	0.429	0.190	0.422	0.598
NA	3.00	0.3	3	10.0	0.656	0.454	0.185	0.434	0.390
NA	3.00	0.4	3	10.0	0.626	0.452	0.167	0.405	0.230
NA	3.00	0.6	3	10.0	0.477	0.434	0.182	0.551	0.155

Table 2.6. Packing Densities at Pt(111)

Compound[a]	$-log\ C$	*Electrode Potential* (*V*)	*Adsorbed* pH	*Rinse* pH	Γ_C (nmol cm^{-2})
pyr	6.00	−0.1	7	7.0	1.56
pyr	5.00	−0.1	7	7.0	2.20
pyr	4.00	−0.1	7	7.0	2.19
pyr	3.00	−0.1	7	7.0	2.30
pyr	2.00	−0.1	7	7.0	2.26
pyr	1.00	−0.1	7	7.0	2.22
pyr	0.00	−0.1	7	7.0	2.26
pyr	Neat	Open			2.31
NA	6.00	−0.2	7	3.3	1.22
NA	5.00	−0.2	7	3.3	1.96
NA	4.00	−0.2	7	3.3	2.05
NA	3.00	−0.2	7	3.3	2.14
NA	2.00	−0.2	7	3.3	2.24
NA	1.00	−0.2	7	3.3	2.25
NA	3.00	−0.3	7	7.0	2.10
NA	3.00	−0.3	7	7.0	2.10
NA	3.00	0.3	3	7.0	2.24
NA	6.00	0.3	7	3.3	0.75
NA	5.00	0.3	7	3.3	1.34
NA	4.00	0.3	7	3.3	1.59
NA	3.00	0.3	7	3.3	1.90
NA	2.00	0.3	7	3.3	2.07
NA	1.00	0.3	7	3.3	2.13
NA	3.00	−0.3	3	10.0	2.41
NA	3.00	−0.2	3	10.0	2.27
NA	3.00	−0.1	3	10.0	2.25
NA	3.00	0.0	3	10.0	2.28
NA	3.00	0.1	3	10.0	2.17
NA	3.00	0.2	3	10.0	2.15
NA	3.00	0.3	3	10.0	2.18
NA	3.00	0.4	3	10.0	2.08
NA	3.00	0.6	3	10.0	1.58

[a]NA = nicotinic acid.

Γ_O (nmol cm^{-2})	Γ_N (nmol cm^{-2})	*Molecular* Γ, Γ_K (nmol cm^{-2})	*Molecular* Γ, *Based on* I_C/I°_{Pt} (nmol cm^{-2})	*Based on* I_{Pt}/I°_{Pt} (nmol cm^{-2})
	0.28		0.31	0.33
	0.31		0.44	0.43
	0.33		0.44	0.46
	0.28		0.46	0.45
	0.25		0.45	0.45
	0.27		0.44	0.47
	0.27		0.45	0.47
	0.33		0.46	0.44
0.63	0.09		0.20	0.21
0.86	0.17		0.33	0.31
0.91	0.20		0.34	0.35
0.84	0.19		0.36	0.36
0.92	0.18		0.37	0.36
0.96	0.18		0.38	0.38
1.03	0.18	0.27	0.35	0.38
1.03	0.18	0.27	0.35	0.38
1.11	0.18	0.02	0.37	0.41
0.38	0.15		0.12	0.11
0.71	0.26		0.22	0.24
0.75	0.27		0.26	0.24
1.00	0.29		0.32	0.32
0.96	0.30		0.34	0.36
1.01	0.31		0.35	0.37
1.08	0.30	0.25	0.40	0.39
1.12	0.32	0.26	0.38	0.40
1.08	0.36	0.29	0.37	0.38
1.09	0.26	0.31	0.38	0.40
1.05	0.27	0.31	0.36	0.37
0.97	0.33	0.20	0.36	0.36
1.03	0.32	0.13	0.36	0.35
1.03	0.29	0.08	0.35	0.37
0.99	0.32	0.05	0.26	0.28

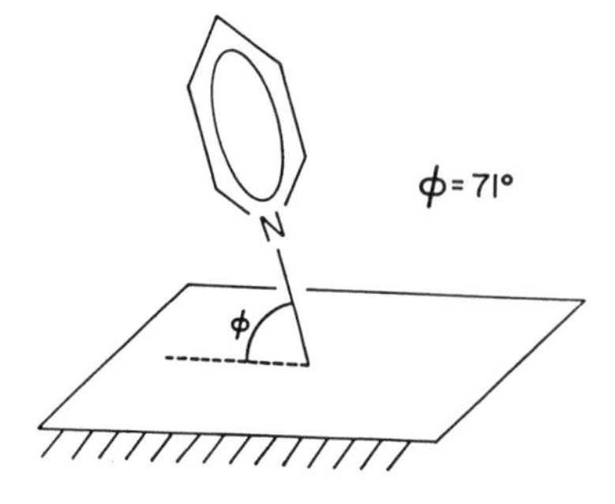

Figure 2.31. Structural model of adsorbed pyr at Pt(111). [Reprinted with permission from D. A. Stern, L. Laguren-Davidson, D. G. Frank, J. Y-P. Gui, C-H. Lin, F. Lu, G. N. Salaita, N. Walton, D. C. Zapier, and A. T. Hubbard, *J. Am. Chem. Soc.*, *111*, 877 (1989). Copyright © (1989) American Chemical Society.]

orientation would have resulted in about 0.384 nmol cm^{-2}, which is smaller than the value observed.

Barradas and Conway (101, 102) calculated that an adsorbed pyr molecule on a silver electrode should occupy a 38-$Å^2$ site in the horizontal orientation.

Figure 2.33 shows the EELS spectrum of pyr adsorbed from aqueous solution at Pt(111). Also shown is the IR spectrum of neat pyr (103). The two spectra are similar except for the Pt—N stretch at 416 cm^{-1}, which is not seen in the IR spectrum. The vibrational assignments for pyr adsorbed on Pt(111) are given in Table 2.7.

Pyridine adsorbed at Pt(111) from aqueous solution exhibited a LEED pattern such as shown in Fig. 2.34*A*. This pattern has an oblique unit mesh. The surface structure is incommensurate; that is, the adsorbed molecules are aligned

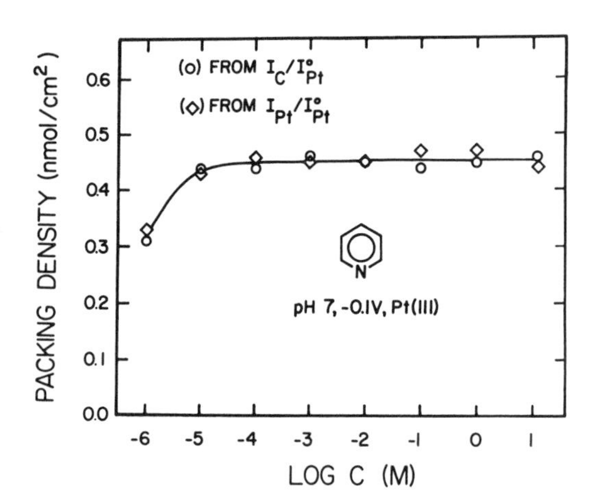

Figure 2.32. Packing density of pyr at Pt(111) versus concentration. Experimental conditions: electrode potential, −0.1 V versus Ag/AgCl reference (1*M* KCl); electrolyte, 10 m*M* KF adjusted to pH 7 with KOH; rinsed with 0.1 m*M* KF (pH = 7); temperature 23 ± 1°C. [Reprinted with permission from D. A. Stern, L. Laguren-Davidson, D. G. Frank, J. Y-P. Gui, C-H. Lin, F. Lu, G. N. Salaita, N. Walton, D. C. Zapien, and A. T. Hubbard, *J. Am. Chem. Soc.*, *111*, 877 (1989). Copyright © (1989) American Chemical Society.]

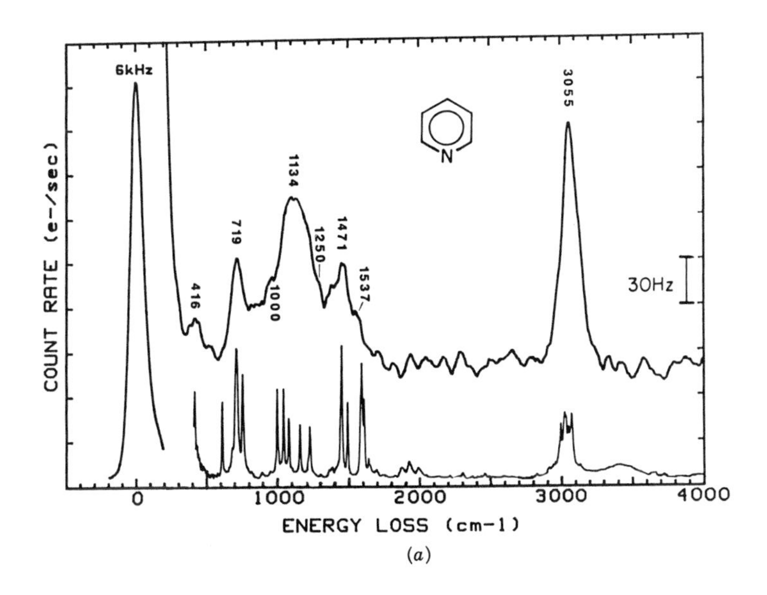

(*a*)

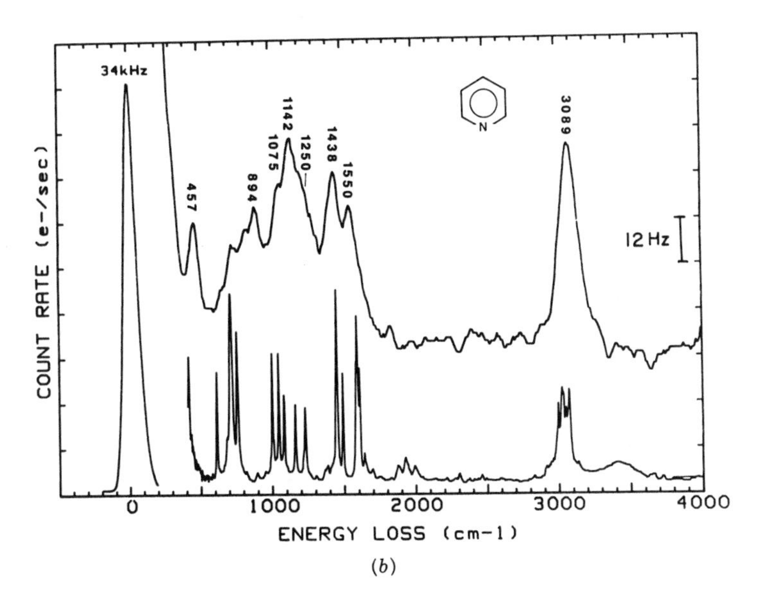

(*b*)

Figure 2.33. Vibrational spectra of pyr. (*A*) EELS spectrum of pyr adsorbed at Pt(111), −0.1 V, pH 3. Lower curve in A and B is mid-IR spectrum of neat pyr. (*B*) EELS spectrum of pyr adsorbed at Pt(111), 0.4 V, pH 3. Experimental conditions: adsorption from 1 m*M* pyr in 10 m*M* KF adjusted to pH 3 with HF, followed by rinsing with 2 m*M* HF (pH 3); other conditions as in Fig. 2.21. [Reprinted with permission from D. A. Stern, L. Laguren-Davidson, D. G. Frank, J. Y-P. Gui, C-H. Lin, F. Lu, G. N. Salaita, N. Walton, D. C. Zapien, and A. T. Hubbard, *J. Am. Chem. Soc.*, *111*, 877 (1989). Copyright © (1989) American Chemical Society.]

Table 2.7. EELS Bands of Absorbed Pyridine and Nicotinic Acid

Compound	*pH/ Electrode Potential* (V)	*Peak Frequency* (cm^{-1})	*Symmetry Species*	*Description*
pyr	3/−0.1	3055	$C_{2v}: A_1, B_2$	C—H stretch
		1537	A_1, B_2	CC stretch
		1471	A_1, B_2	CC, CN stretch
		1250	B_1, B_2	C—H bend
		1134	A_1	C—H bend
		1000	A_1, A_2, B_2	Ring, C—H bend
		719	B_1	Ring, C—H
		416	$B_1; A_1$	Ring bend; Pt—N stretch
pyr	3/+0.4	3089	$C_{2v}: A_1, B_2$	C—H stretch
		1550	A_1, B_2	CC stretch
		1438	A_1, B_2	CC, CN stretch
		1250	B_1, B_2	C—H bend
		1142	A_1	C—H bend
		1075	A_1, A_2, B_2	Ring, C—H bend
		894	B_1	Ring, C—H bend
		457	$B_1; A_1$	Ring bend; Pt—N stretch
NA	3/−0.2	3566	$C_s: A'$	O—H stretch
		3068	A'	C—H stretch

		1748	A'	C═O stretch
		1566	A'	CC stretch
		1368	A'	CC, CN stretch
		1132	A', A''	C—O stretch; C—H, O—H bend
		784	A', A''	OCO, C—H bend
		652	A', A''	CC, OCO bend
		465	$A''; A'$	Ring bend; Pt—N stretch
NA	3/+0.6	3567	$C_s: A'$	O—H stretch (weak)
		3071	A'	C—H stretch
		1733	A'	C═O stretch
		1592	A'	CC stretch
		1388	$A'; A''$	CC, CN stretch; C—H bend
		1192	A'	C—H bend
		1117	A'	C—O stretch
		1007	A', A''	Ring, CH bend
		824	A''	OCO, ring bend
		673	A''	Ring bend
		463	$A''; A'$	Ring bend, Pt—N stretch
		203	A'	Pt—O stretch
NA	10/−0.3	3068	$C_s: A'$	C—H stretch
		1612	A'	CC, CO (asym.) stretch
		1379	$A'; A''$	CC, CN, CO (sym.) stretch; C—H bend
		1136	A'	C—H bend
		815	A', A''	OCO, ring bend
		425	$A''; A'$	Ring bend; Pt—N stretch

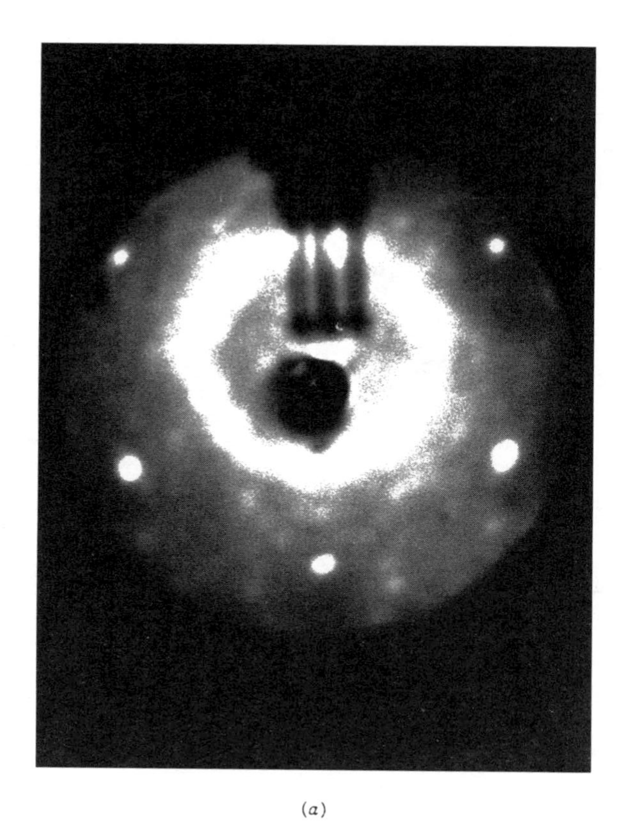

(a)

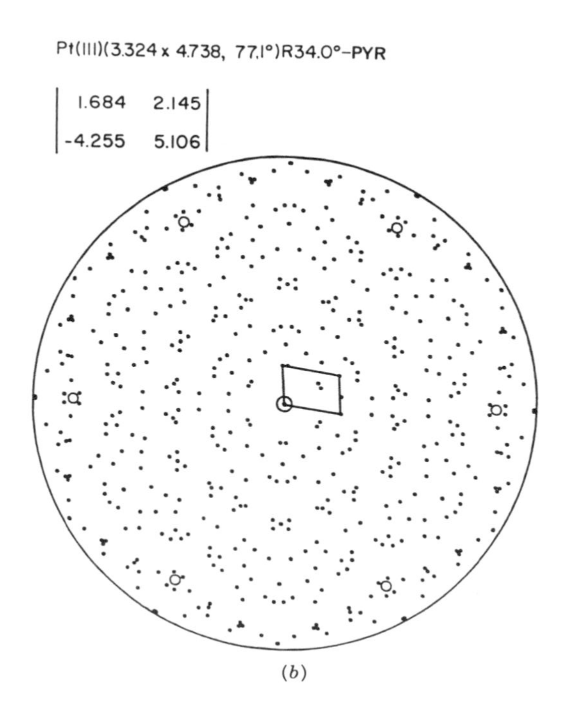

(b)

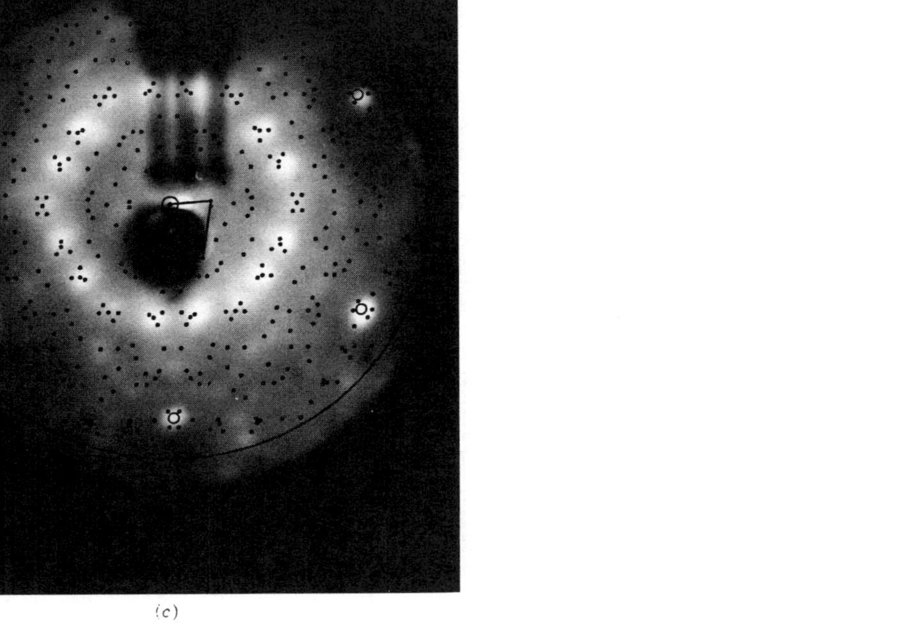

(c)

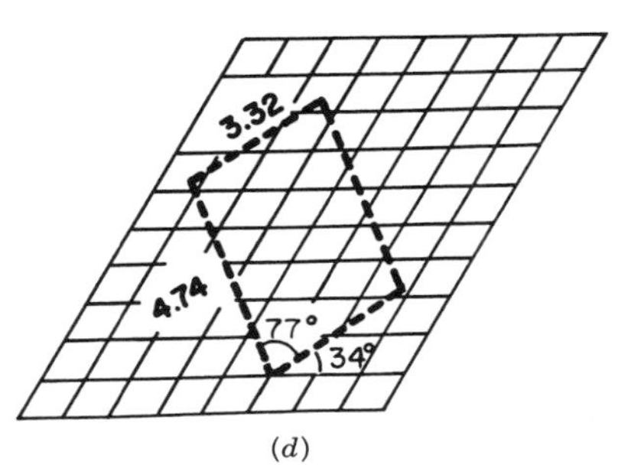

(d)

Figure 2.34. LEED pattern and structure of pyridine at Pt(111). (*A*) LEED pattern of pyr adsorbed at Pt(111), 51 eV. This pattern is Pt(111) (3.324 × 4.738, 77.1°)R34°-pyr. In matrix notation: $\begin{vmatrix} 1.684 & 2.145 \\ -4.255 & 5.106 \end{vmatrix}$ (*B*) Diagram of the LEED pattern in A. (*C*) Comparison of calculated and observed LEED patterns. (*D*) Diagrams of observed and nearest commensurate unit meshes. (*E*) Diagram of calculated LEED pattern corresponding to nearest commensurate structure $(2\sqrt{3} \times \sqrt{21}, 79.1°)R30°$. (*F*) Model of the Pt(111)(3.324 × 4.738, 77.1°)R34°-pyr. The pyr packing density in this model is 0.421 nmol cm^{-2}. Experimental conditions: adsorption from 1 m*M* pyr in 10 m*M* (pH 7) at −0.3 V, followed by rinsing with 0.1 m*M* KF (pH 7). [Reprinted with permission from D. A. Stern, L. Laguren-Davidson, D. G. Frank, J. Y-P. Gui, C-H. Lin, F. Lu, G. N. Salaita, N. Walton, D. C. Zapien, and A. T. Hubbard, *J. Am. Chem. Soc.*, *111*, 877 (1989). Copyright © (1989) American Chemical Society.]

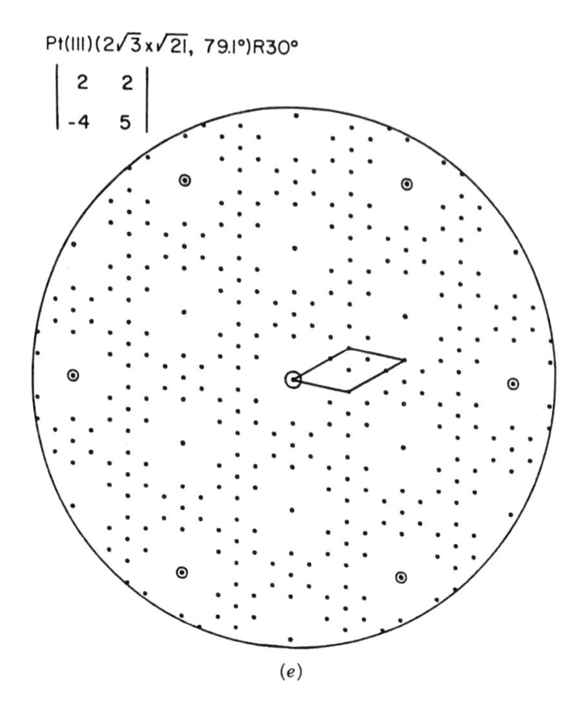

(*e*)

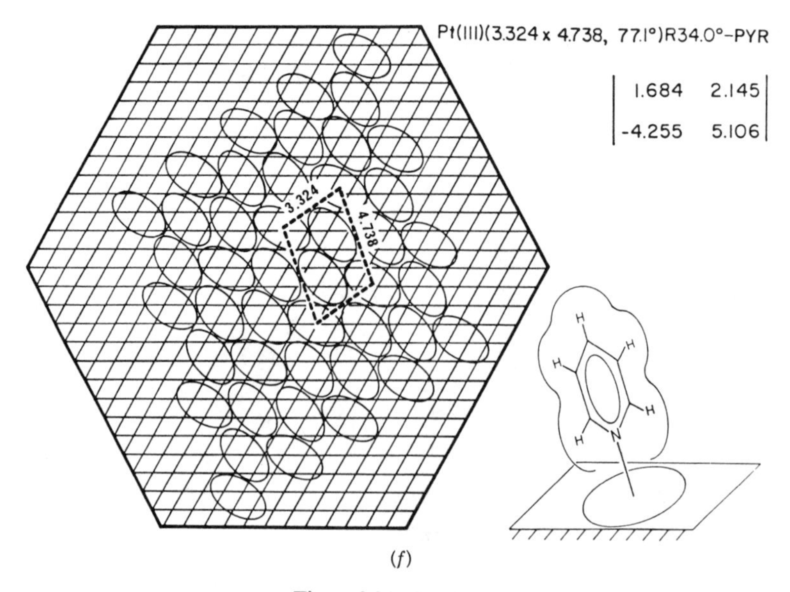

(*f*)

Figure 2.34. Continued.

rotationally but not translationally with respect to the Pt(111) substrate. A diagram of the LEED pattern is shown in Fig. 2.34*B*. The angle of tilt between the plane of the ring and the surface plane is 71°. A simulation of this structure superimposed on the observed LEED pattern appears in Fig. 2.34*C*. The calculated and observed LEED patterns are closely similar. In contrast, the closest commensurate structure, $(2\sqrt{3} \times \sqrt{21}, 79°)R30°$, as shown in Fig. 2.34*D*, would have given a LEED pattern completely different from that observed, Fig. 2.34*E*. With three pyr molecules per $(3.324 \times 4.738, 77°)R34°$ unit cell, the calculated packing density is 0.421 nmol cm^{-2}, in close agreement with the measured packing density, 0.45 nmol cm^{-2}. Therefore, based on observations from EELS, Auger, and LEED, it is proposed that the structure of pyr at Pt(111) from solution is Pt(111) $(3.324 \times 4.758, 77°)R34°$-pyr as shown in Fig. 2.34*F* (see Ref. 12 for further details).

Adsorbed pyr "passivates" the Pt(111) surface somewhat, as demonstrated by cycle voltammetry (solid curve) (Fig. 2.35) both with respect to adsorption of OH (oxidation) and towards adsorption of H atoms.

2.3.5B Nicotinic Acid

Shown in Fig. 2.30*C* is the Auger spectrum of adsorbed NA, also known as "niacin" or "vitamin B_3" (12, 104) adsorbed at a Pt(111) surface from aqueous solution. Packing densities were calculated from the Auger data based on Eqs. 2.58–2.62:

$$\Gamma_C = (I_C/I_{Pt}^{\circ})/[B_C(\tfrac{1}{3} + \tfrac{1}{3}2f)] \tag{2.58}$$

$$\Gamma = \tfrac{1}{6}\Gamma_C \tag{2.59}$$

$$\Gamma_N = (I_N/I_{Pt}^{\circ})/B_N f^2) \tag{2.60}$$

$$\Gamma_O = (I_O/I_{Pt}^{\circ})/[B_O(\tfrac{3}{4} + \tfrac{1}{4}f)] \tag{2.61}$$

$$I_{Pt}/I_{Pt}^{\circ} = (1 - 9K\Gamma) \tag{2.62}$$

Auger data are given in Table 2.8. The molecular packing densities were calculated based on Eqs. 2.58–2.62 and are given in Table 2.6. The measured packing densities are graphed versus concentration in Fig. 2.36. The two Auger methods give virtually identical results over the entire concentration range from $10^{-6}\,M$ to the solubility limit, about $0.1M$. Virtual constancy of the packing density from 10^{-4} to $10^{-1}M$ is indicative of the stability and uniformity of the tilted structure of adsorbed NA at Pt(111) shown in Fig. 2.37. The packing density at the solubility limit ($10^{-1}M$) is 0.38 nmol cm^{-2}. The theoretical limiting packing density for the model shown in Fig. 2.37 is 0.38 nmol cm^{-2} (43.71 Å^2 molecule^{-1}), based on covalent and van der Waals radii (25). The angle between the plane of the ring and Pt surface is 75°, Eq. 2.57 (with $a = 8.43$ and $b = 7.33$). On the other hand, vertically oriented NA would have led to a packing density 0.579 nmol cm^{-2} (28.7 Å^2 molecule^{-1}), much larger than that observed, while a

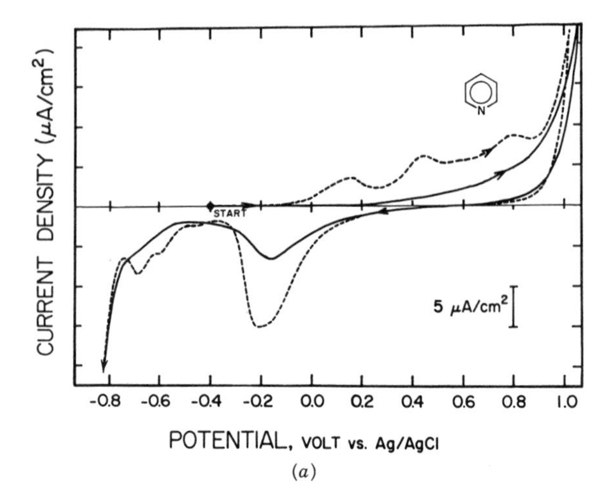

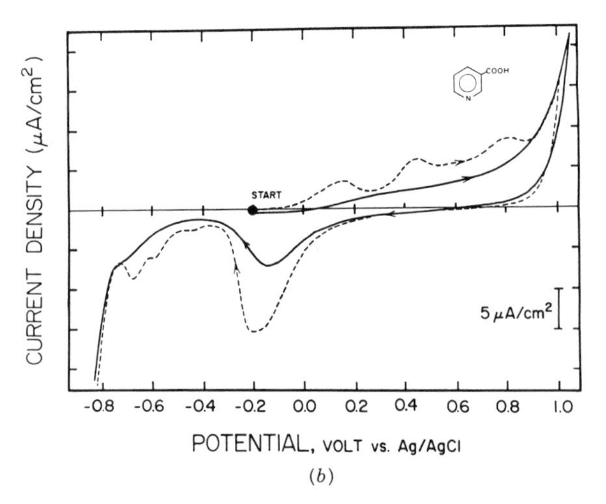

Figure 2.35. Cyclic voltammetry at Pt(111). (*A*) pyr. (*B*) NA. Solid curve (——): adsorbed layer; dashed curve (---): clean Pt(111). Experimental conditions: adsorption from pyr (0.1 m*M*, -0.1 V), NA (1 m*M*, -0.2 V) in 10 m*M* KF (pH 7); rinse and scan in 10 m*M* KF (pH 7); 5 mV s^{-1}. [Reprinted with permission from D. A. Stern, L. Laguren-Davidson, D. G. Frank, J. Y-P. Gui, C-H. Lin, F. Lu, G. N. Salaita, N. Walton, D. C. Zapien, and A. T. Hubbard, *J. Am. Chem. Soc.*, *111*, 877 (1989). Copyright © (1989) American Chemical Society.]

horizontal orientation would have given a much lower packing density, 0.290 nmol cm^{-2} (57.2 Å^2 molecule^{-1}).

Electron energy loss spectra of adsorbed NA at Pt(111) are shown in Fig. 2.38. Also shown are the mid-IR spectra of NA vapor (104) and solid potassium nicotinate (103). Assignments of the EELS bands by analogy with the accepted IR assignments for pyr and derivatives (105) are given in Table 2.7. Figure 2.38*A* shows the EELS spectrum of NA adsorbed at Pt(111) from acidic solutions at a

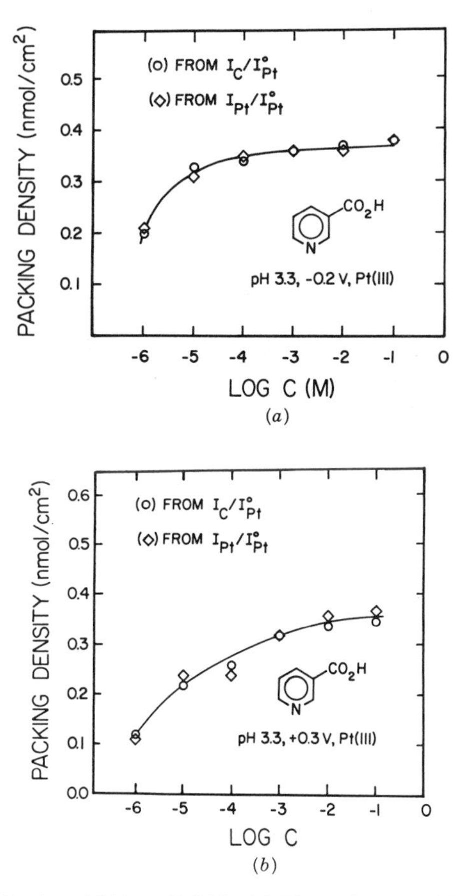

Figure 2.36. Packing density of NA at Pt(111). (*A*) Electrode potential, -0.2 V. (*B*) Electrode potential, $+0.3$ V. Experimental conditions; adsorption from NA solutions containing 10 m*M* KF adjusted to pH 7, followed by rinsing with 1 m*M* HF (pH 3.3); temperature $23 \pm 1^\circ$. [Reproduced with permission from D. A. Stern, L. Laguren-Davidson, D. G. Frank, J. Y-P. Gui, C-H. Lin, F. Lu, G. N. Salaita, N. Walton, D. C. Zapien, and A. T. Hubbard, *J. Am. Chem. Soc.*, *111*, 877 (1989) Copyright © (1989) American Chemical Society.]

relatively negative potential. Under such conditions, the O—H stretching band at $3566\,\text{cm}^{-1}$ is very pronounced. However, this band disappears when the adsorbed layer of NA is rinsed with base (0.1 m*M* KOH) as shown in Fig. 2.38*C*, and becomes apparent again when the surface is rinsed with acid as expected for a pendant carboxylic acid OH group:

$$\text{(N-bound pyridine)}\text{—}CO_2H + KOH \rightleftharpoons \text{(N-bound pyridine)}\text{—}CO_2K + H_2O \qquad (2.63)$$

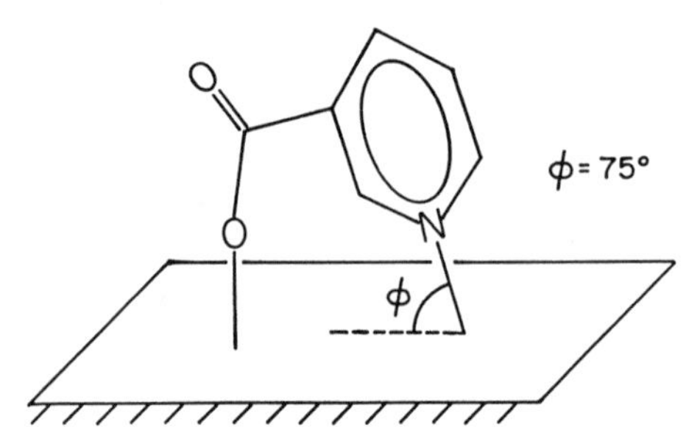

Figure 2.37. Structural model of adsorbed NA (0.6 V) at Pt(111). [Reprinted with permission from D. A. Stern, L. Laguren-Davidson, D. G. Frank, J. Y-P. Gui, C-H. Lin, F. Lu, G. N. Salaita, N. Walton, D. C. Zapien, and A. T. Hubbard, *J. Am. Chem. Soc.*, *111*, 877 (1989). Copyright © (1989) American Chemical Society.]

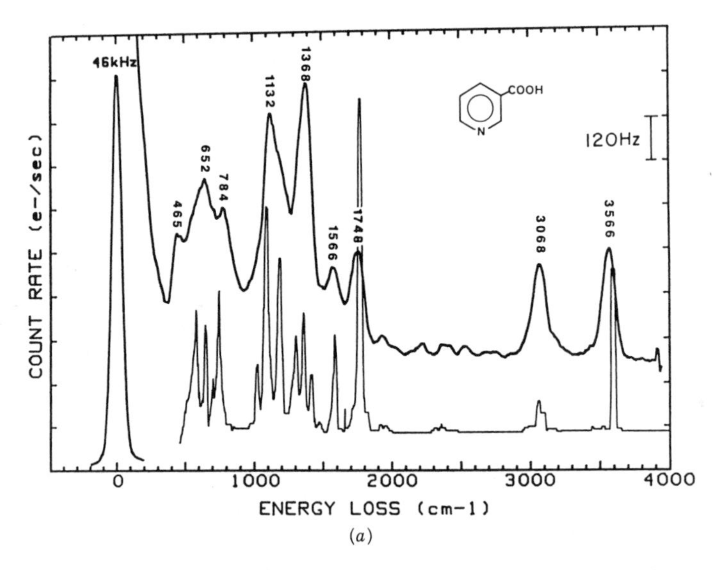

Figure 2.38. Vibrational spectra of NA. (*A*) EELS spectrum of NA adsorbed at Pt(111), −0.2 V, pH 3. Lower curve in *A* and *B* is mid-IR spectrum of NA vapor. (*B*) EELS spectrum of NA at Pt(111), +0.6 V, pH 3. (*C*) EELS spectrum of NA at Pt(111), −0.3 V, rinsed at pH 10. Lower curve is mid-IR spectrum of solid KNA. Experimental conditions: adsorption from 1 m*M* NA in 10 m*M* KF, pH 3 (*A* and *B*) or pH 7 (*C*), followed by rinsing with 2 m*M* HF (*A* and *B*, pH 3) or 0.1 m*M* KOH (pH 10, *C*); other conditions as in Fig. 2.21. [Reprinted with permission from D. A. Stern, L. Laguren-Davidson, D. G. Frank, J. Y-P. Gul, C-H. Lin, F. Lu, G. N. Salaita, N. Walton, D. C. Zapien and A. T. Hubbard, *J. Am. Chem. Soc.*, *111*, 877 (1989). Copyright © (1989) American Chemistry Society.]

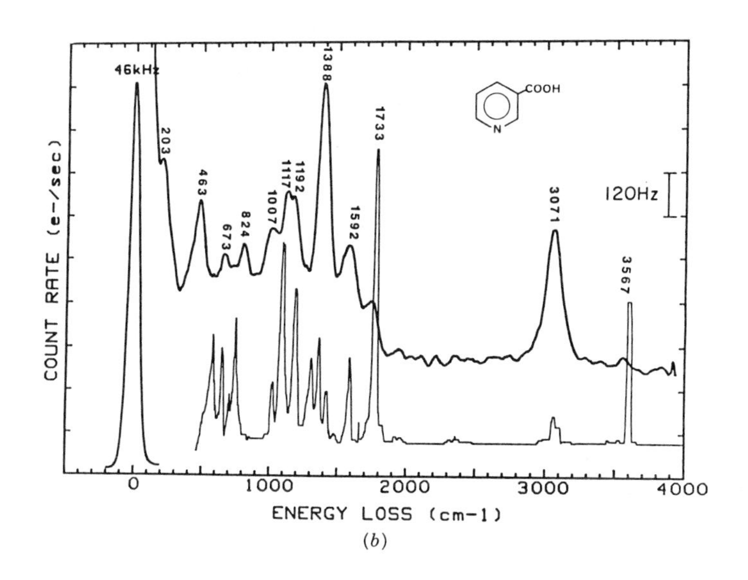

(b)

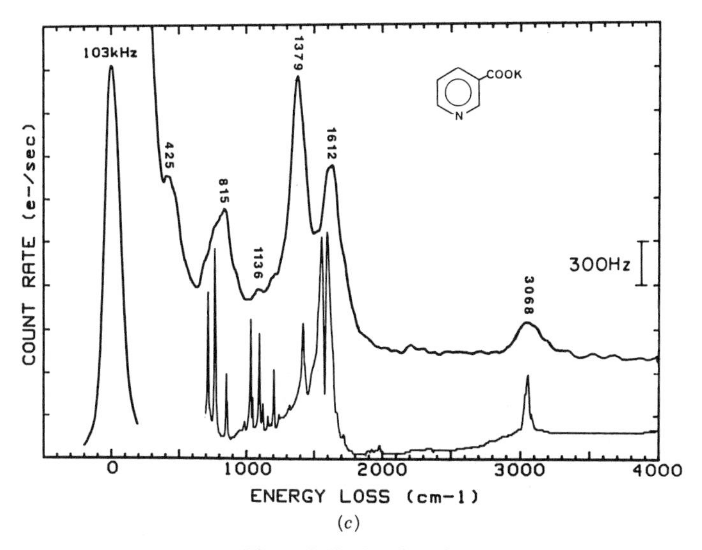

(c)

Figure 2.38. Continued.

Another interesting characteristic of adsorbed NA is that the intensity of the O—H stretching band during adsorption varies from maximum at negative potential to minimum at positive potential as shown in Fig. 2.38 and 2.39. Figure 2.39*A* shows the corresponding variation of rhe EELS O—H stretching peak height (normalized by the C—H intensity) with respect to electrode potential. Also shown in Fig. 2.39*A* is the packing density of adsorbed NA obtained from (I_C/I°_{Pt}). As can be seen, the packing density of adsorbed NA

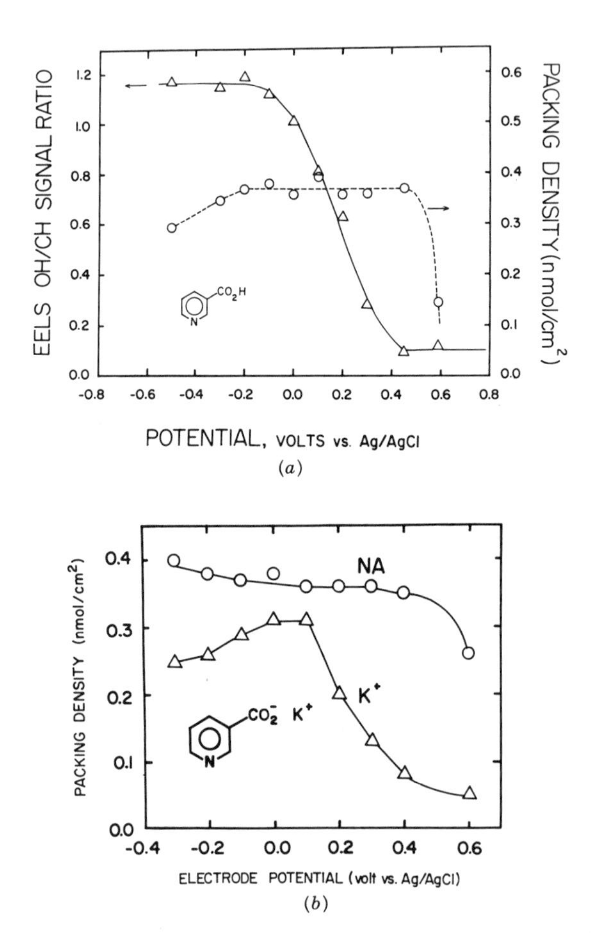

Figure 2.39. The O—H/C—H signal ratio (EELS) and packing density (Auger) of NA at Pt(111) versus electrode potential. (*A*) Ratio of EELS O—H ($3566\,cm^{-1}$) to C—H ($3068\,cm^{-1}$) peak height (△); packing density of NA (○). (*B*) Packing density of potassium ions (△); packing density of NA (○). Experimental conditions: (*A*) adsorption from 1 m*M* NA in 10 m*M* KF at pH 7, followed by rinsing in 2 m*M* HF (pH 3); (*B*) adsorption from 1 m*M* NA in 10 m*M* KF at pH 3, followed by 3 rinses in 0.1 m*M* KOH (pH 10). EELS conditions as in Fig. 2.38. Auger conditions as in Fig. 2.30. [Reprinted with permission from D. A. Stern, L. Laguren-Davidson, D. G. Frank, J. Y-P. Gui, C-H. Lin, F. Lu. G. N. Salaita, N. Walton, D. C. Zapien, and A. T. Hubbard, *J. Am. Chem. Soc.*, *111*, 877 (1989). Copyright © (1989) American Chemical Society.]

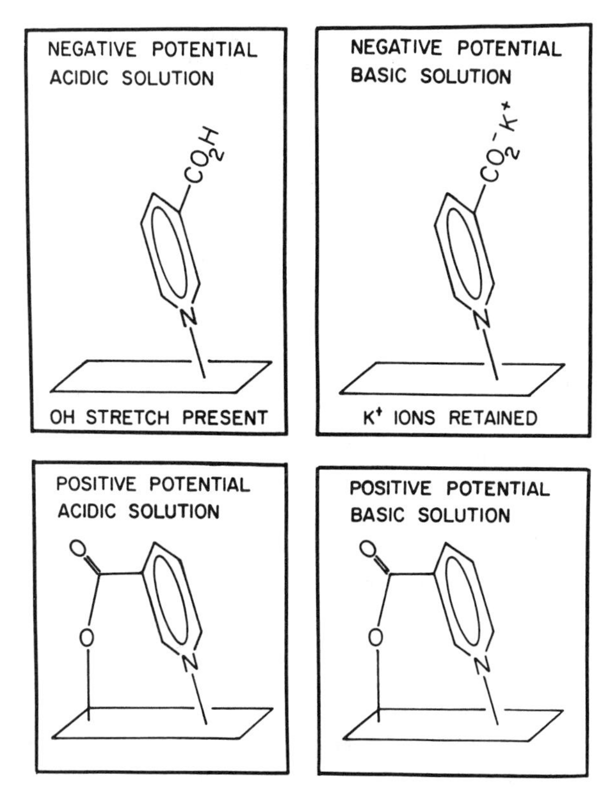

Figure 2.40. Structural variations of NA versus pH and electrode potential. [Reprinted with permission from D. A. Stern, L. Laguren-Davidson, D. G. Frank, J.Y-P. Gui, C-H. Lin, F. Lu, G. N. Salaita, N. Walton, D. C. Zapien, and A. T. Hubbard, *J. Am. Chem. Soc.*, *111*, 877 (1989). Copyright © (1989) American Chemical Society.]

remains constant over the same range of potentials (12). A model of this transition is given in Fig. 2.40. Note that the disappearance of the O—H vibration as the electrode potential becomes more positive suggesting that there is a slight change in the angle of tile and rotation of the pyridine ring allowing the carboxylate group to form a coordinate covalent bond with the Pt surface. The shortest Pt—O distance, which can be reached with only minor changes in bond angles ($\pm 5°$) is about 2 Å. This is in agreement with the Pt—O covalent bond length of Ca (2.05 Å) (11, 12, 25).

Adsorbed NA at sufficiently negative potential favors the pendant carboxylic acid moiety as was shown in Eq. 2.63 above. Figure 2.30*D* clearly shows the retention of the K^+ after rinsing the surface with a basic K^+ solution and keeping the surface at the same negative potential. However, when the potential is made more positive, the adsorbed NA loses its ability to retain potassium ions, Fig. 2.39*B*. These changes are reversed upon returning the potential to more negative values. This trend is not attributable simply to electrostatic interaction between K^+ and the electrode surface, because this behavior is exhibited only by the *meta*-carboxypyridine and not by the *para*-carboxypyridine or pyridine itself (12). The potential dependence of K^+ retention by NA indicates that the coordination of pyridine carboxylate moieties by Pt surface involves charge transfer as in Eq. 2.64.

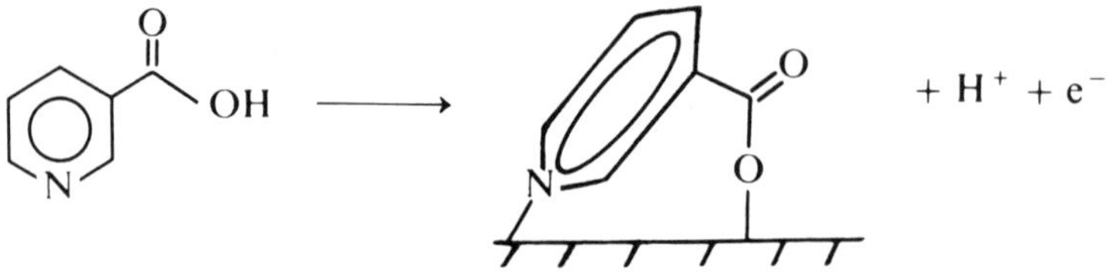

$$+ H^+ + e^- \qquad (2.64)$$

This charge-transfer process occurs gradually as the electrode potential is increased from -0.15 to $+0.45$ V (vs. Ag/AgCl), and the measured anodic charge is about 27 μC cm^{-2} at the limit of full oxidative coordination, based upon subtraction of the negative going voltammetric scan for adsorbed pyr from that for NA and integration of the difference. Models of processes are presented in Fig. 2.40.

Experiments were carried out to demonstrate the reversibility of the potential dependence of adsorbed NA vibrational modes and retention of K^+ ions, as follows: NA was adsorbed at an electrode potential of 0.45 V, followed by a potential step to -0.3 V for 120 s. Then the adsorbed species was characterized by Auger spectroscopy and EELS. A complete equilibration took place as shown in Eq. 2.65.

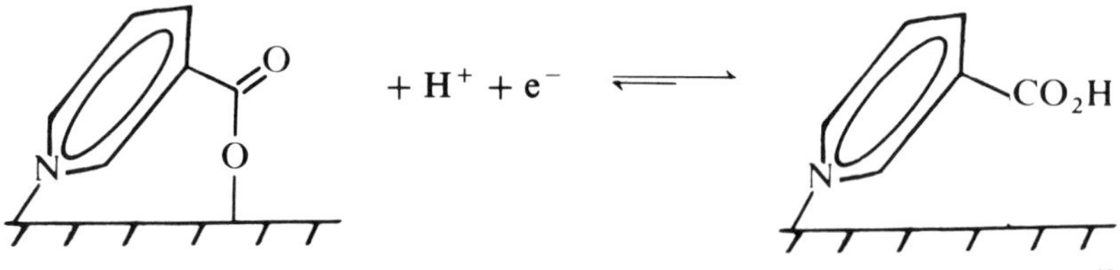

$$+ H^+ + e^- \rightleftharpoons \quad CO_2H \qquad (2.65)$$

That is, the reaction proceeded from the coordinated carboxylate state (at $+0.45$ V) to the pendant-carboxylate state (at -0.3 V). However, when the potential was negative during adsorption (-0.3 V) and adjusted to a positive value afterwards ($+0.45$ V), the reaction proceeded to only about 70% of completion. Such oxidative-coordination reactions of adsorbed anions are typical behavior for Pt (112–114), Ag(114) and probably other metals.

$$M + X \xrightleftharpoons{\text{surface}} MX + e^- \qquad (2.66)$$

Since aliphatic carboxylic acids are not strongly adsorbed at Pt surfaces, carboxylate coordination to Pt in the adsorbed NA layer may be due at least in part to an entropy effect in which adsorption of a pyridine moiety induces a stronger interaction between Pt and the carboxylate group, analogous to that occurring in metal chelates.

REFERENCES

1. J. F. Nicholas, *An Atlas of Models of Crystal Surfaces*, Gordon and Breach, New York (1965); *J. Phys. Chem. Solids*, *21*, 230 (1961).
2. B. Bent, Ph.D. Thesis, University of California, Berkeley (1986).
3. M. W. Roberts and C. S. McKee, *Chemistry of Metal–Gas Interface*, Oxford University Press (1978).
4. G. A. Somorjai, *Chemistry in Two Dimensional Surfaces*, Cornell University Press (1981).
5. F. Lu, G. N. Salaita, L. Laguren-Davidson, D. A. Stern, E. Wellner, D. G. Frank, N. Batina, D. C. Zapien, N. Walton, and A. T. Hubbard, *Langmuir*, *4*, 637 (1988).
6. D. A. Stern, E. Wellner, G. N. Salaita, L. Laguren-Davidson, F. Lu, N. Batina, D. G. Frank, D. C. Zapien, N. Walton, and A. T. Hubbard, *J. Am. Chem. Soc.*, *110*, 4885 (1988).
7. G. N. Salaita, L. Laguren-Davidson, F. Lu, E. Wellner, D. A. Stern, N. Batina, D. G. Frank, C. S. Benton, and A. T. Hubbard, *J. Electroanal. Chem. Interfacial Electrochem.*, *245*, 253 (1988).
8. N. Batina, D. G. Frank, J. Y-P. Gui, B. F. Kahn, C. H. Lin, F. Lu, J. W. McCargar, G. N. Salaita, D. A. Stern, D. C. Zapieu, and A. T. Hubbard, *Electrochem. Acta*, *34*, 1031 (1989).
9. D. A. Stern, G. N. Salaita, F. Lu, J. W. McCargar, N. Batina, P. G. Frank, L. Laguren-Davidson, C. N. Lin, N. Walton, S. Y. Gui, and A. T. Hubbard, *Langmuir*, *4*, 711 (1988).
10. G. N. Salaita, D. C. Zapien, N. Batina, E. Wellner, N. Walton, and A. T. Hubbard, in *Chemically Modified Surfaces*, Vol. 2, D. Leyden (Ed.), Gordon and Breach, New York (1987).
11. Arthur T. Hubbard, "Electrochemistry at Well-Characterized Surfaces," *Chem. Rev.*, *88*, 633 (1988).
12. D. A. Stern, L. Languren-Davidson, D. G. Frank, J. Y-P. Gui, C-H. Lin, F. Lu, G. N. Salaita, N. Walton, D. C. Zapien, and A. T. Hubbard, *J. Am. Chem. Soc.*, *111*, 877 (1989).
13. A. T. Hubbard, D. G. Frank, D. A. Stern, M. J. Tarlov, N. Batina, N. Walton, and E. Wellner, in *Redox Chemistry and Interfacial Behavior of Biological Molecules*, G. Dryhurst and K. Niki (Eds.), Plenum, New York (1988).
14. M. P. Soriaga and A. T. Hubbard, *J. Am. Chem. Soc.*, *104*, 3397 (1982).
15. M. P. Soriaga, J. H. White, and A. T. Hubbard, *J. Phys. Chem.*, *87*, 3048 (1983).
16. V. K. F. Chia, M. P. Soriaga, and A. T. Hubbard, *J. Electroanal. Chem.*, *167*, 97 (1984).
17. T. Solomun, B. C. Schardt, S. D. Rosasco, A. Wieckowski, J. L. Stickney, and A. T. Hubbard, *J. Electroanal. Chem.*, *176*, 309 (1984).
18. D. G. Frank, J. Y. Katekaru, S. D. Rosasco, G. N. Salaita, B. C. Schardt, M. P. Soriaga, D. A. Stern, J. L. Stickney, and A. T. Hubbard, *Langmuir*, *1*, 587 (1985).
19. B. C. Schardt, J. L. Stickney, D. A. Stern, D. G. Frank, J. Y. Katekaru, S. D. Rosasco, G. N. Salaita, M. P. Soriaga, and A. T. Hubbard, *Inorg. Chem.*, *24*, 1419 (1985).
20. S. D. Rosasco, J. L. Stickney, G. N. Salaita, D. G. Frank, J. Y. Katekaru, B. C. Schardt, M. P. Soriaga, D. A. Stern, and A. T. Hubbard, *J. Electroanal. Chem.*, *188*, 95 (1985).
21. B. C. Schardt, J. L. Stickney, D. A. Stern, A. Wieckowski, D. C. Zapien, and A. T. Hubbard, *Langmuir*, *3*, 239 (1987).
22. D. A. Stern, H. Baltruschat, M. Martinez, J. L. Stickney, D. Song, S. K. Lewis, D. G. Frank, and A. T. Hubbard, *J. Electroanal. Chem.*, *217*, 101 (1987).
23. M. P. Soriaga and A. T. Hubbard, *J. Am. Chem. Soc.*, *104*, 2735 (1982).
24. V. K. F. Chia, Ph.D., Thesis, University of California, Santa Barbara (1986).

25. L. C. Pauling, *Nature of the Chemical Bond*, 3rd ed., Cornell University Press, New York, 221 (1960).
26. M. P. Soriaga, P. H. Wilson, A. T. Hubbard, and C. S. Benton, *J. Electroanal. Chem.*, *142*, 317 (1982).
27. M. P. Soriaga and A. T. Hubbard, *J. Electroanal. Chem.*, *137*, 34 (1984).
28. V. K. F. Chia, M. P. Soriaga, and A. T. Hubbard, *J. Electroanal. Chem.*, *167*, 79 (1984).
29. M. P. Soriaga, D. Song, D. C. Zapien, and A. T. Hubbard, *Langmuir*, *1*, 123 (1985).
30. D. Song, M. P. Soriaga, K. L. Vieria, and A. T. Hubbard, *J. Electroanal. Chem.*, *184*, 171 (1985).
31. (a) M. P. Soriaga, J. L. Stickney, and A. T. Hubbard, *J. Mol. Catal.*, *21*, 211 (1983). (b) M. P. Soriaga, J. L. Stickney, and A. T. Hubbard, *J. Electroanal. Chem.*, *144*, 207 (1983). (c) M. P. Soriaga and A. T. Hubbard, *J. Phys. Chem.*, *88*, 1758 (1984). (d) M. P. Soriaga and A. T. Hubbard, *J. Electroanal. Chem.*, *159*, 101 (1983). (e) K. L. Vieira, D. C. Zapien, M. P. Soriaga, A. T. Hubbard, K. P. Low, and S. E. Anderson, *Anal. Chem.*, *58*, 2974 (1986). (f) V. K. F. Chia, J. H. White, M. P. Soriaga, and A. T. Hubbard, *J. Electroanal. Chem.*, *217*, 121 (1987). (g) D. Song, M. P. Soriaga, and A. T. Hubbard, *J. Electroanal. Chem.*, *193*, 255 (1985).
32. M. Boudart and G. Djega-Mariadassou, *Kinetics of Heterogeneous Catalytic Reactions*, Princeton University Press (1984).
33. J. E. Lennard-Jones, *Trans. Faraday Soc.*, *28*, 333 (1932).
34. J. E. Lennard-Jones, *Physicia*, *4*, 941 (1937).
35. A. Wieckowski, S. D. Rosasco, G. N. Salaita, A. T. Hubbard, B. E. Bent, F. Zaera, D. Godkey, and G. A. Somorjai, *J. Am. Chem. Soc.*, *107*, 5910 (1985).
36. J. Horiuti and K. Miyahara, *Hydrogenation of Ethylene on Metallic Catalysis*, NSRDS-NBS-13 (1969).
37. O. Beeck, *Discuss. Faraday Soc.*, *8*, 118 (1950).
38. G. C. A. Schuit, *Discuss. Faraday Soc.*, *8*, 126 (1950).
39. O. Beech, *Rev. Mod. Phys.* *17*, 61 (1945).
40. F. Zaera and G. A. Somorjai, *J. Am. Chem. Soc.*, *106*, 2288 (1984).
41. F. Zaera, Ph.D. Thesis, University of California Berkeley (1984).
42. D. G. Castner and G. A. Somorjai, *Chem. Rev.* *79*, No. 3, 233 (1979).
43. S. Rosasco, Ph.D. Thesis, University of California, Santa Barbara (1986).
44. M. L. Patterson and M. J. Weaver, *J. Phys. Chem.*, *89*, 1331 (1985).
45. M. L. Patterson and M. J. Weaver, *J. Phys. Chem.*, *89*, 5046 (1985).
46. (a) J. E. Demuth, *I.B.B. J. Res. Dev.*, *22*, 265 (1978). (b) T. E. Felter and W. H. Weinberg, *Surface Sci.*, *103*, 265 (1981). (c) P. Skinner, M. W. Howard, I. A. Oxton, S. F. A. Kettle, D. B. Powell, and N. Sheppard, *J. Chem. Soc. Faraday Trans.*, *2*, 77 (1981).
47. L. L. Kesmodel, L. H. Dubois, and G. A. Somorjai, *Chem. Phys. Lett.*, *56*, 267 (1978).
48. L. L. Kesmodel, L. H. Dubois, and G. A. Somorjai, *J. Chem. Phys.*, *70*, 2180 (1979).
49. M. R. Albert, L. G. Sheddon, W. Eberhardt, F. Greuter, T. Gustafsson, and E. W. Plummer, *Surf. Sci.*, *120*, 19 (1982).
50. H. Steininger, H. Ibach, and S. Lehwald, *Surf. Sci.*, *117*, 685 (1982).
51. (a) J. E. Demuth, *Surf. Sci.*, *80*, 367 (1979). (b) M. Salmeron and G. Somorjai, *J. Phys. Chem.*, *86*, 341 (1982).
52. M. Bartok, *Stereochemistry of Heterogeneous Metal Catalysis*, Wiley, New York (1985).
53. L. E. Samuels, *Metallographic Polishing by Mechanical Methods*, Pitman, London (1963).
54. E. A. Wood, *Crystal Orientation Manual*, Columbia University Press, New York (1963).
55. J. L. Stickney, Ph.D. Thesis, University of California Santa Barbara (1984).
56. M.-C. Tsai and E. L. Muetterties, *J. Am. Chem. Soc.*, *104*, 2534 (1982).

57. S. Lehwald, H. Ibach, and J. E. Demuth, *Surf. Sci.*, *78*, 577 (1978).
58. N. V. Richardson, *Surf. Sci.*, *87*, 622 (1979).
59. G. W. Rubloff, N. Luth, J. E. Demuth, and W. D. Grobman, *J. Catal.*, *53*, 423 (1978).
60. T. E. Fischer, S. R. Kelemen, and H. P. Bonzel, *Surf. Sci.*, *64*, 157 (1977).
61. J. M. Baset, G. Dalmai-Imelik, M. Primet, and R. Mutin, *J. Catal.*, *37*, 22 (1975).
62. P. C. Stair and G. A. Somorjai, *J. Chem. Phys.*, *67*, 4361 (1977).
63. J. L. Gland and G. A. Somorjai, *Adv. Colloid Interface Sci.*, *5*, 203 (1976).
64. C. Besoukhanova, J. P. Candy, M. Forissier, *C. R. Acad. Sci. Paris, Ser. C*, 287, 479 (1978).
65. C. P. Radar and H. A. Smith, *J. Am. Chem. Soc.*, *84*, 1443 (1962).
66. M. Abon, J. C. Bertolini, J. Billy, J. Massardier, and B. Tardy, *Surf. Sci.*, *162*, 395 (1985).
67. J. E. Demuth and D. E. Eastman, *Phys. Rev.*, *B13*, 1523 (1976).
68. (a) F. P. Netzer, E. Bertel, and J. A. D. Matthew, *Surf. Sci.*, *92*, 43 (1982). (b) R. W. Wexier and E. L. Mutterties, *J. Am. Chem. Soc.*, *106*, 4810 (1984).
69. C. M. Friend and E. L. Mutterties, *J. Am. Chem. Soc.*, *103*, 733 (1981).
70. M. A. Chesters and G. A. Somorjai, *Surf. Sci.*, *52*, 211 (1975).
71. J. C. Bertolini and J. Roussea, *Surf. Sci.*, *89*, 467 (1979).
72. (a) H. Jobic, J. Tomkinson, J. P. Candy, P. Fouilloux, and A. J. Renouprez, *Surf. Sci.*, *95*, 494 (1980). (b) M. Moskovis and D. P. Dilella, *J. Chem. Phys.*, *73*, 6068 (1980). (c) P. Gao and M. J. Weaver, *J. Phys. Chem.*, *89*, 5040 (1985).
73. J. L. Garnett and W. A. Solich-Baumgartner, *Adv. Catal.*, *16*, 95 (1966).
74. H. A. Smith and W. E. Campbell, *Proc. 3rd Intermat. Cong. Catal.*, Amsterdam, 1373 (1965).
75. A. K. Myers, G. R. Schoofts, and J. B. Benziger, *J. Phys. Chem.*, *91*, 2230 (1987).
76. E. D. Becker, E. Charney, and T. Anno, *J. Chem. Phys.*, *42*, 942 (1965).
77. H. W. Wilson, *Spectrochim. Acta*, *30*, 2141 (1974).
78. H. A. Smith, in *Catalysis*, Vol. 5, P. H. Emmett (Ed.), Reinhold, New York (1957).
79. R. L. Augustine, *Catalytic Hydrogenation*, Marcel Dekker, New York (1965).
80. A. P. G. Kieboom and F. van Rantwijk, *Hydrogenation and Hydrogenolysis in Synthetic Organic Chemistry*, University of Delft, Delft (1977).
81. P. N. Rylander, *Catalytic Hydrogenation in Organic Synthesis*, Academic, New York (1979).
82. H. J. Rimek, in *Methoden der Organischen Chemie*, Houken-Weyl, Stuttgart, (1980).
83. S. Nishimura and H. Taguchi, *Bull. Chem. Soc. Jpn.*, *36*, 353 (1963).
84. F. Zymalkwaski and G. Strippel, *Arch. Pharm.*, *298*, 604 (1965).
85. H. van Bekkum, B. van de Graaf, G. van Minnen-Pathuis, J. A. Peters, and B. M. Wepster, *Rec. Trav. Chim. Pays-Bas.*, *89*, 521 (1970).
86. W. Schneider and R. Dillman, *Chem. Ber.*, *96*, 2377 (1963).
87. S. Yonemoni and M. Noshiro, *Nippon Kagaku Kaishi*, 1980, 1924 (1980), *Chem. Abstr.*, *94*, 1741 (1981).
88. M. Alnot and R. Ducross, *Appl. Surf. Sci.*, *14*, 114 (1982).
89. P. Tetenyi and L. Babernies, *J. Catal.*, *8*, 215 (1967).
90. H. A. Smith and W. E. Campbell, *Proc. 3rd Internat. Cong. Catal.*, Amsterdam, 1373 (1965).
91. G. C. Bond, *Catalysis by Metals*, Academic, London (1962).
92. T. J. Nieuwstad, J. P. Klapwijk, and H. van Bekkum, *J. Catal.*, *29*, 404 (1973).
93. A. T. Hubbard, V. K. F. Chia, D. G. Frank, J. Y. Katekaru, S. D. Rosasco., G. N. Salaita, B. C. Schard, D. Song, M. P. Soriaga, D. A. Stern, J. L. Stickney, J. H. White, K. L. Vieira, A. Wieckowski, and D. C. Zapien, in *New Dimensions in Chemical Analysis*, B. L. Shapiro (Ed.), Texas A & M University Press, College Station, TX (1985).

94. F. Ullman, *Justus Liebigs Ann. Chem.*, *68*, 332 (1904).
95. B. G. Bravo, T. Mebrahtu, M. P. Soriaga, D. C. Zapien, A. T. Hubbard, and J. L. Stickney, *Langmuir*, *3*, 595 (1987).
96. (a) V. H. Grassian and E. L. Muetterties, *J. Phys. Chem.*, *91*, 389 (1987). (b) G. D. Waddill and L. L. Kesmodel, *Phys. Rev. B.*, *31(8)*, 4940 (1985).
97. J. L. Gland and G. A. Somorjai, *Surf. Sci.*, *38*, 157 (1973).
98. G. A. Somorjai, *Adv. Catal.*, *26*, 1 (1977).
99. J. H. S. Green, *Spectrochim Acta*, *33A*, 575 (1977).
100. J. E. Demuth, K. Christmann, and P. N. Sanda, *Chem. Phys. Lett.*, *76*, 201 (1980).
101. R. G. Barradas and B. E. Conway, *J. Electroanal. Chem.*, *6*, 314 (1963).
102. D. L. Jeawmaike and R. P. Van Duyne, *J. Electroanal. Chem.*, *84*, 1 (1977).
103. C. J. Pouchert, *The Aldrich Library of FTIR Spectra*, Aldrich Chemical Co., Milwaukee, WI (1985).
104. *The Interpretation of Vapor-Phase Spectra*, Vol. 2 Sadtler Research Laboratories, Philadelphia (1984).
105. (a) J. H. S. Green, W. Kynaston, and H. M. Paisley, *Spectrochim. Acta*, *19*, 549 (1963). (b) J. H. S. Green, W. Kynastom, and A. S. Lindsey, *Spectrochim. Acta*, *17*, 486 (1961). (c) G. N. Salaita, D. A. Stern, F. Lu, H. Baltruschat, B. C. Schardt, J. L. Stickney, M. P. Soriaga, D. G. Frank, and A. T. Hubbard, *Langmuir*, *2*, 828 (1986). (d) D. A. Stern, H. Baltruschat, M. Martinez, J. L. Stickney, D. Song, S. K. Lewis, D. G. Frank, and A. T. Hubbard, *J. Electroanal. Chem.*, *217*, 101 (1987). (e) F. Lu, G. N. Salaita, H. Baltruschat, and A. T. Hubbard, *J. Electroanal. Chem.*

Chapter III

MODIFICATION OF ELECTRODE SURFACES WITH SELF-ORGANIZED ELECTROACTIVE MICROSTRUCTURES

John S. Facci
Xerox Webster Research Center, 114/39D, Webster, NY

3.1 INTRODUCTION

Electrode surface modification remains the subject of intensive interdisciplinary study beginning with the seminal work of Lane and Hubbard (1) and

Molecular Design of Electrode Surfaces,
Edited by Royce W. Murray. Techniques of Chemistry Series, Vol. XXII.
ISBN 0-471-55773-0

Moses and Murray (2) over a decade ago. Areas such as electrocatalysis, energy conversion, display technology, current rectification, and macromolecular electronics continue to be actively explored as applications for surface modification (3). Practical examples of modified metal and semiconducting surfaces, for example, may be found in the electrophotography literature (4). In this chapter we systematically explore the utility, scope, and limitations of the Langmuir–Blodgett transfer technique in the modification of electrodes with organized electroactive microstructures.

Many electrode modification techniques such as the covalent chemical attachment of redox groups to chemically active electrode sites, dip-, evaporative- and spin-coating, and electropolymerization, result in a random spatial and orientational arrangement of redox centers at the electrode surface. Organization of film molecules, especially electroactive species, is not usually achieved by the application of these techniques. On the other hand, organized, oriented assemblies of molecules can be fabricated by either the Langmuir–Blodgett (L–B) monolayer transfer technique (5, 6) or by molecular self-assembly. The latter was pioneered a decade ago by Sagiv (7). Both techniques are being applied, for example, to construct special molecular architectures to provide molecular level explanations of wetting (8), tribology (9) and electron transfer (10). It is further expected that unique or special device functionality in areas such as nonlinear optics, sensors, molecular electronics, microlithography, and chemically modified electrodes will be enabled by organizing functional materials at surfaces (11).

Numerous monolayer and multilayer films containing electroactive and electroinactive organized molecular assemblies have been fabricated by both the self-assembly and L–B transfer techniques on many different substrates, including electrode surfaces. The interest in these structures emerged in parallel with the ability to characterize them. Surface sensitive molecular characterization techniques such as Fourier transform intrared spectroscopy (FTIR), scanning tunneling microscopy (STM), atomic force microscopy (AFM), electron diffraction techniques, He diffraction, X-ray standing waves, and bulk characterization techniques such as wetting, fluorescence microscopy, ellipsometry, and electrochemistry have been applied to obtain a clearer understanding of the structure–property relationships in organized arrays of molecules (7, 12).

A large body of work, directed toward the modification of electrode surfaces with organized assemblies fabricated by the L–B technique (13–17) and self-assembly (18–30), has clearly demonstrated the potential of these techniques to tailor the spatial and orientational microstructure of electroactive species at electrode interfaces. Much of the early work involved bringing about an understanding of the relationship between surface modification procedure and the resultant monolayer or multilayer structure; clearly, progress in this area will continue. These techniques have emerged as useful protocol for the investigation of interfacial electron-transfer dynamics as they can be used to control substrate-electroactive reagent distance (7a, 10, 16a, d, f) and orientation (31).

The relationship between surface microstructure and electrochemical (10, 24b, 32) and photoelectrochemical (16b) reactivity has begun to unfold. Novel molecular electronic applications based on L–B films have also been reported (33).

3.1.1 Self-Assembled Films

Several groups are actively exploring the microstructures that are formed by molecular self-assembly. The deliberate self-assembly of long-chain siloxane monolayers on glass, quartz, Al, Ge, and ZnSe was first done by Sagiv (7a–e). Film wettability, FTIR, ellipsometry, polarized absorption, and fluorescence spectroscopy were used to demonstrate that such monolayers contain closely packed extended alkyl chains in an all-trans configuration, which are oriented perpendicular to the substrate surface. The orientation of film molecules in monolayer self-assembled films was found to be identical to those in analogous L–B monolayers. Self-assembled films are not perfect. Pinholes arising from incomplete monolayer formation or disorder, however, can be filled. Rubinstein and Sagiv (21) showed that an alkylthiol could be used to "backfill" pinholes both in monolayers of octadecyltrichlorosilane and in monolayers of a metal chelating ligand adsorbed onto Au. Finklea et al. (20c) electropolymerized phenol in order to fill pinhole defects in octadecylthiol monolayers. The latter healed monolayers were employed in the blocking of electrochemical reactions of dissolved redox couples.

Sagiv and his co-workers (7d) also demonstrated that built-up self-assembled films could be fabricated by repetition of a sequence of self-assembly and chemical activation of the outer surface of an alkylsilane monolayer containing a terminal vinyl group. A three monolayer film was fabricated this way but FTIR showed that increasing disorder accompanied each additional monolayer. Ulman et al. (26a) were similarly able to obtain a 25 monolayer film of methyl-23-(trichlorosilyl)tricosanoate on single-crystal Si wafers. However, disorder also tended to increase with each additional monolayer. In addition, they demonstrated that built-up self-assembled films could potentially be prepared on Au by first assembling a hydroxy terminated *n*-alkylthiol monolayer on Au as an adhesive layer for the self-assembly of an alkylsiloxane monolayer.

Other methods of fabricating oriented molecular assemblies involve the adsorption of *n*-alkanoic acids on alumina (25a–b) and the adsorption of *n*-alkylthiols (8a–h, 10, 20b–c, 21, 24, 25c–f, 26e), *n*-alkyldisulfides (25g–i), and sulfides (27) on Au. The rather strong Au-S interaction in the sulfur containing adsorbates leads to the irreversible adsorption of a single monolayer with the hydrocarbon tails oriented nearly perpendicular to the surface in an extended all-trans configuration forming well-ordered, oriented monolayers. These monolayers have excellent versatility in terms of introducing various functional groups at the monolayer–liquid and monolayer–gas interface (8, 24c). These systems are also particularly notable for their ability to block the diffusion of dissolved redox couples to the underlying electrode. Studies of electrode reactions occurring over "long" distances involving dissolved reagents require substantially pinhole-free films for the prevention of physical diffusion of redox couples to the underlying electrode interface. Finklea (20b–c) used alkylthiol and alkylsiloxane oriented monolayers to block Au and Pt electrodes. However, problems with pinholes or stability were reported. Rubinstein (21a, b) employed such pinholes, as arrays of microelectrodes, to obtain heterogeneous rate constants for fast outer-sphere redox couples. Because the alkylthiol chain length is readily varied, self-assembly of this system provides a variable thickness diffusion barrier. Porter et al. (10) demonstrated substantially pinhole-free monolayers in *n*-alkylthiol monolayers for chain lengths greater than C_{10}. Electron transfer to/from solution reagents was dramatically hindered and depended strongly on the spacer thickness. Similar studies on L–B monolayers appear not to have been done.

Several groups have assembled electroactive amphiphiles onto various electrode substrates. Electrochemical characterization of ferrocene (18, 19a), viologen (15a, 18a–e, 19b, 22d, 23b), and quinone (18f) amphiphiles assembled on electrodes have been conducted. Several novel characterization tools were employed. Monolayer self-assembly on porous aluminum oxide templates allowed the measurement of *lateral* electron-transfer rates among organized electroactive sites (18). A particularly powerful tool for characterizing the dynamics of counterion motion within monolayers is the combination of electrochemistry and the quartz crystal microbalance (19). The emphasis in the above studies, however, is on electroanalytical characterization, which provides an important though indirect probe of monolayer structure.

3.1.2 Langmuir–Blodgett Films

While the fabrication of planned multilayered self-assembled films is relatively new, the fabrication of ordered, oriented multilayered L–B films is well established (34). Tremendous advances in L–B film fabrication and characterization (5, 6) have occurred over the last 15 years, since the seminal work of Kuhn et al. (35) and Möbius (36). The interest in these molecular assemblies was driven in part by a need for pinhole-free molecule thick films and by the need for precise control of intermolecular distances (molecular spacers) in studies of electron and energy transfer (35, 36). In the last decade the development and application of surface characterization techniques resulted in detailed understanding of the structure of L–B films (6). Numerous proposed technological applications incorporating L–B films have since been investigated including nanolithography (37), insulating gates (38), materials for LEDs (39), and gas sensors (40), to name a few.

Transfer of monolayers from the water surface need not rely on strong film–substrate interaction or intramonolayer intermolecular interactions, although film stability, as expected, is greatly enhanced by such interactions (41, 42). Interlayer chain interdigitation is not usually encountered in L–B films (see, however, Ref. 43). Molecular intercalation is observed in some self-assembled films (18, 23b). Finally, in an L–B multilayer the deposition of similar or dissimilar multilayers is readily achieved without introducing additional functional groups for chemical activation. This flexibility in multilayer planning is one of the main advantages of L–B film fabrication over self-assembly. It is achieved at the cost of possibly producing metastable structures that may lead to uncertainty in the structure of the overlayers, especially when contacted with a liquid electrolyte. Meaningful case to case comparisons are sparse (7b–d). The L–B film deposition and self-assembly techniques are thus to some extent complementary modification techniques. Selection of one technique over the other should be based on the nature of the phenomena and the system under study. In some cases it is envisioned that a combination of self-assembled and L–B film may prove to be useful.

The formation of molecule thick organized monolayers of ambipolar surfactants, such as *n*-alkanoic acids, on water is a well-known phenomenon, first systematically explored over 50 years ago. Blodgett (34), working with Langmuir, demonstrated that a floating monolayer of barium and cadmium stearate could be transferred from the water surface to a submerged hydrophilic glass substrate upon emersion of the substrate from the water subphase. Repeated sequential immersion and emersion of the substrate through the floating monolayer resulted in the transfer of an ordered monomolecular layer with each pass through the floating monolayer. A multilayer film is built up monolayer by monolayer in precisely defined increments of known thickness. The accuracy and precision of the film thickness, as well as its uniformity over large areas, led to some commercial use of L–B monolayers of barium stearate as a thickness guage (5a, 44).

In practice, the formation of an organized monolayer at the air–water interface requires the correct balance of hydrophobic tail interactions and interaction of the "head group" with the aqueous subphase. The spreading of monolayers at the air–water interface is usually achieved by applying to the water surface an aliquot of a dilute solution containing an accurately known quantity of the surfactant. Ideally, after evaporation, a randomly distributed array of widely separated noninteracting surfactant molecules is left on the surface, as schematically shown in Fig. 3.1 top (see, however, Ref. 15b). Decreasing the area occupied by the surfactant monolayer on the water surface, that is, monolayer compression, can be accomplished, by moving a hydrophobic barrier such as Teflon® over the water surface (Fig. 3.1 bottom). As the surface excess of surfactant at the air–water interface $\Gamma_{a/w}$ increases, the tails can achieve an extended all-trans configuration normal to the water surface, which ideally is completely preserved upon transfer. In addition, the intermolecular attractive forces between surface water molecules are replaced in part by water–head group interactions. Consequently, the water surface tension is lowered. Equation 3.1

$$\pi = \Delta\gamma = \gamma_0 - \gamma \tag{3.1}$$

expresses the monolayer surface pressure π as the lowering of the water surface tension $\Delta\gamma$ from its value γ_0 at a pure monolayer free surface to its value at a

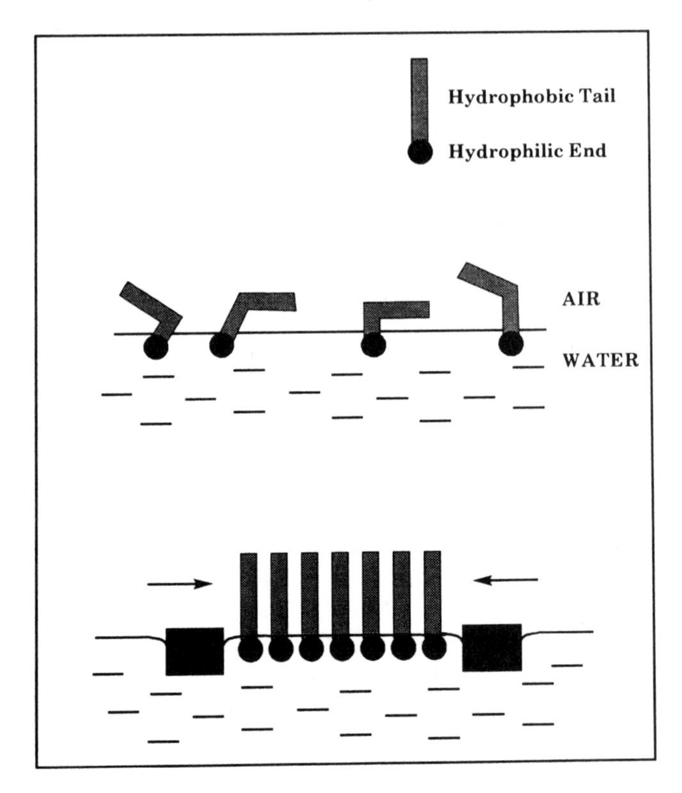

Figure 3.1. Schematic depiction of an insoluble amphiphilic monolayer at the air–water interface prior to compression (top) and after compression (bottom).

monolayer coated surface γ. The surface pressure π can be measured by a Langmuir balance. This is a differential method that requires a floating barrier to separate monolayer coated and monolayer free surfaces. The differential surface tension places a force on the barrier, which is measured in a variety of ways by the Langmuir balance. Surface pressure is also measured directly via a Wilhelmy plate that employs a hydrophilic plate (e.g., glass, Pt, or paper), which is immersed into the subphase. The plate is typically suspended from a sensitive electrobalance and the changes in surface tension are related to weight changes (5a, c). The advantages and disadvantages of both methods are discussed elsewhere (5c).

Film transfer from the air–water interface to a hydrophilic substrate (during emersion) results in the hydrophilic head groups being in contact with the substrate. Immersion of a hydrophobic substrate through a floating monolayer results in the hydrophobic tails touching the substrate. While the substrate is immersed in the subphase at this point, one might attempt to obtain a single tails down monolayer by emersing the substrate through a monolayer-free water surface (after sweeping it clean). Such modification is rather uncertain as it is usually accompanied by the monolayer transferring back to the water surface upon emersion. For this reason it is usually necessary to transfer two layers when the first layer is transferred tails down.

Several features of L–B films make them attractive for electrode modification. Many electroactive groups are easily functionalized by hydrocarbon tails and thus it should be possible to prepare complex organized structures with tailored electrical (45) or electrochemical properties. The L–B films are of uniform thickness over wide areas (± 1–2 Å in our hands) and should be useful in studying long distance electrode reactions. Workers in our laboratory (46) and elsewhere (31) have shown that chromophore and electrophore orientation can be controlled by appropriate hydrocarbon substitution of the parent compound. Thus the nature of the amphiphile, the order of film transfer, the substrate wettability, hydrocarbon chain length, substitution pattern, and other factors may be controlled to obtain flexibility in designing molecular architectures on electrode surfaces.

The electrochemical characterization of electroactive L–B monolayers has, until recently, received sparse attention. The L–B monolayers of amphiphilic $[Ru^{II}(bipy)_3]^{2+}$ (bipy = 2,2'-bipyridine) were assembled on optically transparent SnO_2 and Au electrodes (16a). These monolayers were spaced from the electrode by a variable number of arachidic acid L–B monolayers. The photocurrents, in the presence of a hydroquinone (HQ) sensitizer, and the photoemission spectra were strongly dependent on Ru-electrode distance. The authors suggest, therefore, that electron transfer was nearly completely blocked by a single 25-Å spacer monolayer. Unfortunately the distance resolution was limited to multiples of 25 Å, the arachidic acid chain length. The deliberate spatial manipulation of electroactive species to achieve specific function was elegantly demonstrated by Fujihira (16b) in the fabrication of photodiodes based on L–B layers containing donor, acceptor, and photosensitizer molecules. In addition, cationic viologen L–B films were shown to behave as anion exchange layers incorporating $[PtCl_6]^{2-}$, which could be electrocatalytically reduced by viologen to Pt^0 islands (16c). Electrode reactivity toward solution redox species was modulated by so-called ion gating L–B films. The gating of counterions was modulated by electrochemically changing the oxidation state of the viologen amphiphiles in the outermost L–B monolayers (16d) and by photochemical control of the hydrophilicity of a spiropyran L–B multilayer (16e). Suggestive preliminary electrochemical studies of long-distance electrode reactions (16d–f) claimed electron transfer through seven ostensibly insulating layers of arachidic acid. Electron transfer in this case almost certainly involved defects in the insulating layers possibly due to translayer diffusion (16h) of the electroactive amphiphiles. Several groups have electrochemically characterized incipient L–B monolayers right at the air–water interface. Majda and his co-workers (18f) investigated lateral electron transfer among electroactive ferrocene sites in monolayer films at the air–water interface by floating a Au microband electrode at a ferrocene monolayer coated water surface. Fujihira (16g), using the horizontal touching technique (47), demonstrated that the electroactivity of an anthraquinone monolayer at the air–water interface could be characterized by cyclic voltammetry. Bard and his co-workers (15b) extended this technique to studying the lateral homogeneity of a floating monolayer of amphiphilic $[Ru(bipy)_3]^{2+}$ as function of surface pressure. An excellent comparison of films transferred by vertical and horizontal transfer was provided as well as a comparison (15a) of the electrochemistry of L–B and adsorbed hexadecylviologen monolayer films. Electrogenerated chemiluminescence (ECL) of L–B films of amphiphilic $[Ru(bipy)_3]^{2+}$ was also reported (15c). Tin oxide electrodes were modified with L–B films of amphiphilic $[M(bipy)_3]^{2+}$ (M = Ru Os) and the heterogeneous electron-transfer rate constants theoretically modeled and measured (17b) by Matsuda, Tokuda and their co-workers. Mediated oxidation of Fe^{2+} by these L–B films was also modeled and described (17c). In the above studies of electroactive L–B films, spectroscopic structural characterization of the films is largely absent; they have not received the same degree of structural scrutiny as have self-assembled monolayers. The perturbing effect of the large electroactive head groups on structure has, for example, not been addressed. Clearly this information would be quite valuable.

While the L–B technique offers a high degree of fabrication flexibility, a significant potential, on the other hand, exists for the occurrence of artifacts (48) in the monolayer preparation, transfer, and characterization steps. Slow monolayer collapse at the air–water interface is one of the most insidious sources of artifacts. In addition, floating monolayers often need to be screened for film dissolution at the air–water interface, molecular segregation in mixed monolayers, monolayer respreading during transfer, molecular overturning (49), optimal transfer speeds, and sensitivity to subphase composition. Occasionally, the high degree of molecular exposure at the air–water interface results in decomposition or change in the oxidation state of the surfactant. As mentioned above, electrochemical characterization, especially of multilayered L–B modified electrodes, may potentially introduce artifacts such as film restructuring–respreading when crossing the air–electrolyte interface and the perturbation of molecular microstructure during oxidation state changes.

In order to investigate the scope of electrode modification with L–B films, particularly with respect to interfacial long-range electron transfer, we discuss below the systematic electrochemical characterization of the electroactive surfactant $C_{18}Fc$ (where FC=ferrocene) at the air–aqueous electrolyte interface,

Fe N^+ PF_6^-

N

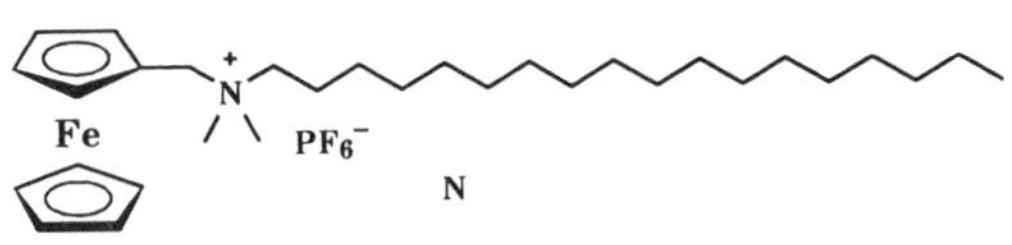

$C_{18}Fc$

and the modification of hydrophilic and hydrophobic Au electrodes with L–B films of $C_{18}Fc$. The electrochemistry of L–B films of $C_{18}Fc$ with ferrocene in direct contact with the electrode is conducted as a function of transfer pressure (molecular area). The modification and electrochemistry of remotely spaced ferrocene is also described. External reflection–absorption FTIR spectroscopy is utilized to characterize the monolayer film state ex situ. Because adsorption and monolayer transfer are not a priori expected to engender the same molecular surfaces, a comparison of the properties and structures of self-assembled and L–B $C_{18}Fc$ films is presented.

3.2 EXPERIMENTAL TECHNIQUES

3.2.1 Monolayer Preparation

The fabrication of good quality monolayer films at the air–water interface requires that adequate care be taken to prevent the introduction of film forming contaminants and deleterious impurities into the experiment (5a, c). Sources of contamination include the trough and associated barriers, subphase water, electrolyte, spreading solvent, and the amphiphile itself. The trough and all barriers must by hydrophobic, relatively inert, and scrupulously clean. The Langmuir trough (Lauda) used in our group employs a modified base milled from solid Teflon®. A trough cleaning procedure employing successive washes with pure ethanol, water, and a mixture of hot concentrated nitric and sulfuric acids, followed by a rinse with monolayer grade water is generally sufficient to yield excellent isotherms, provided that other sources of contamination are precluded. A floating Teflon® barrier is used to separate the monolayer covered water surface from the clean surface. The accuracy of the π measurements is thus directly related to the absence of surface active impurities in the subphase, since their adsorption at the air–water interface lowers the aqueous surface tension relative to a pure subphase.

In our laboratory monolayer grade water is prepared by reverse osmosis, passage through a Millipore filtration train (ion exchange, organic removal), and distillation from alkaline permanganate under purified Ar. A final distillation from a slightly acidic reservior yields "monolayer" water used to prepare the subphase. Other water purification techniques such as catalytic pyrodistillation (50) should also be suitable. Sufficient care should be exercised so that adventitious surface active impurities are not introduced into the subphase. For example, the subphase should not be contacted, even briefly, with plastics that are known to contain plasticizers or, when preparing carboxylate monolayers, with metals such as Cu, Al, Fe, and brass, which readily corrode in water.

Monolayers are spread on the water surface by the dropwise application of a solution containing an accurately known concentration of the surfactant ($\sim 1\ mg\,mL^{-1}$) in a low-boiling high-purity solvent. A micrometer driven syringe, or the like, is used to deliver accurate volumes of the surfactant solution to the water surface. The spreading solvent should be immiscible with water and it should not form a lens on the water surface. Suitable spreading solvents include chloroform, benzene, hexane, and cyclohexane. Mixtures containing a small fraction of water soluble solvents such as ethanol may be used to dissolve sparingly soluble amphiphiles, but these mixed solvents must be used with care in order to avoid dissolution into the subphase of the monolayer. In order to insure that high-purity commercially available solvents are suitable for monolayer work, one typically spreads and evaporates 50–100 μL of the pure solvent on the water surface. An acceptable solvent will yield no discernible surface pressure π at the minimum molecular area of the trough. Generally, 15 min are required for complete evaporation of chloroform from $C_{18}Fc$ monolayers. The evaporation time varies, as expected, depending on the nature of the amphiphile and should be determined experimentally. Incompletely evaporated solvent may be manifested as a sloping isotherm baseline or as an irreproducible isotherm (14c).

We found that ripples on the water surface caused by external vibrations can adversely affect film transfer, particularly during the transfer of "stiff" monolayers (46). In order to minimize this effect, the trough is located on a vibration isolation table. In our laboratory monolayers are prepared in a constant humidity, dust-free (Class 100 clean room) environment. However, filtered air laminar flow hoods and gloveboxes are also used. Pressure–area isotherms are obtained at low areal compression rates under computer control. Monolayers are compressed to the desired molecular area or surface pressure and transferred to hydrophilic electrodes by emersion of the electrode through the floating monolayer at constant surface pressure (π_t). In order to prepare bilayers containing dissimilar monolayers, the first monolayer is transferred tails down on a hydrophobic substrate and left in the subphase while another monolayer is spread. The second monolayer is then transferred on the upstroke. In order to minimize subphase contamination, inert materials are used and washed immediately prior to the beginning of the experiment.

3.2.2 Electrode Preparation

Electrode preparation and subsequent characterization are done in an external Class 1000 area. The preparation of electrodes for film transfer is relatively straightforward. To achieve heads down monolayer transfer, a hydrophilic electrode must be emersed vertically from the subphase. Conversely, tails down transfer is done by vertical immersion of a hydrophobic electrode or transfer by the horizontal touching technique (47). We have had difficulty in transferring heads down $C_{18}Fc$ monolayers onto Pt. More successful are transfers to Au. Solid Au flag electrodes ($2.54 \times 2.54 \times 0.05$ cm) are mechanically polished by standard metallography. Smoother Au electrodes are obtained by electron beam deposition of 100 nm of Au onto a 5-nm Cr adhesion layer on both sides of highly polished $\langle 100 \rangle$ single-crystal Si wafers (Polishing Corporation of America, Santa Clara, CA). In order to obtain meaningful transfer ratios, transfers to wafers coated with Au on one side are avoided as the transfer ratios to Au and Si are not the same (14c). Electrodes of any desired area are obtained by dicing the wafer with a diamond tipped stylus. Hydrophilic Au can be prepared by potential cycling of the Au film electrode in pure 1 $M H_2SO_4$, or by immersing Au flags in 0.1M ethanolic KOH. A more consistent technique involves immersion of Au films into hot 1:3 H_2O_2 (30%)–H_2SO_4 for 2–5 min. Cleanliness in the preparation of Au electrodes cannot be overemphasized. Clean hydrophilic Au electrodes are rinsed with distilled water. While drying, it is critical that the electrode contact only scrupulously clean surfaces. In our experience, ignoring this step inevitably results in hydrophobic areas on the electrode surface to which the monolayer does not transfer (14c). Hydrophobic electrodes are prepared via the adsorption of sulfur containing small molecules on clean Au (see below). Monocrystalline Pt bead electrodes for adsorption experiments are prepared from a 0.5-mm diameter Pt wire (99.99%) by the procedure of Clavilier et al. (51).

3.2.3 Materials

The synthesis of $C_{18}Fc$ from 1-iodooctadecane and (ferrocenylmethyl)-dimethylamine was previously described (14b). The compound $C_{18}Fc$ was recrystallized three times from methanol–water. High purity chloroform (Fisher LC Grade) containing pentene stabilizer was used as received. High purity methanol (Burdick & Jackson) and absolute undenatured ethanol were used throughout. The sulfur containing adsorbates thiolacetic acid, 3-methyl-thiophene, 3-thiophenecarboxylic acid, 3-thiopheneacetic acid, di-*tert*-butyl-disulfide, dimethyldisulfide, and *tert*-butyl mercaptan were purified prior to use. 3-Methylthiophene was distilled under Ar and di-*tert*-butyldisulfide was twice vacuum distilled. 3-Thiopheneacetic acid was twice recrystallized from cyclo-hexane. Octadecylmercaptan was twice recrystallized from absolute ethanol. Adsorption of these species is described in Section 3.3.2. Stearic acid was recrystallized three times from ethanol. Cadmium chloride (Aldrich, Gold Label) was recrystallized from water. Sulfuric and perchloric acids (G. F. Smith) were doubly distilled from Vycor. Tetraethylammonium iodide (TEAI) and KI were used as received. Argon, used in blanketing and degassing of electrolyte solutions, was purified via an Ace–Burlitch inert atmosphere system.

3.2.4 Apparatus

Conventional Princeton Applied Research electrochemical instrumentation was used for voltammetric studies. Electrochemical cells were degassable and of conventional design. A large area Pt gauze counterelectrode encircled the working electrode. All potentials are referenced to a sodium chloride saturated calomel electrode (SSCE) of conventional design. Cyclic voltammetry was performed immediately following monolayer transfer to Au or adsorption on Pt. A Digilab 15C FTIR spectrometer was used to obtain reflection–absorption IR spectra of monolayers. Ellipsometric film thickness measurements (52) on Au film electrodes were done on a Rudolph Research Model ELIV ellipsometer using 70° incident 632.8 nm radiation. An uncoated Au film electrode was used as a reference to determine the bare substrate ellipsometric constants Δ_r and Ψ_r averaged over 3–6 spots. In order to obtain the best reference, the reference Au surface was sectioned from the same wafer as the sample and its surface treated in the same manner as the monolayer coated electrode. Ellipsometric constants

Δ_f and Ψ_f were determined for several spots on the film coated substrate and film thicknesses calculated from values of $\delta\Delta = \Delta_f - \Delta_{r,avg}$ and $\delta\Psi = \Psi_f - \Psi_{r,avg}$. A refractive index of 1.5 was assumed consistent with literature values of aromatic surfactants (26a). Using a value of 1.4 increased the film thickness by nearly 2 Å.

3.3 RESULTS AND DISCUSSION

In Sections 3.3.1 and 3.3.2 we describe the systematic modification of Au electrodes with L–B films of C_{18}Fc heads down on Au and describe several schemes for remotely spacing ferrocene a uniform distance from the electrode. These structures are electrochemically characterized. In Section 3.3.3 the preparation and characterization of self-assembled C_{18}Fc monolayers adsorbed from aqueous acid electrolytes is described and compared with L–B films.

3.3.1 Heads Down C_{18}Fc Monolayers on Gold Electrodes

3.3.1A C_{18}Fc Behavior at the Air–Electrolyte Interface

Before undertaking the modification of electrode surfaces with C_{18}Fc L–B monolayers and bilayers, it is necessary to first understand and control the film state at the air–electrolyte interface. Monolayer films of C_{18}Fc are spread at the air–water interface by evaporation of a dilute chloroform or 50:50 chloroform–benzene solution of an accurately known concentration of the surfactant. The monolayer properties of C_{18}Fc are obtained from the measurement of surface pressure versus molecular area at constant temperature (π–A isotherm). Shown in Fig. 3.2*a* is a typical first cycle compression expansion π–A isotherm for C_{18}Fc on a 0.1*M* Na_2SO_4 subphase when the film is compressed to about 35 Å^2 molecule^{-1}. The shape of the curve and absence of well-defined transitions imply that the film exists in a two-dimensional liquid state, loosely analogous to a three-dimensional liquid, over the range of molecular areas indicated in the figure. When the film is compressed below 24–27 Å^2 molecule^{-1} ($\pi > 42\,\mathrm{mN\,m^{-1}}$), film collapse occurs as indicated in Fig. 3.2*b*. A somewhat different isotherm is obtained on a 0.1*M* $NaClO_4$ subphase (18f). Extrapolation of the π–A curve in the high π region to zero surface pressure (dashed line) yields a measure of the projected molecular area, $A_{lim} = 51$ Å^2 molecule^{-1}, of the C_{18}Fc head group. Molecular models indicate a molecular area of 49 Å^2. In the case of liquid films, A_{lim} may be an inexact measure of molecular area. The possibility of head groups slipping past one another at high π introduces another uncertainty in the interpretation of A_{lim}. Under the above conditions, however, A_{lim} appears to be a useful approximation of head group molecular area.

The kinetics of the molecular processes associated with film compression and expansion are slow relative to the experimental (e.g., compression–expansion) timescale. That the film is not in equilibrium with the barrier movement is shown by the observation that surface pressure decreases slightly during film

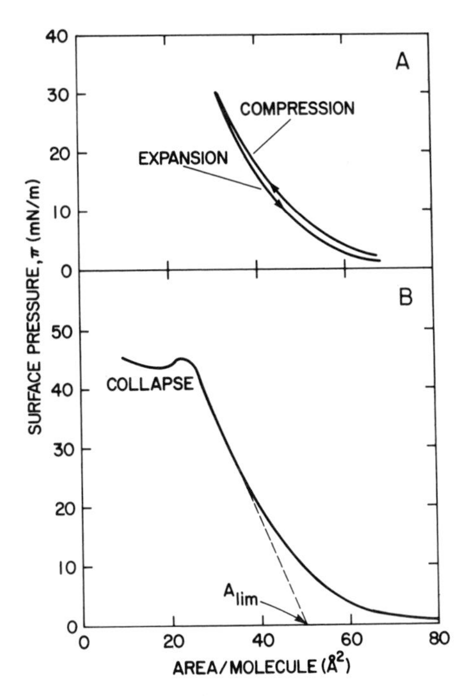

Figure 3.2. (A) Surface pressure versus molecular area isotherm of $C_{18}Fc$ at the air—$0.1M$ Na_2SO_4 interface; (B) estimation of the molecular limiting area (A_{lim}) from the high π region of the curve, prior to collapse of the film. [Reprinted with permission from Facci, J. S. et al., *Langmuir*, **1986**, *2*, 732. Copyright © (1986) American Chemical Society.]

compression when the moving barrier motion is halted and *increases* when the barrier movement is halted during expansion. The decrease is due to slow accommodation of molecules to a lower surface area, while the increase is due to slow molecular expansion to accommodate a larger surface area. The effect is more pronounced at higher π. At the air–electrolyte interface these films, therefore, behave as viscous liquidlike films.

Because of the charge on the surfactant, the subphase ionic strength significantly influences the limiting molecular area of $C_{18}Fc$ at the air–aqueous electrolyte interface. The limiting molecular area passes through a minimum near $0.1M$ Na_2SO_4 as the concentration of Na_2SO_4 is varied from 10^{-4} to $0.3M$. A $0.1M$ Na_2SO_4 subphase is therefore employed throughout this study as it minimizes intermolecular coulombic repulsion. In addition, it provides electrolytic conductivity for voltammetric studies, which are done in situ in the trough subphase. The monolayer and voltammetry are done at the same electrolyte concentration since the microstructure of $C_{18}Fc$ is sensitive to the nature (53) and concentration of the electrolyte counterion (see Section 3.3.3).

The compound $C_{18}Fc$ spontaneously forms an organized, nearly close packed monolayer in equilibrium with its *crystalline* form. A monolayer film can be formed at the air–electrolyte interface by placing a crystal of $C_{18}Fc$ on the subphase. As the monolayer spreads from the crystal the surface pressure increases until an equilibrium value π_e, the equilibrium spreading pressure, is achieved. On a pure water subphase $\pi_e = 18\,mN\,m^{-1}$ corresponding to a molecular area $A = 40–50\,Å^2$ molecule^{-1} ($4.2–3.3 \times 10^{-10}\,mol\,cm^{-2}$), or approximately A_{lim}. Compression of a monolayer above $18\,mN\,m^{-1}$ does not necessarily imply immediate film collapse, however, since bulk crystalline $C_{18}Fc$ nuclei or nucleating centers are not present during compression at the air–water interface under our experimental conditions (54). A length of time substantially longer than the transfer time may be needed to form these nuclei; other film area loss processes such as a slow film collapse or two-dimensional crystallization (18f) are likely to occur first.

3.3.1B Transfer to Au Electrodes

We now describe the preparation of organized monolayers of $C_{18}Fc$ oriented heads down on Au electrodes. In order to achieve this, a hydrophilic electrode surface must be presented to the floating monolayer upon electrode emersion from the subphase. Gold electrodes are made hydrophilic by soaking in either 0.1*M* ethanolic KOH or acidified peroxide. Gold working, (SSCE) reference and Pt auxiliary electrodes are placed in the trough subphase and a $C_{18}Fc$ film is spread and compressed at an areal compression rate of $5–8\,Å^2$ molecule^{-1} min^{-1}. Because of the film viscosity described above, the monolayers were annealed at constant pressure for 3–5 min. During this time the rate of film contraction not related to transfer (e.g., reorganization and crystallization) is monitored and calculated. The Au electrode is emersed from the subphase at $1.0\,mm\,min^{-1}$ at constant pressure.

As expected, all L–B films emerged dry from the subphase (because of electrolyte drainage during transfer) and were hydrophobic because of the extended alkyl tails. The advancing water contact angle was 90–95°, which is significantly less than that for a highly ordered monolayer of *n*-alkyl chains. This suggests that the alkyl tail region of the monolayer is somewhat disordered (55). These L–B films were durable and could not be rinsed from the electrode with a stream of water or repeated immersion through the air–water interface. It is thought that their physical stability accrues from the organization of the hydrocarbon tails, which imparts a relatively high activation barrier for film dissolution, analogous to the high dissolution activation barrier for organized stearic acid molecules at the air–water interface (5a). In addition the film molecules may be electrostatically anchored to the surface by an adsorbed monolayer of OH^-. If this is the case, Na^+ cations associated with Au/OH^- and SO_4^{2-} anions associated with $C_{18}Fc$ are removed as the subphase is drained during transfer. Cation induced adsorption (ads) resulting from OH^-_{ads} at Au was previously reported (56); in addition, the voltammetry of Au in 0.1*M* NaOH shows evidence for the OH^-_{ads} species on Au (56).

In order to understand the influence of film structure on the electrochemistry, films were transferred to Au at several different surface pressures. The π–A

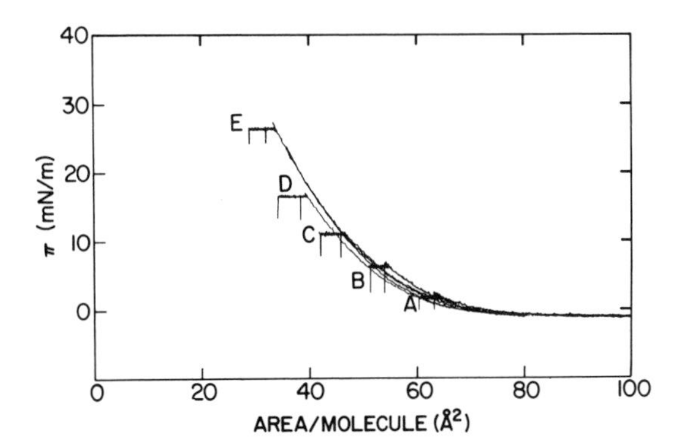

Figure 3.3. Pressure–area isotherms for $C_{18}Fc$ monolayers transferred to hydrophilic Au at $\pi_t = 3$–$27\,mN\,m^{-1}$. [Reprinted with permission from Facci, J. S. et al., *Langmuir*, **1986**, *2*, 732. Copyright © (1986) American Chemical Society.]

curves recorded during transfer from a $0.1M$ Na_2SO_4 subphase at $\pi_t = 2.7$, 7.2, 12.1, 17.4, and $26.9 \pm 0.3\,mN\,m^{-1}$ are shown in Fig. 3.3 (curves A–E, respectively). The right- and left-hand tick marks define the beginning and end of film transfer, and the molecular areas associated with them are denoted by A_{init} and A_{fin}. The total film contraction (A_{loss}) unrelated to transfer is calculated from the constant π region in the annealing portion of the curve. The number of moles n of $C_{18}Fc$ transferred to the Au electrode is calculated from Eq. 3.2, where K is the number of molecules spread and N_A is Avogadro's number. Transfer ratios, defined as the ratio

$$n = K(A_{init} - A_{fin} - A_{loss})/(N_A A_{init}) \tag{3.2}$$

of the area of the film transferred from the water surface to the projected or geometric area of the substrate, is calculated at each π_t and summarized in Table 3.1. Corrected transfer ratios lie close to unity except for the one at $22\,mN\,m^{-1}$. Although transfers at $\pi_t = 22$ and $27\,mN\,m^{-1}$ were done at $\pi > \pi_e$, these films are kinetically stable to collapse at short times, that is, less than 15 min. All variables related to film structure and transfer (electrode emersion and reimmersion rates, nature of electrode, and electrolyte concentration) are held constant in order to eliminate as many experimental variables as possible.

3.3.1C Voltammetry of Heads Down $C_{18}Fc$ L–B films

Immediately after monolayer transfer, cyclic voltammetry is done in the trough subphase by removing most of the film from the water surface, reexpanding the film to the maximum trough area, and reimmersing the electrode in the subphase. Upon reimmersion the $C_{18}Fc$ film did not appear to respread to the water surface. Fig. 3.4 curves A–C show representative cyclic voltammograms at $200\,mV\,s^{-1}$ of L–B films of $C_{18}Fc$ on Au flags at 2.7, 7.2, and $27\,mN\,m^{-1}$, respectively. The voltammograms display the characteristics of

Table 3.1. Summary of Transfer Parameters for the Deposition of L–B Monolayers of $C_{18}Fc$ from a $0.1M\ Na_2SO_4$ Subphase at Various Surface Pressures

A_{init}[a]	π_t	$10^3 \times A_{loss}$ ($Å^2$ molecule^{-1} min^{-1})	*Transfer Ratio*[b]	$10^9 N$ (mol)[c]
63.5	2.7	1.5	1.04	2.72
54.2	7.2	1.9	1.01	3.09
46.0	12.1	3.8	1.01	3.65
38.6	17.4	8.3	0.92	3.97
37.4	22.0	5.4	0.73	3.27
32.4	27.0	5.6	0.96	4.91

[a]Molecular area at beginning of transfer.
[b]Ratio of the electrode projected area to the area of film lost from water surface.
[c]Moles transferred.

surface confined electroactive reagents (57). These films show no degradation to potential cycling in $0.1M\ Na_2SO_4$ and $1M\ H_2SO_4$ for 30 min at $100\,mV\,s^{-1}$. This observation is somewhat surprising in view of the known chemical instability of the ferrocenium cation toward water (58). Similar stability is observed below (Section 3.3.3) for self-assembled films of $C_{18}Fc$. The peak potential difference (ΔE_p) is 20–30 mV and the anodic peak current ($i_{p,a}$) is linear with sweep rate. The surface formal potential $E^{\circ\prime}_{surf} = 0.52$ V, determined from the average of the anodic and cathodic peak potentials, is somewhat more positive than the solution formal potential $E^{\circ\prime}_{soln} = 0.36$ V of the model compound (ferrocenylmethyl)trimethylammonium sulfate in $0.1M\ Na_2SO_4$. The term $E^{\circ\prime}_{soln}$ is taken as the formal potential of $C_{18}Fc$ in the absence of adsorption

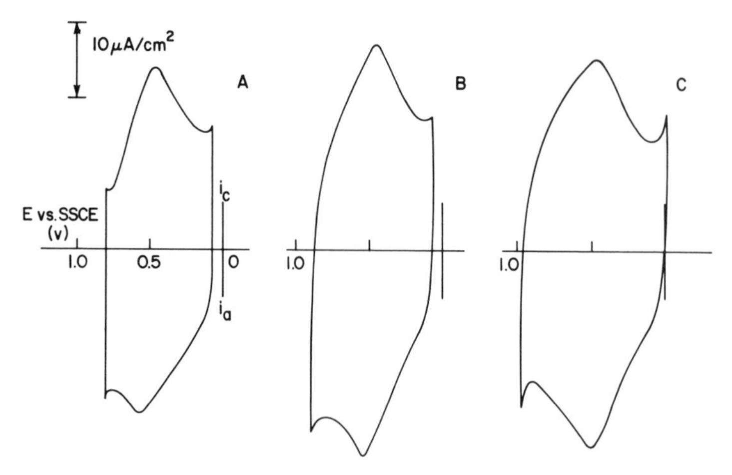

Figure 3.4. Representative cyclic voltammograms ($0.1M\ Na_2SO_4$, $200\,mV\,s^{-1}$) of L–B films of $C_{18}Fc$ transferred to hydrophilic Au flags at 2.7, 7.2, and $22\,mN\,m^{-1}$.

or ion pairing. The difference between the solution and surface formal potentials (-0.14 V) may be interpreted in terms of the differences in the strength of adsorption (57) at Au of the reduced (Red) and oxidized (Ox) forms of $C_{18}Fc$. In the reaction scheme in Eqs. 3.3a–c,

$$\mathrm{Red_{soln}} \leftrightarrows \mathrm{Red_{ads}} \tag{3.3a}$$

$$\mathrm{Red_{ads}} - e^- \rightarrow \mathrm{Ox_{ads}} \tag{3.3b}$$

$$\mathrm{Ox_{ads}} \leftrightarrows \mathrm{Ox_{soln}} \tag{3.3c}$$

soln and ads refer, respectively, to the solution and adsorbed forms of $C_{18}Fc$. The term $\Delta G^{\circ}_{\mathrm{R,ads}}$ and $G^{\circ}_{\mathrm{Ox,ads}}$ are defined as the free energies of adsorption of Red and Ox from solution. $\Delta G^{\circ}_{\mathrm{Red,\ ads}}$ and $\Delta G^{\circ}_{\mathrm{Ox,ads}}$ refer to Eq. 3.3a and the reverse of Eq. 3.3c, respectively. $\Delta E^{\circ\prime}$ is then expressed in terms of the oxidation half-wave potentials $E^{\circ\prime}_{\mathrm{surf}}$ and $E^{\circ\prime}_{\mathrm{soln}}$ (Eq. 3.4).

$$\Delta E^{\circ\prime} = E^{\circ\prime}_{\mathrm{soln}} - E^{\circ\prime}_{\mathrm{surf}} = (\Delta G^{\circ}_{\mathrm{Red,ads}} - \Delta G^{\circ}_{\mathrm{Ox,ads}})/\mathscr{F} \tag{3.4}$$

Using $\Delta E^{\circ\prime} = -0.14$ V, one finds $\Delta G^{\circ}_{\mathrm{Red,ads}} - \Delta G^{\circ}_{\mathrm{Ox,ads}} = \Delta(\Delta G^{\circ}_{\mathrm{ads}}) = -3.2$ kcal mol^{-1}. Thus Red is more strongly adsorbed to Au than is Ox. Alternatively, Red is stabilized relative to Ox by being incorporated in a hydrophobic environment created by the tails.

The charge under the anodic wave Q_a ($\pm 15\%$) decreases slightly with sweep rate for all $C_{18}Fc$ L–B films. The charge Q_a becomes independent of v for $v = 10$–$20\,\mathrm{mV\,s^{-1}}$ at high π_t (27 mN M^{-1}). The sweep rate dependence of Q_a at these relatively slow sweep rates is thought to be a consequence of the organized nature and packing of the film hydrocarbon tails. The ferrocene film sites (Fc) in heads down L–B monolayers are shielded from the aqueous electrolyte by a hydrophobic hydrocarbon layer while the oxidation reaction $Fc \rightarrow Fc^+ + e^-$ necessitates counterion movement through this layer to accommodate the incipient increase in film charge. However, because of the hydrocarbon tails, ion–solvent transfer across the hydrophobic tail region may be hindered on the cyclic voltammetric timescale. It is postulated that the sweep rate dependence of Q_a is a consequence of a slight diffusion barrier created by the hydrocarbon chains. A similar though more pronounced sweep rate dependence of Q_c was observed during the reduction of anthraquinone phospholipid L–B monolayers on Au (14c). The effect is believed to be more pronounced in the phospholipid L–B monolayers because of the better chain packing.

The number of moles of $C_{18}Fc$ in the L–B film ($Q_a/n\mathscr{F}$) and apparent surface coverage, $\Gamma_{\mathrm{app}} = Q_a/n\mathscr{F}A$ (A = projected electrode area) are determined at slow sweeps (10–20 mV s^{-1}). The projected area rather than true area, is used since L–B films bridge over surface imperfections (59). Figure 3.5 top plots Γ_{app} versus $\Gamma_{\mathrm{a/w}}$ and A_{init}, where $\Gamma_{\mathrm{a/w}}$ is the surface excess at the air–water interface during transfer. Initially Γ_{app} increases with $\Gamma_{\mathrm{a/w}}$ and π_t but they attains a constant value of $\Gamma_{\mathrm{app}} \sim 2.7 \times 10^{-10}\,\mathrm{mol\,cm^{-2}}$ when $\Gamma_{\mathrm{a/w}} = 3.6 \times 10^{-10}\,\mathrm{mol\,cm^{-2}}$. Figure

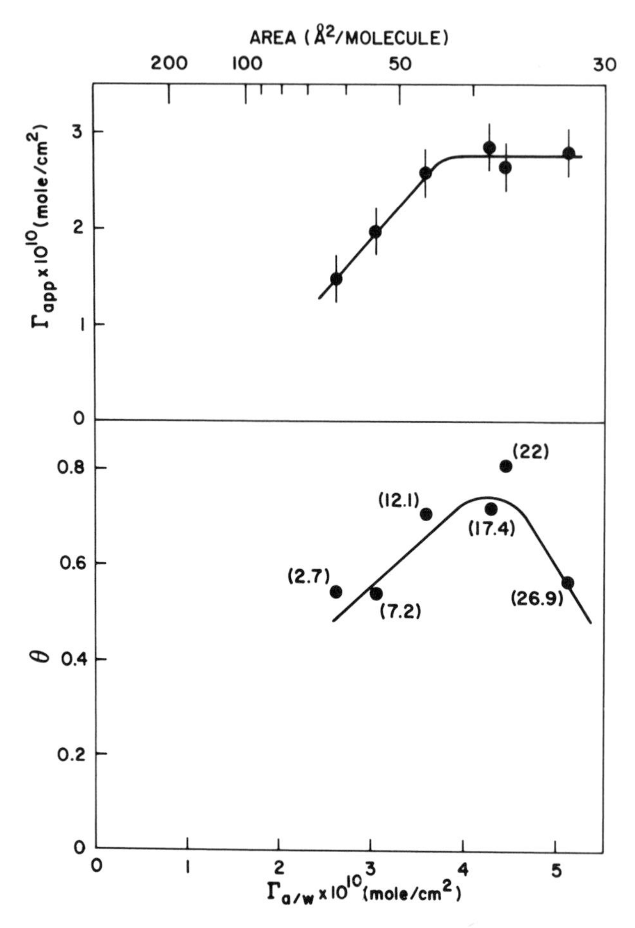

Figure 3.5. (Top) Plot of apparent surface coverage Γ_{app} determined cyclic voltammetrically versus the water surface coverage $\Gamma_{a/w}$ prior to transfer; (Bottom) plot of the fraction of transferred sites that are electrochemically active versus $\Gamma_{a/w}$. Numbers in parentheses are values of π_t. [Reprinted with permission from Facci, J. S. et al., *Langmuir*, **1986**, *2*, 732. Copyright © (1986) American Chemical Society.]

3.5 bottom plots θ, the fraction of transferred ferrocene sites that are redox active as a function of $\Gamma_{a/w}$. Values of θ are in all cases less than unity and exhibit a maximum near $\Gamma_{a/w} = 4 \times 10^{-10}$ mol cm^{-2} (41 Å^2 molecule^{-1}).

Systematic experimental errors such as leakage behind trough barriers during transfer, subphase impurities, and monolayer respreading to the water surface during reimmersion are readily eliminated as an explanation for $\theta < 1$. In addition, the disorder in the tails precludes substantial film passivation by the blocking of counterion diffusion through the hydrocarbon tails. Finally, molecular overturning (49) is not thought to occur. Film bridging over flaws in the Au substrate would lead to only a fraction of the ferrocene sites being in direct physical contact with the electrode. Those sites not in contact can only become oxidized by lateral electron self-exchange with neighboring ferrocene sites. Full participation of all electroactive sites may not occur. A similar explanation was reported by Matsuda and his co-workers (17b) in L–B films of surfactant derivatives of $[Ru(bipy)_3]^{2+}$ and $[Os(bipy)_3]^{2+}$ on optically transparent SnO_2 electrodes. Partial film dissolution or film expansion on the Au surface may be involved (15b). However, recent indirect coulometric measurements by Majda (60) of C_{18} viologen surfactant L–B films on Au showed that all sites which were transferred could be detected electrochemically by indirect coulometric assay, while fewer sites were directly engaged in electron transfer with the underlying electrode. It is unfortunately not possible at present to discern which of these explanations dominate.

The full width at half-maximum (fwhm) of the cyclic voltammetric wave at half-height (E_{fwhm}) for C_{18}Fc L–B monolayers increases linearly from 285 to 375 mV as $\Gamma_{a/w}$ increases from 2.6 to 5.1×10^{-10} mol cm^{-2}. These values are significantly larger than the ideal $E_{fwhm} = 90.6$ mV for a one electron transfer (57, 61) implying a strongly repulsive interaction parameter $r \ll 0$. Although a molecular significance has not been attributed to r, it is tempting to speculate that the increasing E_{fwhm} is related to greater steric crowding since E_{fwhm} correlates much better with $\Gamma_{a/w}$ than with Γ_{app}. The term E_{fwhm} is dramatically larger in C_{18}Fc L–B monolayers than in adsorbed C_{18}Fc monolayers (Section 3.3.3) *even at comparable coverages* suggesting profound differences in microstructure. A somewhat more appropriate model for explaining cyclic voltammetric peak widths and peak shape has been proposed by Tokuda and Matsuda and their co-workers (17b) in which a regular arrangement of electroactive sites can give rise to broad and even split waves when the interaction energy is sufficiently negative (repulsive). In this model broad cyclic voltammetric waves result from the greater stability of Red/Ox pairs than Red–Red or Ox–Ox pairs. The unusual increase in E_{fwhm} suggests an increasing C_{18}Fc coverage as $\Gamma_{a/w}$ increases.

3.3.1D Surface FTIR Spectroscopic Characterization

Grazing angle reflection–absorption Fourier transform infrared (RA–FTIR) spectroscopy is a very useful tool for the investigation of L–B monolayer (62) and self-assembled film (7, 8, 25) structures. The use of polarized IR radiation enables the study of the anisotropic structure of oriented films. Finite absorption of IR radiation will occur only for those vibrations that have a transition dipole moment with a component perpendicular to the plane of the substrate (62f). The RA–FTIR spectroscopy is done on C_{18}Fc films on Au that was electron beam deposited onto highly polished single-crystal ⟨100⟩ Si substrates. These Au substrates were considerably smoother than Au flags used in the voltammetric studies just described. The root-mean-square-smoothness of the vapor deposited Au electrodes was 21 Å as measured by Mireaux inter-

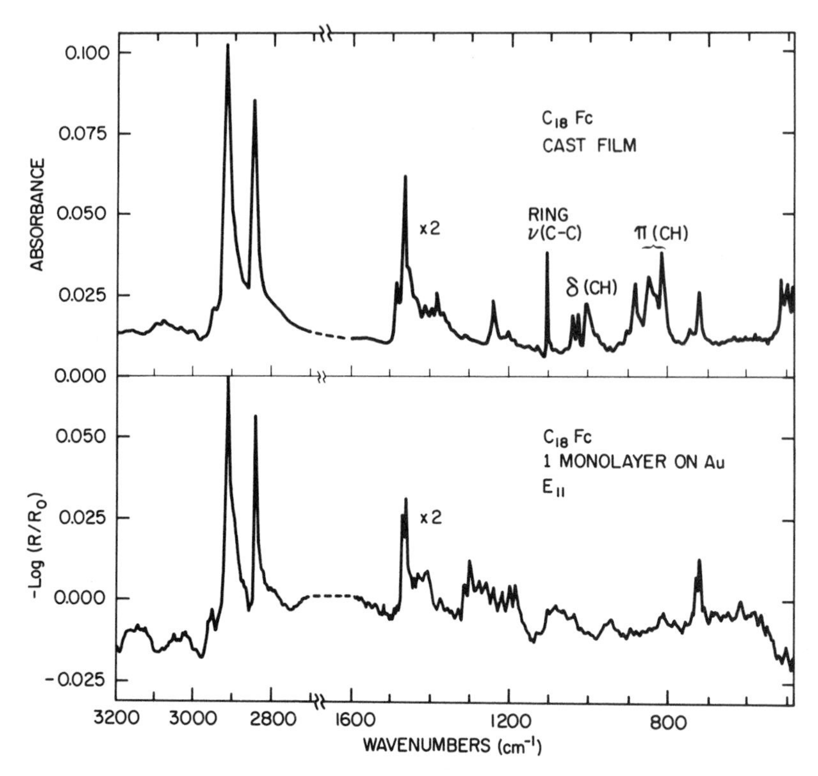

Figure 3.6. Comparison of the grazing incidence reflection–absorption FTIR spectra (p-polarized incident radiation) of bulk $C_{18}Fc$ (top) and a single monolayer of $C_{18}Fc$ transferred at $\pi_t = 22\,mN\,m^{-1}$ (bottom).

ferometry (63) employing a $1\,\mu m$ focused laser beam. Scanning tunneling microscopy, which has a much better lateral resolution, shows the presence of Au islands about 300 Å in diameter and about 150 Å high.

The RA–FTIR spectrum of a single monolayer of $C_{18}Fc$ transferred at $22\,mN\,m^{-1}$ heads down on Au (after voltammetry) is shown in Fig. 3.6 bottom and is compared with a RA–FTIR of bulk $C_{18}Fc$ in Fig. 3.6 top on an identically prepared substrate. Ellipsometry indicates a film thickness of 23 ± 2 Å. Differences between the cast film and the monolayer film are evident below $1600\,cm^{-1}$. A progression of closely spaced CH_2 wagging, twisting, and rocking absorbances observed at $1330–1170\,cm^{-1}$ corresponds to a methylene chain in an all-trans zigzag configuration (62). The monolayer CH_2 rocking vibration at 719 and $730\,cm^{-1}$ is consistent with an orthorhombic subcell packing (62b) as is the splitting of the CH_2 bending vibration at 1475 and $1460\,cm^{-1}$. The corresponding region of the bulk spectrum contains only a single adsorption at $1250\,cm^{-1}$ and a complex pattern at $1475–1490\,cm^{-1}$. Peaks corresponding to aliphatic CH_2 and CH_3 stretches are observed in the $2965–2825\text{-}cm^{-1}$ region appear similar to those in the bulk spectrum. The asymmetric and symmetric CH_2 stretching bands at 2920 and 2850 cm^{-1} are slightly different from bulk CH_2. A monolayer containing chains perpendicular to the surface shows CH_2 stretching bands of much lower intensity (8, 62b–c) while tilted chains would show a greater CH_2 absorption (62e). The ratio of the absorbances of the asymmetric and symmetric CH_2 stretching band relative to the asymmetric CH_3 stretch at 2963 cm^{-1} further indicate that the chains are not perpendicular to the surface and/or somewhat disordered. The monolayer CH_2 stretching region appears similar to cadmium arachidate monolayers just prior to melting (62c). These results suggest that the tails are disordered, which is consistent with the liquid state of the monolayer at the air–water interfaces and the difference in the head and tail cross-sectional areas. Ferrocene ring C—C stretch at 1100 cm^{-1}, out of plane C—H bend (1000–1040 cm^{-1}) and in plane C—H bend (800–850 cm^{-1}) observed in the RA–FTIR of the bulk $C_{18}Fc$ are absent in the spectrum of the monolayer. This is consistent with (though not proof of) ferrocene ring orientation perpendicular to the surface.

3.3.2 Remote Spacing of Ferrocene Sites

It is of theoretical and experimental interest to be able to space electroactive sites a remote but uniform distance from the electrode surface in terms of understanding the details of interfacial electron transfer. Previous studies also examined electroactive groups remotely spaced from atomically smooth Hg (32) and Au (10, 24b) electrodes. These strategies involved adsorption at electrodes of molecular spacer segments, which stiffly hold the electroactive moiety a uniform distance from the electrode. Previously, as here, spacer segments were close packed organized hydrocarbon chains. Weaver and his co-workers (32) employed short chains and found an exponential decrease in the rate of heterogeneous electron transfer as a function of electrode–redox site distance. Porter et al. (10) demonstrated that electron transfer to/from dissolved redox couples dramatically depended on the thickness of an essentially pinhole-free *n*-alkylthiol spacer layer.

This section describes an alternate approach, namely, that of utilizing L–B films for remotely spacing electroactive reagents. The four microstructures, which are built toward this end, are presented in Schemes 1–4 in Fig. 3.7. Scheme 1 results from transfer of two monolayers of $C_{18}Fc$ with the first layer transferred tails down to hydrophobic Au. Scheme 2 depicts a bilayer modified Au electrode in which a monolayer of cadmium stearate is used as a spacer for heads down $C_{18}Fc$. Schemes 3 and 4 depict bilayer modified electrodes in which $C_{18}Fc$ is introduced into the spacer layer.

In order to transfer L–B monolayers tails down (onto smooth electron beam deposited Au), hydrophobic Au surfaces are needed. Advantage was taken of the fact that sulfur containing small molecules irreversibly adsorb onto Au. The thiol, disulfide, and thiol acid adsorbates shown in Table 3.2 are therefore expected to orient methyl groups toward the aqueous phase, conferring hydrophobicity to the Au surface. Adsorption of these species begins by electrochemically precleaning Au in either $1M\,H_2SO_4$ or $0.1M$ KOH by 5–7

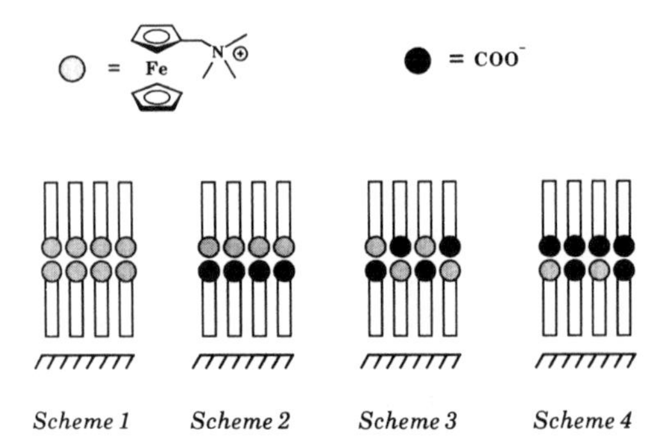

Figure 3.7. Schematic depiction of four schemes employing L–B bilayers used to achieve remote spacing of ferrocene groups from hydrophobic Au interfaces. Dark circles and open circles represent the ferrocene and carboxylate head groups, respectively.

potential scans between oxide formation and the onset of hydrogen evolution. (The potential limits in 0.1*M* KOH, e.g., are +0.5 and −1.1 V.) Electrodes are then immediately rinsed in monolayer grade H_2O and immersed for 5 min in a 1–2 vol% ethanolic solution of the adsorbate. Monolayer coated electrodes were then rinsed successively with ethanol and H_2O and the advancing contact angle measured in ambient air. The results are summarized in Table 3.2. Gold electrodes modified with 3-methylthiophene (Au/3MT), di-*tert*-butyldisulfide [Au/$(t\text{-BuS})_2$] and *tert*-butylmercaptan (Au/*t*-BuSH) had the highest hydrophobicity with a contact angle of 85–90°. Higher contact angles (113°) are observed at Au coated with a self-assembled monolayer of octadecylmercaptan, known to contain closest packed methyl groups at the monolayer surface. This implies that closest packing of methyl groups in these adlayers does not occur. Despite this fact, they are still useful.

In addition to its hydrophobic character, the adsorbate layer must not exhibit electrochemical waves and must be stable toward potential cycling in pH 6.6, 0.1*M* $NaClO_4$ electrolyte. Gold coated with thiolacetic acid exhibited a faradaic

Table 3.2. Water Contact Angles (Degrees) at Vapor Deposited Au Films Treated with Various Sulfur Containing Adsorbates

Electrolyte[a]	CH_3COSH	3-MT[b]	$(t\text{-BuS—})_2$	$(MeS—)_2$	*t*-BuSH	$Me_3Si—$[c]	TPA[d]
1*M* H_2SO_4	75	90	85–88	84–85	86		
1*M* H_2SO_4			80–81				
0.1*M* KOH	65	90	86–88	84	87–90	NR	88–89

[a]Electrolyte used to electrochemically clean Au.
[b]3-Methylthiophene.
[c]1-(Trimethylsilyl)imidazole.
[d]Thiopivalic acid, $(CH_3)_3CC(O)SH$.

wave at 0.5 V due to thiol oxidation or desorption and was therefore not useful. The electrodes Au/3MT, Au/(*t*-BuS)$_2$, and Au/*t*-BuSH exhibited the onset of background current at 0.7–0.75 V and were suitable as hydrophobic substrates. Experimentally, however, Au/3MT is found to yield superior transfer of tails down monolayers and is thus preferred over Au/(*t*-BuS)$_2$ and Au/*t*-BuSH despite the similarity in contact angles. Immersion of Au/3MT through C_{18}Fc monolayers and C_{18}Fc–stearic acid mixed monolayers is accompanied by a 90° meniscus contact angle while immersion of Au/(*t*-BuS)$_2$ through the same floating monolayer results in only about 70° meniscus contact angle and low transfer ratios. The reason for this behavior is not clear. Thus, Au/3MT substrates are used for the transfer of tails down monolayers to Au.

In principle, the configuration in Scheme 1 results in the spacing of ferrocene units about 25 Å from the electrode. However, the transfer of the first monolayer of C_{18}Fc tails down even on Au/3MT was invariably accompanied by transfer ratios significantly less than one and could therefore not be investigated in detail. It was therefore necessary to improve the transfer ratio. It was felt this could be best accomplished by improving the molecular packing of the tails by resorting to mixed monolayers containing stearic acid. The monolayer transfers in Schemes 2–4 were, in fact, accompanied by transfer ratios near unity.

In Scheme 2, a monolayer of cadmium stearate is first transferred tails down to Au/3MT, followed by heads down transfer of a C_{18}Fc monolayer. The presence of Cd^{2+} in the subphase stabilizes the tails down $C_{18}H_{37}CO_2^-$ monolayer. The Fig. 3.8 inset shows the transfer of a monolayer of cadmium stearate. The steep rise in the isotherm is characteristic of a close packed monolayer in the solid state. The deposition ratio is 1.04 indicating an ideal transfer. The stearate modified Au/3MT electrode is then totally submerged in the subphase, the surface swept clean of the floating monolayer and a monolayer of C_{18}Fc spread. As the Au/3MT/$C_{18}H_{37}CO_2^-$ substrate is withdrawn through the C_{18}Fc monolayer at 17 mN m^{-1}, the meniscus contact angle is nearly 0°, as expected, indicating that the tails down stearate monolayer did not desorb from the surface while submerged in the subphase. The transfer of C_{18}Fc (not shown) is similar to those shown in Fig. 3.3. A near unity deposition ratio is again observed. At pH > 6, stearate film anions are not protonated lending electrostatic cohesiveness between adjacent but oppositely charged first and second monolayers. The overall thickness of this two-monolayer surface structure is about 55 Å from molecular models. A 50–55-Å film thickness range is obtained ellipsometrically, assuming a refractive index of 1.50 ± 0.05. On the basis of the ellipsometry and transfer ratios, it is concluded that Scheme 2 depicts reasonably well the actual surface structure.

Figure 3.8 shows the cyclic voltammetry of the resulting Au/3MT/$C_{18}H_{37}CO_2^-$/C_{18}Fc structure in 0.1*M* Na_2SO_4 at pH 6.6. Cyclic voltammetric waves are not observed even at sweep rates as slow as 10 mV s^{-1}. Note the absence of background reactions for the oxidation of 3-MT in the potential window at this pH. These results indicate that the ferrocene head groups do not react at the Au/3MT interface and hence do not penetrate the stearate

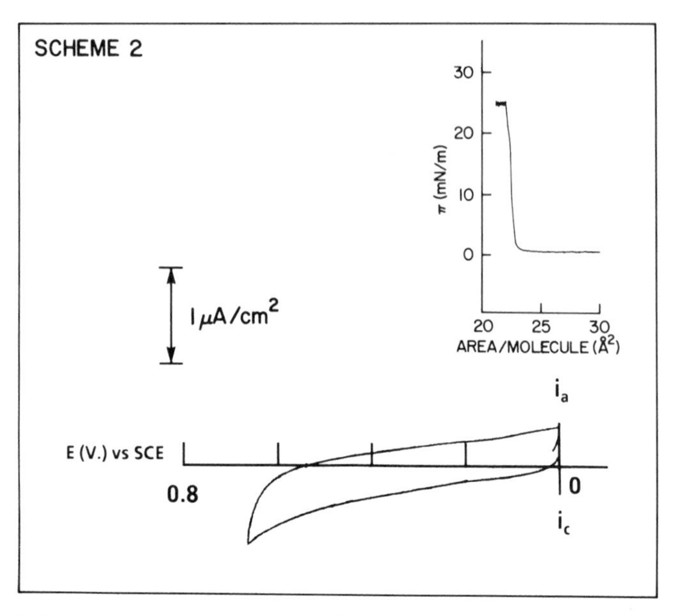

Figure 3.8. Cyclic voltammetric waves (20 mV s^{-1}, pH 6.6, 0.1 *M* Na_2SO_4) of a L–B film, fabricated as in Scheme 2, in Fig. 3.7, on hydrophobic Au (Au/3MT/$C_{17}H_{35}CO_2^-$/C_{18}Fc). Inset: π–A isotherm for the transfer of the stearate spacer layer on Au/3-MT.

monolayer. From the background current the interfacial capacitance is estimated to be 12–20 μF cm^{-2}. In contrast, the capacitance of virtually pinhole-free octadecylthiol monolayers is nearly 1 μF cm^{-2} (10). Thus there must certainly be pinholes in both the 3MT layer and in both monolayers. It appears that the cohesiveness of L–B monolayers and the ability of these monolayers to bridge defects prevents interlayer diffusion even in the presence of pinholes in the inner layer. The fact that ellipsometry shows no change in bilayer thickness after cyclic voltammetry demonstrates that the outer ferrocene containing layer has not desorbed when contacted with electrolyte. The other, perhaps more important, implication of these results is that electron transfer does not occur on the timescale of these experiments over a distance of about 25 Å. This is in contrast with the voltammetry of ferrocene, which is remotely spaced by the self-assembly of 18-(ferrocenyl)octadecylthiol (24b). In the latter system, well-developed cyclic voltammetric waves are observed and a measurable heterogeneous rate constant is found. One of the underlying differences that may account for the difference between these two systems is that the ferrocene sites in the latter study are connected via chemical bonds to the Au substrate. In the L–B bilayer films direct bonding does not exist so a through bond electron-transfer pathway (super exchange) is not possible.

In Scheme 3 of Fig. 3.7 C_{18}Fc molecules are introduced into the spacer layer. Here two mixed monolayers containing a 1:1, 1:2, and 1:4 molar ratio of C_{18}Fc and $C_{18}H_{37}CO_2^-$, respectively, are transferred by sequential immersion and

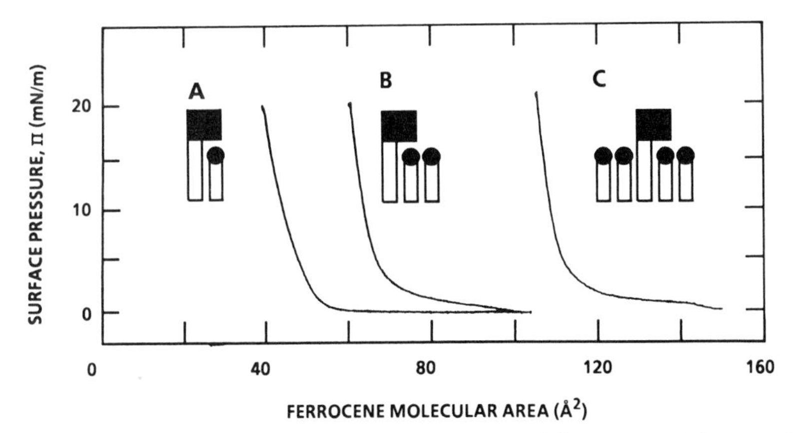

Figure 3.9. Pressure–area isotherms of mixed monolayers containing $C_{18}Fc$ and $C_{17}H_{35}CO_2^-$ in 1:1, 1:2, and 1:4 molar ratios at *A*, *B*, and *C*, respectively. Subphase composition: pH 6.8 $NaHCO_3$ buffer, $10^{-4}M$ $CdCl_2$. Symbols indicate schematic representation of molecular packing. Filled circles and rectangles represent carboxylate and ferrocene head groups, respectively.

emersion through the mixed monolayer. Figure 3.9*A–C* shows the π–A isotherms for mixed monolayers of $C_{18}Fc$ and $C_{18}H_{37}CO_2^-$ in 1:1 and 1:2 and 1:4 ratios, respectively, on a pH 6.8, $10^{-4}M$ $CdCl_2$ subphase. The molecular areas are calculated on a per $C_{18}Fc$ basis and include a contribution from stearate chains. The area occupied per ferrocene molecule is 46, 65, and 106 Å^2 for the 1:1, 1:2, and 1:4 mixed monolayers, respectively. The molecular area per ferrocene in the 1:1 mixed monolayer is similar to that in a pure monolayer on $0.1M$ Na_2SO_4 (51 Å^2 molecule^{-1}) in which molecular packing is maximized. In $C_{18}Fc$ the molecular area of the ferrocene head group is about 50 Å^2 molecule^{-1} while 21 Å^2 molecule^{-1} is attributed to the hydrocarbon tail. This approximate 30 Å^2 molecule^{-1} disparity in the head-and-tail areas implies that a second alkyl chain can be accommodated in this space in a 1:1 mixed monolayer. Incorporation of a stearate molecule thus leads to improved chain packing and organization. This is shown in the idealized structure Fig. 3.9*A*. The ferrocene molecular area in the 1:2 mixed monolayer is then easily explained in terms of an additional 21 Å^2 molecule^{-1} for the additional stearate in the 1:1 unit cell. A molecular area of 67 Å^2 molecule^{-1} for $C_{18}Fc$ in the 1:2 mixed monolayer would be predicted, while 65 Å^2 molecule^{-1} is in fact observed. Analogously, 109 Å^2 molecule^{-1} is predicted for a 1:4 mixed monolayer by the addition of three stearate anions to the 1:1 unit cell. The prediction of 109 Å^2 molecule^{-1} agrees well with the 106 Å^2 molecule^{-1} observed. While it is possible to conclude that stearate and $C_{18}Fc$ are molecularly miscible in 1:1 monolayers, it is not possible to conclude that the 1:2 and 1:4 monolayers are not composed of aggregates of 1:1 stearate–$C_{18}Fc$ and cadmium stearate because of the linear additivity of molecular areas (5a).

Figure 3.10*A* shows a family of cyclic voltammograms for Au/3MT modified with two layers of the 1:1 mixed monolayer fabricated according to Scheme 3.

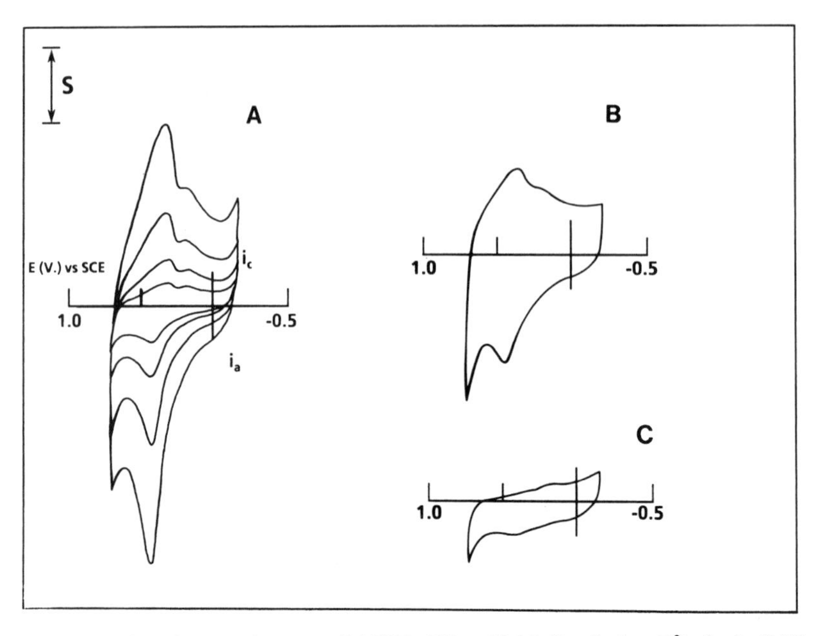

Figure 3.10. Cyclic voltammetric curves (0.1*M* Na_2SO_4, pH 6.6, S = 7 $\mu A\ cm^{-2}$) of a L–B bilayer film composed of varying mole ratios of $C_{18}Fc$ and $C_{17}H_{35}CO_2^-$ on Au/3MT as indicated in Scheme 3. Curves at *A*, *B*, and *C* are for 1:1, 1:2, and 1:4 molar ratios, respectively. A = 50, 100, 200, 400 $mV\ s^{-1}$; B = 400 $mV\ s^{-1}$; C 100 $mV\ s^{-1}$.

The deposition ratio of the inner layer is 0.7 while the transfer ratio for the outer layer was unity. At Au/3MT electrodes, a single anodic wave at +0.40 V is observed. Peak currents $i_{p,a}$ for the oxidation of ferrocene to ferrocenium were found to be linear with sweep rate. The coverage calculated from the charge under the anodic wave is 0.6×10^{-10} mol cm^{-2}, much less than the number of moles transferred (6.1×10^{-10} mol cm^{-2}). This corresponds to oxidation of 0.099 of the available ferrocene sites. Expecially intriguing are the *two* cathodic waves that are invariably observed in Scheme 3 modification. The more negative reduction wave is referred to with an asterisk. Note that $E_{p,a} = E_{p,c} = 0.4$ V and $E^*_{p,c} = 0.15$ V. The dependence of $i_{p,c}$ and $i^*_{p,c}$ on sweep rate is less than linear while the ratio of the two peak currents $i_{p,c}/i^*_{p,c}$ is strongly dependent on the sweep rate, decreasing as the sweep rate increases. Discussion of these results is postponed until later in this section.

Figure 3.10*B* shows a cyclic voltammogram at 400 $mV\ s^{-1}$ of the 1:2 bilayer film in Scheme 3. The wave at $E^{\circ\prime} = 0.4$ V is the dominant feature of voltammograms and the peak currents $i_{p,a}$ are linear with sweep rate. Much less pronounced is the wave at $E^*_{p,c} = 0.15$ V relative to the 1:1 mixed bilayer. The area under the anodic wave corresponds to about 0.3×10^{-10} mol cm^{-2}, which is 0.067 of the available 4.5×10^{-10} mol cm^{-2} of the ferrocene sites transferred from the air–water interface. Figure 3.10*C* shows a relatively featureless cyclic

Table 3.3. Fraction of Injecting Electroactive Sites as a Function of Ferrocene Site Dilution in Bilayers of Mixtures of $C_{18}Fc$ and Stearate Anions

Film[a]	$10^{10} \times \Gamma_{app}$[b] (mol cm^{-2})	$10^{10} \times \Gamma_0$[c] (mol cm^{-2})	Θ[d]
1:1	0.6 ± 0.1	6.1	0.099
1:2	0.3	4.5	0.067
1:4	0.1	2.7	0.037

[a]Ratio of $C_{18}Fc$ to $C_{17}H_{35}CO_2^-$.
[b]Electrochemically measured coverage.
[c]Moles transferred from the air–water interface.
[d]Γ_{app}/Γ_0.

voltammogram wave (100 mV s^{-1}) for the 1:4 mixed bilayer. A rough estimate of the charge under the anodic wave indicates about 4% of the available 2.7×10^{-10} mol cm^{-2} ferrocene sites transferred undergo direct electron transfer with the electrode. The electroactivity ratios θ are summarized in Table 3.3. The ratio θ decreases smoothly as the mixed bilayer becomes enriched in stearate film molecules. Thus, electron exchange between the underlying electrode and most of the film ferrocene sites is blocked at all dilutions. However, charge blocking is enhanced at higher $C_{18}Fc$ dilution. In view of the results of the pure stearate spacer, it is felt that the main wave at 0.4 V in the 1:1 and 1:2 mixtures is associated with the leakage of charge from the Fc sites at the monolayer–monolayer interface arising from physical defects in the spacer layer.

Finally, Fig. 3.11*A* and *B* illustrate a series of cyclic voltammograms between 20 and 800 mV s^{-1} for a Au electrode modified as shown in Scheme 4 in Fig. 3.7. In contrast to Scheme 3 modification, this bilayer exhibits split cathodic and anodic waves. At low sweep rates the wave at $E^*_{p,c} = 0.19$ V is dominant while at higher sweep rates the wave at $E_{p,c} = 0.4$ V becomes dramatically prominent. The coverage calculated from the anodic wave, 2×10^{-10} mol cm^{-2}, is sweep rate independent and represents about 60% of transferred ferrocene sites.

Scheme 2 modification leads to complete blocking of electron exchange between ferrocenes located about 25 Å from the electrode and the underlying electrode. Replacing some of the spacer stearate molecules, as shown in Scheme 3 with $C_{18}Fc$ results in a small fraction of charge exchange with the remote ferrocene sites. The presence of $C_{18}Fc$ seems to perturb the organization of the spacer layer relative to a pure stearate monolayer. This is readily apparent from the dilution study. It is thought that film defects (pinholes) lead to two paths for electron mediation, namely, molecular floppiness (trans-layer motion of ferrocene sites) and lateral electron transfer from ferrocene sites in contact with the electrode to ferrocene sites at the monolayer–monolayer interface. These

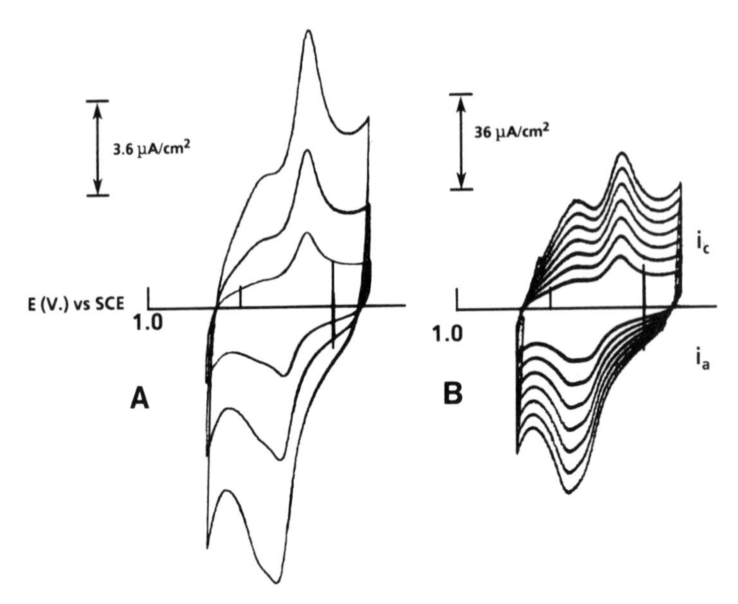

Figure 3.11. Cyclic voltammetric curves (pH 6.6, $0.1M$ Na_2SO_4) for a L–B bilayer film composed of an inner layer of 1:1 $C_{18}Fc$–$C_{17}H_{35}CO_2^-$ and an outer layer of pure $C_{17}H_{35}CO_2^-$. $A = 20$, 50, $100\,mV\,s^{-1}$; $B = 200$–$800\,mV\,s^{-1}$ in $100\,mV\,s^{-1}$ increments.

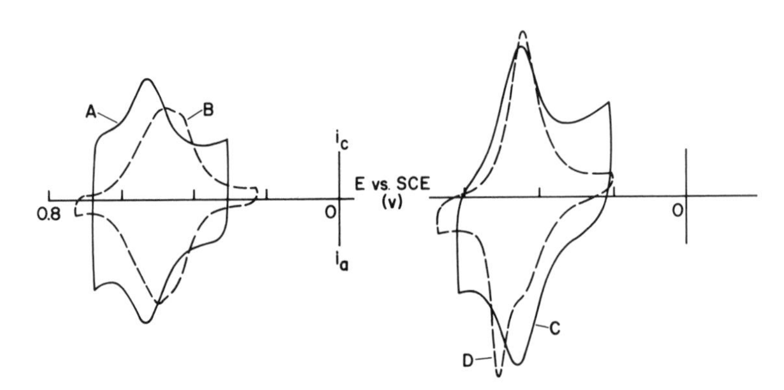

Figure 3.12. Cyclic voltammetric waves in $1M\,H_2SO_4$ (curves A and B) and $1M\,HClO_4$ (curves C and D) for $C_{18}Fc$ adsorbed from $1M\,H_2SO_4$ (A, B) and $1M\,HClO_4$, (C, D). In curves A and B, $c = 0.03\,\mu M$ and $0.2\,\mu M$ and $\Gamma = 0.33 \times 10^{-10}$ and $1.5 \times 10^{-10}\,mol\,cm^{-2}$, respectively. In curves C and D, $c = 0.12$ and $1.5\,\mu M$ and $\Gamma = 0.26 \times 10^{-10}$ and $1.7 \times 10^{-10}\,mol\,cm^{-2}$.

defects can be healed as the $C_{18}Fc$ sites are diluted with stearic acid and the nature of the spacer layer approaches that of a pure stearate layer (Scheme 2 in Fig. 3.7). Also contributing to the rapid drop in electron mediation with increased dilution is the lower concentration of ferrocene sites at the monolayer–monolayer interface. The variability in electron blocking–mediation in bilayers fabricated by Schemes 1–4 demonstrates the need to exercise care in the design of electroactive architectures on electrodes built up from L–B films.

The broadening and splitting of cyclic voltammetric peaks in organized films composed of a regular array of electroactive species was theoretically treated by Matsuda and his co-workers (17b). Other explanations include different ferrocene environments or geometrical arrangements, or ion pairing with stearate. We do not as yet have a satisfactory explanation of these results. We have observed similar multiple peak response during the reduction–oxidation of well-characterized L–B films of anthraquinone functionalized phospholipids in buffered aqueous chloride electrolyte (14c) that do not appear to be related to site periodicity. In the latter system the multiple peaks were attributed to the difficulty in counterion permeation through the tails in aggregated and expanded microheterogeneous phases of the transferred monolayer.

3.3.3 Adsorption from Aqueous Media

3.3.3A Adsorption from 1 M H_2SO_4 and 1 M $HClO_4$

All adsorption experiments are done on Pt bead and planar Pt electrodes modified with an adsorbed monolayer of iodine (Pt/I) as previously described (14b). Formation of a Pt–I surface is confirmed by X-ray photoelectron spectroscopy (XPS). The roughness factor of the Pt bead is 1.11 as determined from the charge under hydrogen underpotential deposition waves. Mettalographically polished Pt, by comparison, has a roughness factor of 1.3–1.4. Relative to unmodified Pt electrodes, Pt–I electrodes exhibit greatly suppressed background current, oxide formation and reduction, and underpotential hydrogen deposition in 1 M H_2SO_4 and 1 M $HClO_4$. Adsorption is conducted in a separate cell by incrementally dissolving methanolic aliquots of $C_{18}Fc$ in a 1 M H_2SO_4 or 1 M $HClO_4$ electrolyte. Rapid and irreversible adsorption of $C_{18}Fc$ at Pt–I occurs upon immersion of Pt–I into the $C_{18}Fc$ containing electrolyte.

Figure 3.12*A* illustrates the cyclic voltammetric response for the adsorption of submonolayer $C_{18}Fc$ ($\Gamma = Q_a/n\mathcal{F}A_t = 3.3 \times 10^{-11}$ mol cm^{-2}) in 1 M H_2SO_4 ($E^{\circ\prime}_{surf} = 0.52$ V) prepared from 0.030 μM $C_{18}Fc$ in 1 M H_2SO_4. Such monolayers are denoted $C_{18}Fc$–H_2SO_4. The $C_{18}Fc$–H_2SO_4 films are durable: No observable loss of ferrocene coverage occurs after 100 cycles in surfactant-free 1 M H_2SO_4; extended cycling (15,000 cycles) results in a loss of about 70% of film electroactivity due to the slow reaction of oxidized ferrocene with water (58). The appearance and behavior of all such submonolayer voltammetric waves are distinctly more ideal (57, 61, 64) than those observed in L–B films of $C_{18}Fc$ in 0.1 M Na_2SO_4 or in 1 M H_2SO_4. Increasing the surfactant concentration to

0.2 μM results in a higher $C_{18}Fc$ coverage ($\Gamma = 1.5 \times 10^{-10}$ mol cm^{-2}, Fig. 3.12*B*). The appearance and behavior of the higher coverage surface voltammetric waves is substantially different from the submonolayer waves as shown in curve B. These films are characterized by a second wave (Wave A), that is, anodic and cathodic shoulders superimposed on the negative potential side of the low coverage wave.

Films of $C_{18}Fc–HClO_4$, similarly prepared by adsorption from 1*M* $HClO_4$, display dramatically different voltammetric behavior from their $C_{18}Fc–H_2SO_4$ counterparts. Figure 3.12*C* presents the cyclic voltammetric response of a submonolayer $C_{18}Fc–HClO_4$ film ($\Gamma = 2.6 \times 10^{-11}$ mol cm^{-2}) adsorbed from 0.12 μM solution. These waves exhibit ideal surface behavior with coverage independent $E^{\circ\prime}_{surf} = 0.44$ V. Adsorption from 1.5 μM $C_{18}Fc$ yields a second anodic peak (Wave B) with no cathodic counterpart (Fig. 3.12*D*, $\Gamma = 1.7 \times 10^{-10}$ mol cm^{-2}). The peak potential $E_{p,a}$ of Wave B is 0.50 V, 60 mV positive of the formal potential of the low coverage wave and $i_p \propto v^{0.75}$.

Similar to L–B films of $C_{18}Fc$, the formal potentials of the various $C_{18}Fc–HClO_4$ and $C_{18}Fc–H_2SO_4$ waves are distinctly different from the $E^{\circ\prime}_{soln} = 0.36$ V of the model compound (ferrocenylmethyl)trimethylammonium sulfate, taken to be the formal potential of unadsorbed $C_{18}Fc$. The formal potential $E^{\circ\prime}_{surf}$ of $C_{18}Fc–H_2SO_4$ ($\Gamma = 1 \times 10^{-11}$ mol cm^{-2}) and submonolayer $C_{18}Fc–HClO_4$ are 160 and 90 mV positive of this reference value, respectively. The formal potential differences $\Delta E^{\circ\prime} = E^{\circ\prime}_{soln} - E^{\circ\prime}_{surf}$ (Eq. 3.4) are interpreted as above in terms of the reaction scheme in Eqs. 3.3a–c. For $C_{18}Fc–H_2SO_4$, $\Delta E^{\circ\prime} = -0.16$ V and thus $\Delta G^{\circ}_{R,ads} - \Delta G^{\circ}_{Ox,ads} = -3.7$ kcal mol^{-1}. Analogously $\Delta G^{\circ}_{Red,ads} - \Delta G^{\circ}_{Ox,ads} = -1.4$ kcal mol^{-1} for Wave A. For the $C_{18}Fc–HClO_4$ submonolayer, $\Delta E^{\circ\prime} = -0.09$ V and $\Delta G^{\circ}_{Red,ads} - \Delta G^{\circ}_{Ox,ads} = -2.0$ kcal mol^{-1}, while $\Delta E^{\circ\prime} = -150$ mV and $\Delta G^{\circ}_{Red,ads} - \Delta G^{\circ}_{Ox,ads} = -3.5$ kcal mol^{-1} for Wave B. Submonolayer $C_{18}Fc–H_2SO_4$ is therefore more strongly adsorbed to the surface than $C_{18}Fc–H_2SO_4$ in Wave A, assuming the strength of adsorption of Ox is the same. The opposite holds for $C_{18}Fc–HClO_4$. Submonolayer $C_{18}Fc–HClO_4$ is less strongly adsorbed than $C_{18}Fc–HClO_4$ in Wave B, again assuming the strength of adsorption of Ox is the same in both waves. Finally, Red is more strongly adsorbed than Ox in both $C_{18}Fc–H_2SO_4$ and $C_{18}Fc–HClO_4$ films. This may be attributed to a solvation effect if Red is located in a hydrophobic environment. It may also be rationalized in terms of a larger intermolecular coulombic repulsion between divalent oxidized ferrocene sites and a higher solubility of Ox than Red. Coulombic repulsion is thought to be responsible for the disruption of micelles of $C_{12}Fc$, a dodecyl(ferrocenylmethyl)dimethylammonium cation, in aqueous electrolyte when they are oxidized (65).

3.3.3B Coverage Dependent Cyclic Voltammetric Behavior

The full width at half-height E_{fwhm} and $E^{\circ\prime}_{surf}$ of $C_{18}Fc–H_2SO_4$ and $C_{18}Fc–HClO_4$ are dependent on coverage and electrolyte anion. The variation of $E^{\circ\prime}_{surf}$ in 1*M* H_2SO_4 is shown in Fig. 3.13*A*. The parameter $E^{\circ\prime}$ varies from +0.53 to

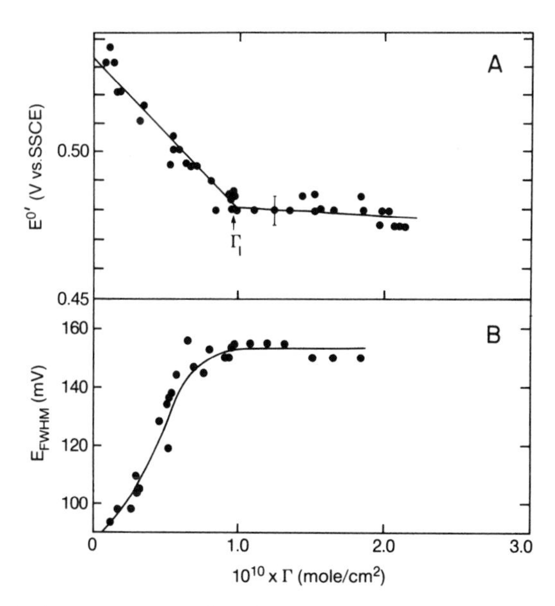

Figure 3.13. Plot of $E^{o'}_{surf}$ versus coverage (Panel *A*) and E_{fwhm} versus coverage (Panel *B*) for $C_{18}Fc-H_2SO_4$ films on Pt–I electrodes. The coverage Γ_1 corresponds to the knee of both curves. [Reprinted with permission from Facci, J. S., *Langmuir*, **1986**, *3*, 525. Copyright © (1986) American Chemical Society.]

$+0.48$ V as Γ increases from 1×10^{-11} to 1×10^{-10} mol cm^{-2} and levels off at higher coverages. The relative stabilization of Red decreases from -3.9 to -2.8 kcal mol^{-1} over this coverage range. Wave A is observed only when $\Gamma > 1 \times 10^{-10}$ mol cm^{-2}. In contrast, $E^{o'}_{surf}$ of submonolayer $C_{18}Fc-HClO_4$ is coverage independent. Similar to Wave A, Wave B is observed only when $\Gamma > 1 \times 10^{-10}$ mol cm^{-2}.

Specific anion effects are observed in the E_{fwhm}. Figure 3.13*B* shows the dependence of E_{fwhm} for $C_{18}Fc-H_2SO_4$. The term E_{fwhm} rises monotonically from a near ideal value of 94 mV at 1×10^{-11} mol cm^{-2} to 150 mV at about 1×10^{-10} mol cm^{-2} and levels off at 150–160 mV when $\Gamma > 1 \times 10^{-10}$ mol cm^{-2}. Estimates of E_{fwhm} at higher coverages are subject to uncertainty because of the overlapping Wave A shoulder. The parameter E_{fwhm} for $C_{18}Fc-HClO_4$ shows no significant systematic dependence on Γ. The knee of the curves in Fig. 3.13 are denoted by $\Gamma_1 \simeq 1 \times 10^{-10}$ mol cm^{-2}.

3.3.3C Adsorption Isotherm Behavior

In order to better understand the molecular structural states associated with Waves A and B, adsorption isotherms for $C_{18}Fc-H_2SO_4$ and $C_{18}Fc-HClO_4$ (Fig. 3.14*A*) were collected and interpreted in connection with contact angle data and the coverage-dependent behavior just described. Inspection of the figure readily reveals that the adsorption behavior and hence, structure is markedly

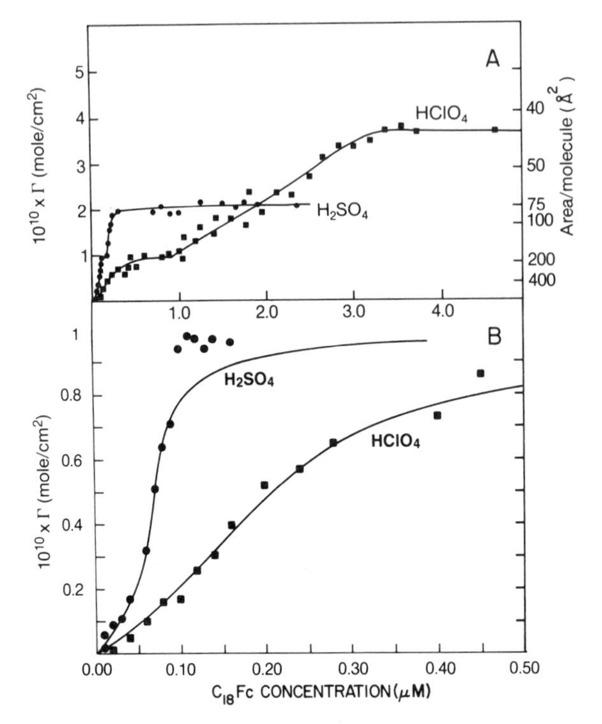

Figure 3.14. Panel *A*: Adsorption isotherm for $C_{18}Fc{-}H_2SO_4$ and $C_{18}Fc{-}HClO_4$ films. Panel *B*: Frumkin isotherm (Eq. 6) fit to the submonolayer data of Panel A. The solid line is the fit to the data points (■, ●). [Reprinted with permission from Facci, J. S., *Langmuir*, **1986**, *3*, 525. Copyright © (1986) American Chemical Society.]

dependent on the nature of the electrolyte anion. In the $C_{18}Fc{-}H_2SO_4$ isotherm, two concentration independent coverages are observed at $\Gamma_1 = 0.98 \times 10^{-10}$ and $\Gamma_2 = 2 \times 10^{-10}$ mol cm^{-2}, that is $\Gamma_2 \simeq 2\Gamma_1$. The short plateau at Γ_1 was highly reproducible. The two steps in coverage for $C_{18}Fc{-}H_2SO_4$ from the origin to about Γ_1 and from Γ_1 to Γ_2 are of similar slope. Interestingly Γ_1 corresponds to the knee of the curves in Fig. 3.13. The two steps in coverage for $C_{18}Fc{-}HClO_4$ from the origin to Γ_1 and from Γ_1 to $\Gamma_3 = 3.7 \times 10^{-10}$ mol cm^{-2} are much less steep than those in the $C_{18}Fc{-}H_2SO_4$ isotherm.

Molecular models suggest that $C_{18}Fc$ lies flat on the electrode surface at $\Gamma_1 = 169$ Å^2 molecule^{-1}. Interestingly, π–A isotherms of stearic acid at the air–Hg interface demonstrated that stearic acid occupies a large molecular area on mercury, consistent with the molecule lying flat on the Hg surface (66). Interactions of the hydrocarbon tail with these hydrophobic surfaces predominate at low coverage. The threshold for the observation of "Wave A" is Γ_1.

Below this coverage only the monolayer wave is observed. Between Γ_1 and Γ_2, the height of Wave A becomes more pronounced and the peak current for two waves become approximately equal at Γ_2. The ratio of Wave A to the submonolayer wave is independent of sweep rate unlike the multiple wave voltammograms in Schemes 2–4 above and $C_{18}Fc{-}HClO_4$ (see below). The similar appearance of the $C_{18}Fc{-}H_2SO_4$ isotherm between the origin and Γ_1 and from Γ_1 to Γ_2 suggest that a second monolayer is deposited over the first. The hydrocarbon tails in the overlayer should also be flatly oriented on the electrode based on the differential coverage $\Gamma_2 - \Gamma_1$. However, water contact angles with $C_{18}Fc{-}H_2SO_4$ films ($\Gamma > 1.5 \times 10^{-10}$ mol–cm^{-2}) indicate a hydrophilic surface at higher coverage, which would not be consistent with a two monolayer model.

A full monolayer of flatly oriented $C_{18}Fc$ is adsorbed onto Pt–I from $1 M$ $HClO_4$ at $\Gamma = \Gamma_1$. On the other hand, molecular models and the π–A isotherm behavior indicate that $C_{18}Fc$ is oriented perpendicular or nearly perpendicular to the surface at $\Gamma_3 = 45$ Å^2 molecule^{-1} since this closely corresponds to the projected area of the ferrocene head group (cf. Sections 3.3.1 and 3.3.2). Thus the $C_{18}Fc{-}HClO_4$ reorientation from flat to perpendicular is likely occurring between Γ_1 and Γ_3. The charged polar head group is likely oriented toward the aqueous phase as the resultant surface is hydrophilic. Wave B is thought to correspond to an organized phase in which $C_{18}Fc$ is more strongly adsorbed to the surface than the unorganized phase, as described above. Wave B becomes more pronounced as Γ approaches Γ_3, while the peak at 0.45 V diminishes. The organized phase at 0.52 V and the unorganized phase at $E^{\circ\prime}_{surf} = 0.45$ V may be in equilibrium with one another. Oxidation of the ferrocene film may disrupt the organized phase as the peak potential, shape and sweep behavior of the reduction wave is independent of coverage over the entire coverage range. As the sweep rate decreases, Wave B becomes more pronounced so that at very low sweeps ($v < 10$ mVs^{-1}) and $\Gamma \sim \Gamma_3$, the wave at 0.45 V is inconspicuous. During steady state fast sweeps insufficient time exists for reorganization of the disrupted phase and the film appears in its unorganized phase. With slow potential sweeps reformation of an organized phase is more possible and peak currents for Wave B predominate. A similar faradaically induced disruption was observed in dodecyl(ferrocenylmethyl)dimethylammonium micelles in water (65).

The steeper rise in the $C_{18}Fc{-}H_2SO_4$ isotherm relative to $C_{18}Fc{-}HClO_4$ stems from the larger equilibrium constant K for the reaction $Red_{soln} \rightleftharpoons Red_{ads}$ in $1 M$ H_2SO_4. It is thought that this is a result of the approximately 10-mM SO_4^{2-} present from dissociation of HSO_4^- in $1 M$ H_2SO_4. Divalent SO_4^{2-} electrostatically compensates two adsorbate molecules, whereas ClO_4^- charge compensates only one. Counterion effects on the stability of ionized surfactant films adsorbed at the air–water interface have long been known (5a). For example, ionized *n*-octadecylammonium monolayers at the air–water interface are stabilized by the presence of SO_4^{2-} relative to Cl^- ($10^{-3} M$ H_2SO_4) (68). In order to quantify the intermolecular interactions in these films, the data were fit to the Langmuir equation, Eq. 3.5:

$$\theta(1 - \theta)^{-1} = Kc \tag{3.5}$$

where θ is defined as Γ/Γ_1 and c is the solution concentration. Plots of $\theta/(1 - \theta)$ versus concentration were nonlinear, increasing monotonically with concentration implying attractive intermolecular interactions. One measure of intermolecular interaction in this case is obtained by assuming a coverage dependent equilibrium constant that varies exponentially with increasing coverage as embodied in the Frumkin isotherm (57), Eq. 3.6.

$$Kc = [\Gamma/(\Gamma_s - \Gamma)]\exp(2g\Gamma/RT) \tag{3.6}$$

In Eq. 3.6, the saturation coverage $\Gamma_s = \Gamma_1$, K is the equilibrium constant in the absence of intermolecular interactions (i.e., extrapolated to zero coverage) and g is a measure of the coverage dependence of the adsorption energy. Our results are expressed as g' values, where $g' = 2g\Gamma_s/RT$. Negative and positive values of g' signify attractive and repulsive intermolecular interactions, respectively. The fit of the submonolayer $C_{18}Fc{-}H_2SO_4$ and $C_{18}Fc{-}HClO_4$ data is very good and is shown in Fig. 3.14*B*. The data for adsorption from $1M\ H_2SO_4$ was best fit by $K = 3.2 \pm 0.4 \times 10^6 M^{-1}$ and $g' = -2.84 \pm 0.08$ while adsorption from $1M\ HClO_4$ was best fit by $K = 1.6 \pm 0.2 \times 10^6 M^{-1}$ and $g' = -1.78 \pm 0.20$ (Table 3.4). The zero coverage equilibrium constant for the reaction $Red_{soln} \rightleftharpoons Red_{ads}$ is twice as large for adsorption from $1M\ H_2SO_4$ and g' is distinctly more negative implying stronger attractive intermolecular interactions. Also summarized in Table 3.4 as a function of electrolyte are the free energies of adsorption for $Red_{soln} \rightleftharpoons Red_{ads}$ (in the absence of intermolecular attractions) determined from $\Delta G^\circ_{Red,ads} = -RT\ln K$. Similarly, the free energy of adsorption $\Delta G^\circ_{Ox,ads}$ for the reaction $Ox_{soln} \rightleftharpoons Ox_{ads}$ is determined from the expression $\Delta G^\circ_{Ox,ads} = \Delta G^\circ_{Red,ads} - \Delta(\Delta G^\circ_{ads})$. Values of $\Delta G^\circ_{Ox,ads}$ for adsorption from $1M\ H_2SO_4$ and $1M\ HClO_4$ are -5.0 and $-6.6\ \text{kcal mol}^{-1}$, respectively

Table 3.4. Summary of Adsorption Parameters for the Adsorption of $C_{18}Fc$ at Pt/I Electrodes

Electrolyte	g'^{a}	$K\,(M^{-1})$	$\Delta G^\circ_{Red,ads}$ [b]	$\Delta(\Delta G^\circ_{ads})$ [c]	$\Delta G^\circ_{Ox,ads}$ [d]
$1M\ H_2SO_4$	-2.84 ± 0.08	$3.2 \pm 0.4 \times 10^6$	-8.9	-3.9	-5.0
$1M\ HClO_4$	-1.78 ± 0.20	$1.6 \pm 0.2 \times 10^6$	-8.5	-2.0	-6.6

[a] $g' = 2g\Gamma_s/RT$ (see Eq. 3.5).
[b] $\Delta G^\circ_{Red,\ ads} = -RT\ln K$. (All ΔG values in kcal mol^{-1}.)
[c] $\Delta(\Delta G^\circ_{ads}) = E^{\circ\prime}_{soln} - E^{\circ\prime}_{ads}$.
[d] $\Delta G^\circ_{Ox,ads} = \Delta G^\circ_{Red,ads} - \Delta(\Delta G^\circ_{ads})$.

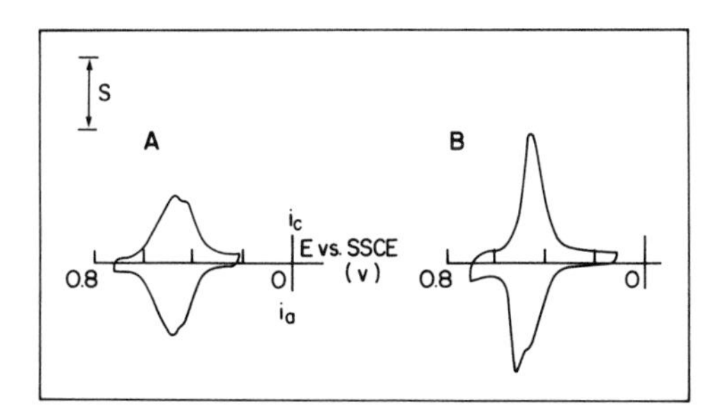

Figure 3.15. *A*: Cyclic voltammetric waves for $C_{18}Fc-H_2SO_4$ ($\Gamma = 1.8 \times 10^{-10}\,mol\,cm^{-2}$, in $1M\,H_2SO_4$. *B:* Cyclic voltammetric waves for the $C_{18}Fc-H_2SO_4$ film in *A* when transferred to $1M\,HClO_4$. $S = 3.8\,\mu A\,cm^{-2}$, $50\,mV\,s^{-1}$.

(Table 3.4), significantly larger than the difference between $\Delta G^{\circ}_{Red,ads}$ for $C_{18}Fc-H_2SO_4$ ($-8.9\,kcal\,mol^{-1}$) and $C_{18}Fc-HClO_4$ ($-8.5\,kcal\,mol^{-1}$).

As noted above, $E_{fwhm} > 90.6\,mV$ for all adsorbates studies because of surface activity (61, 64) or organizational (67) effects. Negligible electroactive center interactions ($E_{fwhm} = 90.6/n\,mV$) are observed in the limit of zero coverage for $C_{18}Fc-H_2SO_4$. Although it is not possible to unambigously interpret these results at a molecular level, they are consistent with the notion that electrostatic binding strongly influences the nature of $C_{18}Fc-H_2SO_4$ films: E_{fwhm} is generally higher in $C_{18}Fc-H_2SO_4$ than in $C_{18}Fc-HClO_4$, indicating a larger average intermolecular repulsive interaction (61) or higher degree of organization (67). The dichotomy of repulsive ferrocene–ferrocene interactions and the attractive tail interactions ($g' < 0$), although not understood in detail, appear to result from molecular organization that hold both the ferrocene sites and the alkyl chains in close proximity.

3.3.3D Electrolyte Induced Surface Reorganization

Desorption of $C_{18}Fc$ films is slow on the cyclic voltammetric timescale. Hence, they can be removed from the surfactant containing electrolyte and transferred to a separate electrolyte solution containing the same anion but no surfactant with no discernible loss in coverage. The cyclic voltammograms of $C_{18}Fc-H_2SO_4$ and $C_{18}Fc-HClO_4$ are surface morphology specific. Surprisingly, film reorganization is found when a $C_{18}Fc-H_2SO_4$ film is transferred to pure $1M\,HClO_4$ electrolyte. At $\Gamma > \Gamma_1$, for example, the cyclic voltammogram of a $C_{18}Fc-H_2SO_4$ film (Fig. 3.15*A*) electrode converts immediately to that associated with a $C_{18}Fc-HClO_4$ film (Fig. 3.15*B*). When the electrode is then transferred back to H_2SO_4, the voltammetry remains that associated with $C_{18}Fc-HClO_4$. Similarly, when $C_{18}Fc-HClO_4$ is transferred to surfactant free $1M\,H_2SO_4$ the voltammetry does not change to that associated with $C_{18}Fc-H_2SO_4$. Thus interconversion between the two surface structures does not occur. The structure of $C_{18}Fc-H_2SO_4$ is easily converted to a partially organized phase at coverages $\Gamma > \Gamma_1$. However, the organized structure of

$C_{18}Fc-HClO_4$ does not convert to the $C_{18}Fc-H_2SO_4$ structure at any coverage. The activation energy for the latter process must be prohibitive.

3.4 SUMMARY

Organized films of $C_{18}Fc$, an electroactive surfactant, were prepared on Au electrodes using the Langmuir–Blodgett transfer technique. The L–B technique was used to prepare monolayers in which the electroactive head groups were in contact with the electrode and the density of packing was controlled by the surface pressure at transfer. The electrochemistry was shown to be sensitive to the molecular packing density and not all sites that were transferred were electroactive. The L–B technique was also used in investigations involving controlled molecular orientation. Mixed L–B bilayers of several compositions were prepared on hydrophobic Au electrodes in which the ferrocene group is spaced nominally by 25 Å from the surface. The cyclic voltammetric response of the bilayer is strongly dependent on its microstructure. For example, electron transfer is shut down in an L–B bilayer composed of an inner layer of tails down stearate and an outer layer of heads down $C_{18}Fc$. Defects in other bilayers that contain $C_{18}Fc$ in the spacer layer may be responsible for electron mediation and a complex voltammetric response. This uncertainty in the structure of L–B films, especially when contacted by an electrolyte solution, is perhaps one of the major drawbacks of this technique.

The amphiphile $C_{18}Fc$ was adsorbed from aqueous acidic solution. The thermodynamics of adsorption and monolayer structure were investigated in detail. The voltammetry of adsorbed $C_{18}Fc$ was remarkably sensitive to the nature of the electrolyte, the level of adsorption, and sweep rate. Adsorption from $1 M$ H_2SO_4 appears to yield self-organized monolayers whose structure is electrolyte dependent.

Investigating heterogeneous electron-transfer reactions over distances less than 25 Å is probably best done by utilizing a variable length spacer (i.e., a self-assembled film) and an immobilized overlayer. The use of an immobilized electroactive overlayer results in simplification of the interpretation of electrode kinetics (69). Possible microstructures toward this end could involve either a combination of self-assembly and L–B deposition or attachment of a terminal electroactive species to a self-assembled monolayer (24b). We envision the use of self-assembled layers as providing a variable length ultrathin spacer (inner layer) in combination with a cohesive electroactive L–B film outer layer. Such molecular architecture could be structurally advantageous over the use L–B films or self-assembled films alone. Such studies are in progress in our laboratory.

ACKNOWLEDGMENTS

The support and encouragement of Dr. T. W. Smith and Dr. W. M. Prest, Jr., during the course of this work is very gratefully acknowledged. This work owes much to the experimental efforts of my colleagues, Dr. C. C. Chen and Dr. J. M. Gold, and Mr. R. Mosher and Mr. P. A. Falcigno.

3.5 REFERENCES

1. (a) Lane, R. F.; Hubbard, A. T. *J. Phys. Chem.* **1974**, *77*, 1405; (b) Lane, R. F.; Hubbard, A. T. *J. Phys. Chem.* **1974**, *77*, 1411.
2. Moses, P. R.; Murray, R. W. *J. Am. Chem. Soc.* **1976**, *98*, 7435.
3. Murray, R. W. in *Electroanalytical Chemistry*, Bard, A. J. (Ed.), Marcel Dekker: New York (1984), Vol. 13, p. 191, and references cited therein.
4. (a) Scharfe, M. *Electrophotographic Principles and Optimization*, Wiley: New York (1984); (b) Mort, J. *Science* **1980**, *20*, 819; (c) Mort, J.; Pai, D. M. (Eds.), *Photoconductivity and Related Phenomena*, Elsevier: Amsterdam (1976).
5. (a) Gaines, G. L., Jr.; *Insoluble Monolayers at Liquid–Gas Interfaces*, Interscience: New York (1966); (b) Vincett, P. S.; Roberts, G. G. *Thin Solid Films* **1980**, *68*, 135; (c) Roberts, G. G. *Langmuir–Blodgett Films*, Plenum: New York (1990); (d) Roberts, G. G. *Adv. Phys.* **1985**, *34*, 475; (e) Knobler, C. M., in *Advances in Chemical Physics*, Prigogine, I; Rice, S. A. (Eds.), Wiley: New York (1990) 397.
6. For a review of activity in this field see the Proceedings of the 1st, 2nd, 3rd, and 4th International Conferences on Langmuir–Blodgett Films: (a) *Thin Solid Films* **1980**, *68*; (b) *Thin Solid Films* **1983**, *99*; (c) *Thin Solid Films* **1985**, *132*–134; (d) *Thin Solid Films* **1988**, *160*.
7. (a) Sagiv, J. *J. Am. Chem. Soc.* **1980**, *102*, 92; (b) Maoz, R.; Sagiv, J. *J. Colloid Interface Sci.* **1984**, *100*, 465; (c) Gun, J.; Iscovici, R.; Sagiv, J. *J. Colloid Interface Sci.* **1984**, *101*, 201; (d) Netzer, L.; Iscovici, R.; Sagiv, J. *Thin Solid Films*, **1983**, *100*, 67; (e) Cohen, S. R.; Naaman, R.; Sagiv, J. *J. Phys. Chem.* **1986**, *90*, 3054; (f) Maoz, R.; Sagiv, J. *Langmuir* **1987**, *3*, 1034; (g) Maoz, R.; Sagiv, J. *Langmuir* **1987**, *3*, 1045; (h) Sabatani, E.; Rubinstein, I.; Maoz, M.; Sagiv, J. *J. Electroanal. Chem.* **1987**, *219*, 365.
8. (a) Troughton, E. B.; Bain, C. B.; Whitesides, G. M.; Nuzzo, R. G.; Allara, D. L.; Porter, M. D. *Langmuir* **1988**, *4*, 365; (b) Bain, C. D.; Biebuyck, H. A.; Whitesides, G. M. *Langmuir* **1989**, *5*, 723; (c) Bain, C. D.; Whitesides, G. M. *Langmuir* **1989**, *5*, 1370; (d) Bain, C. D.; Whitesides, G. M. *Science* **1988**, *240*, 62; (e) Bain, C. D.; Whitesides, G. M. *J. Am. Chem. Soc.* **1988**, *110*, 3665; (f) Bain, C. D.; Whitesides, G. M. *J. Am. Chem. Soc.* **1988**, *110*, 5897; (g) Bain, C. D.; Whitesides, G. M. *J. Am. Chem. Soc.* **1988**, *110*, 6560; (h) Bain, C. D.; Troughton, E. B.; Tao, Y.-T.; Evall, J.; Whitesides, G. M.; Nuzzo, R. G. *J. Am. Chem. Soc.* **1989**, *111*, 321; (i) Strong, L.; Whitesides, G. M. *Langmuir* **1988**, *4*, 546; (j) Wasserman, S. R.; Tao, Y.-T.; Whitesides, G. M. *Langmuir* **1989**, *5*, 1074.
9. Novotny, V.; Swalen, J. D.; Rabe, J. P. *Langmuir* **1989**, *5*, 485.
10. Porter, M. D.; Bright, T. B.; Allara, D. L.; Chidsey, C. E. D. *J. Am. Chem. Soc.* **1987**, *109*, 3559.
11. Swalen, J. D.; Allara, D. L.; Andrade, J. D.; Chandross, E. A.; Garoff, S.; Israelachvilli, J.; McCarthy, T. J.; Murray, R.; Pease, R. F.; Rabolt, J. F.; Wynne, K. J. *Langmuir* **1987**, *3*, 932.
12. See Ref. 6 for an excellent overview of the application of these and other characrerization techniques to LωB films.
13. Faulkner, L. R. *Chem. Eng. News* **1984**, 28.
14. (a) Facci, J. S.; Falcigno, P. A.; Gold, J. M. *Langmuir* **1986**, *2*, 732; (b) Facci, J. S. *Langmuir* **1987**, *3*, 525; (c) Liu, M. D.; Leidner, C. F.; Facci, J. S. *J. Phys. Chem.*, in press; (d) Facci, J. S.; Chen, C. C., unpublished results.
15. (a) Lee, C.-W., Bard, A. J. *J. Electroanal. Chem.* **1988**, *239*, 441; (b) Zhang, X.; Bard, A. J. *J. Am. Chem. Soc.* **1989**, *111*, 8098; (c) Zhang, X.; Bard, A. J. *J. Phys. Chem.* **1988**, *92*, 5566.
16. (a) Fujihara, M.; Aoki, K.; Inoue, S.; Takemura, H.; Muraki, H.; Aoyagui, S. *Thin Solid Films* **1985**, *132*, 221; (b) Fujihara, M.; Nishiyama, K.; Yamada, H. *Thin Solid Films* **1985**, *132*, 77; (c)

Fujihira, M.; Poosittisak, S. *J. Electroanal. Chem.* **1986**, *199*, 481; (d) Fujihara, M.; Araki, T. *J. Electroanal. Chem.* **1986**, *205*, 329; (e) Nishiyama, K.; Fujihara, M. *Chem. Lett.* **1987**, 1443; (f) Fujihira, M.; Araki, T. *Bull, Chem. Soc. Jpn.* **1986**, *59*, 2375; (g) Fujihira, M.; Araki, T. *Chem. Lett.* **1986**, 921; (h) Fujihira, M. Nishiyama, K.; Hamaguchi, Y. J. *Chem. Soc., Chem. Commun.* **1986**, 823.

17. (a) Daifuku, H.; Aoki, K.; Tokuda, K.; Matsuda, H. *J. Electroanal. Chem.* **1985**, *183*, 1; (b) Aoki, K.; Tokuda, K.; Matsuda, H. *J. Electroanal. Chem.* **1986**, *199*, 69; (c) Daifuku, H.; Yoshimura, I.; Hirata, I.; Aoki, K.; Tokuda, K.; Matsuda, H. *J. Electroanal. Chem.* **1986**, *199*, 47.

18. (a) Miller, C. J.; Majda, M. *J. Am. Chem. Soc.* **1986**, *108*, 3118; (b) Miller, C. J.; Widrig, C. A.; Charych, D. H.; Majda, M. *J. Phys. Chem.* **1988**, *92*, 1928; (c) Widrig, C. A.; Majda, M. *Anal. Chem.* **1987**, *59*, 754; (d) Miller, C. J.; Majda, M. *Anal. Chem.* **1988**, *60*, 1168; (e) Widrig, C. A.; Majda, M. *Langmuir* **1989**, *5*, 689; (f) Goss, C. A.; Miller, C. J.; Majda, M. *J. Phys. Chem.* **1988**, *92*, 1937; (g) Widrig, C. A; Miller, C. J.; Majda, M. *J. Am. Chem. Soc.* **1988**, *110*, 2009.

19. (a) Donohue, J. J.; Buttry, D. A. *Langmuir* **1989**, *5*, 671; (b) De Long, H. C.; Buttry, D. A. *Langmuir* **1990**, *6,* 1319.

20. (a) Finklea, H. O.; Robinson, L. R.; Blackburn, A.; Richter, B.; Allara, D.; Bright, T. *Langmuir* **1986**, *2*, 239; (b) Finklea, H. O.; Avery, S.; Lynch, M.; Furtsch, T. *Langmuir* **1987**, *3*, 409; (c) Finklea, H. O.; Snider, D. A.; Fedyk, J. *Langmuir* **1990**, *6*, 371.

21. (a) Sabatani, E.; Rubinstein, I. *J. Phys. Chem.* **1987**, *91*, 6663; (b) Rubinstein, I.; Steinberg, S.; Tor, Y.; Shanzer, A.; Sagiv, J. *Nature (London)* **1988**, *332*, 426; (c) Sabatani, E.; Rubinstein, I.; Maoz, M.; Sagiv, J. *J. Electroanal. Chem.* **1987**, *219*, 365.

22. (a) Quintela, P. A.; Kaifer, A. E. *Langmuir* **1987**, *3*, 769; (b) Quintela, P. A.; Diaz, A.; Kaifer, A. E. *Langmuir* **1988**, 4, 663; (c) Garcia, O. J.; Quintela, P. A.; Kaifer, A. E. *Anal. Chem.* **1989**, *61*, 979; (d) Diaz, A; Kaifer, A. E. *J. Electroanal. Chem.* **1988**, *249*, 333.

23. (a) Bunding-Lee, K. A.; Mowrey, R.; McLennan, G.; Finklea, H. O. *J. Electroanal. Chem.* **1988**, *246*, 217; (b) Bunding-Lee, K. A. *Langmuir* **1990**, *6*, 709.

24. (a) Harris, A. L.; Chidsey, C. E. D.; Levinos, N. J.; Loiacono, D. N. *Chem. Phys. Lett.* **1987**, *141*, 350; (b) Chidsey, C. E. D. *Science*, in press; (c) Chidsey, C. E. D.; Loiacono, D. N. *Langmuir* **1990**, *6*, 682.

25. (a) Allara, D. L.; Nuzzo, R. G. *Langmuir* **1985**, *1*, 45; (b) Allara, D. L.; Nuzzo, R. G. *Langmuir* **1985**, *1*, 52; (c) Allara, D. L.; Nuzzo, R. G. *Langmuir* **1985**, *1*, 45; (d) Allara, D. L.; Nuzzo, R. G. *Langmuir* **1985**, *1*, 52; (e) Nuzzo, R. G.; Dubois, L. H.; Allara, D. L. *J. Am. Chem. Soc.* **1990**, *112*, 558; (f) Dubois, L. H.; Zegarski, B. R.; Nuzzo, R. G. *J. Am. Chem. Soc.* **1990**, *112*, 570; (g) Nuzzo, R. G.; Allara, D. L. *J. Am. Chem. Soc.* **1983**, *105*, 4481; (h) Nuzzo, R. G.; Zegarski, B. R.; Dubois, L. H. *J. Am. Chem. Soc.* **1987**, *109*, 733; (i) Nuzzo, R. G.; Fusco, F. A.; Allara, D. L. *J. Am. Chem. Soc.* **1987**, *109*, 2358.

26. (a) Tillman, N.; Ulman, A.; Schildkraut, J. S.; Penner, T. L. *J. Am. Chem. Soc.* **1988**, *110*, 6136; (b) Ulman, A.; Eilers, J. E.; Tillman, N. *Langmuir* **1989**, *5*, 1147; (c) Tillman, N.; Ulman, A.; Penner, T. L. *Langmuir* **1989**, *5*, 101; (d) Tillman, N.; Ulman, A.; Elman, J. F. *Langmuir* **1989**, *5*, 1020; (e) Ulman, A.; Tillman, N. *Langmuir* **1989**, *5*, 1418.

27. Barner, B. J.; Corn, R. M. *Langmuir* **1990**, *6*, 1023.

28. Diem, T.; Czajka, B.; Weber, B.; Regen, S. L. *J. Am. Chem. Soc.* **1986**, *108*, 6094.

29. (a) Yokota, T.; Itoh, K.; Fujishima, A. *J. Electroanal. Chem.* **1987**, *216*, 289; (b) Okahata, Y.; Yokobori, M.; Ebara, Y.; Ebato, H.; Ariga, K. *Langmuir* **1990**, *6*, 1148.

30. (a) Lee, H.; Kelpley, L. J.; Hong, H. G.; Akhter, S.; Mallouk, T. E. *J. Phys. Chem.* **1988**, *92*, 2597; (b) Putvinski, T. M.; Schilling, M. L.; Katz, H. E.; Chidsey, C. E. D.; Mujsce, A. M.; Emerson, A. B. *Langmuir* **1990**, *6*, 1567.

31. (a) Fukuda, K., Nakahara, H., Kato, T. *J. Colloid Interface Sci.* **1976**, *54*, 430; (b) Fukuda, K.; Nakahara, H.; Kato, T. *J. Colloid Interface Sci.* **1979**, *69*, 24; (c) Nakahara, H.; Fukuda, K. *J. Colloid Interface Sci.* **1981**, *83*, 401; (d) Nakahara, H.; Fukuda, K. *J. Colloid Interface Sci.* **1983**, *93*, 530; (e) Nakahara, H.; Fukuda, K.; Sato, M. *Thin Solid Films* **1985**, *133*, 1.

32. Li, T. T.-T.; Weaver, M. J. *J. Am. Chem. Soc.* **1984**, *106*, 6107.
33. (a) Roberts, G. G.; Twigg, M. V. *Br. Patent* 8,129,018, **1981**; (b) Zhang, X.; Bard *U.S. Patent*, in preparation.
34. (a) Blodgett, K. B. *J. Am. Chem. Soc.* **1934**, *56*, 495; (b) Blodgett, K. B. *J. Am. Chem. Soc.* **1935**, *57*, 1007.
35. (a) Kuhn, H.; Möbius, D.; Bucher, G. in Weissberger, A; Rossiter, B. W.(Eds.), *Physical Methods of Chemistry*, Vol. 1, Part 3B, Wiley, New York: 1972, p. 597; (b) Kuhn, H. *Thin Solid Films* **1983**, *99*, 1; (c) Kuhn, H. *J. Photochem.* **1979**, *10*, 111; (d) Kuhn, H. Möbius, D. *Angew. Chem. Int. Ed. Engl.* **1971**, *10*, 620.
36. (a) Möbius, D. *Acc. Chem. Res.* **1981**, *14*, 63; (b) Möbius, D. *Ber. Bunsenges. Phys. Chem.* **1978**, *82*, 848.
37. (a) Barraud, A. *Thin Solid Films* **1983**, *99*, 317; (b) Broers, A. N.; Pomerantz, M. *Thin Solid Films* **1983**, *99*, 323.
38. (a) Roberts, G. G.; Pande, K. P.; Barlow, W. A. *Proceedings IEEE Part I, Solid State Electron Devices* **1978**, *2*, 169; (b) Fung, D. D.; Larkins, G. L. *Thin Solid Films* **1985**, *132*, 33.
39. Roberts, G. G.; Petty, M. C.; Dharmadasa, I. M. *IEE Proc. Part I* **1981**, *128*, 197.
40. (a) Baker, S.; Roberts, G. G.; Petty, M. C. *Proc. IEE Part I* **1983**, *130*, 260; (b) Roberts, G. G.; Petty, M. C.; Baker, S.; Fowler, M. T.; Thomas, N. J. *Thin Solid Films* **1985**, *132*, 113.
41. Barraud, A.; Rosilio, C.; Ruaudel-Teixier, A. *Thin Solid Films* **1980**, *68*, 7.
42. Barraud, A.; Rosilio, C.; Ruaudel-Teizier, A. *J. Colloid Interface Sci.* **1977**, *62*, 509.
43. (a) Belbeoch, B.; Roulliay, M.; Tournarie, M. *Thin Solid Films* **1985**, *134*, 89; (b) Vincett, P. S.; Barlow, W. A. Thin Solid Films **1980**, *71*, 305.
44. Blodgett, K. B. *U.S. Patent 2,587,282,* **1952**.
45. Sugi, M.; Sakai, K.; Saito, M.; Kawabata, Y.; Iizima, S. *Thin Solid Films* **1985**, *132, 69.*
46. Law, K. Y.; Chen, C. C. *J. Phys. Chem.* **1989**, *93*, 2533.
47. Gaines, G. L., Jr.; *Insoluble Monolayers at Liquid–Gas Interfaces*, New York (1966), Chapter 8.
48. Gaines, G. L. *Thin Solid Films* **1980**, *68*, 1.
49. Ter Minassian-Saraga, L. *J. Chim. Phys.* **1955**, *52*, 181.
50. Conway, B. E.; Angerstein-Kozlowska, H.; Sharp, W. B. A.; Criddle, E. *Anal. Chem.* **1973**, *45*, 1331.
51. Clavilier, J.; Faure, R.; Guinet, G.; Durand, R. *J. Electroanal. Chem.* **1980**, *107*, 205.
52. Gottesfeld, S., in *Electroanalytical Chemistry, A Series of Advances*, Bard, A. J. (Ed.); Vol. 15, Marcell Dekker: New York (1989), Chapter 2.
53. Majda reported that $C_{18}Fc$ monolayers on a $NaClO_4$ subphase may crystallize (18f).
54. For comparison, note that the equilibrium spreading pressure π_e of stearic acid is about $5\,mN\,m^{-1}$, yet π–A isotherms can be obtained up to $40\,mN\,m^{-1}$, well above π_e (5a).
55. Closest packed, ordered alkylthiol and alkylsiloxane monolayers show an advancing water contact angle of 110–114° (7, 8). Introduction of disorder in the alkyl chains exposes the CH_2 groups and lowers the contact angle (7d, 26c).
56. (a) Valles, E.; Gomez, E.; Felieu, J. M.; Aldaz, A. *J. Electroanal. Chem.* **1984**, *190*, 95; (b) Cordova, R.; Martins, M. E.; Arvia, A. J. *Electrochim. Acta* **1980**, *25*, 453; (c) Kirk, D. W.; Foulkes, F. R.; Graydon, W. F. *J. Electrochem. Soc.* **1980**, *127*, 1069.
57. Bard, A. J.; Faulkner, L. R.; *Electrochemical Methods, Fundamentals and Applications*, Wiley: New York, **1980**, Chapter 12.
58. Szentrimay, R.; Yeh, P.; Kuwana, T. in *Electrochemical Studies of Biological Systems* Sawyer, D. T. (Ed.), American Chemical Society, **1977**.
59. Gaines, G. L., Jr., *Insoluble Monolayers at Liquid–Gas Interfaces*, Interscience, New York (1966) p. 330.
60. Majda, M., private communication.

61. Brown, A. P.; Anson, F. C. *Anal. Chem.* **1977**, *49*, 1589.
62. (a) Ohnishi, T.; Ishitani, A.; Ishida, H.; Yamamoto, N.; Tsubomura, H. *J. Phys. Chem.* **1978**, *82*, 1989; (b) Rabolt, J. F.; Burns, F. C.; Schlotter, N. E.; Swalen, J. D. *J. Chem. Phys.* **1983**, *78*, 946; (c) Naselli, C.; Rabolt, J. F.; Swalen, J. D. *J. Chem. Phys.* **1985**, 82, 2136; (d) Rabe, J. P.; Rabolt, J. F.; Brown, C. A.; Swalen, J. D. *Thin Solid Films* **1985**, *133*, 153; (e) Allara, D. L.; Swalen, J. D. *J. Phys. Chem.* **1982**, *86*, 2700; (f) Greenler, R. G. *J. Chem. Phys.* **1966**, *44,* 310; (g) Chollet, P. A. *Thin Solid Films* **1978**, *52*, 343.
63. (a) Zavislan, J. M. Ph.D. Thesis, University of Rochester, Rochester, NY, 1987; (b) Hartman, J. S.; Gordon, R. L.; Lessor, D. L. *Appl. Opt.* **1981**, *20*, 2665; (c) Jabr, S. N. *Opt. Lett.* **1985**, *10*, 526.
64. Laviron, E. *J. Electroanal. Chem.* **1974**, *52*, 395.
65. (a) Saji, T.; Hoshino, K.; Aoyagui, S. *J. Am. Chem. Soc.* **1985**, *107*, 6865; (b) Sakai, K.; Saito, M.; Sugi, M.; Iizima, S. *Jpn. J. Appl. Phys. Part 1* **1985**, *24*, 865.
66. Ellison, R. H. *J. Phys. Chem.* **1962**, *66*, 1867.
67. (a) Matsuda, H.; Aoki, K.; Tokuda, K. *J. Electroanal. Chem.* **1987**, *217*, 1; (b) Matsuda, H.; Aoki, K.; Tokuda, K. *J. Electroanal. Chem.* **1987**, *217*, 15.
68. Davies, J. T. *Proc. R. Soc. London A* **1951**, 208, 224.
69. (a) Hupp, J. T.; Weaver, M. J. *J. Electroanal. Chem.* **1983**, *145*, 43; (b) Hupp, J. T.; Weaver, M. J. *J. Phys. Chem.* **1985**, *89*, 2795.

Chapter IV

DYNAMICS OF ELECTRON TRANSPORT IN POLYMERIC ASSEMBLIES OF REDOX CENTERS

Marcin Majda
Department of Chemistry, University of California at Berkeley, Berkeley, California

---

Molecular Design of Electrode Surfaces,
Edited by Royce W. Murray. Techniques of Chemistry Series, Vol. XXII.
ISBN 0-471-55773-0 © 1992 John Wiley & Sons, Inc.

Modification of electrode surfaces with thin polymer films carrying redox sites has become a major area of electrochemistry for more than a decade (for a review, see Refs. 1–6). Such rapid growth can be understood in view of its many potential applications. The ones most frequently mentioned are electrocatalysis, electrochromic devices, coatings on semiconducting electrodes with photosensitizing and anticorrosive properties, electrochemical switching devices, and electrochemical sensors. In virtually all of these applications, the functioning of microstructural systems depends on the efficiency of the charge transport across the electrode film. It is, therefore, not surprising that the understanding of charge transport and related processes has always been emphasized in this area of research. Indeed, the large body of research on chemical modification of electrodes that has been generated over the years is concerned primarily with two issues. The first deals with the synthesis, preparation, and thermodynamic redox characteristics of the electrode coatings. The second is concerned with the dynamics of charge transport and to some extent addresses the problem of mass transport in the electrode films.

In assessing accomplishments in this field, it seems fair to say that the main progress has been made not in meeting the practical goals mentioned above, but rather in discovering and building the understanding of some basic principles governing the dynamic behavior of these complex multicomponent systems. It appears, in fact, that the single most important benefit of the vast research carried out in this area is the invention of a novel class of chemical systems. These molecular systems, often referred to as molecular assemblies or microstructures, consist of a complex network of redox sites often dispersed in a polymeric or a solid state matrix. In most cases, an interpenetrating solvent is also an integral part of the system. All of these systems are designed to perform a specific function. One particular function common to all such systems is the ability to transport charge.

In view of the multimolecular–multicomponent character of these interfacial structures, it is not surprising that their properties, and specifically the dynamics of charge transport, depend strongly not only on their chemical composition but also on their physical structure and the extent of internal organization. The relationship of structure and function in multimolecular assemblies, as opposed to the classically chemical problem of the structure versus function relationship at the single molecule level, has attracted chemists' attention only recently. Yet, by addressing this problem, chemists can create and investigate many systems mimicking some elementary functions of biological processes and thus broaden and enrich the scope of chemical research and narrow the gap between chemistry and biology. The molecular machinery of natural photosynthesis and multicomponent assemblies of enzymatic catalysis are just two examples of a number of biological systems where the function of an organized molecular assembly depends primarily on its supermolecular structure and organization.

This chapter is written with two related goals in mind. The first is to introduce and discuss in general terms the basic phenomena and processes involved in electron transport in electrode films. The second goal is to present a few selected cases of different types of electrode coatings as an illustration of the basic phenomena, and in order to point out and to analyze the interdependence between the dynamics of electron transport and the chemical and physical structure of the electrode films.

4.1 THE ELEMENTS OF ELECTRON TRANSPORT

Electron transport in microstructural films is a process involving dynamics of several elementary events that overlap and cross influence one another. Electron transfer between redox centers, motion of polymer strands, diffusion of the redox centers and counterions, uptake and expulsion of solvent molecules are the main elements whose dynamics are mutually interrelated. The intention of this section is to describe electron transport, beginning with a simple picture of the system and its elementary processes, and then increasing the complexity of the problem, by showing the additional processes and interrelationships existing in real multicomponent electrochemical systems.

4.1.1 Description of a Model System

Let us consider initially a generic system consisting of a thin, homogeneous polymeric film deposited at a planar electrode surface in contact with an electrolyte solution. The electrode film contains randomly distributed redox sites that are confined exclusively within the film, so that the electrolyte solution is free of the electroactive species. The mode of immobilization of the electroactive species in the film may involve either chemical bonding to the polymer chains of the film matrix, or it may rely on some other, for example, electrostatic, interaction between the redox centers and the polymer matrix[1]. In the former case, the redox sites are immobile beyond a close vicinity of their attachment to the matrix. Otherwise, their mobility is unrestricted within the electrode film.

We will assume that the electrode film is swollen to some extent with a solvent, and that it is permeated with the ions of the supporting electrolyte, which can diffuse both within the film and also across the film–solution interface. The ion permeability of the electrode film is thus an intrinsic parameter of the system and is related primarily to the type of the polymer and to the extent of its swelling in a particular solvent. The mobility of the electroactive sites in the matrix is related to matrix fluidity in that it involves, at least indirectly, segmental motion of the polymer strands.

Let us also assume that the properties of the system are independent of the redox state of the electroactive centers, and that the dynamics of electron transport are independent of the counterion motion in the film. Finally, we want to assume that there is no electric field in the film, except in a double layer region at the electrode–film interface, which is negligibly thin compared to the film thickness.

4.1.2 Electron Hopping and Physical Diffusion of Redox Species

We shall consider an experiment that is initiated by the application of a potential to the electrode surface, and which results in the reduction of the redox species in the electrode film. Electron transport in the polymer film amounts to the propagation of charge through the film until all of the redox sites change their oxidation state (1, 7). This process may involve either one, or both of the following two mechanisms: (a) electron hopping between the redox sites and (b) their physical diffusion through the film. Both processes are shown schematically in Fig. 4.1. Because of the strong dependence on the electrode potential, the kinetics of the electron transfer at the electrode–film interface, which initiates electron transport in the film, are not rate limiting (7). The overall process, and thus the observed current, is then controlled by the kinetics of the electron transport in the film.

Following the reduction of the electroactive centers immediately adjacent to the electrode surface (Fig. 4.1), the concentration gradient of the oxidized species develops. This fuels the physical diffusion of the oxidized species towards the electrode surface, and also initiates the electron hopping. In cases where the redox centers are not bound rigidly to the polymer matrix, physical diffusion alone provides sufficient mechanistic basis for the complete reduction of all the sites within the electrode film. When, however, the redox sites cannot diffuse freely through the electrode film, the reduction of all the sites can take place only via electron hopping. This mechanism involves a series of the electron-transfer reactions between the redox sites in the film. In most cases, this requires diffusional collision of the neighboring centers. Thus some mobility of the centers may be required. Complete lack of matrix fluidity may entirely restrict electron transport.

Physical diffusion and electron hopping can proceed simultaneously. In this case, the rate of electron transport is the sum of the rates of both processes.

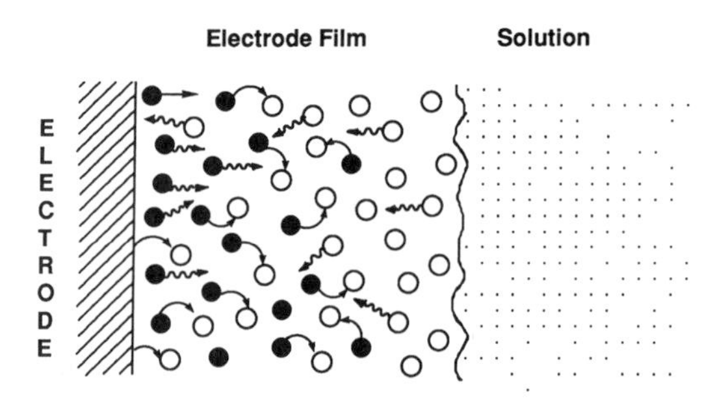

Figure 4.1. A schematic representation of an electrode coated with a polymer film incorporating redox species. Reduction of the oxidized sites (○) is initiated at the electrode surface and propagates diffusively through the entire film. This involves electron transfer between the reduced (●) and oxidized sites, depicted as smooth curved arrows, and diffusion of the oxidized sites directly to the electrode surface, depicted with wavy arrows.

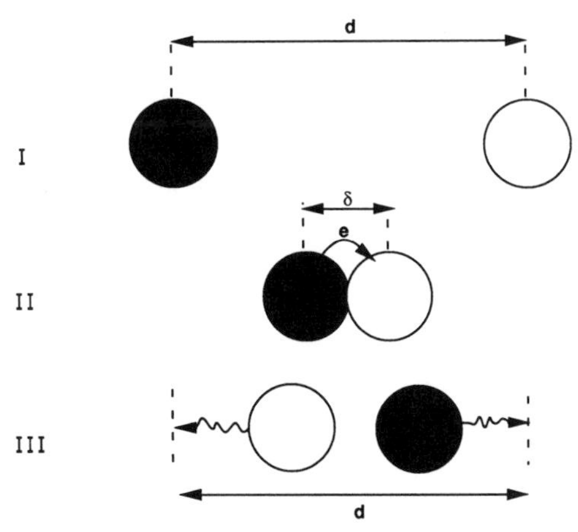

Figure 4.2. A schematic microscopic picture of the events associated with an electron transfer in a polymer film. Stage I shows the two redox sites in their average equilibrium positions. Stage II is the electron transfer that follows the formation of a precursor complex. Stage III depicts separation of the reactants following electron transfer.

Regardless of the mechanism, the rate of the electron transport can be characterized by an observable parameter, the apparent diffusion coefficient, D_{app}. The literature has demonstrated that the processes described here are indeed diffusive in nature for a large number of different electrode film systems, by fitting the observed current–time transients to the Cottrell equation, derived from Fick's laws of diffusion:

$$i = n\mathscr{F}AC_0^*D_{app}/\sqrt{\pi t} \tag{4.1}$$

where n is the number of electrons exchanged in the reduction of an oxidized center at the electrode surface, $\mathscr{F}$ is the Faraday constant, A is the electrode surface area, and C_0^* is the total concentration of the redox species in the electrode film (1, 8, 9).

4.1.3 Electron Hopping

Let us consider in more detail the electron-hopping process and its depiction in Fig. 4.1. It is easy to realize that the situation presented there is inaccurate, in that the redox centers exchanging the electron are drawn at their equilibrium positions rather than at a collision distance where the electron transfer really takes place. Figure 4.2 shows schematically a microscopic picture of the events involved in a single electron-transfer step. These are the diffusional collision of the centers, leading to the formation of a precursor complex, electron transfer, and the diffusional separation of the centers to their initial equilibrium position. The overall effect is transport of electron over a distance d, equal to the average distance between the redox centers in a matrix. Part of this was accomplished by diffusion. The remaining fraction of this distance (δ) was accomplished by the electron transfer (step II in Fig. 4.2). Thus, we conclude that the electron-transfer supplements diffusion, in that it is equivalent to the translational advancement of an electron carrier.

This concept, sometimes referred to as electron-transfer diffusion, has been a subject of considerable interest. Dahms (10) considered the contribution of the electron exchange reaction

$$M(1)^{n+} + M(2)^{(n-1)+} \rightarrow M(1)^{(n-1)+} + M(2)^{n+} \tag{4.2}$$

to the electronic conduction of a solution in which a linear concentration gradient of the reactants exists in the direction of the current flow (his treatment assumes, again, no electric field gradient):

$$\frac{dC_{\mathrm{Ox}}}{dx} = \frac{-dC_{\mathrm{Red}}}{dx} = \frac{dC}{dx} \tag{4.3}$$

The indixes (1) and (2), above, indicate the positions of the reactants and products collinear with the gradient of their concentrations (C). It is apparent that the electron-transfer contribution to the solution conductivity is analogous to the situation in Fig. 4.2. In view of this model, Dahms obtained expressions for the forward and reverse current densities due to Reaction 2:

$$i_{\mathrm{f}} = \mathscr{F}k\delta C_{\mathrm{Ox}}\left(C_{\mathrm{Red}} + \delta\frac{dC_{\mathrm{Red}}}{dx}\right) \tag{4.4}$$

$$i_{\mathrm{b}} = \mathscr{F}k\delta C_{\mathrm{Red}}\left(C_{\mathrm{Ox}} + \delta\frac{dC_{\mathrm{Ox}}}{dx}\right) \tag{4.5}$$

where δ is the quantity introduced in Fig. 4.2 and k is the site-to-site electron exchange rate constant. Since the net current density, $i_{\mathrm{e}} = i_{\mathrm{f}} - i_{\mathrm{b}}$, one obtains the final expression using Eqs. 4.3–4.5:

$$i_{\mathrm{e}} = \mathscr{F}k\delta^2 C\frac{dC}{dx} \tag{4.6}$$

where $C = C_{\mathrm{Ox}} + C_{\mathrm{Red}}$, the total concentration of the redox species. The purely diffusive contribution to the net current density can be obtained from Fick's first law:

$$i_{\mathrm{d}} = \mathscr{F}D_{\mathrm{p}}\frac{dC}{dx} \tag{4.7}$$

Here D_{p} is the diffusion coefficient of both D_{Ox} and D_{Red}. It is clear then that the equation for the total current will contain the sum of the diffusion coefficients representing the electron transfer (D_e) and physical diffusion (D_p):

$$i_t = \mathscr{F}(D_e + D_p)\frac{dC}{dx} \tag{4.8}$$

$$D_e = k\delta^2 C \tag{4.9}$$

Later, the same expression for D_e was obtained independently by Laviron (11) and by Andrieux and Savéant (12) who considered electron transport in polymer films via electron hopping. Andrieux and Savéant derived the corresponding Fick's second-law expression based on the rate of electron transfer between the immobile redox centers (12). In this treatment, the electron hopping was considered as a unidimensional, bimolecular process coincidental with the direction of the concentration gradient. The situation in question is shown in Fig. 4.3. The electron exchange reactions take place across the imaginary planes perpendicular to the concentration gradient. For example,

$$A_{j-1} + B_j \underset{k}{\overset{k}{\rightleftharpoons}} B_{j-1} + A_j \tag{4.10}$$

The starting point in this derivation is the expression for the rate of concentration change of redox species in an infinitesimal (monomolecular) segment of a polymer film parallel to the electrode surface.

$$dC_{A_j}/dt = -kC_{A_j}C_{B_{j-1}} + kC_{A_{j-1}}C_{B_j} - kC_{A_j}C_{B_{j+1}} + kC_{A_{j+1}}C_{B_j} \tag{4.11}$$

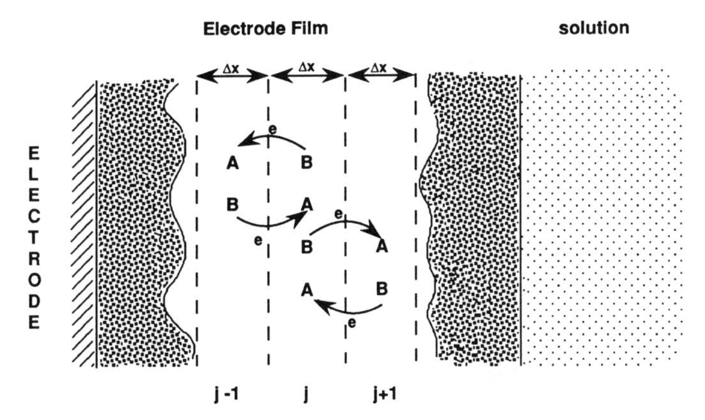

Figure 4.3. An illustration of an approximate model of one-dimensional electron hopping in a polymer film, incorporating fixed redox sites (see text and Eqs. 4.10–4.14). The polymer film is divided into a number of parallel layers, each containing fixed redox sites **A** and **B**. Electron transport is initiated at the electrode surface (**A** + e^- → **B**) and propagates through the electrode film via electron hopping only. Individual electron-transfer reactions (Eq. 4.10) can only occur between sites in the adjacent planes across distance Δx.

Since the sum of concentrations of the redox species in any layer is

$$C_{A_j} + C_{B_j} = C \tag{4.12}$$

then Eq. 4.11 can be rewritten as:

$$dC_{A_j}/dt = kC(C_{A_{j-1}} - 2C_{A_j} + C_{A_{j+1}}) \tag{4.13}$$

The bracketed term on the right-handside of Eq. 4.13 divided by Δx^2 ($\Delta x = \delta$, the distance over which the electron transfer takes place) is the finite difference expression for d^2C_A/dx^2. By making this substitution, Eq. 4.13 becomes

$$\frac{dC_A}{dt} = kC\delta^2 \frac{d^2C_A}{dx^2} \tag{4.14}$$

Thus, the expression for the diffusion coefficient obtained here is the same as in Eq. 4.9.

In both of these derivations leading to Eqs. 4.9 and 4.14, the electron transfer between adjacent sites in the precursor complex is considered to take place only along a single dimension coincidental with the direction of the concentration gradient. Considering the fact that the system in question is three dimensional and isotropic, the rate constant used in Eqs. 4.9 and 4.14 equals $\frac{1}{6}$ the bimolecular rate constant usually used to describe kinetics of electron transfer, k_{ex}. Thus, Eq. 4.9 can be rewritten as:

$$D_e = \tfrac{1}{6}k_{ex}\delta^2C \tag{4.15}$$

A different approach to the same problem was taken by Ruff and Friedrich and their co-workers (13, 14) who wanted to include in their derivation a geometric factor stemming from the three-dimensional orientation of the transition state complex with respect to the vector of the concentration gradient. Their equation for D_e introduces, apparently erroneously, a constant $\pi/4$ to Eq. 4.9.

$$D_e = \frac{\pi}{4}k\delta^2C \tag{4.16}$$

This equation has often been used in the area of electrochemistry concerned with polymer coated electrodes (15–20). Later, Ruff and Botar (21, 22) published a corrected derivation where factor $\frac{1}{6}$ appears in Eq. 4.9. This result is consistent with a stochastic formulation of a diffusion coefficient (21, 22). In this approach, diffusing sites are placed in a cubic lattice with a spacing equal to δ, the center-to-center distance the electron jumps in an electron transfer that takes place within the precursor complex. Naturally, from the point of view of purely physical diffusion, the definition of the magnitude of δ is purely arbitrary and has no reflection in physical reality. It is defined as described here simply in order to include the possibility of electron transfer as an equivalent elementary step in the overall diffusive motion of sites. Botar and Ruff (21, 22) demonstrated that this approach indeed leads to Eq. 4.15, relating the apparent diffusion coefficient and the second-order electron-transfer rate constant. Factors $\frac{1}{4}$ and $\frac{1}{2}$ were obtained for the two-dimensional and one-dimensional cases, respectively. The same result was recently favored by Buck in his discussions of the electron-transfer diffusion (23, 24).

As a direct result of this discussion, we can conclude that the rate constant (k_{ex}) in Eq. 4.15 is not a subject to mass transport limitation. The value of the rate constant can be expressed by

$$k_{ex} = K_p k_{et} \tag{4.17}$$

where K_p is the precursor complex equilibrium constant and k_{et} is the unimolecular rate constant of electron transfer (25–27):

$$k_{et} = \kappa_{el} \nu_n \exp(-\Delta G^*/RT) \tag{4.18}$$

Here, κ_{el} is the electronic transmission coefficient, $\nu_n(s^{-1})$ is the nuclear frequency factor, and ΔG^* is the activation free-energy barrier. The pre-equilibrium constant can be expressed by a product of the term describing the work expanded to overcome electrostatic repulsion in the formation of the preequilibrium complex, and the statistical probability of forming such a complex. Thus, with some approximations, for a pair of spherical reactants, we may write:

$$K_p = 4\pi 10^{-3} N_A R^2\, \delta R \exp(-w_e/RT) \tag{4.19}$$

where N_A is Avogadro's number, R is the average separation of the reacting centers upon electron transfer, and δR is the effective reaction zone thickness (25). It is apparent that R is the same parameter as δ in Eq. 4.15. The term δR expresses a range of nuclear separations within which electron transfer proceeds with a particular rate constant expressed by Eq. 4.18. Hence, the magnitude of δR is linked to the electronic transmission coefficient. Its value is larger for strongly adiabatic reactions, where $\kappa_{el} = 1$, than for nonadiabatic reactions, where $\kappa_{el} \ll 1$ and δR is about 0.5 Å (25–27). The distance dependence of the electron transfer will be considered in more detail in Section 4.4.1. The role of the nuclear frequency factor should also be mentioned here. It describes dynamics of all nuclear motions involved in the formation of the transition state. In cases where so-called "inner-shell" reorganization energy (stemming from the bond distortions of the reactants) is small compared to the "outer-shell" reorganization energy, ν_n describes the solvent dynamics (28). Thus, it should be expected that electron transfer involving redox centers imbedded in a polymer matrix may proceed with a rate constant substantially different from that known for a particular redox reaction observed in a well-characterized solvent. Hence, Eqs. 4.15 and 4.21 (below) have limited predictive power. They may be used, however, to measure the apparent rate constant under a particular set of conditions and to study its dependence on the environmental parameters.

Electron hopping as a mechanism of electron propagation in polymer films was first proposed by Kaufman and his co-workers (29), in their investigations of polystyrene resins incorporating pyrazoline moieties and in the case of poly(phenoxytetrathiafulvalane) (30). The evidence of electron hopping in polymer films was later convincingly demonstrated by Facci and Murray (15). They analyzed the shapes of the voltammetric waves for $[Fe(CN)_6]^{3-/4-}$ and $[IrCl_6]^{2-/3-}$ redox couples simultaneously ion exchanged into copolymer film of vinylpyridine and (γ-methacryloxypropyl)trimethoxysilane. Two sets of voltammograms shown in Fig. 4.4 correspond to the potential scans that were adjusted to encompass each of the redox couples separately (set A), or both together (set B). In the latter case it is apparent that the electrogenerated $[IrCl_6]^{2-}$ enhances transport of $[Fe(CN)_6]^{4-}$ to the electrode surface via electron hopping.

$$[IrCl_6]^{2-} + [Fe(CN)_6]^{4-} = [IrCl_6]^{3-} + [Fe(CN)_6]^{3-} \tag{4.20}$$

Similarly, the onset of $[Fe(CN)_6]^{3-}$ reduction in the cathodic half of the cycle in Fig. 4.4*B* leads to the acceleration of the $[IrCl_6]^{2-}$ flux towards the electrode surface via the same reaction. In this example, the rate constant of electron hopping in Eq. 4.15 refers, naturally, to the cross-exchange electron transfer.

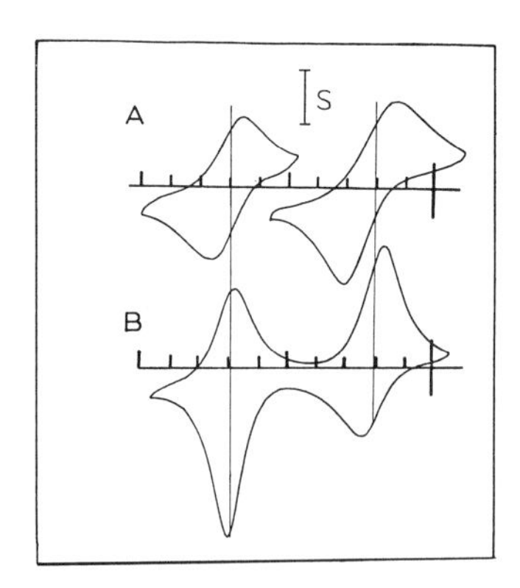

Figure 4.4. Cyclic voltammograms of $[Fe(CN)_6]^{3-/4-}$ (right-hand curves) and $[IrCl_6]^{2-/3-}$ (left-hand curves) redox couples illustrating the presence of electron hopping. The redox species are immobilized in a cationic polymer film at high concentrations of about 0.3*M* each. The two diffusion-controlled cyclic voltammograms shown in set **A** were recorded separately and independently of one another. During a single voltammetric scan, shown in set **B**, the enhancement of the $[IrCl_6]^{3-}$ oxidation current and $[Fe(CN)_6]^{3-}$ reduction current indicates charge transport via electron hopping (Eq. 4.20). $S = 300\,\mu A\,cm^{-2}$. [Reprinted with permission from J. Facci and R. W. Murray, *J. Phys. Chem.*, 1981, **85**, 2870. Copyright © (1981) American Chemical Society.]

Similar cases of electron-hopping enhancement of the electron-transport processes in electrode films were also analyzed by Buttry et al. (31).

In the following section, we consider several examples of polymer films containing redox sites and focus on the concentration dependence of the apparent diffusion coefficient. The latter can be expressed by the following:

$$D_{app} = D_p + D_e \tag{4.21}$$

In view of the concentration dependence of D_e, the experimental verification of the concentration dependence of D_{app} becomes a key element in the experimental strategy designed to obtain rate constants of electron hopping.

4.1.4 Concentration Dependence of D_{app}: Polymer Films with Freely Diffusing Redox Centers

We deal first with those cases where the electroactive species are not covalently bound within the polymer matrix, and thus, physical diffusion of the redox species alone could account for the global change of their oxidation state in an electrode film. As an illustration of this type of behavior, let us consider electron transport in a Nafion film loaded with $[Co(bpy)_3]^{2+}$ (bpy = 2,2–bipyridine) (16, 17). Nafion is a polymer containing sulfonate groups, and thus ion-exchange interactions account for the incorporation of $[Co(bpy)_3]^{2+}$ up to high concentration levels. At the same time, the electroactive centers are free to diffuse through the electrode film. Buttry and Anson (16, 17), who studied this system, investigated the concentration dependence of D_{app} values reflecting a one electron reduction or oxidation of the $[Co(bpy)_3]^{2+}$ sites in a Nafion film (17). The solution value of the electron exchange rate constant of the $[Co(bpy)_3]^{2+/+}$ couple is about $10^8 M^{-1}s^{-1}$ and thus the electron hopping in the electrode film could have a pronounced effect on D_{app}. In contrast, the one-electron oxidation of $[Co(bpy)_3]^{2+}$ in Nafion should proceed via physical diffusion because of the negligibly low value ($2M^{-1}s^{-1}$) of the $[Co(bpy)_3]^{3+/2+}$ electron exchange rate constant.

The dependence of both diffusion coefficients, $D_{2/1}$ and $D_{2/3}$, measured by Buttry and Anson (17) are reproduced in Fig. 4.5*A*. We see that $D_{2/1}$ indeed increases with the $[Co(bpy)_3]^{2+}$ concentration in the Nafion film in accord with Eq. 4.15. However, $D_{2/3}$, which represents a purely diffusive process, shows a small decrease over the same concentration range. This could not be predicted by our current model. The source of this discrepancy was related by the authors to the high concentration of $[Co(bpy)_3]^{2+}$ in the film. In their diffusive motions, cobalt complex residence sites are limited to those in the proximity of the sulfonate groups. At high cobalt complex concentration, the competition for the residence sites might decrease $D_{2/3}$. This phenomenon was referred to as "single file diffusion" (17). A decrease of the observed diffusion coefficients with concentration of the charged species incorporated electrostatically in ion-exchange type polymer films has been observed in several other cases and was related to electrostatic cross-linking (see Section 4.5.1).

The ability to measure D_p independently ($D_{2/3}$ in Fig. 4.5*A*) and the sum of $D_p + D_e$ ($D_{2/1}$ in Fig. 4.5*A*) for $[Co(bpy)_3]^{2+}$ presented a unique opportunity to extract the true dependence of D_e on concentration. The result is shown in Fig. 4.5*B* (17). The evaluation of the slope of the plot in view of the Ruff–Friedrich equation (Eq. 4.16) yielded the rate of electron exchange of $2 \times 10^3 M^{-1}s^{-1}$. If we used, instead, Eq. 4.15, we would obtain the apparent rate constant of the electron exchange of $1.5 \times 10^4 M^{-1}s^{-1}$. Both of these values are significantly smaller than the rate constant observed in homogeneous aqueous solution of about $10^8 M^{-1}s^{-1}$. This, however, does not represent the diffusion limitation of the reaction kinetics as derived by von Smoluchowski (32), but rather, is a pronouncement of strong matrix effects on the electron-transfer kinetics. It can be hypothesized that the observed decrease in Nafion films reveals inability of the reagents to approach each other at a contact distance necessary for an adiabatic electron transfer observed in homogeneous solutions.

The somewhat unexpected decrease of $D_{2/3}$ on the $[Co(bpy)_3]^{2+}$ concentration is another type of matrix effect. However, knowing that the solution value of the rate constant for $[Co(bpy)_3]^{3+/2+}$ is very small, we would not expect in this case to see any enhancement of physical diffusion due to electron hopping. There are several other systems that conform with this class of behavior: $[Ru(NH_3)_6]^{3+/2+}$ and (trimethylammino)methylferrocene ($TMAFc^{2+/+}$) in Nafion will be described briefly. The rate constant of electron exchange for the $[Ru(NH_3)_6]^{3+/2+}$ couple is $4 \times 10^3 M^{-1}s^{-1}$ (33) and, as expected (Eqs. 4.15 and 4.21), its diffusion coefficient in Nafion was invariant with concentration (16). The same result was also obtained when White et al. (34) investigated the concentration dependence of $TMAFc^{2+/+}$ diffusion in Nafion. An average value of D_{app} of $1.7 \pm 0.5 \times 10^{-10} cm^2 s^{-1}$ over the concentration range $7 \times 10^{-3} - 1.2M$ was obtained. That this value of D_{app} does indeed reflect the physical diffusion was demonstrated by the direct measurement of the rate of incorporation of the ferrocene species into the Nafion films. This experiment involved a measurement of an oxidation current transient at a Nafion coated electrode, initially devoid of the electroactive species, upon its immersion into a $TMAFc^+$ solution. Evaluation of the current in terms of the $TMAFc^+$ diffusion across a Nafion film gave essentially the same value of the diffusion coefficient as D_{app} mentioned above (34). A theoretical prediction of D_e for this case based on Eq. 4.15 is in the range 1.3×10^{-12}–1.3×10^{-9} for the Nafion ferrocene concentrations from 1 m*M* to 1*M* (with $\delta = 7$ Å, $k_{ex} = 1.6 \times 10^6 M^{-1}s^{-1}$). These values of D_e are indeed smaller than D_{app}, with the possible exception of the highest concentrations above approximately 0.13*M*. Because the electron-hopping enhancement to D_{app} was not observed suggests that the apparent electron-exchange rate constant of the ferrocene couple in Nafion is lower than $1.6 \times 10^6 M^{-1}s^{-1}$ assumed here based on solution data (35).

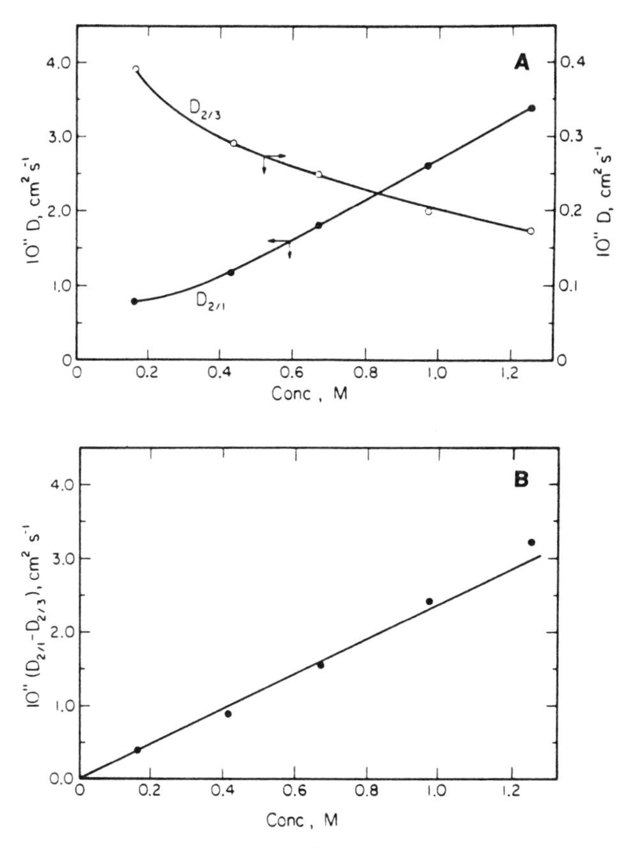

Figure 4.5. Concentration dependence of the diffusion coefficients of $[Co(bpy)_3]^{2+}$ in Nafion coatings obtained by chronocoulometry. The $D_{2/3}$ data were obtained by oxidizing the cobalt complex to its 3+ state. Since the electron-exchange kinetics of the 3+/2+ cobalt couple are negligibly slow, these data are not affected by electron hopping and represent the D_p component of the electron transport. The $D_{2/1}$ plot was obtained by electroreduction of $[Co(bpy)_3]^{2+}$. The increase of $D_{2/1}$ with concentration reveals electron-transport enhancement due to electron hopping. $D_{2/1} = D_p + D_e$. The plot in part *B* shows concentration dependence of D_e obtained as the difference $D_{2/1} - D_{2/3}$. [Reprinted with permission from D. A. Buttry and F. C. Anson, *J. Am. Chem. Soc.* 1983, **105**, 685. Copyright © (1983) American Chemical Society.]

4.1.5 Concentration Dependence of D_{app}: Polymer Films with Covalently Bound Redox Centers

Let us now consider a class of systems where the electroactive sites are covalently attached to polymer backbone and thus unable to diffuse physically through a polymer film. Several examples of this class of polymers are shown in Fig. 4.6. Thin films of these materials can be cast directly at the electrode surface (Fig. 4.6*a–c*), electroprecipitated (Fig. 4.6*d*), or deposited at the electrode surface by one of several electrochemical (Fig. 4.6*e,f*) or organosilane (Fig. 4.6*g*)

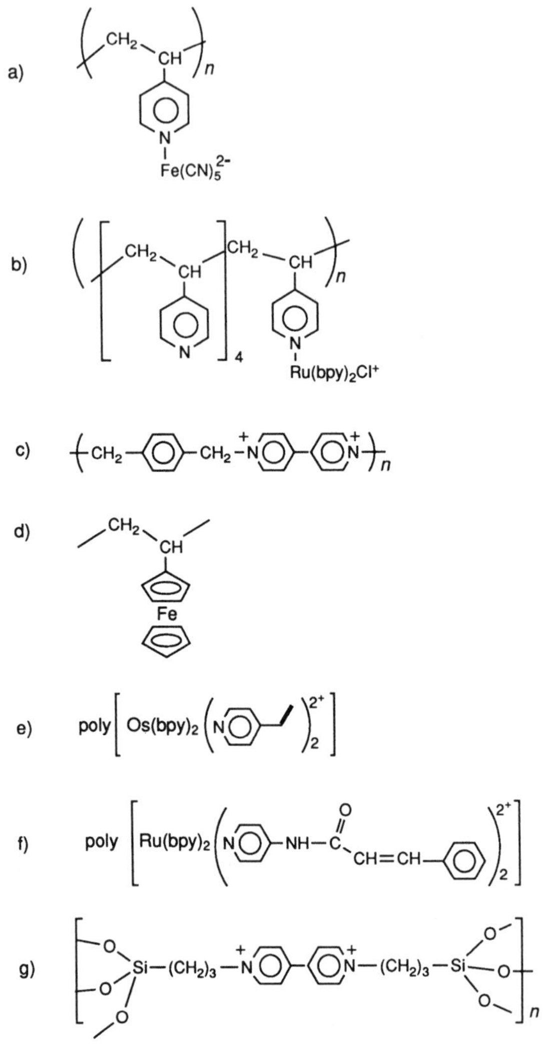

REFS.

a) 36

b) 37, 38

c) 39, 40

d) 41, 42

e) 43, 44

f) 43, 45

g) 46, 47

Figure 4.6. Examples of polymers with covelently bound redox species.

polymerization schemes. In all of these cases and other systems belonging to this class of polymers, it is difficult to vary the concentrations of the electroactive sites in polymer films. This task becomes particularly difficult to accomplish in a way that preserves, at the same time, the morphology and other physical properties of the electrode film. As a result little experimental work is available in the literature reporting this type of measurement.

In view of the electron-hopping model discussed in Section 4.1.3, the diffusion coefficient of electron transport is directly related to the apparent rate of electron exchange by Eq. 4.15. However, in the absence of physical diffusion of the redox sites, the effective range of concentrations where this equation is applicable is rather limited. As we will see below, this model does not provide a good description of the electron hopping in polymer films with covalently bound redox centers under even moderately broad range of redox sites concentrations.

Let us first consider poly$[Os(bpy)_2(vpy)_2]^{2+}$ (vpy = 4-vinylpyridine) electrode coatings developed by Murray and his co-workers (43, 44). This is a highly cross-linked polymer with a large concentration of the $[Os(bpy)_2(py)_2]^{3+/2+}$ (py = pyridine) redox sites ($\sim 1M$). The closely related $[Os(bpy)_3]^{3+/2+}$ redox couple has the electron-exchange constant in a homogeneous solution $k_{ex} = 2.2 \times 10^7 M^{-1} s^{-1}$ (48). A direct interpretation of the measured value of $D_{app}(8 \times 10^{-9} cm^2 s^{-1})$ (49) in view of Eq. 4.15 gives the apparent rate constant of almost an order of magnitude lower, $2.8 \times 10^6 M^{-1} s^{-1}$. In the calculation $d = 13$ Å was used somewhat arbitrarily. The true separation of the osmium centers in the film might be somewhat larger; this would also account for the lower observed value of the rate constant hinting a less adiabatic process (27). The D_{app} measurements were also made for the electron transport supported by the Os(II/I) and Os(I/O) states of the polymer (49). Their values, D_{app} (II/I) $= 2.4 \times 10^{-8} cm^2 s^{-1}$ and D_{app}(I/O) $= 2 \times 10^{-7} cm^2 s^{-1}$ yield the rate constants of $8.5 \times 10^6 M^{-1} s^{-1}$ and $7.1 \times 10^7 M^{-1} s^{-1}$, respectively. It is interesting to note that the latter of these values actually exceeds the homogeneous solution rate constant for the Os(I/O) couple, which is estimated to be about $5 \times 10^6 M^{-1} s^{-1}$ (49). This is one example of an apparent rate enhancement derived from confinement of the redox sites in a polymeric matrix.

It should be mentioned that this type of data (see above) have been often interpreted in the literature in view of the diffusion limited rate constant described by the Smoluchowski equation. Such incorrect interpretation might lead, occasionally, to a "surprising" result that the rate constant calculated from the measured D_{app} value based on Eq. 4.15 is higher than the "diffusion limited" value obtained based on known diffusion coefficients in a particular polymeric matrix. Since, as described in Section 4.1.3, k_{ex} values obtained from Eq. 4.15, based on experimentally determined diffusion coefficients, describe the "site-to-site" electron-transfer rate constants, their interpretation must be done on purely kinetic basis.

Let us come back again to the analysis of Eq. 4.15. The predicted linear dependence of D_e on the concentration of redox centers has to be taken with considerable caution. Specifically, one cannot realistically expect such dependence to hold below a certain minimum concentration level where the electroactive sites are so diluted that their collisions, and thus also electron exchange, are physically impossible. This point was addressed by Shigehara et al. (36), who studied poly(4-vinylpyridine) films with $Fe(CN)_5$ redox centers coordinatively linked to the polymer pyridine groups (Fig. 4.6*a*). They found sharply decreasing D_{app} values when the concentration of the $Fe(CN)_5$ sites in the polymer film decreased below a critical value. In the concentration range above the critical value, the authors did not observe, however, the expected linear increase of D_{app} with concentration. In this case the electron-transport processes were complicated by other phenomena related to the nature of the polymer film. Strong pH effects as well as partial and variable electroactivity of the $Fe(CN)_5$ sites made the full analysis of the transport events difficult (36).

Similar studies were also conducted by Murray and his co-workers (50, 51) on a copolymer film prepared by electropolymerization of $[Os(bpy)_2(4\text{-}pyNHCOCH{=}CHPh)_2]^{2+}$ (where py = pyridine, and Ph = phenyl) and its ruthenium analogue. This method leads to the formation of highly cross-linked polymer films with high density of the redox sites. The rate of electron transport supported by the Os(III/II) redox species was examined as a function of their concentration in polymer films (50). The concentration of the Os sites was varied by changing the ratio of the Ru and Os sites, keeping their total concentration constant. This approach gave the best assurance that the results were not affected by the polymer structural features related to the variable level of loading, since both monomers, their rate of polymerization, and incorporation into the film network were essentially identical. The variation of D_{app} with Os species concentration was obtained by chronoamperometry and is reproduced in Fig. 4.7. Here again, Eq. 4.15 does not find reflection in the experimental

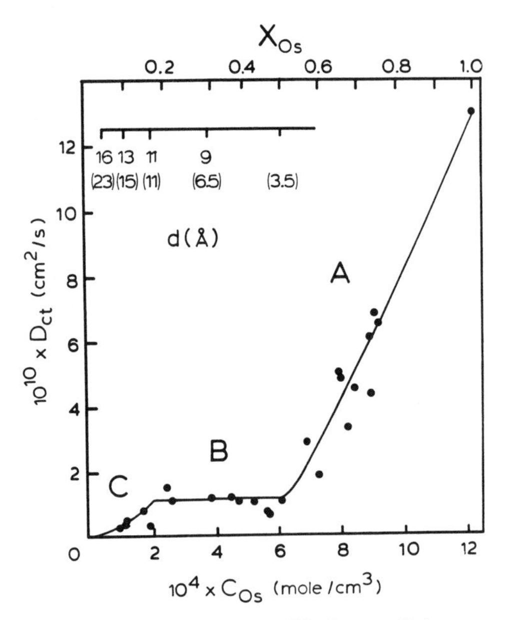

Figure 4.7. Dependence of the electron-transport diffusion coefficient on the mole fraction of osmium sites (top abscissa) and their concentration in a copolymer poly$[Ru/Os(bpy)_2(4\text{-}pyNHCOCH{=}CHPh)_2^{2+}]$ (structure *f* in Fig. 4.6). The calculated edge-to-edge distance *d* between the Os complex sites at various concentrations. [Reprinted with permission from J. S. Facci, R. H. Schmehl, and R. W. Murray, *J. Am. Chem. Soc.* 1982, **104**, 4959. Copyright © (1982) American Chemical Society.]

reality. The authors managed to find a reasonable fit of the data in the region of high Os site concentration (region A in Fig. 4.7) by adopting a random walk model for the electron hopping between stationary nearest neighbors (represented by the dashed line) (50). To account for the plateau in region B, the effect of polymer segmental motion was postulated to create additional Os/Os nearest neighbors. Finally, in the region of the lowest Os redox sites concentration, the polymer self-diffusion is the necessary element of the electron transport in the absence of the Os/Os nearest neighbors pairs (Os centers are diluted to the point of having only Ru-type nearest neighbors). In this region, the observed D_{app} dependence on concentration could either reflect the prediction of Eq. 4.15, or the limiting situation of the physical break-down of the electron-hopping process described in the previous example of poly(4-vinylpyridine). An entirely different interpretation of the data in Fig. 4.7 involving migration effects will be discussed in Section 4.3.4.

An entirely different situation is encountered in polymeric networks in which redox species have relatively high mobility around their point of immobilization. Electron transport combines, then, random oscillations of the electroactive sites and the electron-exchange reaction. In these cases D_{app} also equals to D_e, and thus D_{app} should again be proportional to the concentration of the redox species in the polymer. As an example, we will consider one such system where, because of the particular structural features of the polymer network, this prediction is not fulfilled.

The polymer film in question, investigated by Moran (52), is an agarose gel whose physical and chemical structure are shown in Fig. 4.8. The 25-μm thick agarose films are prepared by spin coating of a viscous, hot solution of agarose at the electrode surface, which is followed by a spontaneous gelation of a thin film of the polymer as the temperature decreased below about 40°C (52, 53). The agarose gels consist of rigid, tightly coiled polymer bundles and purely aqueous domains, as depicted schematically in Fig. 4.8. Water constitutes about 92% of its volume. The immobilization of the electroactive moieties in the gel involved chemical activation of the agarose hydroxyl groups located on the surface of the polymer bundles according to the procedure in Scheme 1.

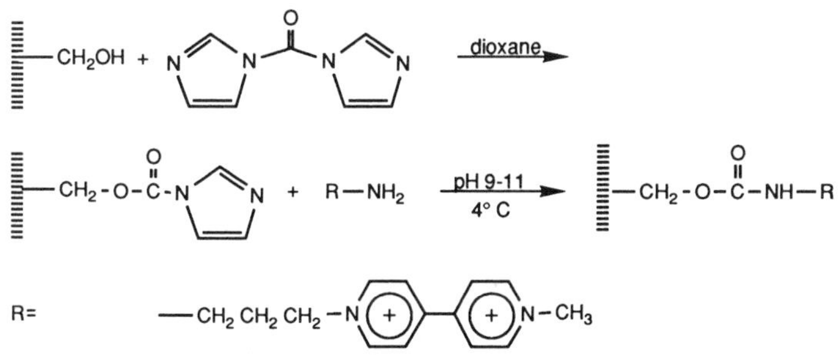

—CH_2OH + (imidazolyl)–C(=O)–(imidazolyl) —dioxane→

—CH_2-O-C(=O)-(imidazolyl) + R—NH_2 —pH 9-11, 4° C→ —CH_2-O-C(=O)-NH—R

R= —$CH_2CH_2CH_2$—N(+)(pyridinium)–(pyridinium)(+)N—CH_3

Scheme 1

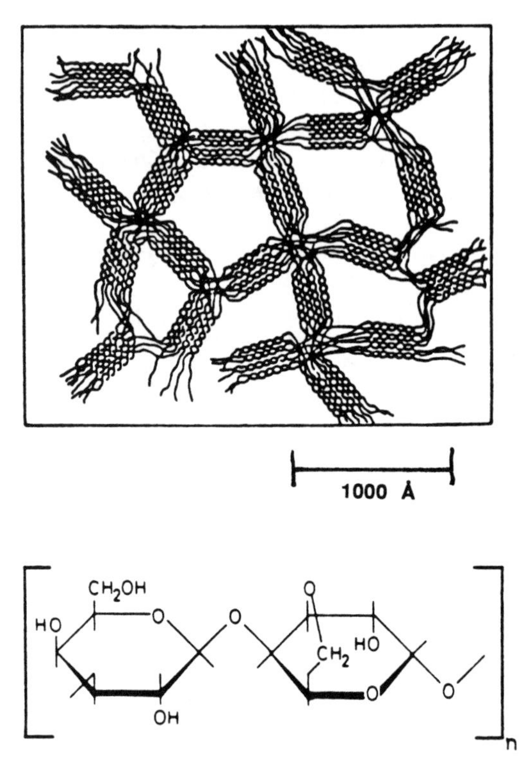

Figure 4.8. Schematic chemical and physical structure of agarose gel. Individual polymer bundles that define the gel structure and aqueous pools consist of several to several hundred double-stranded helices of hydrogen-bonded polysaccharide chains (from Ref. 52).

As a result of the length of the spacer arms, the viologen groups are positioned in the aqueous domains and thus reside in a low viscosity environment of the gel interior. The concentration of the electrochemically active viologen groups in the film (calculated as a ratio of the electrochemically measured quantity of the viologen sites density and the measured swelled film thickness) was varied in the range 0.4–3.2mM. In this range of concentrations, the electron-transport diffusion coefficient $D_{app} = 1.8 \pm 0.4 \times 10^{-7}\,cm^2\,s^{-1}$ did not depend on the viologen concentration in the film (52).

The high value of D_{app} is very surprising. If we assume $\delta = 7\,Å$ and use the highest concentration above, this value of D_{app} would correspond to k_{ex} of $6.9 \times 10^{10} M^{-1} s^{-1}$. This value is three orders of magnitude higher than the literature estimate for the dimethylviologen couple in acetonitrile and simply does not seem to be plausible (54). To account for this surprising result, one has to realize that the system is clearly nonhomogeneous and that the local concentration of the viologen groups is much higher than the value obtained by dividing the quantity of the redox species by the electrode film volume.

To assess the local concentration of the viologen groups on the internal surfaces of the polymer bundles, we noticed first that the total concentration of the viologen species, which was determined electrochemically following an induced structural collapse of the gel film, was much higher, about 5.0–13mM. This means that the electrochemically active viologen groups in a fully swollen film constitute only a small fraction of all the immobilized viologen centers (8–23%). Note also that the increase of the total concentration of the viologen groups from 5 to 13mM (less than threefold) resulted in almost eightfold increases of the electroactive viologen concentration (0.4–3.0mM). These data suggest that the viologen groups are dispersed on the surfaces of the agarose bundles so that only a fraction of these centers have a near neighbor at a sufficiently close distance to exchange electrons. We can estimate this distance to be about 30 Å, the double length of the spacer arm (8 Å) and the viologen molecule itself (7 Å). Thus, the surface concentration of the viologen groups in the electroactive domains is about 2.1×10^{-11} mol cm^{-2}. Since the electron transport proceeds laterally along the polymer bundles, this corresponds to the local concentration of the viologen groups in an imaginary 10-Å layer around the polymer bundles of about 0.2M. A schematic drawing reflecting this situation is given in Fig. 4.9.

The second important point to be made here has to do with the effective distance δ over which the electron is advanced upon single electron transfer (see Figs. 4.2 and 4.9). Since the motion of the viologen groups in a aqueous environment is fast compared to the measured D_{app}, the frequency of collisions of the adjacent redox centers is larger than the frequency of electron transfer. Thus, a single *effective* collision advances the electron over $\delta = 30$ Å, an average distance between the attached viologen groups. If these values are used to

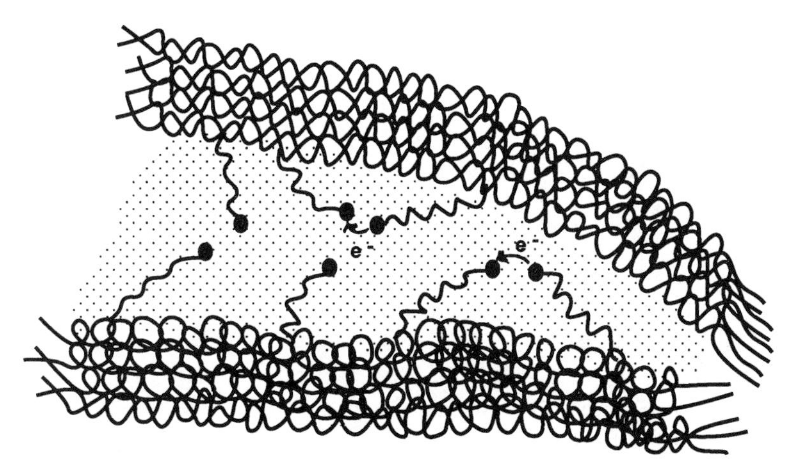

Figure 4.9. A schematic microscopic picture of electron transport in agarose gel derivatized with viologen centers (see Scheme 1). The length of the spacer arm, and the fact that the viologen moieties reside in a medium of low viscosity, lead to efficient encounters of the neighboring centers and electron exchange, despite relatively large distances between their points of attachment.

calculate the apparent rate constant of electron exchange, one obtains $k_{ex} = 3.8 \times 10^7 M^{-1}s^{-1}$. This value is five times higher than the available literature value (54). Since in this case, one could assume that the kinetics of the system is similar to that for dimethylviologen, the reasonably good agreement between the apparent rate constant and its literature value supports the model that was devised here to understand the effect of the structural features of the system on its behavior. The lack of concentration dependence is probably related to two factors. On the one hand, the increase of the viologen sites population appears to contribute primarily to the increase of the level of electroactivity of the system and not to the increase of the viologen concentration in the electroactive domains. The second reason is related to the fact that C and δ are inversely related and thus D_e should, at most, be only weakly dependent on concentration.

There are two important messages stemming from this example. (1) The heterogeneous nature of some polymer films may lead to errors in the assessment of the effective concentration of the electroactive species participating in the electron transport. (2) The distance δ in Eq. 4.15 does not always correspond to the sum of the radii of the redox centers involved in the electron transfer.

A different illustration of the second point was provided in a recent report by He and Chen (55), who investigated electron-transport rate in Nafion films loaded with ruthenium and osmium bipyridyl complexes. Chronocoulometric and ac impedance measurements showed that D_{app} increases significantly faster than the concentration of the redox species. The authors proposed that the increase of the redox sites concentration may lead to a situation where more than two redox centers are instantaneously within electron-transfer distance. This enables a succession of several electron-transfer events to take place upon a single collisional event. This is equivalent to an increase of the electron-hopping distance.

4.2 MIGRATION OF COUNTERIONS

So far, we have considered electron hopping and diffusion of the electroactive centers as the only elements of the electron-transport process. Reexamination of Fig. 4.1, leads, however, to a realization that in the transient experiments such as chronoamperometry, a net charge flow across electrode–film interface requires a concurrent flow of ions across both the film and the film–solution interface. Thus, during the reduction of the redox species in Fig. 4.1, cations of the supporting electrolyte have to migrate into the electrode film and/or anions initially present in the film have to migrate out of the film, in order to maintain its electroneutrality. In many cases, this counterion motion may be the rate-limiting step of the electron transport and thus determines the value of measured diffusion coefficient.

4.2.1 Transient versus Steady State Measurements

It is apparent from the preceding discussion that evaluation of the D_{app} values obtained by transient techniques should involve a determination of the rate-limiting step, in order to properly interpret the measurements. To overcome this uncertainty in the determination of the rate-limiting step of electron transport, Majda and Faulkner (56, 57) developed a luminescence technique in order to determine unambiguously the rate of electron transport unperturbed by counterion effects in a polymer film. Their measurements concerned poly(styrenesulfonate) (PSS) films containing $[Ru(bpy)_3]^{2+}$ ions incorporated into the film via ion exchange. The chronocoulometric measurements of D_{app}, which involved one-electron oxidation of the ruthenium centers, gave $D_{app} = 1.3 \times 10^{-9}\,cm^2\,s^{-1}$ (58). This value was independent of the $[Ru(bpy)_3]^{2+}$ concentration in the film in the range 0.08–0.38M.

The strategy of the luminescence experiment was to obtain D_{app} of electron diffusion under steady state conditions and thus in the absence of the net charge flow across the film. The dynamics of electron diffusion is thus decoupled from counterion migration. In the experiment, the diffusion coefficient of electron propagation was related to the rate of collisional quenching of the excited state, $[{}^*Ru(bpy)_3]^{2+}$ by $[Ru(bpy)_3]^{3+}$ {$[{}^*Ru(bpy)_3]^{2+}$ represents the excited state}. The latter was generated in the electrode film galvanostatically and maintained at a low steady state level. A luminescence–time-decay curve and the accompanying galvanostatic potential–time curve obtained in such experiments are shown in Fig. 4.10 (57). The quenching process is readily apparent from the sharp decay of the luminescence intensity in the initial stage of the experiment, where only about 10% of the $[Ru(bpy)_3]^{2+}$ sites were oxidized. The data analysis leads to the Stern–Volmer plot reproduced in Fig. 4.11 (57). Here, the ratio of luminescence intensities in the absence of the quencher and that following generation of the quencher is plotted as a function of quencher concentration. The linearity of the plot indicates that the quenching process involves biomolecular collisions of $[{}^*Ru(bpy)_3]^{2+}$ and $[Ru(bpy)_3]^{3+}$.

The bimolecular rate constant, which can be obtained from the slope of the Stern–Volmer plot, describes the rate of electron exchange in the absence of counterion migration. Its value, $k_{ex} = 1.2 \times 10^8 M^{-1} s^{-1}$, was found independent of the ruthenium complex concentration in the film. Since the quenching process in the PSS films is diffusion controlled, the D_p obtained, based on k_{ex} and the Smoluchowski equation, was $1.1 \times 10^{-7}\,cm^2\,s^{-1}$. This experiment demonstrated, therefore, that when electron diffusion, which could take place either via electron hopping or physical diffusion of the redox species, is decoupled from counterion migration, its rate is about two orders of magnitude faster than that observed in the transient chronocoulometric experiments. This indicates that in the latter case, the electron-transport process is limited by the counterion migration or by other phenomena related to counterion migration.

In the absence of nonelectroactive ions, the requisite ion migration in electrode films has to be supported by the ionic redox species actively

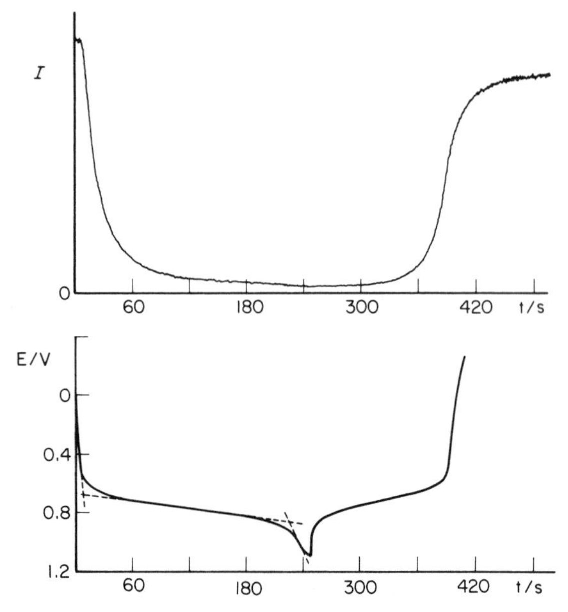

Figure 4.10. Luminescence intensity, **I**, and electrode potential, E, transients recorded simultaneously during a galvanostatic ($i = 3.5\,\mu A\,cm^{-2}$) oxidation and rereduction of $[Ru(bpy)_3]^{2+}$ centers in a poly(styrenesulfonate) film on a Pt electrode. The rapid decay of luminescence intensity is indicative of the effective quenching of the excited state ruthenium complex in its reduced form in presence of only a few percent of the electrogenerated 3+ species. (from Ref. 57).

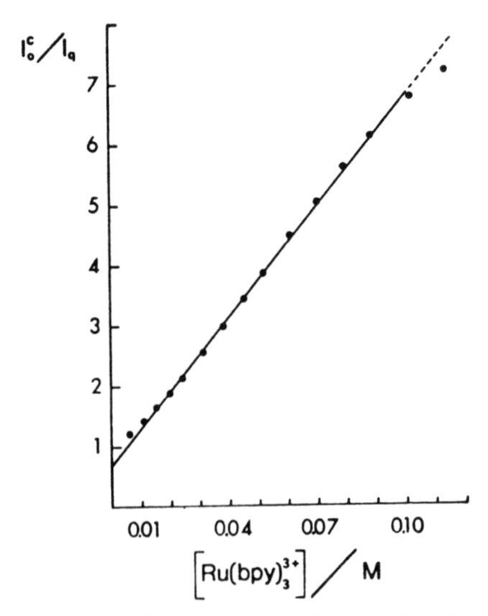

Figure 4.11. Stern–Volmer plot obtained based on the data in Fig. 4.10. The term I_0^c is the luminescence intensity corrected for the depletion of the $[Ru(bpy)_3]^{2+}$ concentration in the course of the galvanostatic oxidation. (Reprinted with permission from Ref. 57.)

participating in the electron-transport process. One such case was described by Doblhofer and co-workers (59–62) who studied quaternized poly(vinylpyridine) films loaded via ion exchange with ferricyanide ions. Their experiment was carried out at a rotating ring-disk electrode with only the disk surface coated with the polymer film (59). An independent control of the disk and ring potential allowed the authors to monitor electrochemically any hexacyanoferrate species leaving the electrode film that was overcoating the disk electrode during the chronoamperometric reduction of $[Fe(CN)_6]^{3-}$ at the disk.

The key results are reproduced in Fig. 4.12 (59). The large dashed-line cathodic transient represents the reduction of the film $[Fe(CN)_6]^{3-}$ at the disk at the beginning of the experiment. If the potential of the ring electrode is set at +0.4 V, the rate of arrival of $[Fe(CN)_6]^{4-}$ ions, the product of the disk reaction can be monitored (curve *b*). The onset of the ring current at this potential is delayed by about 0.3 s, which corresponds to the sum of the intrinsic disk–ring transit time (0.08 s) and the time required for the ferrocyanide ions to diffuse across the electrode film. In contrast, when the ring potential is set at −0.2 V to monitor the arrival of ferricyanide ions, the delay time of the onset of the ring current (curve *a*) is only about 0.1 s. This demonstrates that ferricyanide ions leave the film immediately, after application of the negative potential to the disk

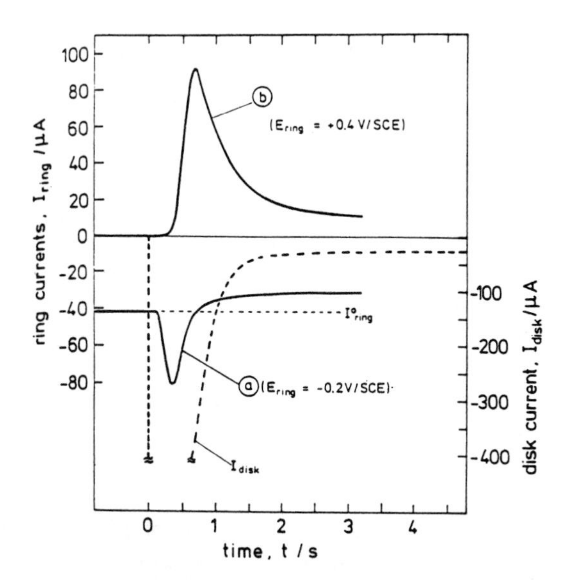

Figure 4.12. Current transients recorded simultaneously at a rotating disk electrode (I_{disk}), and a ring electrode (continuous lines). The disk electrode is coated with a poly(4-vinylpyridine) film loaded with ferricyanide ions. The electrolyte solution is $2 \times 10^{-4} M$ $K_3Fe(CN)_6$ and $0.1 M$ KCl. Potential of the ring electrode can be set at +0.4 V (curve *b*) or −0.2 V (curve *a*) to monitor the rate of arrival of either ferrocyanide or ferricyanide ions, respectively. The shorter delay time of the ring transient *a* compared to *b* suggests that ferricyanide ions are driven by an electric field out of the film as a result of the application of a negative potential to the disk electrode (see text). [Reprinted with permission from Ref. 59.]

electrode. Since there is no concentration gradient of $[Fe(CN)_6]^{3-}$ ions in the outer layer of the film (near the film–solution interface) at the beginning of the disk experiment, the expulsion of ferricyanide ions is driven by the electric field created in the film at the onset of the electroreduction. It can, therefore, be classified as migration. These results suggest that, despite the fact that the experiments were carried out in $0.1M$ KCl solution, $[Fe(CN)_6]^{3-}$ are the only mobile ions in the polymer film. The Donnan exclusion breaks down when the KCl electrolyte concentration is raised to $1M$, when both K^+ and Cl^- ions enter the film and assume the role of the migrating species (59). Similar migration effects combined with Donnan exclusion were also reported by Elliott and Redepenning (63) in thick Nafion coatings.

4.2.2 Electrochemical Steady State Methods

Some of the deficiencies of the transient electrochemical techniques of the D_{app} measurements, such as chronocoulometry and chronoamperometry, can be overcome by the electrochemical steady state techniques. This approach, in general, involves a bipotentiostatic system where current is passed across a polymer film sandwiched between two working electrodes. Under limiting steady state conditions (i.e., the electroactive species in the film are fully reduced and oxidized, respectively, at the cathode and the anode), the magnitude of the current density is proportional to the diffusive flux of electrons, which can be expressed in terms of Fick's first law (in the absence of the migration effects):

$$i = n\mathscr{F}CD_{app}/d \tag{4.22}$$

Here d is the thickness of the polymer film. Murray and his co-workers (64, 65) developed two experimental approaches where D_{app} can be measured at steady state in a bipotentiostatic system. The design of the experimental set-up in these methods is schematically illustrated in Fig. 4.13. The first one is a true sandwichlike assembly, where a polymer film at a metal electrode is overcoated with about a 400-Å layer of gold in vacuum (64). The Au layer plays the role of a second electrode. It is sufficiently thin to be completely permeable to the solvent and ions, and thus it does not interfere physically with the electrochemical behavior of the system.

The second design is an array of narrowly spaced microelectrodes arranged in an interdigitated pattern of two independent electrodes (65). Figure 4.13*B* shows a cross section of a single pair only. Arrays of microelectrodes were first introduced by Wrighton (5, 66–68) in studies of conducting polymers such as poly(pyrrole) and poly(aniline). The polymer film is deposited over the entire array. At steady state, electron transport proceeds laterally through the polymer in between two adjacent electrodes. Conceptually, this arrangement functions in an analogous fashion to the "sandwich" dual-electrode system; thus, it is often referred to as an "open sandwich" arrangement.

Using these methodologies, Murray and his co-workers (49, 65, 69) examined electron transport in a series of electropolymerized electrode coatings

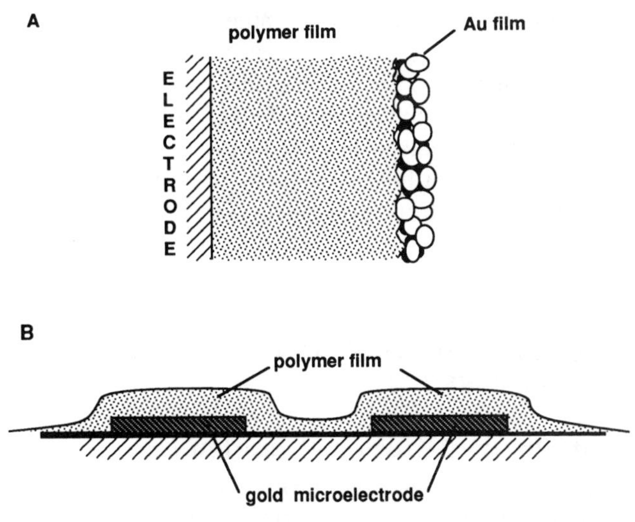

Figure 4.13. A schematic representation of two experimental approaches to steady state measurements of charge-transport processes in polymer films. (*A*) A "sandwich" electrode, where a polymer film deposited at the electrode surface is overcoated with about 400-Å gold layer that remains permeable and can function as a second electrode. (*B*) An "open sandwich" scheme, involving an interdigitated array of microelectrodes (only a single pair is shown here) overcoated with a polymer film (see text and Refs. 64 and 65).

based on such monomers as $[Os(bpy)_2(vpy)_2]^{2+}$, $[Os(bpy)_2(p\text{-cinn})_2]^{2+}$ (64), $[Ru(vbpy)_3]^{2+}$ (49), and $[Fe(vbpy)_3]^{2+}$ (64), where p-cinn = *N*-(4-pyridyl)cinnamamide (4-py-NHCOCH=CHPh) and vbpy = 4-vinyl-4′-methyl-2,2′-bipyridine (see Fig. 4.6). Some transient electrochemical measurements of D_{app} for the poly$[Os(bpy)_2(p\text{-cinn})_2^{2+}]$ were presented in the previous section (see Fig. 4.7 and the related discussion). In all of these cases, the authors found that both the transient and the steady state measurements led essentially to the same values of D_{app} (49, 50, 64). This result demonstrates that the mobility of the counterions is greater than the "mobility" of electrons in these polymers. Since the electron transport is limited in these cases by the kinetics of the electron transfer, it appears that despite extensive cross-linking, the polymer network is fairly permeable to small molecules and ions.

Ion permeability of this class of redox polymers is rarely encountered in other cases where, consequently, counterion migration imposes limitation on the rate of electron transport. In those instances, the transient and the steady state determinations of D_{app} yield significantly different results. A notable example of such behavior was recently presented by Faulkner and his co-workers (70), who investigated the dynamics of charge transport in copolymeric films of styrene and vinylpyridine cross-linked in situ by reaction of the pyridine groups with 1,2-dibromoethane. The electrode films were loaded by ion exchange with $[Os(bpy)_3]^{2+}$. The value of D_{app} obtained by a dual-electrode steady state method analogous to Murray's "open sandwich" arrangement was almost two orders of magnitude higher than the one obtained chronocoulometrically. The higher, steady state value of D_{app}, which describes the rate of electron transport decoupled from the counterion migration effects, is also the value relevant in the mediation and catalytic applications of the polymer coated electrodes. Considering the scheme involved in electrocatalysis shown in Fig. 4.14, one realizes that the electron shuttling is a steady state process analogous to that encountered in the dual-electrode experiment. Thus, the steady state techniques are indispensable in the measurements of the electron transport rate in electrode films.

4.3 THEORETICAL TREATMENTS OF THE COUNTERION MIGRATION EFFECTS

4.3.1 Nernst–Planck Diffusion Migration Law

The description of several experimental cases in the previous section demonstrated the involvement of counterion migration in electron transport. These effects are particularly visible in the nonsteady state potential step experiments. Even though these types of effects can be severe, it is not well understood exactly how the electric field generated in the electrode coatings influences the charge-transport process. In addition to counterion migration–diffusion dynamics, one should mention here the effect of the electric field on the heterogeneous electron transfer and the kinetics of ion transfer across the film–solution interface, as well as the thermodynamics of ion partitioning and

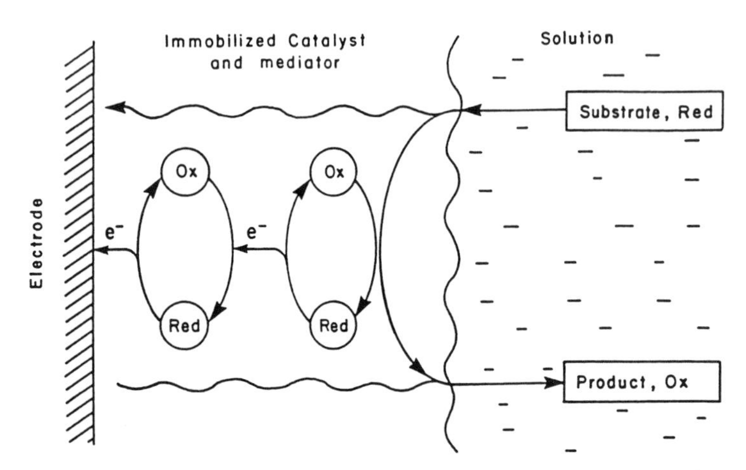

Figure 4.14. A schematic representation of an electrocatalytic scheme involving a polymer coated electrode. An electron-transfer catalyst or mediator, Ox/Red is immobilized in a polymer film. Substrate diffuses into the electrode film where it is converted to product in an electron-transfer reaction with the oxidized catalyst. Electron transport between the individual catalyst sites and the electrode surface involves diffusion and electron hopping. When the flux of substrate towards the electrode film is at steady state, the electron transport also assumes a steady state rate.

Donnan exclusion processes. In view of the complexity of these effects, theoretical predictions and modeling become necessary to sort out the relative significance of various factors, and to provide the experimental criteria that would aid in the design of experiments.

Several theoretical approaches have appeared recently to address these issues (71–81). Recent results of Savéant (74–79) and Buck (80, 81) are particularly interesting both because of the nature of the predictions they make and because their formulation can be used as a basis of experimental verification. In order to treat the motion of nonelectroactive ions in an electric field, the Nernst–Plank equation is used.

$$J_{\mathrm{I}} = -D_{\mathrm{I}}\left[\frac{dC_{\mathrm{I}}}{dx} - \left(\frac{z_{\mathrm{I}}\mathscr{F}}{RT}\right)C_{\mathrm{I}}\frac{d\Phi}{dx}\right] \tag{4.23}$$

The unidimensional flux of ions, J_{I}, is proportional to the gradient of concentration and the electric field in the polymer film. To describe the effect of an electric field on electron propagation, Buck proposed the same equation, but with C_{I} replaced by C_{E}, the total concentration of the redox sites (80). This is equivalent to treating electrons as mobile "ions" hopping between fixed oxidized redox sites.

This approach was criticized by Savéant (76), who derived, and later implemented in a number of cases discussed below, an equivalent Nernst–Planck diffusion–migration equation for the electron hopping, in which the bimolecular character of the electron hopping is taken into account (74) (see also erratum to Ref. 74 in Ref. 75). The electric field induced electron-transfer rate constant takes the following form, where the activation energy barrier was assumed to be symmetrical and thus $\alpha = 0.5$.

$$k = k^{\circ}\exp[-n\mathscr{F}(\Phi_2 - \Phi_1)/2RT] \tag{4.24}$$

Here k° is the electron-exchange rate constant, and the potential difference decreases between the sites *1* and *2*. Following rhe derivation scheme analogous to that for the electron-hopping diffusion equation (Eq. 4.11–4.14 above) (12), the combined diffusion–migration flux expression is the following:

$$\begin{aligned} J_{\mathrm{E}} &= -D_{\mathrm{E}}\left[\frac{dC_{\mathrm{A}}}{dx} + \frac{n\mathscr{F}}{RT}C_{\mathrm{A}}\left(1 - \frac{C_{\mathrm{A}}}{C^{\circ}}\right)\frac{d\Phi}{dx}\right] \\ &= D_{\mathrm{E}}\left[\frac{dC_{\mathrm{B}}}{dx} - \frac{n\mathscr{F}}{RT}C_{\mathrm{B}}\left(1 - \frac{C_{\mathrm{B}}}{C^{\circ}}\right)\frac{d\Phi}{dx}\right] \end{aligned} \tag{4.25}$$

where C_{A} and C_{B} are the concentrations of the oxidized and the reduced species, while C° is their total concentration (74, 75). As in the original derivation, $D_{\mathrm{E}} = k^{\circ}C^{\circ}\delta^2$ (see Eqs. 4.10–4.14 in Section 4.1.3).

4.3.2 Steady State Diffusion and Migration in Counterion Conservative Systems

The results stemming from both types of formalisms (Eq. 4.23—this case is termed "electron displacement" and Eq. 4.25—this case is referred to as electron hopping) were obtained and compared by Savéant in the analysis of electron transport across two types of polymer films (76). The systems were considered under dual-electrode steady state conditions (see Fig. 4.13*A*). In the first case, the immobilized redox couple has $3+/2+$ charge, which is matched exactly by the singly charged mobile counterions. No other ions are present in the film. This case can be considered as a model of Murray's poly[$Os(bpy)_2(vpy)_2^{2+}$] system (43, 44, 49). The second case models Nafion-like films with $[Ru(bpy)_3]^{3+/2+}$ species (17, 33). An important assumption, however, is made that electrons can propagate only via an electron-hopping mechanism. The singly charged and immobile sulfonate groups are the only anions. Their charge is matched exactly by the ruthenium species and mobile sodium ions.

In both cases the initial "redox composition factor" or the ratio of the oxidized (A) and reduced (B) forms of a redox couple is fixed and kept constant. This fixes the total charge of the redox species and hence the concentration of the counterions. At steady state, which is the object of the theoretical treatment, no ions can enter or leave the polymer film. The potential difference applied between the electrodes sandwiching the film is composed of the interfacial potentials obeying the Nernst equation and of the potential drop in the film itself. Interfacial electron transfer is assumed to be fast. At steady state, the diffusion and migration of the mobile, electroinactive counterions compensate each other as expected from the Nernst–Planck equation. Charge neutrality is obeyed everywhere in the film. For the sake of simplicity, the activity and ion-pairing effects were not taken into account (76).

The results obtained for the first type of polymer predict that the maximum steady state current should be about 7 and 22% higher (for the electron-hopping and the "electron displacement" cases, respectively) than the current due to electron hopping across such films in the absence of the electric field (see Fig. 4.15) (76). The maximum current is expected at a higher than 1:1 ratio of A–B, namely, 1.0051 and 1.1445 for the two versions of the Nernst–Planck equation. Finally, only the "electron displacement" formalism predicts slightly nonlinear concentration profiles of A, B and of the nonelectroactive counterions across the electrode film (see Fig. 4.16) (76).

More striking predictions are made for the second type of polymer, which models Nafion-type films. Here, the concentration of the electroinactive mobile ions can be very small, depending on the ratio of the electroactive and bound nonelectroactive ions. In such cases, a significant electric field exists in the film leading to large increases of the limiting current compared to its value in the absence of the electron-hopping migration effect. A representative set of results corresponding to both the electron-hopping and "electron displacement" formalism is reproduced in Fig. 4.17 (76). The ratio of the steady state

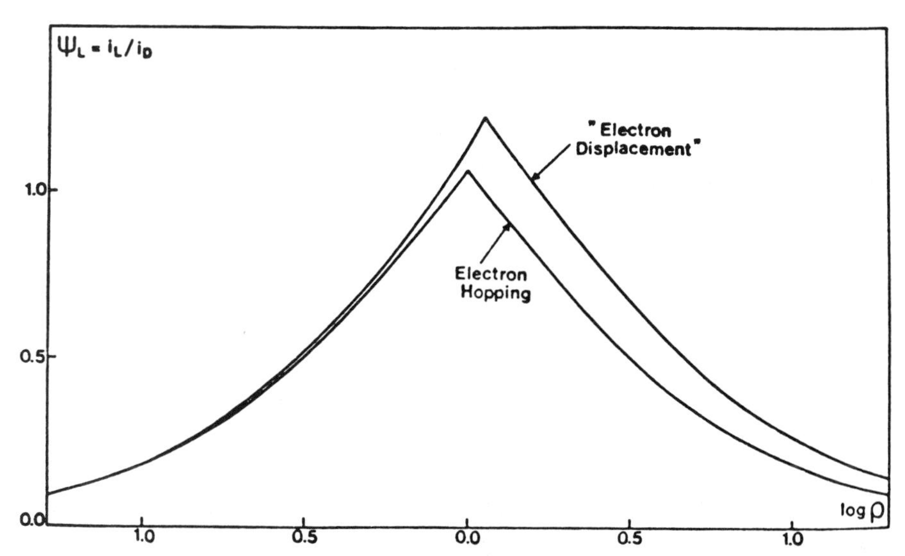

Figure 4.15. Theoretically predicted ratio of a limiting steady state current (under two-electrode "sandwich" configuration of Fig. 4.13*A*) in the presence (i_L) and in the absence (i_D) of the "migration effects" as a function of the initial concentration ratio, C_{Red}/C_{Ox} in the electrode film. Ox/Red is a 3+/2+ type redox couple covalently bound within the polymer matrix. Singly negatively charged mobile counterions are assumed to be present only at the level required by charge neutralization (see text. [Reprinted with permission from Ref. 76.]

migration–diffusion and purely diffusion controlled currents is plotted as a function of log ρ ($\rho = C_{Red}/C_{Ox}$). Individual curves correspond to different values of $f^\circ = C_F/C^\circ$, the concentration ratio of the fixed electroinactive and electroactive ions. This parameter determines the concentration of the mobile counterions through the electroneutrality rule. For example, $f^\circ = 2.75$ and $\rho = 1$ represent a case where $C^\circ = 0.364$, $C_{Ox} = C_{Red} = 0.182$, and the concentration of mobile cations is 0.091. (All the concentrations are expressed in dimensionless form scaled by the concentration of the fixed counterions in the film.) When $f^\circ = 3$, the initial redox composition factor cannot span the range of $\rho < 1$, since the total charge of the electroactive 3+/2+ species would exceed the charge of the bound anions. The $f^\circ = \infty$ curve corresponds to abundance of mobile counterions and is thus free of electric field effects. In addition to the steady state current densities, also included in the paper are the dimensionless concentration and electric field profiles, current potential curves, and "zero-current" conductance curves calculated as a function of ρ and f°.

As one can readily appreciate from Fig. 4.17 and Eq. 4.25, the electric field leads to an enhancement of the electron transport. The magnitude of the enhancement increases as the polymer film becomes more counterion deficient. The electron-hopping formalism leads to less drastic electric field effects than those predicted when the "electron displacement" approach is adopted. In both

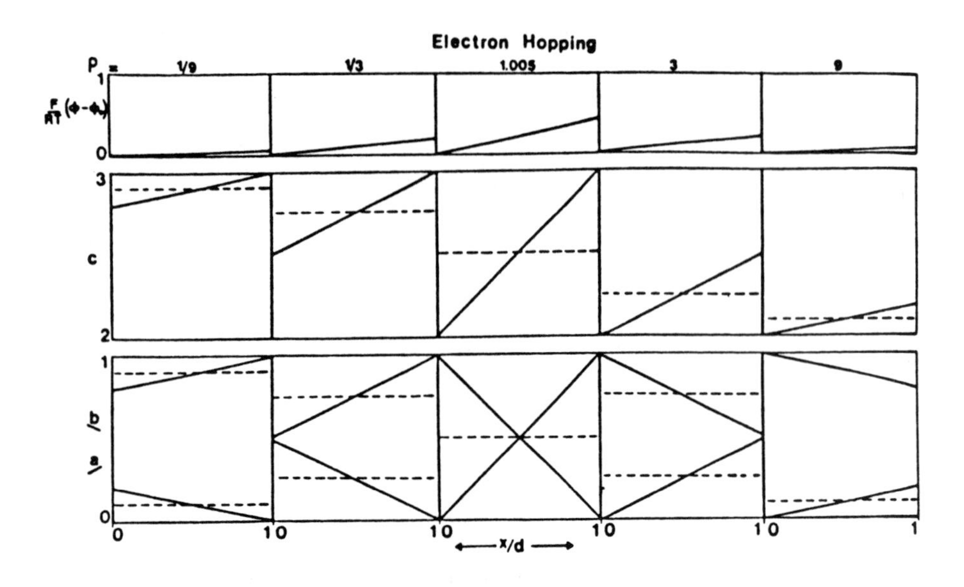

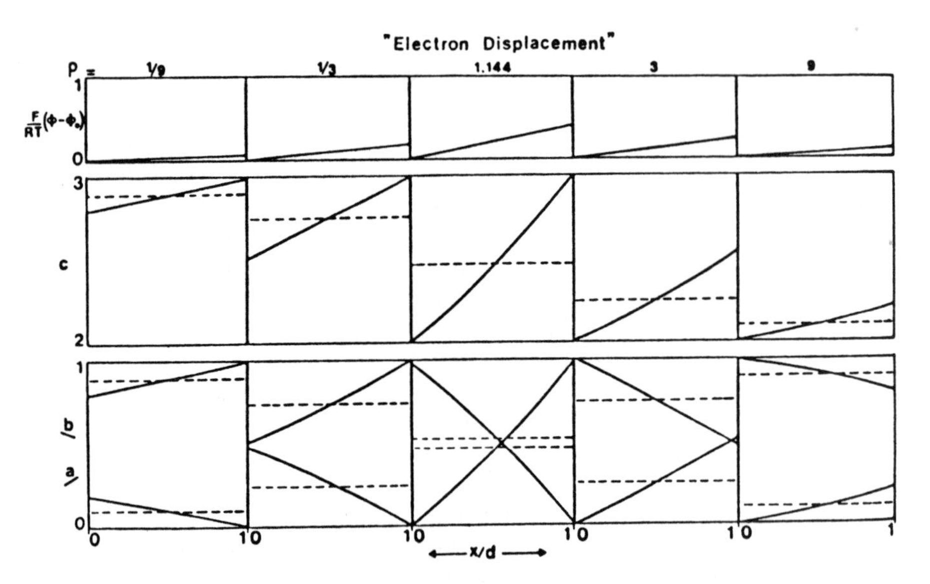

Figure 4.16. Diagrams illustrating theoretical predictions for the shape of electric field and concentration profiles in the polymer film under conditions described in Fig. 4.15. The values of ρ are 1.005 and 1.144 and they refer to the maximum current conditions in Fig. 4.15. a, b, and c refer to the concentrations of Ox, Red, and counterions, respectively, normalized with respect to the total concentration of the redox species (see text). [Reprinted with permission from Ref. 76.]

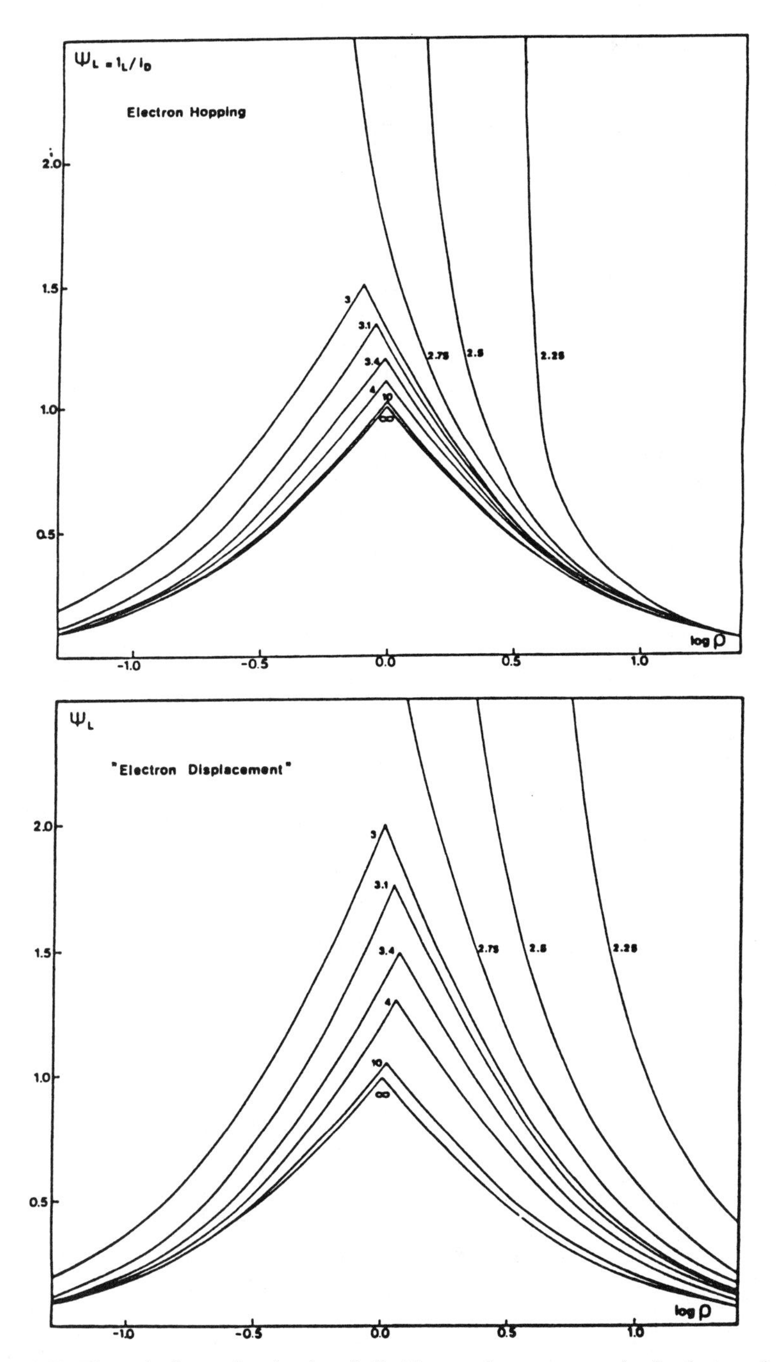

Figure 4.17. Theoretically predicted ratios of a limiting steady state current (under the two-electrode "sandwich" configuration of Fig. 4.13*A*) in the presence (i_L) and in the absence (i_D) of the "migration effects" as a function of the initial concentration ratio, C_{Red}/C_{Ox} in the polymer film. Calculations were done for a $[Ru(bpy)_3]^{3+/2+}$ in a Nafion-type case (except that redox sites were assumed to be immobile). Various plots correspond to various concentration levels of mobile counterions (e.g., Na^+) expressed as $f°$, the ratio of their concentration and the total concentration of the redox species (see text). [Reprinted with permission from Ref. 76.]

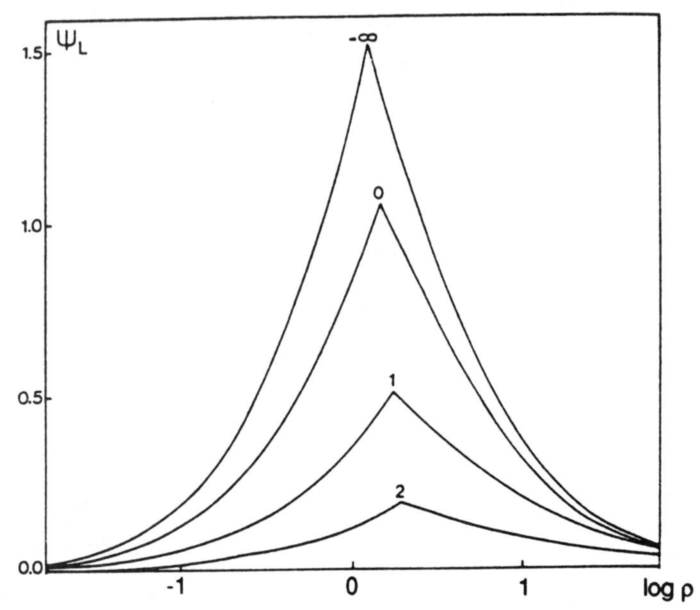

Figure 4.18. The effect of ion-pairing equilibria between singly charged Ox and counterions C^- (Eqs. 4.26 and 4.27 in the text) on the dimensionless current (under steady state "sandwich" conditions) defined in Figs. 4.15 and 4.17. The four plots were obtained for various values of a dimensionless parameter, κ expressing the magnitude of the ion-pairing equilibrium constant. [Reprinted with permission from J.-M. Savéant, *J. Phys. Chem.* 1988, **92**, 4526. Copyright © (1988) American Chemical Society.]

cases, the steady state currents are independent of the nonelectroactive counterion mobility. The latter, as pointed out above, remain macroscopically immobile.

4.3.3 Effect of Ion-Pairing Equilibria

Equally detailed treatment was given, in Savéant's subsequent paper, to a similar steady state cases where mobile electroinactive ions can engage in ion pairing with the bound electroactive ions (78). The following equilibria involving A^+/B redox couple and C^- counterions were considered:

$$A^+(1) + B(2) = B(1) + A^+(2) \tag{4.26}$$

$$A^+ + C^- = AC \tag{4.27}$$

Only the free A^+ ion was considered to be electroactive, while its ion pair, AC cannot participate directly in the electron transport. The ion-pairing equilibrium was represented by a dimensionless parameter κ, the product of the equilibrium constant, and the total film concentration of C^-. Other constraints and assumptions are the same as in the previous case. Some of the most notable results are shown in Fig. 4.18, where again a dimensionless limiting steady state current, $\Psi = i_L/i_D$, is plotted as a function of the log ρ for different values of κ (78). As could be expected, the limiting current decreases drastically as the ion pairing becomes more extensive. The curve for the case with negligible ion pairing, $\kappa = -\infty$, exhibits only the electric field enhancement of the electron hopping as discussed above.

A different aspect of the ion-pairing effect is shown in Fig. 4.19. In this case, the system in question contains also bound electroinactive cations, F^+ (78). Their total concentration equals C_F. A dimensionless diffusion coefficient of electron-hopping bearing migration effect normalized by its purely diffusive value is plotted versus mole fraction of the electroactive ions $X_E = C_E/(C_E + C_F)$ for various values of κ. (The variation of X_E was accomplished by changing C_E/C_F with $C_E + C_F$ remaining constant.) Again, one observes a decrease of D_{app} with the increase of the extent of ion pairing. More interesting, however, is the shape of the individual curves predicting the dependence of D_{app} on the concentration of the electroactive species. Even in the case of negligible ion pairing, the dependence is nonlinear. This is a result of the "migration" effect. The extent of nonlinearity increases, however, as the ion pairing becomes more significant (78). Such nonlinear dependence of D_{app} on concentration was reported in several recent experimental studies (55, 82, 83).

4.3.4 Migration Effects Under Transient Conditions

As the last theoretically treated case of electric field effects on electron transport, we will consider a system under transient, chronoamperometric conditions, presented recently by Andrieux and Savéant (79) (also see Ref. 77). The system is essentially analogous to that considered above under steady state conditions (78). Here again, the authors do not consider ion pairing and activity

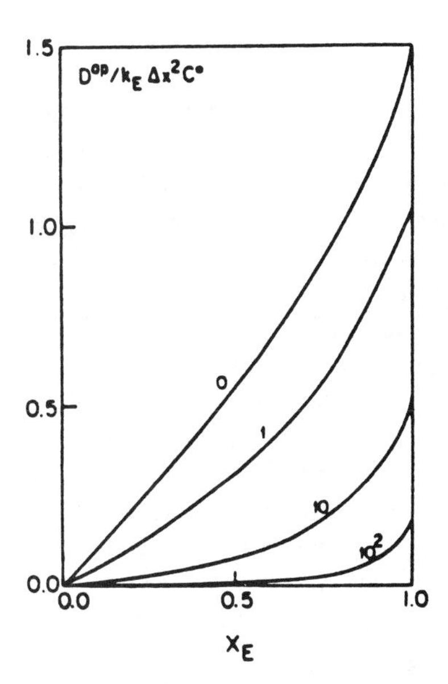

Figure 4.19. The effect of ion-pairing equilibria on dimensionless diffusion coefficient of electron hopping (normalized by its purely diffusive value) as a function of electroactive ions concentration. The latter is expressed here as a mole fraction of all bound ions, x_E. Numerical values next to the individual plots represent κ, a dimensionless parameter proportional to the ion-pairing equilibrium constant (see Eq. 4.27 in the text). [Reprinted with permission from J.-M. Savéant, *J. Phys. Chem.* 1988, **92**, 4526. Copyright © (1988) American Chemical Society.]

effects. The following parameters have to be introduced in order to understand the presentation of some key results. The redox sites A/B are bound to a polymer matrix. The total concentration of A + B is C_E°. Electron hopping is the sole mechanism for electron transport. In addition, two types of counterions are also present in the film: C—mobile and F bound ions. The charges of the ionic species are z_A, z_B, z_C, and z_F. The relative concentration of the bound nonelectroactive ions is expressed by γ, a dimensionless parameter that depends on the charges of the ionic species in the film: $\gamma = (z_A - n + z_F f^\circ)/n$. The term f° is the concentration ratio C_F°/C_E° and n is a number of electrons exchanged in an elementary step defined as positive for reduction. An important new variable is σ, the ratio of ion and electron intrinsic mobilities defined by $\sigma = D_I/D_E$.

The authors solved a set of Fick's second law equations containing the second-order migration term as described above in the flux expression (Eq. 4.25). Semiinfinite conditions and film electroneutrality were assumed in the derivation. A limiting case characteristic for a large potential step experiment was considered, that is, electron and ion transfer at the electrode–film and film–solution interfaces, respectively, are fast. The authors refer to these conditions as "plateau-current response", which carries a reference to the steady state conditions but describes a transient $i \propto 1/\sqrt{t}$ type current decay. These characteristics do indeed prevail in the presence of the electric field effects.

The most interesting and intuitively unexpected result is the prediction that the migration effects lead to current enhancement compared to purely diffusive conditions, regardless of whether electroactive cations or anions are being reduced in the film (79). This would not be observed if the electron transport involved physical diffusion of the redox ions. The following argument provides a qualitative explanation of this effect. Regardless of the charge of the mobile counterions, the electroreduction in the film creates a potential, which increases from the electrode surface towards the interior of the film, as the mobility of the counterions lags the electron hopping. This results in a migrational enhancement of electron hopping. If, however, electron transport involves physical diffusion of the charged redox species, the current enhancement by the electric field is expected only in the case of the reduction of cationic species, which would be attracted towards the electrode surface. The same argument predicts cathodic current depression for the anionic redox species. In all of these cases, the extent of the migration effects increases as the relative intrinsic mobility of the counterions (expressed by σ) and their relative population (expressed by γ) decrease. A representative set of data is shown in Fig. 4.20 (71).

Also important and interesting are the predictions of the dependence of the observed diffusion coefficient, D_{app}, on the concentration of the redox species, C_E°. The answer to this question is not straightforward because the variation in C_E° has to be expressed in terms of the changes in γ and thus C_F°. Also, different relationships between D_{app} and D_E are obtained, depending on the magnitude of σ. Overall, the theory calls for different behaviors ranging from a close to linear to cubic dependence of D_{app} on C_E°, when σ decreases from ∞ to 0.

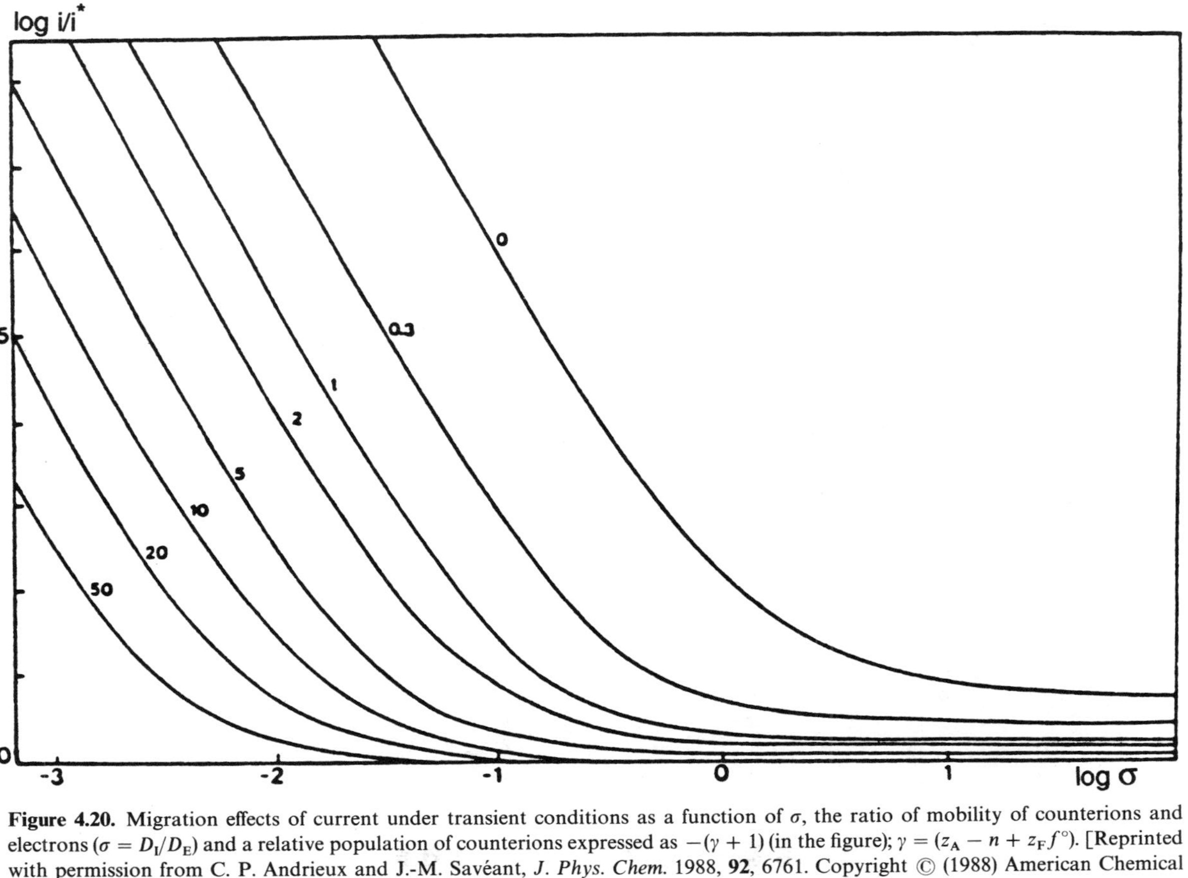

Figure 4.20. Migration effects of current under transient conditions as a function of σ, the ratio of mobility of counterions and electrons ($\sigma = D_I/D_E$) and a relative population of counterions expressed as $-(\gamma + 1)$ (in the figure); $\gamma = (z_A - n + z_F f^\circ)$. [Reprinted with permission from C. P. Andrieux and J.-M. Savéant, *J. Phys. Chem.* 1988, **92**, 6761. Copyright © (1988) American Chemical Society.]

An interesting illustration of this prediction is a fit of the experimentally obtained dependence of D_{app} on the redox sites concentration obtained by Facci et al. (50) to one of the limiting cases (79). The polymer system in question, a mixture of the electropolymerized $Os(bpy)_2(p\text{-}cinn)_2^{2+}$ and its ruthenium analogue, was described above (Fig. 4.7 and the related discussion in Section 4.15) (50). In this case, the Ru sites are considered as bound electroinactive ions, that is, $z_F = +2$. If the diffusion coefficient of ClO_4^-, the mobile counterion, is assumed to be very small relative to D_E so that $\sigma \to 0$, than a good agreement between the theory and the experimental data can be obtained as shown in Fig. 4.21. From the fit, the authors obtained $D_{app}/D_e = 5.5$. The observed diffusion coefficient is five times larger than its true value under conditions unperturbed by the electric field.

The analysis of data in Fig. 4.21 is an example of the applicability of Savéant's theoretical calculations to experimental verifications. Since all his results are presented in the form of dimensionless plots and equations, this should not be difficult. What may be more difficult is reproduction of the often stringent physical assumptions imposing difficulties on the experimental reality. Considering the case just discussed, Murray and his co-workers (50) demonstrated that diffusion of anions like ClO_4^- in similar polymer films is rather fast. Thus, the assumption made above that $D_I/D_E < 10^{-2}$ may not be justified. On the other hand, Murray's experiments were carried out in the presence of $0.1M\ Et_4NClO_4$ and no effort was made to exclude or to measure the concentration of the nonelectroactive ions in the film. The calculated behavior assumes a "counterion conservative" system, where no excess of electrolyte in the film is present.

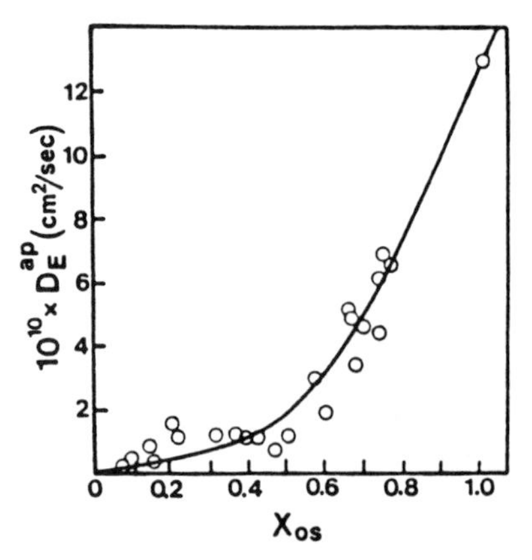

Figure 4.21. Simulation of the experimental conditions existing in the Os/Ru copolymer described in Figs. 4.6*f* and 4.7, with the model of electron hopping bearing migration effects induced by low mobility of counterions relative to electron transport. [Reprinted with permission from C. P. Andrieux and J.-M. Savéant, *J. Phys. Chem.* 1988, **92**, 6761. Copyright © (1988) American Chemical Society.]

4.4 ELECTRON TRANSPORT IN RIGID SYSTEMS

4.4.1 Extended Electron Transfer

Electron hopping has been treated so far as a bimolecular process taking place upon collision of neighboring redox centers. This requires a certain fluidity of the polymer matrix. In cases of rigid networks of redox species, where a combination of a lack of solvent swelling, high extent of cross-linking and high degree of loading may prevent bimolecular collisions, it becomes important to consider "extended" electron transfer. This involves electron hopping between rigidly held nearest neighbor centers, over distances exceeding their combined radii. Theoretical modeling of the concentration dependence of D_{app} in these types of systems was recently reported by Fritsch–Faules and Faulkner (84). Formal description of the electron-transfer process involves unimolecular kinetics. The rate constant depends exponentially on the edge-to-edge separation of the donor and acceptor.

$$k_{et} = k_0 \exp[-\beta(r_{NN} - r^{\circ})] \tag{4.28}$$

k_0 is a rate constant describing an intrinsic kinetic facility of a redox couple, r_{NN} is the average near neighbor distance, r° is the diameter of the redox center, and $1/\beta$ is a characteristic distance describing the extent of electronic overlap between the donor and acceptor. The expression for electron-transport diffusion coefficient stems from the stochastic nature of electron hopping. For a three-dimensional case this is

$$D = k_{et} r_{NN}^2 / 6 \tag{4.29}$$

The authors combined Eq. 4.28 and 4.29 to obtain the dependence of D on concentration. As could be expected, D increases exponentially with concentration in relatively dilute systems. The rise of D becomes less sharp and approaches a limiting value when the concentration reaches its packing limited value. It would be tempting to use this model to explain the data of Facci et al. (Fig. 4.7 in Section 4.1.5). Their polymer film is highly cross-linked and thus might reflect the assumptions of this model. Also, the results of He and Che (55) discussed in Section 4.1.5 showed a similar sharp rise of D_{app} with the concentration of $[Os(bpy)_3]^{2+}$ in Nafion films.

4.4.2 Electric Field Driven Electron Transport

Perhaps the best example of a rigid system, where extended electron transfer could be in operation, was presented recently by Jernigan, Murray and their co-workers (69, 85–87). They investigated steady state electron transport in poly$[Os(bpy)_2(vpy)_2]^{2+}$ type polymers in the absence of a solvent and at low temperatures. There are two very interesting aspects of these measurements. One of them deals with the solvent effects on the electron transport rate (69).

This also encompasses a question of whether solvent is required at all for the steady state electron transport to take place. We will deal with these issues in Section 4.5.

The second issue of interest is the driving force in the electron transport (85–87). So far, we have considered systems where the electron transfer at the electrode–film interface generated a concentration gradient of redox species. This provided for the directional flow of charge across the electrode film Jernigan and Murray (85, 87) showed that the electric field alone applied across a redox polymer film can result in specific cases in the same directional flow of charge. The effect of the electric field is therefore equivalent to the concentration gradient in fueling charge transport.

Consider a poly[$Os(bpy)_2(vpy)_2(ClO_4)_2$] film, 100–500 nm thick, sandwiched between two parallel electrodes as discussed above (see Fig. 4.13*A* and the related discussion in Section 4.3.2). Applying potentials of +1.2 and 0.0 V to the two electrodes in a 0.1 *M* Et_4NClO_4 acetonitrile solution, the redox composition of the polymer was brought to the 50:50 oxidized–reduced state. Before the potential was disconnected, the sandwiched electrode was removed from the electrolyte solution, rinsed with CH_3CN, and dried with N_2. The potential bias between the electrodes, ΔE, was then brought to zero and the concentration profiles existing in the film allowed to relax. With the assembly still in dry N_2, a linearly increasing voltage bias was applied between the electrodes. Depending on the voltage sweep rate, two types of current–voltage curves were obtained as shown in Fig. 4.22 (87).

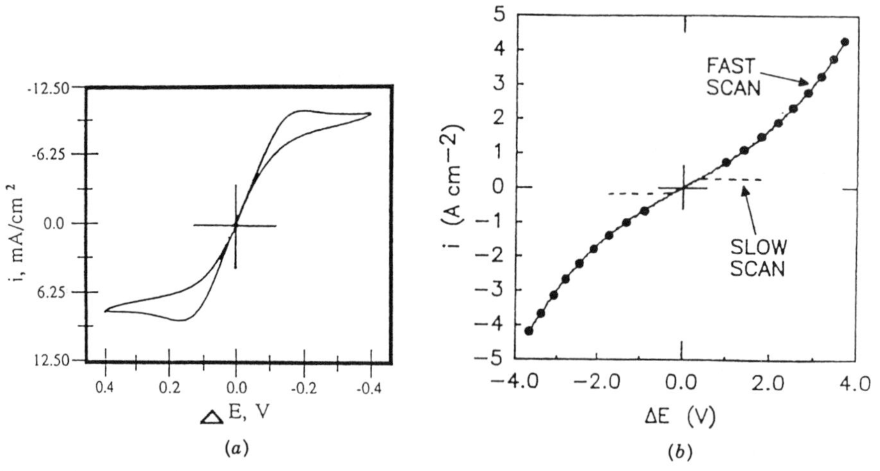

Figure 4.22. Two-electrode ("sandwich" configuration) current versus voltage bias curves for poly[$Os^{III/II}(bpy_2(vpy)_2(ClO_4)_{2.5}$] in dry N_2 at room temperature. *A* is a slow ΔE scan; current is driven by the concentration gradients of Os(III) and Os(II), which can develop across the polymer film under these polarization conditions. *B* is a fast ΔE scan; in the absence of the concentration gradient of redox sites, current is a result of a site-to-site voltage gradient. [Reprinted with permission from J. C. Jernigen, N. A. Surridge, M. E. Zvanut, M. Selver, R. W. Murray, *J. Phys. Chem.* 1989, **93**, 4620. Copyright © (1989) American Chemical Society.]

A slow voltage sweep of about $8\,\mathrm{mV\,s^{-1}}$ produced a sigmoidal response (frame A) characteristic for a steady state response driven by a concentration gradient developed across the film. Despite the dry state of the film bathed in a stream of N_2, the slow voltage bias resulted in a spatial redistribution of the charges to establish a linear and mutually opposing concentration gradients of the Os(II) and the Os(III). Furthermore, ClO_4^- ions migrated towards the oxidized side of the assembly in fulfillment of the charge neutrality requirement. This was possible since the prior equilibration of the system as poly$[Os(bpy)_2(vpy)_2 \cdot (ClO_4)_{2.5}]$ did not require any net charge to enter or leave the polymer film to reach the final steady state.

A different current voltage profile was observed when the experiment was carried out at $1000\,\mathrm{V\,s^{-1}}$ (Fig. 4.22*B*) (87). Under these conditions, the voltage bias was imposed too fast for the system to respond by establishing concentration gradients as it did previously. Instead, the observed current is driven by the voltage gradient applied across the film. In microscopic terms, ΔE of 1 V corresponds to Φ of about 4.4 mV per 12-Å intersite distance. The observed exponential dependence of current on voltage bias can be expressed as:

$$i = i^\circ[\exp(-n\mathscr{F}\rho\Phi/2RT) - \exp(n\mathscr{F}\rho\Phi/2RT)] \tag{4.30}$$

where i° is the exchange current corresponding to $\Phi = 0$:

$$i^\circ = 10^3 n\mathscr{F} C_{\mathrm{Os(II)}} C_{\mathrm{Os(III)}} k_{\mathrm{ex}} \delta \tag{4.31}$$

and ρ is a parameter introduced to account for nonideal behavior of the system. Here, k_{ex} is the bimolecular rate constants (see Section 4.1.3 and Eqs. 4.17–4.19). The fast scan voltage bias experiments were repeated in the temperature range of 295–83 K. The analysis of these results uncovered participation of electron tunneling at the low-temperature limit in the steady state electron transport (87).

The discussion of extended electron transfer, voltage driven electron transport, and electron tunneling mechanism is intended to broaden the range of physical phenomena, which become relevant in the studies of electron-transport processes in supermolecular systems. It is clear that the electrochemical methods can be used successfully in addressing a number of fundamental questions ranging from such issues as the distance dependence of the electron-transfer kinetics and electron tunneling, to conductivity in solid state materials (88).

4.5 POLYMER MORPHOLOGY EFFECTS ON ELECTRON TRANSPORT

The final section of this chapter (Section 4.5) deals briefly with the effects of polymer morphology on electron transport. In general, these effects are very strong and are present in essentially all types of polymer films. This makes it difficult to identify some general patterns. Indeed, one might be led to conclude that each particular type of polymer film behaves completely unlike any other.

4.5.1 Matrix–Redox Sites Interactions

It is apparent from the discussion in the previous section that the factors affecting polymer film fluidity such as molecular cross-linking and the extent of solvent swelling have a strong influence on the dynamics of electron transport. The extent to which polymers swell in a particular solvent is not just an intrinsic property of a polymer. Solvent sorption may also depend on the oxidation state of the incorporated redox species and the type of counterions partitioned from the electrolyte solution. The latter effect was recently documented by Oh and Faulkner (89), who noticed strong differences in the diffusion coefficients, permeability, and activation energies of electron diffusion measured in the presence of either nitrate, *p*-toluenesulfonate, or perchlorate background electrolytes.

The extent of interactions of the electroactive species with the polymer matrix can also imprint their own effects on the electron transport. Electrostatic interactions of the charged electroactive sites with the charged groups of the polymer network have a large effect, referred to as electrostatic cross-linking. These effects depend usually on the concentration of the charged redox centers, and lead to a decrease of the electron-transport rates with increasing concentration of redox sites. Diffusion of ferricyanide and other multiply charged anions in poly(vinylpyridine) films, investigated by Oyama et al. (90), is a good example of such behavior. Representative data showing a decrease of D_{app} with increasing concentration of the electroactive ions in the film is reproduced in Fig. 4.23 (90).

Frequently, one observes large differences in polymer matrix–redox center interactions depending on the oxidation state of the redox species. This may lead

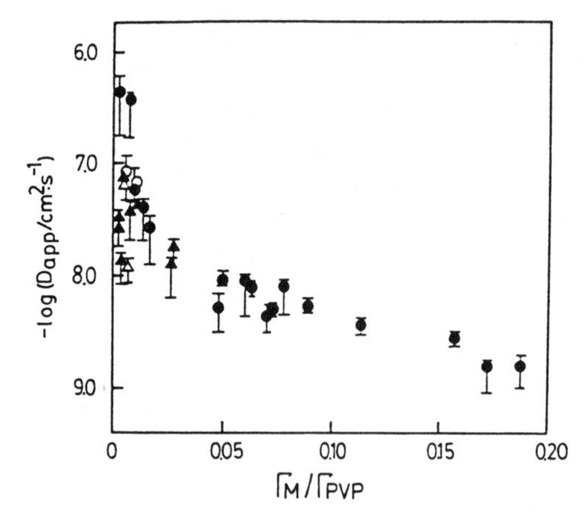

Figure 4.23. Concentration dependence of $[Fe(CN)_6]^{3-}$ diffusion coefficient, D_{app} (●) in a protonated poly(4-vinylpyridine) film. Ferricyanide ion concentration is expressed as a mole fraction of the pyridine moieties. The decrease of D_{app} is indicative of electrostatic cross-linking effect. [Reprinted with permission from N. Oyama, T. Ohsaka, M. Kaneko, K. Sato, and H. Matsuda, *J. Am. Chem. Soc.* 1983, **105**, 6003. Copyright © (1983) American Chemical Society.]

to large shifts in the redox potential of the immobilized species when adjustment of polymer conformation and its interactions with the redox sites in a particular oxidation state is slow compared to electron transport. Such phenomena were observed by Murray and his co-workers (9, 91), Peerce and Bard (92) in the case of poly(vinylferrocene) films, Bowden et al. (93), and by Majda and Faulkner (58) in their investigations of $[Ru(bpy)_3]^{3+/2+}$ in poly(styrenesulfonate) films.

4.5.2 Macroscopically Heterogeneous Systems

A special class of morphological effects influencing electron transport is encountered in cases where electroactive species reside in two distinctly different domains of a structurally heterogeneous polymer film. An example of this type of situation is Nafion, which contains microscopic (~50 nm), interconnected aqueous domains, where most of the sulfonate ion-exchange groups are located, dispersed in a hydrophobic perfluorinated matrix inaccessible to charged species (17, 19, 94–96). These two types of domains are separated by an "interfacial region," which is a zone of intermediate characteristics. The dynamics of electron transport in structurally heterogenous films depend on the rate of electron propagation in the individual domains of the film, on the thermodynamics of partitioning of the redox species between the different domains, and on the kinetics of redox equilibrium between species in bordering domains (17).

Consider electron transport in a polymer film composed of two bicontinuous domains or phases. Redox species partition between these two domains in which the diffusion coefficients of electron transport are D_1 and D_2. Depending on the kinetics of the cross-domain (cross-phase) electron exchange or mass exchange, two limiting cases can be distinguished. In the case of negligibly slow cross-phase exchange, the electron transport will proceed independently in both domains. Under transient conditions, the diffusion zone developed in a domain of large D (D_1) will be broader than that in the domain where a slower diffusion prevails as shown in Fig. 4.24, Scheme A. The observed current is equal to the sum of fluxes developed in each phase. Thus, the measured diffusion coefficient can be expressed by the following:

$$D_{app} = (x_1 D_1^{1/2} + x_2 D_2^{1/2})^2 \tag{4.32}$$

where x_1 and x_2 are the mole fractions of the electroactive species in each type of domain. In the second case, characterized by very fast cross-phase equilibration, the diffusion zones will expand at the same rate in both phases as shown in Fig. 4.24, Scheme B. Electron transport in the domain of lower D, is accelerated or mediated by the faster electron transport carried out in the adjacent domains via rapid electron and/or mass transfer across the domain boundaries. In this case, Andrieux et al. (97) showed that D_{app} is the following function of the mole fractions and D_1 and D_2:

$$D_{app} = x_1 D_1 + x_2 D_2 \tag{4.33}$$

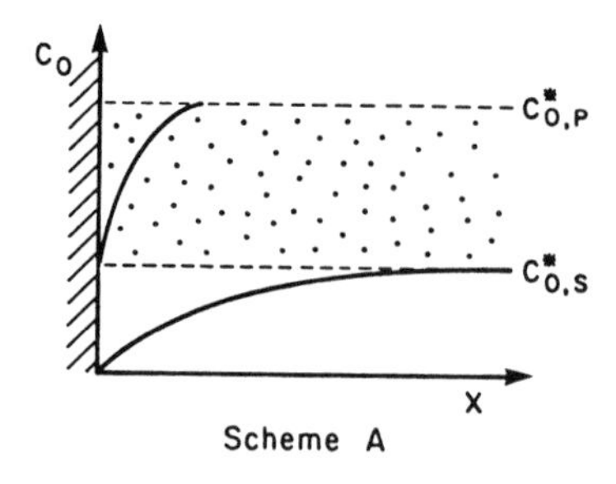

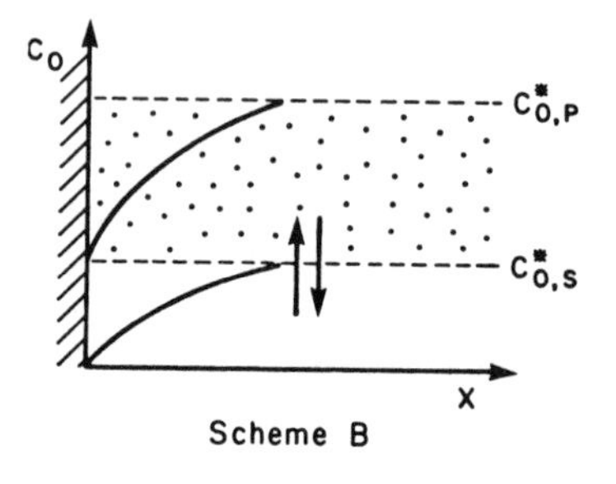

Figure 4.24. Diagrams of concentration profiles during a potential-step experiment in two-domain (S and P) type, heterogeneous polymer films in which different diffusion coefficients characterize transport in each domain. *Scheme A* is a case of negligibly slow kinetics of cross-domain exchange processes. *Scheme B* is a case of fast cross-domain exchange kinetics (see Eqs. 4.32 and 4.33 in text). [Reprinted with permission from Ref. 100.]

As expected, for any case of a two-domain polymer, D_{app}, under conditions of fast exchange kinetics is larger than its value obtained in the absence of the exchange process. Also, the coupling of the diffusional pathways by fast exchange kinetics can lead to a situation where a small fraction of electroactive species present in regions of high electron transport can effectively accelerate the electron transport in the entire polymer film.

4.5.3 Examples of Biphasic Systems

This type of case, involving Nafion films, was described by Buttry and Anson (17) (also see Section 4.1.4). The authors compared electron transport in these films exchanged with $[Co(bpy)_3]^{3+/2+}$ or $[Ru(bpy)_3]^{3+/2+}$ couples. In both cases, only a small fraction of the metal complex resides in the aqueous domains where $D = 9 \times \delta 10^{-7} cm^2 s^{-1}$. Most of the complex is partitioned into interfacial regions of the film, where diffusion coefficients are several orders of magnitude lower. Coupling of the diffusional processes evolving in the two types of domains is possible only in the case of $[Ru(bpy)_3]^{3+/2+}$ due to the fast electron exchange kinetics of this redox couple (so that Eq. 4.33 applies). The rate of electron exchange for the $[Co(bpy)_3]^{3+/2+}$ is negligibly slow, and thus no coupling exists (Eq. 4.32). As a result, the measured value of D_{app} for the ruthenium system in Nafion ($5 \times 10^{-10} cm^2 s^{-1}$) is much larger than its value for the cobalt trisbipyridyl system ($2 \times 10^{-12} cm^2 s^{-1}$).

The coupling of the diffusional pathways in the $[Ru(bpy)_3]^{2+}$ case explains also the lack of concentration dependence of D_{app} (17). The value of D_{app} in this case is essentially dominated by the diffusion coefficient of $[Ru(bpy)_3]^{2+}$ in the aqueous domains, where its value, as shown above, is too large for electron hopping to contribute significantly. Recently, structural effects on electron-transport dynamics in Nafion films were reviewed in a detailed discussion by Whiteley and Martin (96).

Nafion is certainly not the only type of polymer film where morphological effects have a dominating effect on the electron-transport processes. Anson, Savéant and their co-workers (98, 99) investigated in great detail electron transport in poly(*l*-lysine) films. The structure of swollen films of this polymer, shown schematically in Fig. 4.25, also consists of two types of regions: aqueous pools and Donnan domains (98). The latter is comprised of the region along the polymer bakbone where the protonated amine groups reside. In this region, Donnan exclusion equilibria prevail and thus only the anionic species Feedta$^{-/2-}$ redox couple) (edta = ethylenediamine-*N*,*N*,*N*′,*N*′-tetraacetate) are confined. Based on the steady state and transient electrochemical measurements, the authors demonstrated that charge transport is carried in both types of domains, and that fast electron-exchange and place-exchange reactions couple both transport pathways together (98). One interesting and rather unexpected result came from the comparison of the diffusion coefficient values describing charge transport in both domains. The value of $D_E = 6.7 \times 10^{-6} cm^2 s^{-1}$, describing diffusion of Feedta$^{-/2-}$ species along the Donnan domains, is larger than the diffusion coefficient of the redox species in the aqueous domains, $D_S = 2.0 \times 10^{-6} cm^2 s^{-1}$. The large value of D_E may reflect a type of "ion-hopping" mechanism involving the counterion diffusion along the Donnan domains in an environment where the content of solvating water molecules could be lower, due to intrinsic hydrophobicity of the polymer backbone (98).

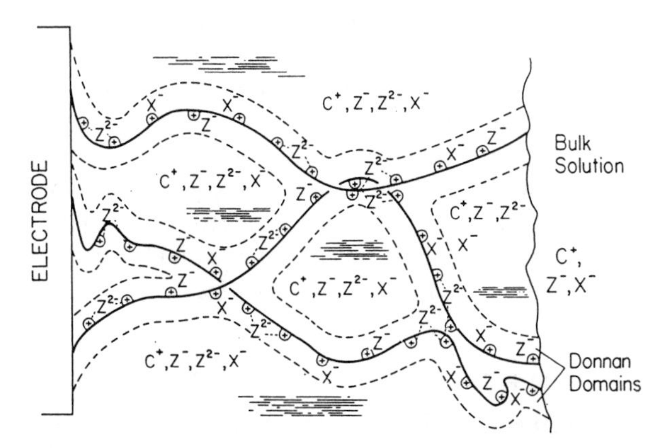

Figure 4.25. Schematic model of a swollen poly(*l*-lisine) film on an electrode surface. The film has been equilibrated with a solution containing an electroinactive electrolyte, CX, and an electroactive couple, Z^-/Z^{2-}. Solid lines represent the polyelectrolyte chains. The dashed lines indicate the regions of extension of the "Donnan domains" containing the electrostatic fields generated by the cationic sites bound to the polymer chains. The purely aqueous domains comprise the remaining volume of the film. [Reprinted with permission from F. C. Anson, J.-M. Savéant, and K. Shigehara, *J. Am. Chem. Soc.* 1983, **105**, 1096. Copyright © (1983) American Chemical Society.]

Several other types of composite polymer films and their electrode behavior have been reported (53, 100–106). In most cases, the coupling between diffusional processes taking place in two different domains of a film results in an overall improvement of the electron transport. In addition, improved permeability and ion retention have been observed. The last point is particularly noteworthy for the class of ion-exchange polymers in which rapid diffusion of the exchanged electroactive ions, a desirable characteristic, is commonly accompanied by their poor retention, a disadvantageous feature. The combination of both high diffusion coefficients and prolonged retention of multiply charged redox species were successfully combined in a new class of composite polymers described in a series of reports by Anson and his co-workers (103–106). For example, casting a polymer film from a solution of a ternary polymer (containing two types of hydrophilic cationic groups and styrene moieties) and a cationic polyelectrolyte [e.g., poly(ethyleneimine) and poly(vinylpyridine)] results in self-segration of the hydrophilic and hydrophobic fragments of the polymers into separate domains (103). Diffusion of the electroactive ions $[Fe(CN)_6]^{3-/4-}$ proceeds only within the hydrophilic domains. Here again, the authors postulate rapid diffusion along the Donnan domains of similar type to those described above for the case of poly(*l*-lysine) to account for unusually hogh ($1-5 \times 10^{-6}\,cm^2\,s^{-1}$) diffusion coefficients (98). At the same time, because the hydrophilic domains are dispersed in a hydrophobic matrix leads to a remarkably high retention of the redox species in the films (103).

The fact that in these types of polymers only hydrophilic domains are involved in the electron-transport processes leads to an apparent dependence of the experimentally obtained diffusion coefficients on the extent of film swelling. This may be difficult to explain on a microscopic level. As pointed out by Inoue and Anson (105), the observed variation of D_{app} with swelling does not reflect differences in transport kinetics within hydrophilic domains where, regardless of the overall extent of film swelling, the rate of diffusion is constant. Instead, the effect stems from averaging the concentration of redox species over the entire film volume, including both the hydrophilic and hydrophobic domains and reflects variations in swelling of the hydrophobic domains. Thus, in cases of macroscopically heterogeneous films in which electroactive species occupy only one type of domain, their concentration cannot be reliably obtained as a ratio of surface density, Γ ($mol\,cm^{-2}$), and the film thickness.

4.6 CONCLUDING REMARKS

One of the objectives of this chapter was to show the complexity of phenomena related to electron transport. Interdependence of many factors governing the dynamics of transport processes has been the most challenging aspect of research in this area. Growth and development of this field so far has primarily involved building and expanding upon a variety of chemical systems and structures used in electrode modification. The few examples of these systems mentioned in this chapter cannot possibly address the breadth and diversity of polymer modified electrodes now existing in the literature. Further progress will probably be best served by more systematic investigations of the individual elements of electron transport. One important aspect requiring systematization is the kinetics of electron transfer in molecular assemblies, where the medium now plays a primary role in determining dynamics of the process. To this end, creating molecular systems in which one could systematically vary the concentration of redox sites while simultaneously controling the morphology of the system, as well as vary the chemical character of the medium independent of the redox sites' concentration, appear to be very much needed and important goals. Polymeric systems are not the only class of molecular assemblies where new effort should be extended. Work involving solid state materials and amphiphilic molecular assemblies has been underway in several laboratories. In these types of systems, a more accurate control of the molecular structure and organization of all the components affecting the system's dynamics is certainly a substantial asset.

ACKNOWLEDGMENTS

I would like to thank Professors Johna Leddy and Jean-Michel Savéant for numerous bitnet communications, which led to sorting out some aspects of electron-hopping kinetics. I also wish to thank Professor Etienne Laviron for his comments on the same subject. Preparation of this article as well as experimental work in my research group reviewed here were supported by grants CHE-8504368 and CHE-8807846 from the National Science Foundation.

REFERENCES

1. R. W. Murray, in *Electroanalytical Chemistry*, A. J. Bard (Ed.), Marcel Dekker, New York, 1984, Vol. 13, p. 191.
2. L. R. Faulkner, *Chem. Eng. News* 1984, **62 (9)**, 28.
3. R. W. Murray, *Annu. Rev. Mater. Sci.* 1984, **14**, 145.
4. C. E. D. Chidsey and R. W. Murray, *Science*, 1986, **231**, 25.
5. M. S. Wrighton, *Science* 1986, **231**, 32.
6. M. Kaneko and D. Worhle, *Adv. Polym. Sci.* 1988, **14**, 142.
7. A. J. Bard and L. R. Faulkner, *Electrochemical Methods. Fundamentals and Applications.* Wiley, New York, 1980, Chapter 5.
8. P. Daum, J. R. Lenhard, D. R. Rolison, and R. W. Murray, *J. Am. Chem. Soc.*, 1980, **102**, 4649.
9. P. Daum and R. W. Murray, *J. Phys. Chem.* 1981, **85**, 389.
10. H. Dahms, *J. Phys. Chem.*, 1968, **72**, 362.
11. E. Laviron, *J. Electroanal. Chem.*, 1980, **112**, 1.
12. C. P. Andrieux and J.-M. Savéant, *J. Electroanal. Chem.* 1980, **111**, 377.
13. I. Ruff and V. J. Friedrich, *J. Phys. Chem.* 1971, **75**, 3297.
14. I. Ruff, V. J. Friedrich, K. Demeter, and K. Csillag, *J. Phys. Chem.* 1971, **75**, 3303.
15. J. Facci and R. W. Murray, *J. Phys. Chem.*, 1981, **85**, 2870.
16. D. A. Buttry and F. C. Anson, *J. Electroanal. Chem.*, 1982, **130**, 333.

17. D. A. Buttry and F. C. Anson, *J. Am. Chem. Soc.* 1983, **105**, 685.
18. J. G. Gaudiello, P. K. Ghosh, and A. J. Bard, *J. Am. Chem. Soc.* 1985, **107**, 3027.
19. I. Rubinstein, *J. Electroanal. Chem.* 1985, **188**, 227.
20. N. Oyama, T. Ohsaka, H. Yamamoto, and M. Kaneko, *J. Phys. Chem.* 1986, **90**, 3850.
21. I. Ruff and L. Botar, *J. Chem. Phys.* 1985, **83**, 1292.
22. L. Botar and I. Ruff, *Chem. Phys. Lett.* 1986, **126**, 348.
23. R. P. Buck, *J. Electroanal. Chem.* 1988, **243**, 279.
24. R. P. Buck, *J. Phys. Chem.* 1983, **92**, 4296.
25. Y. T. Hupp and M. J. Weaver, *J. Electroanal. Chem.* 1983, **152**, 1.
26. N. Sutin, *Prog. Inorg. Chem.* 1983, **30**, 441.
27. M. D. Newton and N. Sutin, *Annu. Rev. Phys. Chem.* 1984, **35**, 437.
28. M. J. Weaver and G. E. McManis III, *Acc. Chem. Res.* 1990, **23**, 294.
29. F. B. Kaufman and E. M. Engler, *J. Am. Chem. Soc.* 1979, **101**, 547.
30. F. B. Kaufman, A. M. Schroeder, E. M. Engler, S. R. Kramer, and J. Q. Chambres, *J. Am. Chem. Soc.* 1980, **102**, 483.
31. D. A. Buttry, J.-M. Savéant, and F. C. Anson, *J. Phys. Chem.* 1984, **88**, 3086.
32. M. von Smoluchowski, *J. Phys. Chem.* 1917, **92**, 129.
33. T. J. Meyer and H. Taube, *Inorg. Chem.* 1968, **7**, 2369.
34. H. S. White, J. Leddy, and A. J. Bard, *J. Am. Chem. Soc.* 1982, **104**, 4811.
35. E. S. Young, M. S. Chan, and A. C. Wahl, *J. Phys. Chem.* 1980, **84**,, 3094.
36. K. Shigehara, N. Oyama, and F. C. Anson, *J. Am. Chem. Soc.* 1981, **103**, 2552.
37. O. Haas and J. G. Vos, *J. Electroanal. Chem.* 1980, **113**, 139.
38. O. Haas, M. Kriens, and J. G. Vos, *J. Am. Chem. Soc.* 1981, **103**, 1318.
39. H. D. Abruna and A. J. Bard, *J. Am. Chem. Soc.* 1981, **103**, 6898.
40. R. J. Mortimer and F. C. Anson, *J. Electroanal. Chem.* 1982, **138**, 325.
41. A. Mertz and A. J. Bard, *J. Am. Chem. Soc.* 1978, **100**, 3222.
42. P. J. Peerce and A. J. Bard, *J. Electroanal. Chem.* 1980, **112**, 97.
43. J. M. Calvert, R. H. Schmehl, B. P. Sullivan, J. S. Facci, T. J. Meyer, and R. W. Murray, *Inorg. Chem.* 1983, **22**, 2151.
44. C. R. Leidner and R. W. Murray, *J. Am. Chem. Soc.* 1984, **106**, 1606.
45. T. Ikeda, R. Schmehl, P. Denisevich, K. Willman, and R. W. Murray, *J. Am. Chem. Soc.* 1982, **104**, 2683.
46. J. A. Bruce and M. S. Wrighton, *J. Am. Chem. Soc.* 1982, **104**, 74.
47. D. C. Bookbinder and M. S. Wrighton, *J. Electrochem. Soc.* 1983, **130**, 1080.
48. M.-S. Chan and A. C. Wahl, *J. Phys. Chem.* 1978, **82**, 2542.
49. P. G. Pickup, W. Kutner, C. R. Leidner, and R. W. Murray, *J. Am. Chem. Soc.* 1984, **106**, 1991.
50. J. S. Facci, R. H. Schmehl, and R. W. Murray, *J. Am. Chem. Soc.* 1982, **104**, 4959.
51. R. H. Schmehl and R. W. Murray, *J. Electroanal. Chem.* 1983, **152**, 97.
52. K. D. Moran, Ph.D. Thesis, University of California at Berkeley, 1987.
53. K. D. Moran and M. Majda, *J. Electroanal. Chem.* 1986, **207**, 73.
54. C. R. Bock, J. A. Conner, A. R. Gutierrez, T. C. Meyer, D. G. Whitten, B. P. Sullivan, and J. K. Nagle, *Chem. Phys. Lett.* 1979, **61**, 522.
55. P. He and X. Chen, *J. Electroanal. Chem.* 1988, **256**, 353.
56. M. Majda and L. R. Faulkner, *J. Electroanal. Chem.* 1982, **137**, 149.
57. M. Majda and L. R. Faulkner, *J. Electroanal. Chem.* 1984, **169**, 97.

58. M. Majda and L. R. Faulkner, *J. Electroanal. Chem.* 1984, **169**, 77.
59. K. Doblhofer, H. Braun, and R. Lange, *J. Electroanal Chem.* 1986, **206**, 93.
60. K. Niwa and K. Doblhofer, *Electrochim. Acta* 1986, **31**, 549.
61. R. Lange and K. Doblhofer, *J. Electroanal. Chem.* 1987, **216**, 241.
62. K. Doblhofer and R. D. Armstrong, *Electrochim. Acta.* 1988, **33**, 453.
63. C. M. Elliott and J. G. Redepenning, *J. Electroanal. Chem.* 1984, **181**, 137.
64. P. G. Pickup and R. W. Murray, *J. Am. Chem. Soc.* 1983, **105**, 4510.
65. C. E. D. Chidsey, B. J. Feldman, C. Lungren, and R. W. Murray, *Anal. Chem.* 1986, **58**, 601.
66. H. W. White, G. P. Kittlesen, and M. S. Wrighton, *J. Am. Chem. Soc.* 1984, **106**, 5375.
67. G. P. Kittlesen, H. W. White, and M. S. Wrighton, *J. Am. Chem. Soc.* 1984, **106**, 7389.
68. E. W. Paul, A. J. Ricco, and M. W. Wrighton, *J. Phys. Chem.* 1985, **89**, 1441.
69. J. C. Jernigan and R. W. Murray, *J. Am. Chem. Soc.* 1987, **109**, 1738.
70. X. Chen, P. He, and L. R. Faulkner, *J. Electroanal. Chem.* 1987, **222**, 223.
71. R. Lange and K. Doblhofer, *J. Electroanal. Chem.* 1987, **237**, 13.
72. R. Lange and K. Doblhofer, *Ber. Bunsenges. Phys. Chem.* 1988, **92**, 578.
73. W. T. Yap, R. D. Durst, E. A. Blugangh, and D. D. Blugangh, *J. Electroanal. Chem.* 1983, **144**, 69.
74. J.-M. Savéant, *J. Electroanal. Chem.* 1986, **201**, 211.
75. J.-M. Savéant, *J. Electroanal. Chem.* 1987, **227**, 299.
76. J.-M. Savéant, *J. Electroanal. Chem.* 1988, **242**, 1.
77. J.-M. Savéant, *J. Phys. Chem.* 1988, **92**, 1011.
78. J.-M. Savéant, *J. Phys. Chem.* 1988, **92**, 4526.
79. C. P. Andrieux and J.-M. Savéant, *J. Phys. Chem.* 1988, **92**, 6761.
80. R. P. Buck, *J. Electroanal. Chem.* 1987, **219**, 23.
81. R. P. Buck, *J. Phys. Chem.* 1988, **92**, 6445.
82. M. Sharp, B. Lindholm, and E.-L. Lind, *J. Electroanal. Chem.* 1989, **274**, 35.
83. F. C. Anson, D. N. Blanch, J.-M. Savéant, and C.-F. Shu, *J. Am. Chem. Soc.* 1991, in press.
84. I. Fritsch-Faules and L. R. Faulkner, *J. Electroanal, Chem.* 1989, **263**, 237.
85. J. C. Jernigan and R. W. Murray, *J. Phys. Chem.* 1987, **91**, 2030.
86. K. Wilbourn and R. W. Murray, *J. Phys. Chem.* 1988, **92**, 3642.
87. J. C. Jernigen, N. A. Surridge, M. E. Zvanut, M. Silver, and R. W. Murray, *J. Phys. Chem.* 1989, **93**, 4620.
88. M. J. Pinkerton, Y. Le Mest, H. Zhang, M. Watanabe, and R. W. Murray, *J. Am. Chem. Soc.* 1990, **112**, 3730.
89. S. M. Oh. and L. R. Faulkner, *J. Electroanal. Chem.* 1989, **269**, 77.
90. N. Oyama, T. Ohsaka, M. Kaneko, K. Sato, and H. Matsuda, *J. Am. Chem. Soc.* 1983, **105**, 6003.
91. K. W. William, R. D. Rocklin, R. Nowak, K. Kuo, F. A. Schultz, and R. W. Murray, *J. Am. Chem. Soc.* 1990, **102**, 7629.
92. P. J. Peerce and A. J. Bard, *J. Electroanal. Chem.* 1980, **114**, 89.
93. E. F. Bowden, M. F. Dautartus, and J. F. Evans, *J. Electroanal. Chem.* 1987, **219**, 91.
94. S. R. Lowry and K. A. Mauritz, *J. Am. Chem. Soc.* 1980, **102**, 4665.
95. M. N. Szentirmay, N. E. Prieto, and C. R. Martin, *J. Phys. Chem.* 1985, **89**, 3017.
96. L. D. Whiteley and C. R. Martin, *J. Phys. Chem.* 1989, **93**, 4650.
97. C. P. Andrieux, P. Hapiot, and J.-M. Savéant, *J. Electroanal. Chem.* 1984, **172**, 49.

98. F. C. Anson, J.-M. Savéant, and K. Shigehara, *J. Am. Chem. Soc.* 1983, **105**, 1096.
99. F. C. Anson, T. Ohsaka, and J.-M. Savéant, *J. Phys. Chem.* 1983, **87**, 640.
100. C. J. Miller and M. Majda, *J. Electroanal. Chem.* 1986, **207**, 49.
101. J. E. Van Koppenhagen and M. Majda, *J. Electroanal. Chem.* 1987, **236**, 113.
102. C. Liu and C. R. Martin, *J. Electroanal. Chem.* 1990, **137**, 510.
102. D. D. Montgomery and F. C. Anson, *J. Am. Chem. Soc.* 1985, **107**, 3431.
104. K. Sumi and F. C. Anson, *J. Phys. Chem.* 1986, **90**, 3845.
105. T. Inoue and F. C. Anson, *J. Phys. Chem.* 1987, **91**, 1519.
106. E. Tsuchida, H. Nishide, N. Ishimaru, D. D. Montgomery, and F. C. Anson, *J. Phys. Chem.* 1987, **91**, 2898.

Chapter **V**

CATALYSIS AT REDOX POLYMER COATED ELECTRODES

C. P. Andrieux and J.-M. Savéant
Laboratorie D'Electrochimie, Universite de Paris 7, 2, Place Jussieu, 75251 Paris Cédex 05, France

---

5.1 INTRODUCTION

Redox polymers are systems resulting from the incorporation of redox centers into polymeric structures. In most cases, these redox centers are organic

Molecular Design of Electrode Surfaces,
Edited by Royce W. Murray. Techniques of Chemistry Series, Vol. XXII.
ISBN 0-471-55773-0

molecules or transition metal complexes that can be reversibly reduced or oxidized by a single electron exchange, both members of the redox couple being chemically stable (1). Some redox polymers however, contain reversible redox couples, the two halves of which interconvert by the concomitent exchange of several electrons and patrons. Quinone–hydroquinone couples are prototypes of such systems (2). The various redox polymers fall into three categories according to the mode of attachment of the redox centers to the polymer backbone, covalent, coordinative, or electrostatic, as sketched and illustrated by a series of examples in Table 5.1.

From the very beginning, a strong incentive for the development of redox polymer coatings on electrode surfaces was their possible use as catalysts for electrochemical reactions. The principle of such applications as it relates to catalysis is sketched in Fig. 5.1. In the uncatalyzed reaction, the substrate (A) diffuses towards the electrode surface, where it is reduced or oxidized into an intermediate (B) that may further undergo a series of reactions leading to a set of final products (C). In homogeneously catalyzed processes, the catalyst is introduced into the solution, together with the substrate (A), under its unreactive form P (P is the oxidized half of the catalyst couple, if the overall reaction is a reduction, and its reduced half, if the overall reaction is an oxidation). In the case of a reduction, the catalyst couple P–Q, is selected so as to have its standard potential positive to the reduction potential of the substrate and vice versa in the case of an oxidation reaction. The reaction is carried out at the potential where

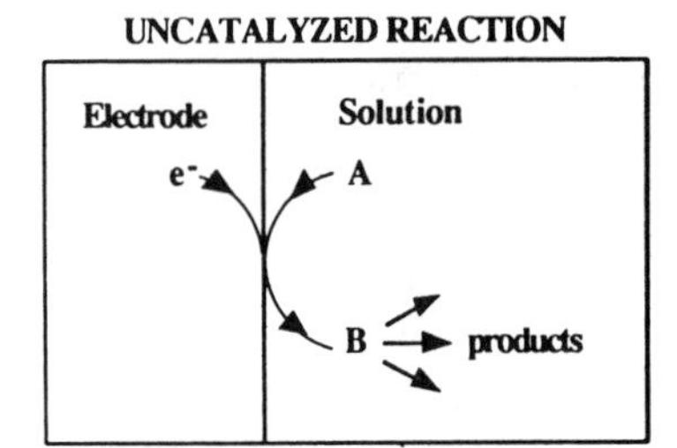

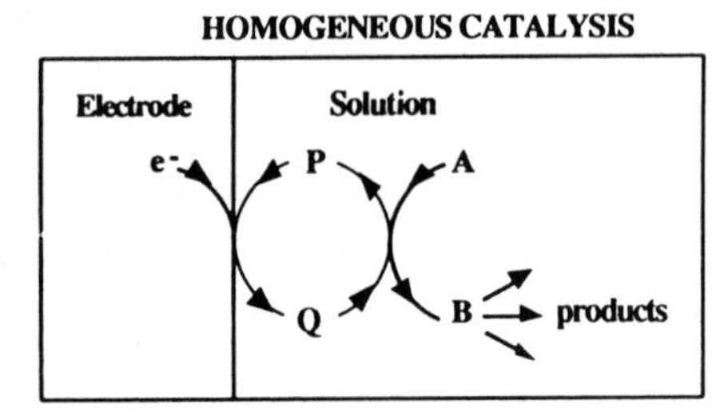

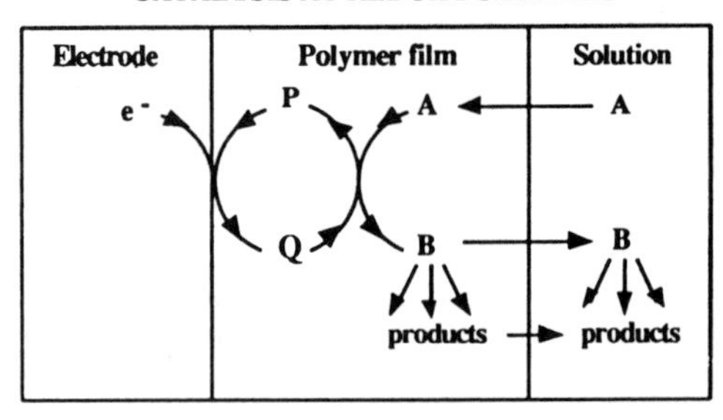

Figure 5.1. Catalysis of electrochemical reactions by redox polymer coatings.

Table 5.1. Redox Polymers

Covalent Attachment	Coordinative Attachment	Electrostatic Attachment	
P P P	L P L′ L P L′ L P L′	− P+ − P+ − P+	+ P− + P− + P−
Poly(acrylamide) substituted with antraquinone (2)	Ruthenium tris vinylbipyridine (14) and analogues (15)	Nafion (19), (20)	Protonated poly(vinylpyridine) (23)
Poly(vinyl) substituted with -ferrocene (3, 4) -*p*-nitrobenzene (5) -*p*-viologen-benzene (6) -naphthalene (7) -anthracene (7) -9–10 diphenyl anthracene (8)	Poly(vinylpyridine) backbone (16)	Poly(styrene sulfonate) (21)	Quaternized poly(vinylpyridine) (24)
Poly(pyrrole) *N* substituted with -viologen (9) -anthraquinone (10) -porphyrin (11) -nitrobenzene (12) -ferrocene (13)	*p*-chlorosulfonated poly-(styrene) backbone (17)	Deprotonated polyacrylic acid (22)	Poly-*l*-lysine (25)
	Poly(vinylimidazole) backbone (18)		

the active form of the catalyst couple (Q) is generated. Since the substrate is not directly reduced (or oxidized) at the electrode at this potential, it reacts with the active form of the catalyst instead of reacting at the electrode. The principle of catalysis at redox polymer films is similar except that the redox couple is now confined within the polymer coating instead of being dispersed in the solution. The passage from homogeneous catalysis to the redox polymer catalysis of electrochemical reactions is thus quite similar to the passage from homogeneous catalysis to supported homogeneous catalysis of chemical reactions. The advantage of supported systems over homogeneous systems in synthetic and sensor applications derives from the physical separation of the catalyst from the substrate. Another point of interest is that the reactivity and selectivity of the same molecular catalyst may be different in the environment offered by the redox polymer structure from what they are in solutions. However, as discussed in details in the following, the catalytic response of the redox polymer coating is not only a function of the intrinsic reactivity of the catalyst but also of the rate of permeation of the substrate and of the rate of electron transport through the coating.

These three key factors, the association of which determines the catalytic efficiency of redox polymer coatings, will be individually discussed. Particular emphasis will be put on the description of the catalytic reaction using the distinction between "redox" and "chemical" catalysis of electrochemical reactions. Then, tackling the analysis of overall catalytic responses, the simple case of catalysis by monolayer coatings will be treated before describing how the association of the three key rate-limiting factors, evoked earlier, controls the catalytic responses of redox polymer coatings. Because of its simplicity, rotating disk electrode voltammetry has become the most popular technique for investigating the catalytic responses of monolayer or multilayer electrode coatings. The analyses of the catalytic processes will thus be described in the context of this technique. Extension to other techniques and, in particular, to preparative-scale electrolysis conditions does not require substantial modifications of the reasonings. Section 5.5 will illustrate the theoretical analysis by experimental examples from the literature.

5.2 THE THREE COMPONENTS OF THE CATALYTIC RESPONSE OF REDOX POLYMER COATINGS

The catalytic response of redox polymer coatings is the result of the combination of three processes taking place inside the film, namely, the catalytic reaction, the transport of electrons, the permeation of the substrate and of the diffusion of the substrate from the bulk of the solution to the film–solution interface.

5.2.1 Redox versus Chemical Catalysis

In a number of cases, the redox centers dispersed in the polymeric structure play an ambident role. On one hand they shuttle the electrons between the electrode and the film–solution boundary and on the other they exchange electrons with the substrate.

In polymers where the redox sites are irreversibly bound to the polymer structure, electron hopping from site to site is the preferred mode of electron transport. The latter thus involves a thermally activated single electron exchange of the outer-sphere type between two adjacent molecular sites. In the numerous cases where the electron carrier molecules also serve as the catalyst, their active form reacts as an outer-sphere single electron donor or acceptor with the substrate. It follows that under such conditions the catalytic reaction is of the "redox" type rather than of the "chemical" type.

The distinction between these two kinds of catalysis was first made when discussing homogeneous catalysis of electrochemical reactions (26). In the redox case, the catalytic reaction starts with an outer-sphere electron transfer even though it may involve one or several successive steps. If the latter are so fast as to make the initial electron-transfer step rate determining, the catalytic reaction is subject to Brønsted (or Marcus)-type activation-driving force limitations (the larger the driving force the faster the catalytic reaction). According to the Marcus quadratic relationship, the activation free energy, $\Delta G^{\ddagger}_{\text{hom}}$, varies with the standard potential difference between the catalyst and substrate redox couples as follows (27):

$$\Delta G^{\ddagger}_{\text{hom}} = \Delta G^{\ddagger}_{0,\text{hom}}\left(1 + \frac{E^0_{\text{PQ}} - E^0_{\text{AB}}}{4\Delta G^{\ddagger}_0}\right)^2 \tag{5.1}$$

Where $\Delta G^{\ddagger}_{0,\text{hom}}$ is the intrinsic barrier free energy, E^0_{PQ} and E^0_{AB} the standard potentials of the catalyst and substrate redox couples. The same relationship applies even in the case where the substrate does not react in an outer-sphere manner in the sense that a bond is broken concertedly with electron transfer (28). In order to estimate its efficiency, a catalytic process must be compared with a reference system regarded as an uncatalyzed reaction. This is naturally defined as the same electrochemical reaction taking place at a bare electrode and involving an outer-sphere heterogeneous single electron transfer as the starting step. The latter reaction is also subject to Marcus-type activation–driving force constraints according to a similar relationship:

$$\Delta G^{\ddagger}_{\text{el}} = \Delta G^{\ddagger}_{0,\text{el}}\left(1 + \frac{E - E^0_{\text{AB}}}{4\Delta G^{\ddagger}_{0,\text{el}}}\right)^2 \tag{5.2}$$

($\Delta G^{\ddagger}_{\text{el}}$ = electrochemical free energy of activation, $\Delta G^{\ddagger}_{0,\text{el}}$ = intrinsic barrier free energy of the electrochemical reaction). It follows that no significant redox catalysis is anticipated to occur at an outer-sphere electrode covered with one monolayer of a redox catalyst (29). In other words, the redox catalysis responses that one may observe with redox catalysts in solution or inside polymeric coatings is essentially the result of their three-dimensional dispersion as opposed to the two-dimensional availability of the electrons (or holes) at an electrode surface.

In chemical catalysis, the active form of the catalyst and the substrate undergo stronger mutual interactions during the course of reaction. These interactions may be so strong as to result in the formation of a transient adduct from which the inactive form of the catalyst (to be reactivated at the electrode surface) will be regenerated simultaneously with the transformation of the substrate into the products or into an intermediate en route to the products. However, the formation of a discrete adduct in not a necessary requirement of chemical catalysis. Another possibility is that the initial electron-transfer step possesses an inner-sphere character not only from the standpoint of the substrate but also from that of the active form of the catalyst, thus involving bonded interactions in the transition state. In such circumstances, the above Marcus-type limitations relating activation to the standard potential of the catalyst are removed. The bonded interactions in the transition state then render the catalytic rate-determining step faster than what it would have been in the case of an outer-sphere electron donor (or acceptor) of same standard potential. In this case, however, as in the preceding one, the primarily formed intermediate should regenerate rapidly the active or inactive form of the catalyst.

If such conditions are fulfilled, chemical catalysis is obviously more efficient than redox catalysis, since it allows a larger decrease of the overpotential: Given the electrode potential the current is expected to be larger in the first case than in the second, or, conversely, the electrode potential required for obtaining a given current is less negative for a reduction, (less positive for an oxidation) in the first case than in the second. Another consequence of the more intimate chemical interactions that prevail in chemical catalysis as compared to redox catalyst is the expectation of better selectivities. A typical example illustrating the superiority of chemical catalysis over redox catalysis in term of both efficiency and selectivity is provided by the homogeneous catalysis of the electrochemical reduction of vicinal dihalides into the corresponding alkenes (30, 31):

$$X \!+\!\!\!+ X + 2e^- \rightarrow \rangle\!=\!\langle + 2X^- \tag{5.3}$$

For the same standard potential, the cobalt(I), iron(I), and nickel(I) porphyrins exhibit rate constants of the catalytic reaction that are several orders of magnitude larger that those observed with typical sphere electron donors such as aromatic anion radicals (including the reduced copper and zinc porphyrins) of same standard potentials. On the other hand the reaction is stereospecific (meso- and *d,l*-dibromides yield exclusively the *trans*- and *cis*-alkenes, respectively) in the first case and not in the second. With the aromatic anion radicals the rate-determining step is the transfer of one electron concertedly with the breaking of one carbon–halogen bond:

$$X \!+\!\!\!+ X + Q \rightarrow \cdot\rangle\!\!+ X + X^- + P \tag{5.4}$$

whereas with the Co, Fe, and Ni porphyrins, an E2 elimination reaction, involving the concerted transfer of two electrons and cleavage of two carbon–halogen bonds takes place:

$$X \!+\!\!\!+\! X + Q \rightarrow \rangle\!\!=\!\!\langle + X^- + QX^+ \tag{5.5}$$

Bonded interactions between Q and X stabilize the transition state in Reaction 5.5 as compared to Reaction 5.4, which is subject to the activation–driving force constraints governed by Eq. 5.1. The loss of stereoselectivity observed with the outer-sphere electron donor originates in the rotation of the β-halo- radical, $\cdot\rangle\!\!+\!\times$ around the carbon–halogen bonds that competes with the further reactions it undergoes, whereas this does not occur with the Co, Fe, and Ni porphyrins, because the transfer of the two electrons and the expulsion of the two bromide ions are concerted processes.

It follows from what precedes that chemical catalysis, still evaluated against an outer-sphere electrode as a reference system, may well take place at a monolayer derivatized electrode since the activation energy is lowered by the bonded interactions in the transition state or by the formation of a discrete adduct. The same catalyst molecules dispersed in a solution or within a polymeric structure may give rise to an even stronger catalysis than at a monolayer coated electrode, since the same three-dimensional effect as evoked earlier may arise. However, as discussed in Section 5.4, the rates at which the electrons propagate and/or the substrate diffuses through the film may jointly impose their own limitations. Extreme situations in which the catalytic reaction only involves a single monolayer located at the electrode–film or film–solution interfaces, may well be the result of such limitations.

In the case of chemical catalysis, it is conceivable that the catalyst, now designed in view of particular chemical properties towards the substrate, still serve as a redox exchanger in the electron-transport process. However, one degree of freedom in the design of the catalytic system will be gained if the two functions, catalysis and transport of electrons through the coating, are dissociated, one type of molecule playing the role of the electron carrier and another that of the catalyst, as illustrated by an experimental example in Section 5.5.

5.2.2 Electron Transport through Redox Polymers

A detailed review of the mechanisms of electron transport through redox polymers is beyond the scope of this chapter but can be found in Chapter IV (32). In this section we only stress a few points useful for the foregoing analysis of catalytic processes.

In redox polymers, electron hopping between redox sites is usually accompanied by the displacement of electroinactive counterions so as to maintain electroneutrality. Thus, unless the electroinactive counterions are totally immobile (see Ref. 33 and references cited therein), transport of electrons is mainly governed by chemical potential gradients of the oxidized and reduced halves of the redox couple involved in the electron-hopping process. In catalytic applications of redox polymer coatings, these chemical potential gradients are created by setting the working electrode potential at a value appropriate for creating, at the electrode surface, an excess of one of two forms of the redox couple over the other. Under these conditions, potential-step transient or steady state electrochemical techniques have most often been used to measure the rates of charge transport.

In most cases, the rate of charge propagation, as derived from the variation of the charge (or the current) with time following a potential step, has been found to obey the Cottrell equation, that is, the instantaneous charge or current are proportional to the square root of time or to its inverse, respectively (provided the timescale is short enough so that the region of space adjacent to the electrode surface in which a sizeable concentration gradient exists is small as compared to the film thickness). Such observations led to the notion that charge transport can be regarded as equivalent to a diffusional process to which an apparent diffusion coefficient, D_{app}, applies.

The way in which electron hopping between pairs of attached oxidized and reduced molecules can result in a diffusionlike behavior was first explained by means of a stochastic model in which the redox centers are regarded as randomly distributed over a *fictitious* cubic lattice whose characteristic length is equal to the average optimal hopping distance, δ_E (34, 35). Fick's laws then appear to provide a satisfactory description of the rate of electron hopping under a concentration gradient of the redox centers. According to this model the diffusion coefficient is given by the following equations.

$$D_E = k_E^{pair} \delta_E^2 C_E = \frac{1}{6} k_E \delta_E^2 C_E^0 \qquad (5.6)$$

where k_E^{pair} is the volume rate constant of electron transfer for a pair of adjacent sites, $k_E = 6k_E^{pair}$ is the conventional electron-transfer volume rate constant, taking into account that each node of the lattice is surrounded by six neighbors. The term C_E^0 is the total concentration of redox sites.

This model applies to immobile redox sites. In the case of electrostatically bound redox centers, physical diffusive displacement may make a significant contribution to the charge propagation in addition to electron hopping between pairs of oxidized and reduced centers. The apparent diffusion coefficient in this case is the sum of the physical displacement (pd) and electron-hopping duffusion coefficients (36):

$$D_E^{app} = D_{pd} + \frac{1}{6} k_E \delta_E^2 C_E^0 \qquad (5.7)$$

Attempts to observe variations in the apparent diffusion coefficient with the concentration of redox sites as predicted by Eqs. 5.6 and 5.7 led to a variety of results, only a few of which appeared to obey the above predictions.

Further improvements of the theory were made by recognizing that the maintenance of electroneutrality means that the electron movement is necessarily coupled to the physical displacement of electroinactive counterions. The situation is analogous to that involved with ordinary solutions of electroactive reactants in the presence of little or no supporting electrolyte, where the migration of charged reactants in the electric field affects the charge propagation rates. Theoretical analysis (37, 38) of such systems involves the addition of classical Nernst–Planck terms to the Fick's diffusion laws to obtain equations showing that the smaller the concentration or mobility of the electroinactive ions present, the more pronounced the effect of the migration. However, the "migration" of electrons is not governed by the Nernst–Planck–Fick equation. It instead obeys the following equations (39–43):

$$J_{\mathrm{P}} = -J_{\mathrm{Q}} = -D_{\mathrm{E}}\left[\frac{\partial C_{\mathrm{P}}}{\partial x} + \frac{\mathscr{F}}{RT} C_{\mathrm{P}}\left(1 - \frac{C_{\mathrm{P}}}{C_{\mathrm{E}}^{0}}\right)\frac{\partial \Phi}{\partial x}\right]$$

$$= D_{\mathrm{E}}\left[\frac{\partial C_{\mathrm{Q}}}{\partial x} - \frac{\mathscr{F}}{RT} C_{\mathrm{Q}}\left(1 - \frac{C_{\mathrm{Q}}}{C_{\mathrm{E}}^{0}}\right)\frac{\partial \Phi}{\partial x}\right] \qquad (5.8)$$

(P, Q = oxidized and reduced forms of the redox couple; Φ = potential, J = fluxes, C = concentrations; D = diffusion coefficients), whereas the Nernst–Planck equation:

$$J = -D\left[\frac{\partial C}{\partial x} + \left(\frac{z\mathscr{F}}{RT}\right) C \left(\frac{\partial \Phi}{\partial x}\right)\right] \qquad (5.9)$$

applies for the electroinactive counterions or for the electroactive ions in the case where their physical displacement prevails over electron hopping. Analyses of the responses expected in both steady state (44) and transient (45) experiments showed that the presence of "migration" always enhances the rate of electron hopping and that the enhancement grows as the mobility of the electroinactive counterions decreases. Thus, earlier suggestions (21, 44–47) that charge-transport rates in redox polymers might be controlled by the slower of the two coupled processes of electron hopping and electroinactive counterion displacement seem inappropriate. On the contrary, the slower the movement of the electroinactive ions, the faster the electron hopping and the larger the resulting current density (at least within the limit that the quasielectroneutrality assumption holds). In all cases, the potential-step current or charge shows a Cottrellian behavior from which apparent diffusion coefficients can be evaluated which contain the effect of the field created by any mismatch between the rates of electroinactive counterion displacement and electron hopping (43). These apparent diffusion coefficients increase with the concentration of redox centers more steeply than the simple proportionality indicated in Eq. 5.6. This feature has been used to interpret (43) previous observations made with poly(vinylpyridine) copolymers containing coordinatively attached osmium and ruthenium ions (48).

One drawback of the models recalled above is that they ignore activity effects on the dynamics of electron hopping under concentration gradients and electric fields in that they tacitly assume that the interactions between each ion and all the others remains the same as the ionic composition of the redox polymer is changed. Since redox polymer coatings contain large ion concentrations, of the order of moles or fractions of a mole, such an assumption seems quite unlikely to hold when the total amount of electroactive material and the redox state of the film are changed as occurs in current applications.

In view of the high ionic content of typical redox polymers and of the weak polarity of large sections of their structure, association between ions of opposite charge is expected to be an important mode of ionic interactions inside of them (49–51). A large part of activity effects may thus be treated in terms of ion-association equilibria. Specifically, the interactions between an individual ion and the other ions may be divided in two categories: formation of an ion pair with one ion of opposite charge on one hand and diffuse interaction with all the other ions on the other. The ion pair and the (relatively) "free" ion are then regarded as engaged in a chemical equilibrium governed by an association constant. The influence of the association constant on the dynamics of charge propagation then depends on the relative role of ion-paired and "free" electroactive ions in the electron-hopping process.

The basic relationships governing the ways in which ion-association affects electron-hopping rates in redox polymers in the presence of electric fields have recently been established for steady state (41, 52) and transient potential step (53) regimes. It appears that the activation barriers for electron hopping between fully associated electroactive ions are high because the electron transfer between the oxidized and reduced halves of the redox couple involves the concerted breaking of an ion-pair linkage and the formation of a new one. Predissociation mechanisms in which the fully ion-paired oxidized half of the redox couple dissociate prior the electron-transfer step, which is then of the outer-sphere type, are thus anticipated to be energetically more favorable. Since the fraction of dissociated ion pair increases with the concentration of redox sites, apparent diffusion coefficients that show steep increases with this concentration are expected (52, 53). These predictions have been experimentally verified with systems such as Ru and $Os(bpy)_3$ in Nafion, where the very cause of the irreversible attachment of these hydrophobic ions resides in their strong association with the fixed sulfonate sites.

In the cubic quasilattice models described so far, the system is regarded as a completely disordered structure. Thus, the probability of finding one particular species at one node of the lattice is proportional to its concentration. In such models, a simple estimate of the hopping distance δ_E in Eqs. 5.6 and 5.7 is the separation between the redox centers when the equivalent hard spheres are in contact. It would be possible to allow electron transfer to occur at greater distances in consideration of significant overlap of the electron transferring and receiving orbitals [extended electron transfer (54)] and/or because each redox center may oscillate out of its equilibrium location. In both cases, integration over a distance dependent probability distribution (a decaying exponential would be the simplest model) would lead to an increase of the average hopping distance. Quite different approaches of the mechanism of electron hopping in redox polymers have been recently proposed. In one of them (54) the redox centers are regarded as individually located at the nodes of an actual cubic lattice. The lattice characteristic distance is taken as equal to the average distance between two nearest neighbors and the probability of electron hopping between them as an exponentially decaying function of this distance. The resulting, diffusionlike charge propagation is then predicted to increase with the concentration of redox sites more rapidly than proportionally. However, it does not seem likely that such a perfect crystal ordering of the electroactive sites could exist in redox polymers. For such systems, the stochastic model described thus seems more realistic.

Another (quite different) model that has been recently proposed to explain the sharp rise in the variation of the apparent diffusion coefficient with the concentration of redox sites observed with the $[Os(bpy)_3]^{3+/2+}$ and $[Ru(bpy)_3]^{3+/2+}$ (bpy = 2,2′-bipyridine) couples in Nafion (55) is based on the notion of charge propagation being the result of a diffusion process that combines physical displacement of the redox centers and electron hopping between them. The overall diffusion coefficient is expressed (56) by Eq. 5.10,

$$D_E^{app} = D_{pd} + \tfrac{1}{4}\pi k_E \delta_E^2 C_E^0 \tag{5.10}$$

but the rate constant k_E is regarded, itself, as possibly limited either by electron transfer between redox centers or by their diffusion one toward the other according to:

$$\frac{1}{k_E} = \frac{1}{k_E^{act}} + \frac{1}{k_D} \tag{5.11}$$

where k_E^{act} is the activation-limited electron self-exchange rate constant and k_D is the diffusion-limited bimolecular rate constant. The suggestion was (55) that the kinetic control of electron hopping passes from diffusion to activation, as the concentration of the redox couple in the film increases. However, this reasoning is difficult to understand because both k_E^{act} and k_D are bimolecular rate constants; the combination of these two rate constants in Eq. 5.11 should therefore be independent of the concentration of redox sites. Interestingly, this same problem has been previously addressed using the concept that the diffusion-limited bimolicular rate constant might be increased by electron transfer (56). However, the rate constant k in Eq. 5.11, or better [the numerical coefficient $\pi/4$ obtained with a thermodynamic model (56) has been recently corrected to $\frac{1}{6}$ (36), which agrees with Eqs. 5.6 and 5.7, which were obtained from the stochastic model evoked earlier] in Eq. 5.7, is merely the activation-limited electron-exchange volume rate constant, and the physical displacement diffusion is entirely taken into account by the first term on the right-hand side of Eq. 5.7.

Stochastic models based on a disordered structure conception of redox polymers as described earlier thus appear to provide a reasonable framework for the description of electron-hopping mechanisms in such materials, provided activity effects are properly taken into account. The knowledge of such effects is important for a precise understanding of electron transport through the film, one of the rate-limiting factors in catalytic applications. However, the most important feature of electron transport, when attempting to combine its kinetic contribution with those of the catalytic reaction and of substrate diffusion, is that it can be treated as a diffusion phenomenon even though it is governed by an apparent rather than a true diffusion coefficient. The exact determination of the apparent diffusion coefficient in steady state or transient measurements requires the knowledge of the concentration of redox sites, (C_E^0), within the coating. The number of moles of redox sites per unit surface area of the electrode, (Γ_E^0 is readily accessible by integration of slow scan cyclic voltammetric responses (1), but the actual thickness of the coating (ϕ) is not so easy to determine. As seen in Section 5.4, however, the parameter required to describe the interference of electron transport as a possible rate-limiting factor in catalytic applications is not the diffusion coefficient D_E^{app} but rather the current density i_E defined as:

$$i_E = \frac{\mathscr{F} D_E^{app} C_E^0}{\phi} = \mathscr{F}\left(\frac{D_E^{app}\Gamma_E^0}{\phi^2}\right) \tag{5.12}$$

$D_E^{app}(\Gamma_E^0)^2/\phi^2$ is itself directly obtained from the Cottrell slopes in potential-step experiments. Since Γ_E^0 is readily measurable, i_E can be derived from these experiments without independent determination of the film thickness.

5.2.3 Permeation of Substrates through Redox Polymers

A convenient technique for investigating the permeation of substrates through polymers is rotating disk electrode voltammetry, whereby the disk electrode is coated with the polymer and dipped into an electrolyte solution containing the substrate (A). If there exists a potential region in which the substrate is reducible (or oxidizable) at the electrode surface, the resulting steady state current–potential curve provides information about the physical displacement of the substrate within the coating.

If the permeation is of the membrane type (4), the substrate diffuses through the polymer with a diffusion coefficient D_S, usually smaller than the diffusion coefficient in the solution, D_A. If, in addition, the crossing of the film–solution interface is fast in both directions, its influence on the current response is simply governed by an equilibrium partition coefficient κ. Steady state plateau current densities (corresponding to a concentration of the substrate equal to zero at the electrode surface) are then obtained as follows assuming that diffusion is linear (57–59). The terms $(C_A)_0$, $(C_A)_{\phi-}$, and $(C_A)_{\phi+}$ are the concentrations of substrate at the electrode surface, at the film–solution boundary from the inside and outside, respectively (ϕ is the film thickness). Then, in the framework of the Nernst approximation

$$\left(\frac{dC_A}{dx}\right)_{\phi+} = \frac{C_A^0 - (C_A)_{\phi+}}{\delta_A} \tag{5.13}$$

δ_A = thickness of the diffusion layer and C_A^0 = concentration of the substrate in the bulk of the solution.

In addition,

$$D_S\left(\frac{dC_A}{dx}\right)_{\phi-} = D_A\left(\frac{dC_A}{dx}\right)_{\phi+} \tag{5.14}$$

and

$$(C_A)_{\phi-} = \kappa(C_A)_{\phi+} \tag{5.15}$$

At steady state, the gradient of A is constant throughout the coating and thus the current density i is given by

$$\frac{i}{\mathscr{F}} = D_S\left(\frac{dC_A}{dx}\right)_0 = D_S\left(\frac{dC_A}{dx}\right)_{\phi-} = D_A\left(\frac{dC_A}{dx}\right)_{\phi+} \tag{5.16}$$

Thus, at the plateau, the current density (i_L) is expressed as

$$\frac{1}{i_L} = \frac{1}{i_S} + \frac{1}{i_A} \tag{5.17}$$

where i_S and i_A are two current densities characterizing the diffusion of the substrate in the film and in the solution, respectively,

$$i_S = \mathscr{F}\left(\frac{D_S \kappa C_A^0}{\phi}\right) \tag{5.18}$$

$$i_A = \mathscr{F}\left(\frac{D_A C_A^0}{\delta_A}\right) \tag{5.19}$$

Since the diffusion layer thickness (δ_A) is inversely proportional to the square root of the electrode rotation rate:

$$\delta_A = 4.98 D_A^{1/3} \nu^{1/6} \omega^{-1/2} \tag{5.20}$$

(δ_A in cm; D_A in cm^2 s^{-1}; ν, the kinematic viscosity of the solution, in cm^2 s^{-1}; ω

in rotations per minute), the Levich current, i_A, is proportional to the square root of the rotation rate (60, 61):

$$i_A = 0.2\mathscr{F} D_A^{2/3} \nu^{-1/6} C_A^0 \omega^{1/2} \tag{5.21}$$

(with the same units and with C_A^0 in ML^{-1}). "Koutecky–Levich" plots of the reciprocal of the plateau current density against the reciprocal of the square root of the rotation rate are thus linear with a nonzero intercept (Fig. 5.2). The slope is proportional to C_A^0 and i_S is obtained from the intercept.

What happens if the crossing of the film–solution boundary by the substrate is not as rapid as assumed above (62)? Let χ_f and χ_b be the first-order rate constants for the in- and out-crossing of the film–solution interface, respectively, where $\chi_f/\chi_b = \kappa$. Then,

$$\frac{i}{\mathscr{F}} = D_S\left(\frac{dC_A}{dx}\right)_0 = D_S\left(\frac{dC_A}{dx}\right)_{\phi-} = D_A\left(\frac{dC_A}{dx}\right)_{\phi+} = \chi_f(C_A)_{\phi+} - \chi_b(C_A)_{\phi-} \tag{5.22}$$

Thus, introducing an additional characteristic current density for the penetration of the substrate into the film:

$$i_P = \mathscr{F}\chi_f C_A^0 \tag{5.23}$$

the plateau current is given by

$$\frac{1}{i_L} = \frac{1}{i_A} + \frac{1}{i_P} + \frac{1}{i_S} \tag{5.24}$$

A linear Koutecky–Levich plot is again obtained with a nonzero intercept (Fig. 5.2), which is both a function of the rate of diffusion of the substrate through the film (i_S) and the rate of penetration of the substrate from the solution into the film (i_P). A way of separating i_P from i_S is to investigate the variation of the reciprocal of the intercept with the film thickness (Fig. 5.2): The term i_S is inversely proportional to ϕ, whereas i_P is independent of ϕ. Since ϕ is not easy to determine, the reciprocal of the intercept may be plotted against the total quantity of polymer deposited onto the electrode surface.

A typical experimental example of such analyses is a study of the permeation of benzoquinone in poly(vinylferrocene) from a tetrabutylammonium tetrafluoroborate acetonitrile solution (62, 63), where χ_f was found as equal to about 0.1 cm s^{-1} and $\kappa C_A^0 D_S$ to 7.8 10^{-9} mol cm^{-1} s^{-1} (62). A series of other examples are listed in Table 5.2. The determination of κD_S requires the film thickness (ϕ) to be measured independently. This is not an easy task as noted earlier. In addition, the determination of D_S would require the independent measurement of κ. This is the reason that permeation data are often made available as values of the global quantity $D_S\kappa/\phi$ (64). As far as the role of substrate diffusion in

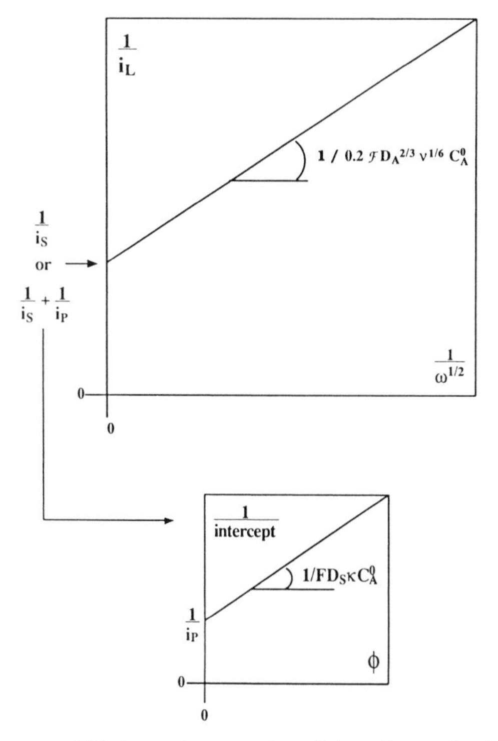

Figure 5.2. Membrane-type diffusion and penetration of the substrate in the coating. Koutecky–Levich plots at the plateau of the substrate reduction (or oxidation) steady state voltammograms.

catalytic coatings is sought this is not a dramatic drawback since the value of i_S, the only necessary parameter, is then obtained (Eq. 5.18).

Whether the diffusion of the substrate in polymer coatings is of the membrane type or if they occur through pinholes in the surface of the film facing the solution is a question that has been discussed in some detail (4, 63). The membrane model was contrasted with a model where the electrode is covered with a thin (as compared to the diffusion layer thickness) insulating film interrupted by a large number of uniformly distributed electrochemically active holes. The conclusion of the discussion, based on potential-step chronoamperometric and rotating disk electrode experiments, is that the permeation of the substrate through the film does not occur through pinholes having dimensions of the same order of magnitude as the solution diffusion layer thickness. In the latter case, rotated disk Koutecky–Levich plots would be predicted not only to be linear but also to pass through the origin at variance with the experimental observations. These conclusions were confirmed, for poly(vinylferrocene) films, by surface techniques such as XPS and SEM (4, 63).

In the membrane model, the molecules diffuse and penetrate in a completely homogeneous phase. At the other extreme, another model consists in envision-

Table 5.2. Summary of Data Taken from the Literature and Analyzed for a Finite Rate of Mass Transfer Across the Polymer–Solution Interface

Polymer	Substrate	$D_S \kappa C_A^0(a)$ (mol cm^{-1} s^{-1})	$D_S \kappa C_A^0(b)$ (mol cm^{-1} s^{-1})	χ_f (cm s^{-1})	Reference
Poly(vinylferrocene)	Benzoquinone	1.3×10^{-9}	7.8×10^{-9}	0.1	63
Poly$[VDQ]^{2+}$(*c*)	Ferrocene	2.2×10^{-12}	3.5×10^{-12}	0.01	64
Poly$[Ru(bpy)_2(p\text{-}cinn)_2]^{2+}$(*d*)	Ferrocene	1.2×10^{-10}	1.3×10^{-10}	>1.1	64
Poly$[Ru(vbpy)_3]^{2+}$(*e*)	$Ru(bpy)_2Cl_2$	1.3×10^{-11}	2.0×10^{-11}	0.05	64
	Benzoquinone	8.7×10^{-11}	7.0×10^{-11}		64
	Ferrocene	4.3×10^{-8}	5.6×10^{-8}	0.01	65
Poly-*l*-lysine/$[Mo(CN)_8]^{3-}$(*f*)	$[Co(tpy)_2]^{2+}$(*g*)	7.9×10^{-10}	7.5×10^{-10}	>1.1	92
Protonated polylysine/ $[Co(C_2O_4)_3]^{2-}$(*f*)	$[Co(C_2O_4)_3]^{2-}$	1.3×10^{-8}	1.3×10^{-8}	>1.1	66

[a]Value found without accounting for a finite interfacial crossing rate, as reported in the reference.
[b]Value found taking amount of a finite interfacial crossing rate.
[c]VDQ^{2+} = vinyl diquat.
[d]p-$(cinn)_2$ = *N*-(4-pyridine) cinnamamide.
[e]vbpy = vinyl analogue of bpy = 2,2′-bipyridine.
[f]pH = 5.5, with 0.2M acetate buffer.
[g]tpy = 2,2′,2″-terpyridine.

ing the polymer coating as composed of two intermingled but sharply different phases. The first phase consists of the macromolecules themselves that are assumed to totally block diffusion and penetration and the other phase is a set of thin, tortuous, solvent-containing channels that would be the only portion of space available for the physical displacement of substrate molecules. Penetration of molecules into the polymer coating would thus occur through the small pinholes connecting the channels to the bathing solution. Similarly, electrode transfer at the electrode would only occur within the areas where the channels communicate with the electrode surface. The question that then arises is whether diffusion and penetration in the polymer coating, according to this mechanism, can be treated formally as membrane-type phenomena and, if so, what are the equivalencies between the parameters characterizing each of the two models. The answers to these questions are the following (67). The surface of the polymer facing the solution, as well as that facing the electrode surface, is regarded as uniformly sprinkled with a large number of circular pinholes of radius R_a, the center of each of them being separated from the center of its nearest neighbor by a distance $2R_0$. The value of R_0 is assumed to be much larger than R_a (say at least 10 times) and R_a is itself small compared to the diffusion layer thickness. How the substrate diffuses in the various sections of space from the electrode to the bulk of the solution is schematically represented in Fig. 5.3. The fractional coverage of the electrode and of the film–solution interface (θ) is such that (68):

$$R_a = R_0(1 - \theta)^{1/2} \tag{5.25}$$

Inside the polymer film, the diffusion of the substrate can be viewed as membranelike with a diffusion coefficient:

$$D_S = \frac{D_C(1 - \theta)}{\tau^2} \tag{5.26}$$

where D_C is the intrinsic average diffusion coefficient in the channels and τ a tortuosity factor that can be defined as the ratio between the average length of the channels and the film thickness. The molecules penetrating the pinholes diffuse in a constrained manner, roughly of the hemispherical type, within a thin layer adjacent to the film–solution interface, thin as compared to the thickness of the solution layer where linear diffusion prevails. The thickness of this thin constrained diffusion layer is denoted μ_{CD}. In this framework, the conservation of matter throughout the various layers composing the system (Fig. 5.3) is expressed by the following set of equations that replaces Eq. 5.20 in the preceding analysis, assuming that the partition of the substrate between the solution and the polymer remains at equilibrium whatever the current flowing through the electrode surface.

$$\frac{i}{\mathscr{F}} = \frac{D_C(1 - \theta)}{\tau^2}\left(\frac{dC_A}{dx}\right)_0 = \frac{D_C(1 - \theta)}{\tau^2}\left(\frac{dC_A}{dx}\right)_{\phi-} = D_A\left(\frac{dC_A}{dx}\right)_{\phi+\mu_{CD}} \tag{5.27}$$

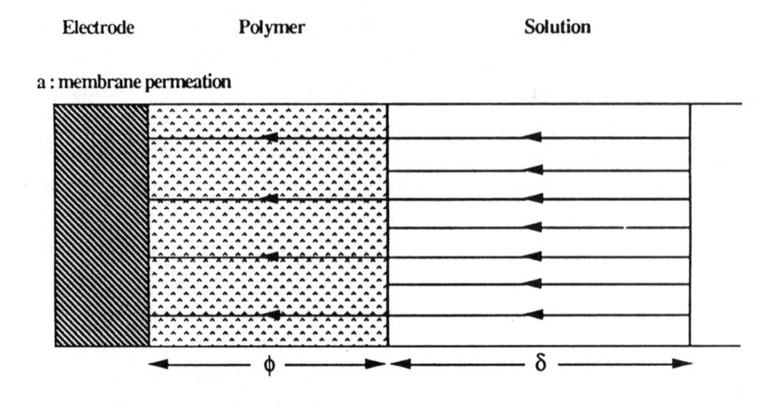

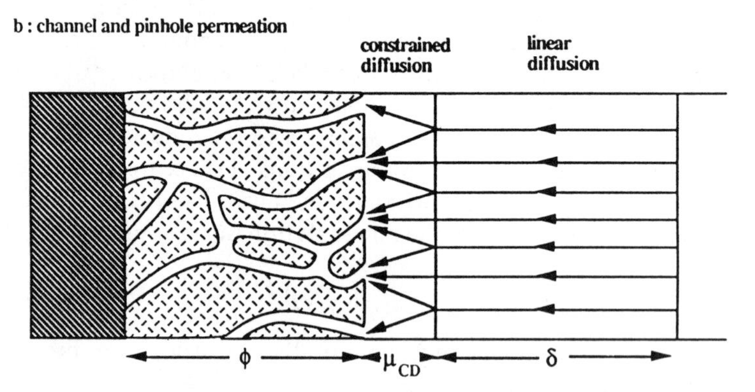

Figure 5.3. Diffusion and penetration of molecules in swollen polymer coatings.

(i, as before, is the current density over the whole geometric area of the electrode). On the other hand

$$\left(\frac{dC_{\mathrm{A}}}{dx}\right)_{\phi+\mu_{\mathrm{CD}}} = \frac{C_{\mathrm{A}}^{0} - (C_{\mathrm{A}})_{\phi+\mu_{\mathrm{CD}}}}{\delta} \tag{5.28}$$

and

$$(C_{\mathrm{A}})_{\phi-} = \kappa(\bar{C}_{\mathrm{A}})_{\phi+} \tag{5.29}$$

There is a slight variation of the concentration of A over the (small) area of each pinhole. For this reason, the bar over C_{A} in the right-hand side of Eq. 5.29, indicates that the concentration of A is averaged over each pinhole area. The partition coefficient (κ) takes account of the possible changes in the interactions between the substrate and its environment when its passes from the solution to the polymer channels. As shown earlier (68), the following relation (Eq. 5.30) provides a reasonably approximate description of constrained diffusion within the thin layer of thickness μ_{CD}.

$$(\bar{C}_A)_{\phi+} = (C_A)_{\phi+\mu_{CD}} - [0.6R_0/(1-\theta)^{1/2}](dC_A/dx)_{\phi+\mu_{CD}} \tag{5.30}$$

At the plateau of the current–potential curve

$$\left(\frac{dC_A}{dx}\right)_0 = \frac{(C_A)_{\phi-}}{\phi} \tag{5.31}$$

Combination of Eqs. 5.25–5.31 finally leads to

$$\frac{1}{i_L} = \frac{1}{i_A} + \frac{1}{i_S} + \frac{1}{i_{CD}} \tag{5.32}$$

where i_S and i_A are the same as defined by Eqs. 5.15 and 5.17.

The new characteristic current density, i_{CD}, is defined as

$$i_{CD} = \mathscr{F}\frac{(1-\theta)^{1/2}D_A C_A^0}{0.6R_0} = \mathscr{F}\left(\frac{R_a}{0.6R_0^2}\right)D_A C_A^0 \tag{5.33}$$

i_{CD} plays formally the same role as the current density i_P characterizing the activated crossing of the film–solution interface in the membrane permeation model described earlier. Constrained diffusion toward pinholes thus amounts to a membrane penetration rate constant.

$$\chi_f = \frac{(1-\theta)^{1/2}D_A}{0.6R_0} = \left(\frac{R_a}{0.6R_0^2}\right)D_A \tag{5.34}$$

In addition, it might be that the passage of the substrate molecules through the pinholes requires an activation barrier if the properties of the solution inside and outside the polymer coating are substantially different one from the other (such activation barriers are, however, likely to be smaller than those controlling the penetration into a homogeneous membrane). In this case, the same characteristic current density as already defined in Eq. 5.23 has to taken into account. It combines with the others in the expression of the plateau current density in the manner shown in Eq. 5.35.

$$\frac{1}{i_L} = \frac{1}{i_A} + \frac{1}{i_{CD}} + \frac{1}{i_P} \tag{5.35}$$

It may thus be concluded that diffusion and penetration of molecules in swollen polymers through small channels and pinholes can be treated formally as membrane-type phenomena with equivalent diffusion coefficients and penetration characteristics given by Eqs. 5.26 and 5.34. The number and size of the pinholes and channels may vary considerably from one polymer to the other and also with the degree of swelling. As an average typical example, let us consider a polymer coating characterized by $R_a = 100\,\text{Å}$, $R_0 = 1000\,\text{Å}$ $(1 - \theta = 10^{-2})$, and $\tau = 3$. Even assuming that the properties of the solution are the same inside the channels and in the outside solution, the above values lead to a decrease of the diffusion coefficient by a factor of 10^3 upon passing from the solution to the interior of the polymer coating. With the same values, the equivalent penetration rate constant (χ_f) would be of the order of 0.1 cm s^{-1} for a diffusion coefficient in the solution of 10^{-5} cm^2 s^{-1}. The experimental values of χ_f, which have been found with several polymer coatings (63), are of the same order of magnitude. Thus the observation of a finite rate of crossing of the film–solution interface may well be the reflection of constrained diffusion towards the pinholes rather than of an activation controlled membrane penetration phenomenon.

In the study of the competition between the electron transfer at the electrode surface and the various transport processes, one has to take account of the fact that a fraction θ of the electrode surface is inactivated by the presence of the polymer. Thus,

$$\frac{i}{\mathscr{F}} = (1 - \theta)k_S \exp[(-/+)(\alpha\mathscr{F}/RT)(E - E^0)]\text{-}(C_A)_0 \tag{5.36}$$

The current density along the wave (i) is then given by

$$\frac{1}{i} = \frac{1}{i_{k_s} \exp[(-/+)(\alpha\mathscr{F}/RT)(E - E^0)]} + \frac{1}{i_L} = \frac{1}{i_{k_s} \exp[(-/+)(\alpha\mathscr{F}/RT)(E - E^0)]} + \frac{1}{i_S} + \frac{1}{i_P} + \frac{1}{i_{CD}} + \frac{1}{i_A} \tag{5.37}$$

i_{k_s} being defined by Eq. 5.32 instead of Eq. 5.15 in the membrane case.

$$i_{k_S} = \mathscr{F} k_S^{\text{app}}(1 - \theta)C_A^0 \tag{5.38}$$

In the case where mass transport is prevalently controlled by the diffusion of the substrate through the polymer coating, the factor that determines the competition between electron and diffusion is

$$\frac{i_{k_S}}{i_S} = k_S^{\text{app}} \Big/ \left(\frac{D_C}{\tau^2}\right)(\phi) \tag{5.39}$$

and not by $k_S^{app}/[(D_C/\tau^2)(1-\theta)\phi]$ as would be obtained by direct application of the membrane permeation model with brute force replacement of D_S by $(D_C/\tau^2)(1-\theta)$. The difference occurs because in the channel–pinhole model only a fraction $(1-\theta)$ of the electrode surface is regarded as electrochemically active, whereas the whole surface is assumed to be active in the membrane model. In the case where diffusion towards the pinholes predominatly controls mass transport the competition with the electron transfer is governed by the ratio

$$\frac{i_{k_S}}{i_{CD}} = \frac{k_S R_a}{0.6 D_A} \tag{5.40}$$

which depends on the size of the pinholes and not upon their number on the surface. This occurs because the pinholes cover a fraction of the film–solution interface that is the same as the electrochemically active area on the electrode surface.

The assumption, made in the above channel-pinhole model, is that the macromolecules totally blocking diffusion and penetration most probably do not apply to all polymer coatings and all substrates. For example, organic substrates are likely to penetrate and diffuse in regions of the coating that are predominantly constituted by intermingled polymer chains and not containing large amounts of solvent. In such cases, the dynamics of permeation through the film would nevertheless be also formally equivalent to those of membrane phenomena.

5.3 CATALYTIC MONOLAYERS

Redox polymer coatings containing several equivalent monolayers are expected to give rise to larger catalytic responses than monolayer coatings of the same redox couple because of the volume effect evoked earlier. We thus analyze first the catalytic responses of monolayer coatings as a reference system for the foregoing analyses of multilayer coatings, bearing in mind that significant catalytic effects at monolayers are expected only if the monolayer works in a chemical catalysis fashion.

In the absence of substrate, a monolayer coating containing a reversible redox couple, P/Q, does not give rise to any steady state current. In the presence of the substrate a wave develops, exhibiting a plateau current that is reached at potentials sufficiently negative (for a reduction, positive for an oxidation) for the concentration of the active form of the catalyst at the electrode surface to equal zero. Then, in the simple case where the catalytic reaction is irreversible:

$$\mathrm{P} \pm \mathrm{e}^- \rightleftarrows \mathrm{Q} \tag{5.41}$$

$$\mathrm{Q} + \mathrm{A} \rightarrow \mathrm{P} + \text{products} \tag{5.42}$$

the steady state current density, i_L, at a rotating disk electrode is given by Eq. 5.43 (29, 69–71)

$$\frac{1}{i_L} = \frac{1}{i_k} + \frac{1}{i_A} \tag{5.43}$$

in which i_A is the Levich current density, defined earlier (Eqs. 5.17 and 5.19). The term i_k is a current density characterizing the catalytic reaction:

$$i_k = \mathscr{F} k \Gamma_E^0 C_A^0 = \mathscr{F} k_m C_E^0 C_A^0 \tag{5.44}$$

where Γ_E^0 is the total surface concentration of catalytic sites, whereas C_E^0 is the equivalent volume concentration. The term k_m is the surface rate constant of the catalytic reaction [dimension = $(\text{mol})^{-1}$ $\text{length}^4 \text{time}^{-1}$] and k the equivalent volume rate constant. Since the Levich current is proportional to $\omega^{-1/2}$, Eq. 5.41 is conveniently displayed from experimental data under the form of a Koutecky–Levich plot of the inverse of the plateau current versus the inverse of the square root of the rotation rate (Fig. 5.4). The plot is expected to be linear. Its intercept provides the inverse of the current density characterizing the catalytic reaction and its slope is proportional to the concentration of substrate in the bulk of the solution.

Equation 5.43 is established as follows. At the electrode surface:

$$\frac{d\Gamma_Q}{dt} = -\frac{d\Gamma_P}{dt} = \frac{i}{\mathscr{F}} - k\Gamma_Q(C_A)_0 \tag{5.45}$$

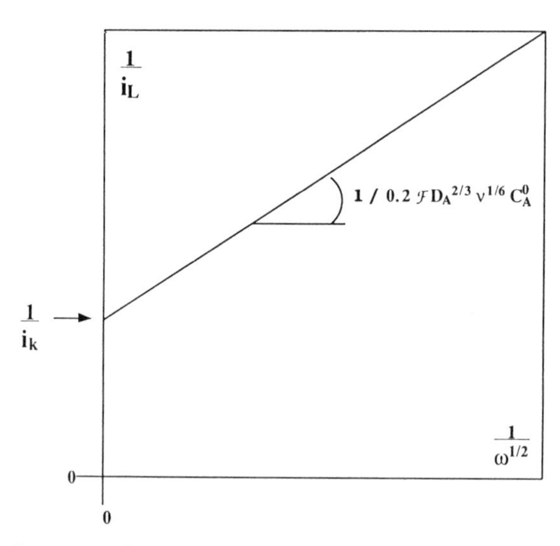

Figure 5.4. Catalysis at a monolayer coated rotating disk electrode "Koutecky–Levich" plot (units for the slope are defined in the text).

(Γ_P and Γ_Q are the surface concentrations of P and Q, respectively). At steady state:

$$\frac{d\Gamma_Q}{dt} = -\frac{d\Gamma_P}{dt} = 0 \tag{5.46}$$

and thus

$$\Gamma_Q + \Gamma_P = \Gamma_E^0 \tag{5.47}$$

since the potential is such that $\Gamma_P = 0$, it follows that $\Gamma_Q = \Gamma_E^0$. Thus,

$$\frac{i}{\mathscr{F}} = k\Gamma_E^0(C_A)_0 = D_A\left(\frac{dC_A}{dx}\right)_0 \tag{5.48}$$

If the diffusion of A in the solution is assumed to be linear:

$$\left(\frac{dC_A}{dx}\right) = \left(\frac{dC_A}{dx}\right)_0 = \frac{C_A^0 - (C_A)_0}{\delta} \tag{5.49}$$

in the framework of the Nernst approximation. Combination of Eqs. 5.48 and 5.49 taking account of the definition of i_A and i_k (Eqs. 5.19 and 5.44) leads to Eq. 5.43.

5.4 IDENTIFICATION AND ASSOCIATION OF THE RATE-CONTROLLING STEPS IN CATALYTIC REDOX POLYMER COATINGS

5.4.1 One-Step Irreversible Catalytic Reactions

The reaction scheme that we analyze now is the same as that discussed in the preceding section (Eqs. 5.41 and 5.42), but the catalytic reaction now takes place in a multilayer coating through which the electrons have to propagate to regenerate the active form of the catalyst and the substrate has to diffuse to reach the catalytic centers. As discussed earlier, these two transport processes can be regarded as diffusional processes governed by diffusion coefficients (D_E^{app} and D_S, respectively). Following preliminary and semiquantitative treatments (72, 73), the problem posed by the combination of the three rate-limiting factors in the film has been analyzed in a more systematic manner (74–76) (see also Ref. 77 where a similar treatment can be found with, however, quite different notations).

As in the case discussed in the preceding section there is no steady state current in the absence of substrate. Upon adding the substrate to the solution one or two waves develop. The objective of the following analysis is to establish the expression of the plateau current density of these waves as a function of the parameters governing the system. Inside the film, that is, for $0 \leqslant x \leqslant \phi$, the concentrations of P, Q and A, C_P, C_Q, C_A, are solutions of following differential equations at steady state (still assuming that diffusion to and from the planar disk electrode is linear).

$$D_{\mathrm{E}}^{\mathrm{app}} \frac{d^2 C_{\mathrm{P}}}{dx^2} + k C_{\mathrm{A}} C_{\mathrm{Q}} = 0 \tag{5.50}$$

$$D_{\mathrm{E}}^{\mathrm{app}} \frac{d^2 C_{\mathrm{Q}}}{dx^2} - k C_{\mathrm{A}} C_{\mathrm{Q}} = 0 \tag{5.51}$$

$$D_{\mathrm{S}} \frac{d^2 C_{\mathrm{A}}}{dx2} - k C_{\mathrm{A}} C_{\mathrm{Q}} = 0 \tag{5.52}$$

From the combination of Eqs. 5.50 and 5.51 it follows that:

$$C_{\mathrm{P}} + C_{\mathrm{Q}} = C_{\mathrm{E}}^{0} \tag{5.53}$$

throughout the film. As the electrode potential is made more and more negative (for a reduction, positive for an oxidation) the first wave reaches a plateau corresponding to

$$(C_{\mathrm{P}})_0 = 0 \quad \text{and thus} \quad (C_{\mathrm{Q}})_0 = C_{\mathrm{E}}^{0} \tag{5.54}$$

and to

$$\left(\frac{dC_{\mathrm{A}}}{dx}\right)_0 = 0 \tag{5.55}$$

The current density is then given by

$$\frac{i_1}{\mathscr{F}} = D_{\mathrm{E}}^{\mathrm{app}} \left(\frac{dC_{\mathrm{P}}}{dx}\right)_0 = -D_{\mathrm{E}}^{\mathrm{app}} \left(\frac{dC_{\mathrm{Q}}}{dx}\right)_0 \tag{5.56}$$

At still more negative (for a reduction positive for an oxidation) potentials, the direct reduction of the substrate comes into play. The boundary conditions are then

$$(C_{\mathrm{Q}})_0 = C_{\mathrm{E}}^{0} \quad \text{and} \quad (C_{\mathrm{A}})_0 = 0 \tag{5.57}$$

and the current density

$$\frac{i_1 + i_2}{\mathscr{F}} = -D_{\mathrm{E}}^{\mathrm{app}} \left(\frac{dC_{\mathrm{Q}}}{dx}\right)_0 + D_{\mathrm{S}} \left(\frac{dC_{\mathrm{A}}}{dx}\right)_0 \tag{5.58}$$

i_1 is the current density corresponding to catalysis while $i_1 + i_2$ represents both the catalytic current and the direct reduction (or oxidation) of the substrate when it is able to reach the electrode surface.

At the film–solution interface ($x = \phi$), the boundary conditions are

$$\left(\frac{dC_P}{dx}\right)_{\phi-} = \left(\frac{dC_Q}{dx}\right)_{\phi-} = 0 \quad (5.59)$$

$$D_S\left(\frac{dC_A}{dx}\right)_{\phi-} = D_A\left(\frac{dC_A}{dx}\right)_{\phi+} \quad (5.60)$$

$$(C_A)_{\phi-} = \kappa(C_A)_{\phi+} \quad (5.61)$$

The subscript $\phi-$ represents the film–solution interface, approached from the film side and $\phi+$ the film–solution interface, approached from the solution side. Equation 5.59 expresses that both members of the catalyst couple are confined within the coating. Conservation of matter (A) leads to Eq. 5.60, and Eq. 5.61 expresses that the activation barriers for the passage of the substrate through the film–solution interface are negligible (κ is the corresponding partition equilibrium constant). The case where this condition is not fulfilled is discussed in the next section.

Besides the Levich current density (i_A), which is a measure of the diffusion kinetics in the solution, the current densities, i_1 and i_2, depend on three current characteristic densities measuring the kinetics of the three rate-limiting processes taking place inside the film.

Transport of electrons $$i_E = \frac{\mathscr{F} D_E^{app} C_E^0}{\phi} = \frac{\mathscr{F} D_E^{app} \Gamma_E^0}{\phi^2} \quad (5.12)$$

Diffusion of the substrate $$i_S = \frac{\mathscr{F} D_S \kappa C_A^0}{\phi} \quad (5.18)$$

Catalytic reaction $$i_k = \mathscr{F} \kappa k C_E^0 C_A^0 \phi = \mathscr{F} \kappa k \Gamma_E^0 C_A^0 \quad (5.62)$$

It is convenient to introduce the following dimensionless variables.

Space $$y = \frac{x}{\phi} \quad (5.63)$$

Concentrations $$q = \frac{C_Q}{C_E^0} \qquad a = \frac{C_A}{\kappa C_A^0} \quad (5.64)$$

The plateau current densities at the first and second waves, i_1 and $i_1 + i_2$ are obtained from the computation of one or the other of the two following set of equations.

In terms of Q:

at the first wave:

$$\frac{d^2q}{dy^2} - \frac{i_k}{i_E} q \left\{1 - \frac{i_E}{i_S}\left[q_1 - q - \left(\frac{dq}{dy}\right)_0 \left(1 + \frac{i_S}{i_A} - y\right)\right]\right\} = 0 \quad (5.65)$$

with

$$q_0 = 1 \qquad \left(\frac{dq}{dy}\right)_1 = 0 \qquad i_1 = -i_E\left(\frac{dq}{dy}\right)_0 \tag{5.66}$$

at the second wave:

$$\frac{d^2q}{dy^2} - \frac{i_k}{i_E} q \left\{ \frac{i_A}{i_A + i_S} \left[1 + \frac{i_E}{i_S}(1 - q_1) \right] y - \frac{i_E}{i_S}(1 - q) \right\} = 0 \tag{5.67}$$

with

$$q_0 = 1 \qquad \left(\frac{dq}{dy}\right)_1 = 0 \qquad i_1 + i_2 = \frac{i_A}{i_S + i_A}[i_S + i_E(1 - q_1)] \tag{5.68}$$

In terms of A:

at the first wave:

$$\frac{d^2a}{dy^2} - \frac{i_k}{i_S} a \left\{ 1 - \frac{i_S}{i_E} \left[\frac{i_A}{i_S}(1 - a_1)y - a + a_0 \right] \right\} = 0 \tag{5.69}$$

with

$$\left(\frac{da}{dy}\right)_0 = 0 \qquad 1 - a_1 = \frac{i_S}{i_A}\left(\frac{da}{dy}\right)_1 \qquad i_1 = i_A(1 - a_1) \tag{5.70}$$

at the second wave:

$$\frac{d^2a}{dy^2} - \frac{i_k}{i_S} a \left[1 - \frac{i_A}{i_E}(1 - a_1)y + \frac{i_S}{i_E} a \right] = 0 \tag{5.71}$$

with

$$a_0 = 0 \qquad 1 - a_1 = \frac{i_S}{i_A}\left(\frac{da}{dy}\right)_1 \qquad i_1 + i_2 = i_A(1 - a_1) \tag{5.72}$$

Although these two systems of equations can be readily computed numerically to obtain i_1 and i_2 as a function of i_A, i_E, i_S, i_k, it is interesting to identify limiting situations in which the plateau currents depend on a lesser number of parameters as the result of a kinetic control by two or even one of the three rate-limiting phenomena taking place inside the film, namely the catalytic reaction (R), the substrate diffusion (S), and the transport of electrons (E). Figure 5.5 displays the various limiting cases together with the concentration profiles of the active form of the catalyst (Q) the substrate (A) in the film normalized against C_E^0 and κC_A^0, respectively. The equations giving directly the plateau current densities at the first and second waves for the various limiting cases are listed in Table 5.3.

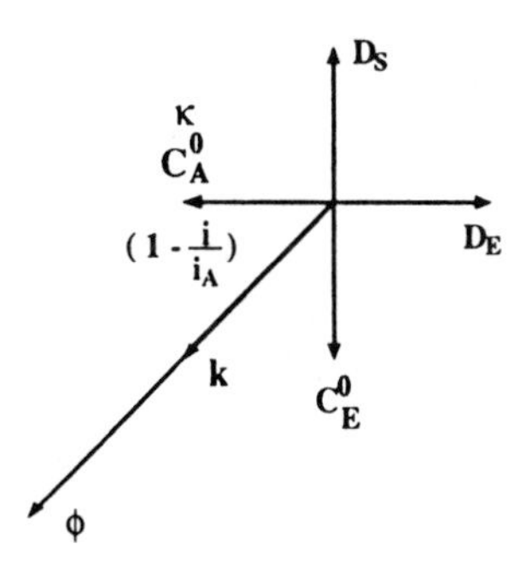

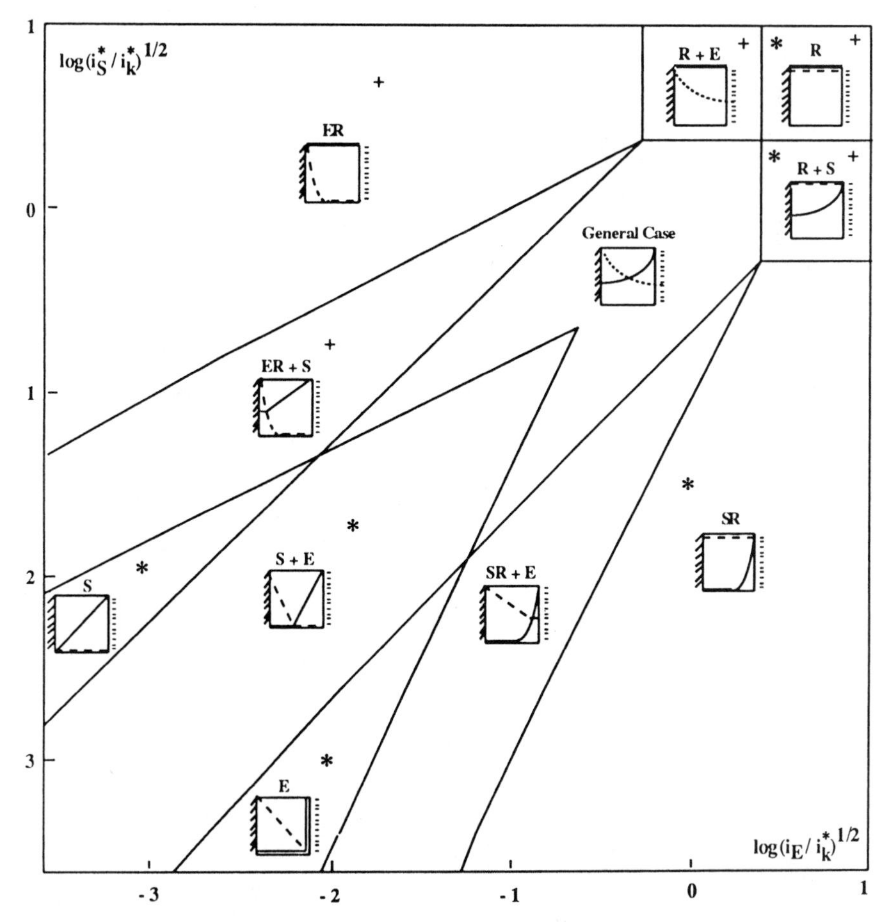

Figure 5.5. Zone diagram for a single-step reversible catalytic reaction. The normalized concentration profiles (comprised between 0 and 1) of the substrate ($C_A/\kappa C_A^0$: ——) and of the active form of the catalyst (C_Q/C_E^0: – – – –) at the plateau of the first wave are exhibited in the inserts. The symbol // on the left-hand side of each insert represents the electrode and the = on the right-hand side represents the solution. An * represents linear Koutecky–Levich plots and a + represents a second wave (in the absence of + there is no second wave). The set of arrows on the top of the diagram represents, in direction and magnitude, the displacement of the system representative point with the various experimental parameters. The boundary lines separating the various zones are calculated on the basis on a 5% uncertainty on the determination of i_1. The symbols are defined in the text and in the glossary. (Reprinted with permission from Ref. 62.)

Table 5.3. Expressions of the Plateau Currents[a]

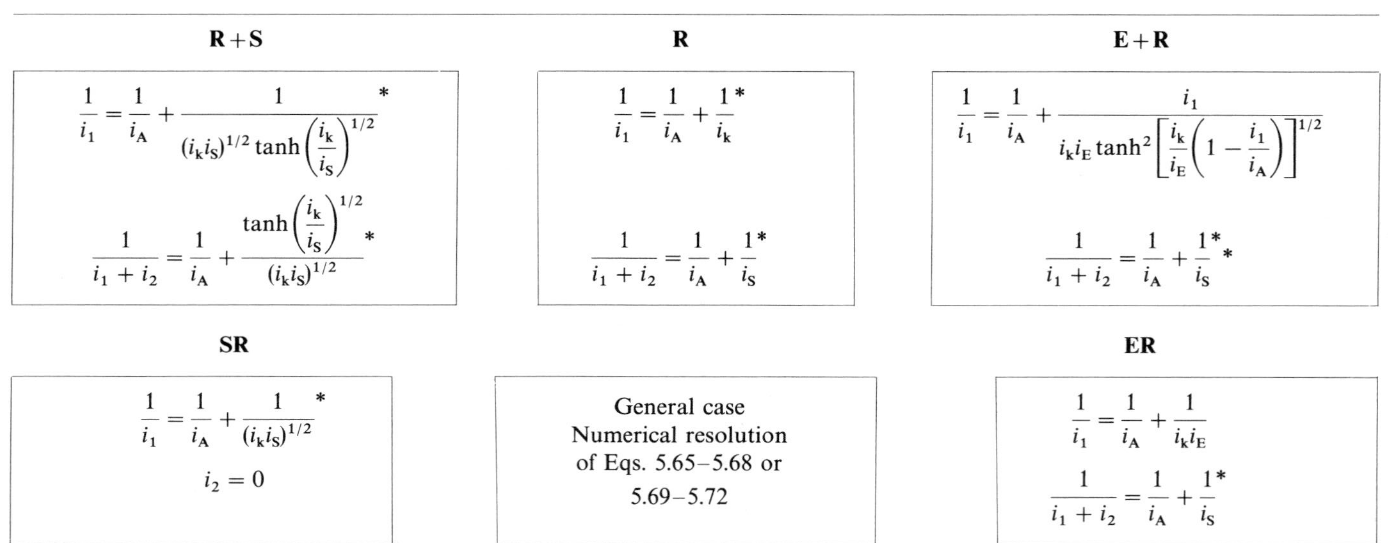

R+S	R	E+R
$\frac{1}{i_1} = \frac{1}{i_A} + \frac{1}{(i_k i_S)^{1/2} \tanh\left(\frac{i_k}{i_S}\right)^{1/2}}$ * $\frac{1}{i_1 + i_2} = \frac{1}{i_A} + \frac{\tanh\left(\frac{i_k}{i_S}\right)^{1/2}}{(i_k i_S)^{1/2}}$ *	$\frac{1}{i_1} = \frac{1}{i_A} + \frac{1}{i_k}$ * $\frac{1}{i_1 + i_2} = \frac{1}{i_A} + \frac{1}{i_S}$ *	$\frac{1}{i_1} = \frac{1}{i_A} + \frac{i_1}{i_k i_E \tanh^2\left[\frac{i_k}{i_E}\left(1 - \frac{i_1}{i_A}\right)\right]^{1/2}}$ $\frac{1}{i_1 + i_2} = \frac{1}{i_A} + \frac{1}{i_S}$ * *
SR		**ER**
$\frac{1}{i_1} = \frac{1}{i_A} + \frac{1}{(i_k i_S)^{1/2}}$ * $i_2 = 0$	General case Numerical resolution of Eqs. 5.65–5.68 or 5.69–5.72	$\frac{1}{i_1} = \frac{1}{i_A} + \frac{1}{i_k i_E}$ $\frac{1}{i_1 + i_2} = \frac{1}{i_A} + \frac{1}{i_S}$ *

SR + E

$$\frac{1}{i_1} = \frac{1}{i_A} + \frac{1}{\left[i_k i_S\left(1 - \frac{1}{i_E}\right)\right]^{1/2}}$$

$$i_2 = 0$$

ER + S

$$\frac{1}{i_1} = \frac{1}{i_A} + \frac{1}{i_S} + \frac{i_1}{i_k i_E}$$

$$\frac{1}{i_1 + i_2} = \frac{1}{i_A} + \frac{1}{i_S}$$

E

$$\frac{1}{i_1} = \frac{1}{i_E}^{*}$$

$$i_2 = 0$$

S + E

$$\frac{1}{i_1} = \frac{i_S}{i_S + i_E}\frac{1}{i_A} + \frac{1}{i_S + i_E}^{*}$$

$$i_2 = 0$$

S

$$\frac{1}{i_1} = \frac{1}{i_A} + \frac{1}{i_S}^{*}$$

$$i_2 = 0$$

[a] * = exhibits linear Koutecky–Levich behavior

It is seen that in some cases the "Koutecky–Levich plots" at the first or the second wave are expected to be linear while in other cases they are not. In some cases a second wave is predicted to follow the first wave, whereas in others there is no second wave. These two features can serve as diagnostic criteria to recognize the particular limiting situation in which the system lies. In the case where a linear "Koutecky–Levich plot" is obtained, the intercept provides the measure of a characteristic film current density or combination of current densities.

How the system passes from one limiting situation to the other upon varying the experimental parameters is summarized in Fig. 5.5. In spite of the large number of intervening parameters, a two-dimensional display of the representative points of the system was made possible by integrating the diffusion of the substrate in the solution into the current densities characterizing the electron transport and the catalaytic reaction in the film. Instead of being referred to the solution bulk concentration of the substrate, these two current densities are now referred to the substrate concentration at the film–solution boundary. Thus, since at the first wave:

$$(C_A)_{\phi+} = C_A^0\left(1 - \frac{i_1}{i_A}\right) \tag{5.73}$$

$$i_S^* = i_S\left(1 - \frac{i_1}{i_A}\right) \tag{5.74}$$

and

$$i_k^* = i_k\left(1 - \frac{i_1}{i_A}\right) \tag{5.75}$$

with these transformations the system only depends on the two parameters i_S^*/i_k^* and i_E/i_k^*. The representation given in Fig. 5.5 shows the direction and magnitude of the effect of varying the experimental parameters.

In this connection, the effect of increasing the film thickness (ϕ) deserves particular attention. One interest of polymer coatings over monolayer coating resides in the expectation of an increased catalytic current caused by the volume effect invoked earlier that would make redox catalysis possible and enhance chemical catalysis. Starting from the situation designated by "R" (Fig. 5.5), increasing ϕ (or equivalently increasing the number of redox sites by unit surface area, Γ_E^0, keeping C_E^0 constant) leads to a proportional increase of the current density i_k (Eq. 5.62). The concentration profiles of both A and Q are then flat throughout the film. The catalytic plateau current (i_1) increases as shown in Fig. 5.6 with, however, a tendency to level off as i_k becomes closer and closer to and eventually overcomes i_A, in which case i_1 becomes equal to i_A (see the R equation in Table 5.3). The increase of the film thickness thus indeed produces the expected increase of the catalytic current but a first limitation is met pertaining to the rate at which the substrate diffused from the bulk of the solution to the film–solution interface. Other limitations may also be encountered either after

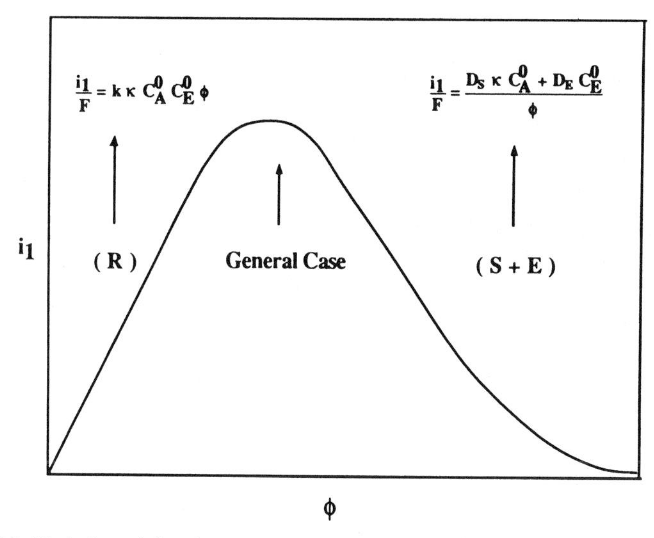

Figure 5.6. Variation of the plateau current density of the first wave, i_1, with thickness, ϕ.

or before the control by i_A has been attained since both the characteristic current densities i_E and i_S decrease as ϕ increases (Eqs. 5.12 and 5.18). The point representing the system in the zone diagram of Fig. 5.5 then tends to come out the R zone and will eventually enter the S + E (or S or E) zone in which the transport of charge and/or the diffusion of the substrate through the film entirely control the kinetics of catalysis. The catalytic reaction is then so fast, relative to the other rate-limiting factors, that it takes place only in a thin reaction layer whose location in the film depends on the relative rates of the E and S processes. If $i_E \gg i_S$ the S zone is reached, the reaction layer being located at the film–solution interface. If, conversely, $i_E \ll i_S$ the E zone is reached, the reaction layer being located at the electrode–film interface. The passage into the S + E (or S or E) zone implies a decrease of the current i_1 as i_S and i_E decreases (see the S + E, E, and S equations in Table 5.3). An example of the increase, leveling off, and decrease of the catalytic current upon increasing the film thickness is shown in Fig. 5.6. The above discussion thus lead to the notion of an optimal thickness of catalytic redox polymer coatings. This optimal thickness as the catalytic reaction gets faster when compared to the transport of charge and the diffusion of the substrate.

The catalytic reaction takes place in a thin reaction layer (thin as compared to the film thickness) also in the ER, ER + S, SR, and SR + E zones, adjacent to the electrode in the first two cases and to the solution in the last ones. Approximate values of the reaction layer thickness (μ) are

in the ER case:

$$\mu = \phi \left[\frac{i_E}{i_k(1 - i/i_A)} \right]^{1/2} \rightarrow \left(\frac{D_E}{\kappa k C_A^0} \right)^{1/2} \tag{5.76}$$

(when $i \ll i_A$)

in the SR case:

$$\mu = \phi\left(\frac{i_S}{i_k}\right)^{1/2} = \left(\frac{D_S}{kC_E^0}\right)^{1/2} \tag{5.77}$$

When the reaction layer is so thin as to reach molecular dimensions, only a single monolayer located either at the electrode surface or at the film–solution interface is catalytically active. Then, in the first case:

$$\frac{1}{i_1} = \frac{1}{i_A} + \frac{1}{i_S} + \frac{1}{i_k} \qquad \frac{1}{i_1 + i_2} = \frac{1}{i_A} + \frac{1}{i_S} \tag{5.78}$$

(linear Koutecky–Levich plots)

and in the second case:

$$\frac{1}{i_1} = \frac{1}{i_A} + \frac{1}{i_k(1 - i_1/i_E)} \qquad i_2 = 0 \tag{5.79}$$

(non-linear Koutecky–Levich plots).

In these expressions i_k is defined as in the discussion of catalysis at monolayer coatings (Eq. 5.44) in the preceding section.

Under such conditions, not only the volume effect enhancement of the catalytic current does not exist but catalysis is generally less efficient than with a monolayer directly deposited onto the electrode surface because of limitations by either the substrate diffusion (the $1/i_S$ term in Eq. 5.78) or the electron transport (the $1 - i_1/i_E$ term in Eq. 5.79).

5.4.2 More Complex Reaction Schemes

5.4.2.1 Effect of Slow Penetration of the Substrate Across the Film–Solution Interface

In the preceding discussion, the penetration of the substrate from the solution into the film and the reverse process were assumed to be so fast as to remain at equilibrium (Eq. 5.61), whatever the magnitude of the current flowing through the electrode surface. A complete treatment of the case where this restrictive assumption is removed has been given for the same reaction scheme (one-step irreversible catalytic reaction) as in the preceding section (63). The results are quite simply obtained by changing $1/i_A$ into $(1/i_A) + (1/i_P)$ systematically in all the equations. The term i_P is the characteristic current density measuring the kinetics of the penetration reaction as defined by Eq. 5.23. If constrained diffusion toward pinholes interfere in the way discussed in Section 2.3, $1/i_A$ should be replaced by $(1/i_A) + (1/i_P) + (1/i_{CD})$ (see Eqs. 5.32–5.35). New versions of Table 5.3, Fig. 5.5, Eqs 5.73 and 5.74 are readily obtained by this transformations, all the other definitions remaining the same [see Table 1 and Fig. 2 in Ref. (62)].

5.4.2.2 Reversible One-Step Catalytic Reaction (78)

The equilibrium constant (K) of the catalytic reaction:

$$A + Q \rightleftarrows B + P \tag{5.80}$$

is related to the difference between the standard potentials of the two redox couple according to

$$\frac{RT}{\mathscr{F}} \ln K = E^0_{AB} - E^0_{PQ} \tag{5.81}$$

The irreversible case treated in Section 5.4.1 corresponds to two possible practical situations. One case is when the intermediate B is chemically unstable, which occurs frequently in the reduction or oxidation of organic molecules. If this deactivation reaction is itself irreversible and fast so as to prevent backward Reaction 5.80 from occuring, the forward Reaction 5.80 is the rate-determining step of the catalytic process. It is then possible to find effective catalysis with P/Q redox couples positive (for a reduction, negative for an oxidation) to the A reduction (or oxidation) wave, forward Reaction 80 being the rate-determining step as in the case treated in Section 5.4.1. In the case where B is chemically stable, Reaction 5.80 is in favor of the right-hand side at a equilibrium only if $E^0_{AB} \gg E^0_{PQ}$. In order to simultaneously have a first wave corresponding to the reduction (or oxidation) of the inactive form of the catalyst, electron transfer to (or from) A has to be slow enough for the A/B wave tod be negative (for a reduction, positive for an oxidation) to the P/Q wave. This situation corresponds to the first case A in Fig. 5.7.

As to the reversibility of the catalytic reaction and the relative order of the P/Q and A/B waves other possibilities exist as summarized in Fig. 5.7 in the case where the intermediate B is chemically stable. The principle of the calculation of the plateau current is the same as in Section 5.4.1, taking account of the change in the boundary conditions. The same characteristic current densities, i_A, i_E, i_S, i_k can be used as parameters together with the equilibrium constant of the catalytic reaction, K [i_k corresponds to the rate constant (k) of the forward Q + A → P + B reaction]. The same type of limiting situations designated by the same combination of the letters R, E, S are reached for extreme values of the parameters. Tables 5.4–5.6 give the expression of the plateau currents for cases A, B, and C as defined in Fig 5.7 for each limiting situation and in the general case.

As expected intuitively, the total height of the two waves is the same in all three cases whatever the rate-limiting situation. What differs from one case to

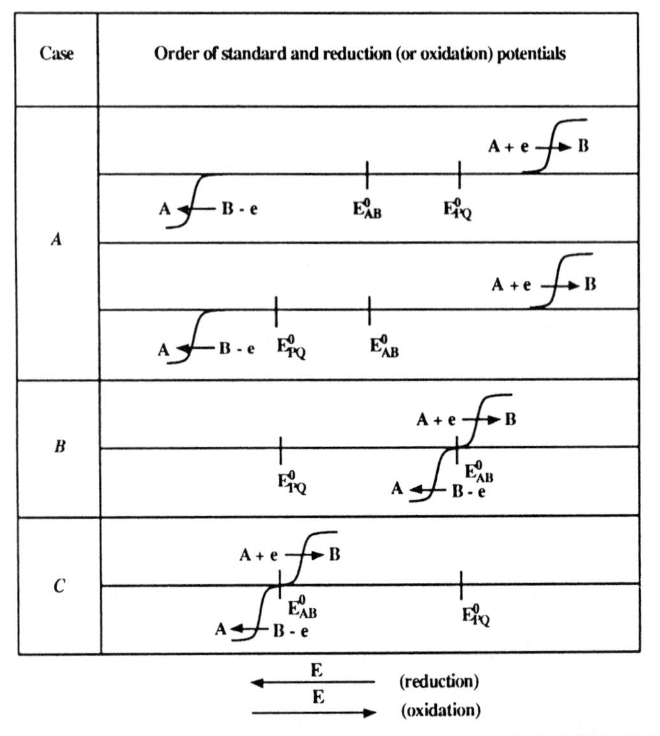

Case	K	First wave		Second wave	
		Nature	Boundary conditions	Nature	Boundary conditions
A	$K \gtrless 1$	Reduction (oxidation) of P	$\left(\frac{dC_A}{dx}\right)_0 = 0$ $(C_Q)_0 = C_E^0$	Reduction (oxidation) of A	$(C_A)_0 = 0$ $(C_Q)_0 = C_E^0$
B	$K \leq 1$	Reduction (oxidation) of P	$(C_A)_0 = C_A^0$ $(C_Q)_0 = C_E^0$	Reduction (oxidation) of A	$(C_A)_0 = 0$ $(C_Q)_0 = C_E^0$
C	$K \geq 1$	Reduction (oxidation) of A	$(C_A)_0 = 0$ $(C_Q)_0 = 0$	Reduction (oxidation) of P	$(C_A)_0 = 0$ $(C_Q)_0 = C_E^0$

Figure 5.7. One-step reversible catalytic reaction. Various redox situations.

Table 5.4. Reversible One-Step Catalytic Reaction: Case A

R

$$\frac{1}{i_1} = \frac{1}{i_A} + \frac{1}{i_k}$$

$$\frac{1}{i_1 + i_2} = \frac{1}{i_A} + \frac{1}{i_S}$$

R + S

$$\frac{1}{i_1} = \frac{1}{i_A} + \frac{1}{(i_k i_S)^{1/2} \tanh\left(\frac{i_k}{i_S}\right)^{1/2}}$$

$$\frac{1}{i_1 + i_2} = \frac{1}{i_A} + \frac{\tanh\left(\frac{i_k}{i_S}\right)^{1/2}}{(i_k i_S)^{1/2}}$$

R + E

$$\frac{1}{i_1} = \frac{1}{i_A}(1 - K^{-1}) + \left[\frac{i_A - (1 - K^{-1})i_1}{i_A - i_1}\right]^2 \frac{i_1}{i_E i_k \tanh^2 \left\{\frac{i_k}{i_E i_A[i_A - (1 - K^{-1})i_1]}\right\}^{1/2}}$$

$$\frac{1}{i_1 + i_2} = \frac{1}{i_A} + \frac{1}{i_S}$$

SR

$$\frac{1}{i_1} = \frac{1}{i_A} + \frac{1}{(i_k i_S)^{1/2}}$$

$$i_2 = 0$$

General case[a]

ER

$$\frac{1}{i_1} = \frac{1}{i_A}(1 - K^{-1}) + \left[\frac{i_A - (1 - K^{-1})i_1}{i_A - i_1}\right]^2 \frac{i_1}{i_E i_k}$$

$$\frac{1}{i_1 + i_2} = \frac{1}{i_A} + \frac{1}{i_S}$$

Table 5.4. (Continued)

SR + E

$$\frac{1}{i_1} = \left\{\frac{1}{i_A} + \frac{i_E^{1/2}}{(i_k i_S)^{1/2}[i_E - (1 - K^{-1})i_1]^{1/2}}\right\}\frac{i_E - i_1(1 - K^{-1})}{i_E - i_1}$$

$$i_2 = 0$$

ER + S

$$\frac{1}{i_1} = \left(\frac{1}{i_A} + \frac{1}{i_S}\right)(1 - K^{-1}) + \frac{i_E i_k}{i_1}\left(\frac{1}{i_1} - \frac{1}{i_A} - \frac{1}{i_S}\right)^2$$

$$\frac{1}{i_1 + i_2} = \frac{1}{i_A} + \frac{1}{i_S}$$

E

$$\frac{1}{i_1} = \frac{1}{i_E}$$

$$i_2 = 0$$

S + E

$$\frac{1}{i_1} = \frac{i_S}{i_S + i_E}\left(\frac{1}{i_A} + \frac{1}{i_S}\right)$$

$$i_2 = 0$$

S

$$\frac{1}{i_1} = \frac{1}{i_A} + \frac{1}{i_S}$$

$$i_2 = 0$$

[a]In terms of A:

at the first wave

$$\frac{d^2a}{dy^2} - \frac{i_k}{i_S}\left(a\left\{1 - \frac{i_S}{i_E}\left[\frac{i_A}{i_S}(1 - a_1)y - a + a_0\right]\right\} - \frac{1 - a}{K}\left[\frac{i_S}{i_E}(1 - a_1)y - a + a_0\right]\right) = 0$$

with

$$\left(\frac{da}{dy}\right)_0 = 0 \qquad 1 - a_1 = \frac{i_S}{i_A}\left(\frac{da}{dy}\right)_1 \qquad i_1 = i_A(1 - a_1)$$

at the second wave:

$$\frac{d^2a}{dy^2} - \frac{i_k}{i_S}\left\{a\left[1 - \frac{i_A}{i_E}(1 - a_1)y + \frac{i_S}{i_E}a\right] - \frac{1-a}{K}\left[\frac{i_A}{i_E}(1 - a_1)y - \frac{i_S}{i_E}a\right]\right\} = 0$$

with

$$a_0 = 0 \qquad 1 - a_1 = \frac{i_S}{i_A}\left(\frac{da}{dy}\right)_1 \qquad i_1 + i_2 = i_A(1 - a_1)$$

In terms of Q:

at the first wave

$$\frac{d^2q}{dy^2} - \frac{i_k}{i_E}\left(q\left\{1 - \frac{i_E}{i_S}\left[q_1 - q - \left(\frac{dq}{dy}\right)_0\left(1 + \frac{i_S}{i_A} - y\right)\right]\right\} - \frac{1-q}{K}\frac{i_E}{i_S}\left[q_1 - q - \left(\frac{dq}{dy}\right)_0\left(1 + \frac{i_S}{i_A} - y\right)\right]\right) = 0$$

with

$$q_0 = 1 \qquad \left(\frac{dq}{dy}\right)_1 = 0 \qquad i = -i_E\left(\frac{dq}{dy}\right)_0$$

at the second wave:

$$\frac{d^2q}{dy^2} - \frac{i_k}{i_E}\left(q\left\{\frac{i_A}{i_A + i_S}\left[1 + \frac{i_E}{i_S}(1 - q_1)\right]y - \frac{i_E}{i_S}(1 - q)\right\} - \frac{1-q}{K}\left\{1 - \frac{i_A}{i_A + i_S}\left[1 + \frac{i_E}{i_S}(1 - q_1)\right]y + \frac{i_E}{i_S}(1 - q)\right\}\right) = 0$$

with

$$q_0 = 1 \qquad \left(\frac{dq}{dy}\right)_1 = 0 \qquad i_1 + i_2 = \frac{i_A i_S}{i_A + i_S}\left[1 + \frac{i_E}{i_S}(1 - q_1)\right]$$

Table 5.5. Reversible One-Step Catalytic Reaction: Case B[a]

R

$$i_1 = 0$$

$$\frac{1}{i_1 + i_2} = \frac{1}{i_A} + \frac{1}{i_S}$$

R+S

$$\frac{1}{i_1} = \left[\frac{1}{i_A} + \frac{\tanh\left(\frac{i_k}{i_S}\right)^{1/2}}{(i_k i_S)^{1/2}}\right]\frac{\cosh\left(\frac{i_k}{i_S}\right)^{1/2}}{\cosh\left(\frac{i_k}{i_S}\right)^{1/2} - 1}$$

$$\frac{1}{i_1 + i_2} = \frac{1}{i_A} + \frac{\tanh\left(\frac{i_k}{i_S}\right)^{1/2}}{(i_k i_S)^{1/2}}$$

R+E

$$i_1 = 0$$

$$\frac{1}{i_2} = \frac{1}{i_A} + \frac{1}{i_S}$$

SR

$$\frac{1}{i_1} = \frac{1}{i_A} + \frac{1}{i_k {i_S}^{1/2}}$$

$$i_2 = 0$$

General case[b]

ER

$$i_1 = 0$$

$$\frac{1}{i_2} = \frac{1}{i_A} + \frac{1}{i_S}$$

SR+E

$$\frac{1}{i_1} = \left\{\frac{1}{i_A} + \frac{i_E^{1/2}}{(i_k i_S)^{1/2}[i_E - (1 - K^{-1})i_1]^{1/2}}\right\}\frac{i_E - i_1(1 - K^{-1})}{i_E - i_1}$$

$$i_2 = 0$$

ER+S

$$i_1 = 0$$

$$\frac{1}{i_2} = \frac{1}{i_A} + \frac{1}{i_S}$$

E

$$i_1 = i_E$$

$$i_2 = 0$$

S+E

$$\frac{1}{i_1 + i_2} = \frac{i_S}{i_S + i_E}\left(\frac{1}{i_A} + \frac{1}{i_S}\right)^c$$

S

$$i_1 = 0$$

$$\frac{1}{i_2} = \frac{1}{i_A} + \frac{1}{i_S}$$

[a] $K \leqslant 1$

[b] In terms of A:

at the first wave:

$$\frac{d^2a}{dy^2} - \frac{i_k}{i_S}\left\{a\left[1 - \frac{i_A}{i_E}(1 - a_1)y - \frac{i_S}{i_E}(1 - a)\right] - \frac{1 - a}{K}\left[\frac{i_A}{i_E}(1 - a_1)y + \frac{i_S}{i_E}(1 - a)\right]\right\} = 0$$

the other are the individual heights of the first and second waves. The equations in Tables 5.4–5.6 allows the calculation of the plateau currents in all cases, whatever the value of the equilibrium constant.

In the particular case of a self-exchange reaction ($K = 1$) the expression of the plateau current is particularly simple (79). There is a single wave, the plateau current density of which is given by

$$\frac{1}{i} = \frac{1}{i_A} + \frac{1}{i_S + i_E}\left[1 + \frac{i_E}{i_S}\frac{\tanh\left[\left(\frac{i_k}{i_S} + \frac{i_k}{i_E}\right)^{1/2}\right]}{\left(\frac{i_k}{i_S} + \frac{i_k}{i_E}\right)^{1/2}}\right] \tag{5.82}$$

In the cases where the catalytic polymer coating functions as a monolayer located at the electrode–film or film–solution interface the expressions of the plateau currents are those listed in Table 5.7 (78).

The location of the first and second waves is also a function of the various characteristic current densities. A detailed analysis of this question can be found in Ref. 78.

Here, as before, the interference of a slow penetration of the substrate into the film and/or of constrained diffusion toward pinholes could be simply taken into account by introducing the i_P and/or i_{CD} characteristic current densities as already discussed in Section 5.4.2.1.

5.4.2.3 Preactivation Mechanisms

In the preactivation mechanism (80) shown in Fig. 5.8, the substrate (C) is converted into an activated form (A) prior to the electron exchange with the active form of the redox catalyst (Q). Such mechanisms have been discussed in homogeneous catalysis and illustrated experimentally by the reduction of

---

with

$$a_0 = 0 \qquad 1 - a_1 = \frac{i_S}{i_A}\left(\frac{da}{dy}\right)_1 \qquad i_1 = i_A(1 - a_1)$$

In terms of Q:

$$\frac{d^2q}{dy^2} - \frac{i_k}{i_E}\left\{q\left[1 - \frac{i_E}{i_S}(1-q) + \frac{i_A}{i_A + i_S}\left(\frac{i_E}{i_S}\right)(1 - q_1)y\right]\right.$$
$$\left. - \frac{1-q}{K}\left[\frac{i_E}{i_S}(1-q) - \frac{i_A}{i_A + i_S}\left(\frac{i_E}{i_S}\right)(1 - q_1)y\right]\right\} = 0$$

with

$$q_0 = 0 \qquad \left(\frac{dq}{dy}\right)_1 = 0 \qquad i_1 = \frac{i_A i_S}{i_A + i_S}\left(\frac{i_E}{i_S}\right)(1 - q_1)$$

The second wave is obtained from the same differential equations and boundary conditions as in Table 5.4.

[c]Single wave.

Table 5.6. Reversible One-Step Catalytic Reaction: Case C[a]

R

$$\frac{1}{i_1} = \frac{1}{i_A} + \frac{1}{i_S}$$
$$i_2 = 0$$

R + S

$$\frac{1}{i_1} = \cosh\left(\frac{i_k}{Ki_S}\right)^{1/2}\left[\frac{1}{i_A} + \frac{K^{1/2}}{(i_S i_k)^{1/2}}\tanh\left(\frac{i_k}{Ki_S}\right)^{1/2}\right]$$
$$\frac{1}{i_1 + i_2} = \frac{1}{i_A} + \frac{\tanh\left(\frac{i_k}{i_S}\right)^{1/2}}{(i_S i_k)^{1/2}}$$

R + E

$$\frac{1}{i_1} = \frac{1}{i_A} + \frac{1}{i_S}$$
$$i_2 = 0$$

SR

$$\frac{1}{i_1} = \cosh\left(\frac{i_k}{Ki_S}\right)^{1/2}\left[\frac{1}{i_A} + \frac{K^{1/2}}{(i_S i_k)^{1/2}}\tanh\left(\frac{i_k}{Ki_S}\right)^{1/2}\right]$$
$$\frac{1}{i_1 + i_2} = \frac{1}{i_A} + \frac{1}{(i_S i_k)^{1/2}}$$

General case[b]

ER

$$\frac{1}{i_1} = \frac{1}{i_A} + \frac{1}{i_S}$$
$$i_2 = 0$$

SR + E

$$\frac{1}{i_1} = \cosh\left(\frac{i_k}{Ki_S}\right)^{1/2}\left[\frac{1}{i_A} + \frac{K^{1/2}}{(i_S i_k)^{1/2}}\tanh\left(\frac{i_k}{Ki_S}\right)^{1/2}\right]$$
$$\frac{1}{i_1 + i_2} = \left\{\frac{1}{i_A} + \frac{i_E^{1/2}}{(i_S i_k)^{1/2}[i_E - (1 - K^{-1})(i_1 + i_2)]^{1/2}}\right\}\frac{i_E - (i_1 + i_2)(1 - K^{-1})}{i_E - (i_1 + i_2)}$$

ER + S

$$\frac{1}{i_1} = \frac{1}{i_A} + \frac{1}{i_S}$$
$$i_2 = 0$$

E	S+E	S
$i_1 = 0$ $i_2 = i_E$	$\dfrac{1}{i_1 + i_2} = \dfrac{i_S}{i_S + i_E}\left(\dfrac{1}{i_A} + \dfrac{1}{i_S}\right)$[c]	$\dfrac{1}{i_1} = \dfrac{1}{i_A} + \dfrac{1}{i_S}$ $i_2 = 0$

[a] $K \geqslant 1$

[b] In terms of A:

at the first wave

$$\frac{d^2a}{dy^2} - \frac{i_k}{i_S}\left\{a\left[\frac{i_S}{i_E}a - \frac{i_A}{i_E}(1 - a_1)y\right] - \frac{1-a}{K}\left[1 - \left(\frac{i_S}{i_E}\right)a + \frac{i_A}{i_E}(1 - a_1)y\right]\right\} = 0$$

with

$$a_0 = 0 \qquad 1 - a_1 = \frac{i_S}{i_A}\left(\frac{da}{dy}\right)_1 \qquad i_1 = i_A(1 - a_1)$$

In terms of Q:

$$\frac{d^2q}{dy^2} - \frac{i_k}{i_E}\left\{q\left[\frac{i_A}{i_A + i_S} + \left(1 - \frac{i_E}{i_S}q_1\right)y + \frac{i_E}{i_S}q\right] - \frac{1-q}{K}\left[1 - \frac{i_A}{i_A + i_S}\left(1 - \frac{i_E}{i_S}q_1\right)y - \left(\frac{i_E}{i_S}\right)q\right]\right\} = 0$$

with

$$q_0 = 0 \qquad \left(\frac{dq}{dy}\right)_1 = 0 \qquad i_1 = \frac{i_A i_S}{i_A + i_S}\left[1 - \left(\frac{i_E}{i_S}\right)q_1\right]$$

The second wave is obtained from the same differential equations as in Table 5.4.

[c] Single wave.

Table 5.7. Expressions of the Plateau Currents of the First and Second Waves When the Film Functions as a Monolayer Located at the Electrode–Film or Film–Solution Interface

Redox Situations	Monolayer Located at the Film–Solution Interface	Monolayer Located at the Electrode–Film Interface
Case A	$\frac{1}{i_1} = \frac{1}{i_A} + \frac{1}{i_E} + \frac{1}{i_k} + \left(\frac{1}{K} - 1\right)\frac{i_1}{i_E i_A}$	$\frac{1}{i_1} = \frac{1}{i_A} + \frac{1}{i_S} + \frac{1}{i_k}$
	$i_2 = 0$	$\frac{1}{i_1 + i_2} = \frac{1}{i_A} + \frac{1}{i_S}$
Case B	$\frac{1}{i_1} = \frac{1}{i_A} + \frac{1}{i_E} + \frac{1}{i_k} + \left(\frac{1}{K} - 1\right)\frac{i_1}{i_E i_A}$	$i_1 = 0$
	$i_2 = 0$	$\frac{1}{i_2} = \frac{1}{i_A} + \frac{1}{i_S}$
Case C	$i_1 = 0$	$\frac{1}{i_1} = \frac{1}{i_A} + \frac{1}{i_S}$
	$\frac{1}{i_1} = \frac{1}{i_A} + \frac{1}{i_E} + \frac{1}{i_k} + \left(\frac{1}{K} - 1\right)\frac{i_2}{i_E i_A}$	$i_2 = 0$

triphenylmethyl-type halides through the prior formation of the carbocation (81). The same system has also been briefly investigated with poly(vinyl-ferrocene) coatings (82). Such reaction schemes are the homogeneous or supported homogeneous catalysis counterparts, of "CE" mechanisms (60) in direct electrochemistry. On the other hand, as will be discussed in the next section, the same formal kinetics is obeyed with systems in which the electron carrier and the catalyst are two different species and in which a chemical catalysis occurs by the uphill formation of an adduct between the active form of the catalyst and the substrate. The treatment of the kinetics of "CE" reaction

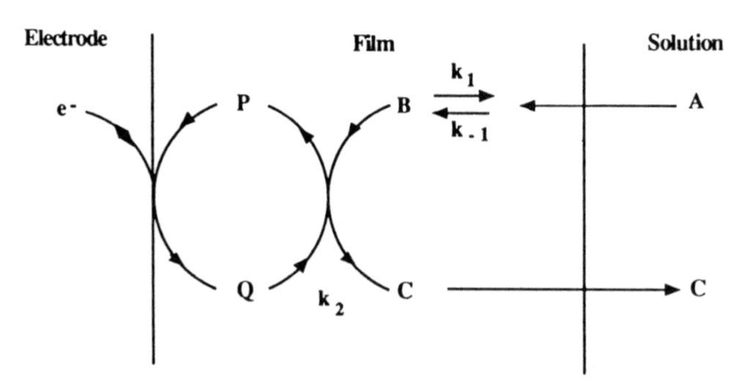

Figure 5.8. Preactivation mechanism in redox polymers.

schemes in redox polymers also provides a starting example for further analyses of more complex systems such as polymer coatings containing immobilized enzymes where the redox couple serve as an immobilized coenzyme.

The kinetics of the system is described, in the general case, by the following set of differential equations and boundary conditions, assuming the steady state assumption holds for the activated form of the substrate, B.

$$D_S \frac{d^2 C_A}{dx^2} - \frac{k_1 k_2 C_C C_A}{k_{-2} + k_2 C_Q} = 0 \tag{5.83}$$

$$D_E \frac{d^2 C_Q}{dx^2} - \frac{k_1 k_2 C_A C_Q}{k_{-1} + k_2 C_Q} = 0 \tag{5.84}$$

with

$$(C_Q)_0 = C_E^0 \tag{5.85}$$

$$\left(\frac{dC_Q}{dx}\right)_{\phi-} = 0 \qquad D_S\left(\frac{dC_A}{dx}\right)_{\phi-2} = D_A\left(\frac{dC_A}{dx}\right)_{\phi+} \qquad (C_A)_{\phi-} = \kappa(C_A)_{\phi+} \tag{5.86}$$

$$i_1 \text{ or } i_1 + i_2 = D_S\left(\frac{dC}{dx}\right)_0 - D_E\left(\frac{dC_Q}{dx}\right)_0 = D_S\left(\frac{dC_A}{dx}\right)_{\phi-} = D_A\left(\frac{dC_A}{dx}\right)_{\phi-} \tag{5.87}$$

and, at the first wave

$$\left(\frac{dC_A}{dx}\right)_0 = 0 \tag{5.88}$$

whereas, at the second wave

$$(C_A)_0 = 0 \tag{5.89}$$

The same dimensionless variables, y, a, and q as previously defined are then introduced (Eqs. 5.63 and 5.64), as well as the same characteristic current densities i_E and i_S for electron transport and substrate diffusion (Eqs. 5.12 and 5.18), respectively. The characteristic current density for the catalytic reaction can be defined either as:

$$i_{k1} = \mathscr{F} k_1 \kappa C_A^0 \phi \tag{5.90}$$

or as:

$$i_k = \mathscr{F}\left(\frac{k_1 k_2}{k_{-1}}\right)\kappa C_A^0 C_E^0 \phi \tag{5.91}$$

with

$$\sigma = \frac{i_k}{i_{k1}} = \frac{k_2 C_E^0}{k_{-1}} \tag{5.92}$$

As compared to the one-step irreversible case discussed in Section 5.4.1, σ is an additional parameter measuring the competition between the electron transfer between the activated form of the substrate and the active form of the redox catalyst on the one hand and the deactivation of the substrate on the other. The equation giving the plateau current densities of the first and second waves in the various limiting cases is given in Table 5.8 together with the equations to be solved numerically in the general case.

When $\sigma \rightarrow 0$, that is, when the electron-transfer reaction is rate determining with the preactivation being at equilibrium, the kinetics is exactly the same as in the case of the one-step irreversible reaction (Section 5.4.1). Thus, taking account of the new expression of i_k (Eq. 5.86), the zone diagram in Fig. 5.5 and the equations displayed in Table 5.3 apply to this case.

The kinetic behavior is different when $\sigma \rightarrow \infty$ that is, when the preactivation of the substrate becomes rate determining. The zone diagram in Fig. 5.9 and the set of equations in Table 5.9 then apply. The appropriate characteristic current density for the catalytic reaction is now i_{kl} (Eq. 5.91).

As in the preceding cases, the interference of low penetration of the substrate into the film and/or of constrained diffusion toward pinholes could be simply taken into account by introducing the corresponding characteristic current densities as described in Section 5.4.2.1.

5.5 EXPERIMENTAL EXAMPLES

A large number of monolayer and multilayer catalytic coatings have been described (for reviews, see Refs. 1, 15, 77, 83–85). Quantitative analysis of the rotating disk electrode data has been attempted in several cases (25, 62, 64, 86–110). We discuss below in some details a few typical examples where the theoretical analyses developed in the preceding section have been applied.

5.5.1 The Incorporated Redox Couple Serves Both as Catalyst and Electron Carrier

Typical eamples of case *A* (Fig. 5.7) with large values of the cross-exchange equilibrium constant κ have found in the oxidation or reduction of substrates that have a large overpotential at the bare electrode such as ferrous ions in acidic aqueous solutions, ascorbate, catechol, and L-DOPA, in oxidation, and dioxygen in reduction.

The oxidation of Fe^{2+} mediated by a protonated poly(vinylpyridine) film loaded with the $[IrCl_6]^{3-/4-}$ couple (71) was one of the first examples quantitatively analyzed by means of the above described models (74). The system then appeared to be under the joint control of the catalytic reaction and of the diffusion of the substrate in the film (a "R + S" situation as defined in Fig. 5.5 and Table 5.3). A more systematic investigation of the same oxidation reaction was carried out using the same redox couple incorporated in a polymer made of protonated poly(vinylpyridine) mixed with 5% protonated poly-*l*-lysine (8). A "R + S" behavior was again found. Its assignment was based on the fact

Table 5.8. Preactivation Mechanism. Mixed Kinetic Control by Reactions 5.1 and 5.2

R

$$\frac{1}{i_1} = \frac{1}{i_A} + \frac{1+\sigma}{i_k}$$

$$\frac{1}{i_1 + i_2} = \frac{1}{i_A} + \frac{1}{i_S}$$

R + S

$$\frac{1}{i_1} = \frac{1}{i_A} + \frac{(1+\sigma)^{1/2}}{(i_k i_S)^{1/2} \tanh\left[\frac{i_k}{i_S(1+\sigma)}\right]^{1/2}}$$

$$\frac{1}{i_1 + i_2} = \frac{1}{i_A} + \frac{(1+\sigma)^{1/2}}{(i_k i_S)^{1/2}} \tanh\left[\frac{i_k}{i_S(1+\sigma)}\right]^{1/2}$$

R + E

$$\frac{d^2q}{dy^2} - \frac{i_k}{i_E} \frac{q\left[1 + \frac{i_E}{i_A}\left(\frac{dq}{dy}\right)_0\right]}{1+\sigma q} = 0$$

$$q_0 = 1 \qquad \left(\frac{dq}{dy}\right)_1 = 0 \qquad i_1 = -i_E\left(\frac{dq}{dy}\right)_0$$

$$\frac{1}{i_1 + i_2} = \frac{1}{i_A} + \frac{1}{i_S}$$

SR

$$\frac{1}{i_1} = \frac{1}{i_A} + \frac{(1+\sigma)^{1/2}}{(i_k i_S)^{1/2}}$$

$$i_2 = 0$$

General case[a]

ER

$$\frac{1}{i_1} = \frac{1}{i_A} + \frac{i_1}{i_k i_E} \frac{\sigma^2}{2[\sigma - \ln(1+\sigma)]}$$

$$\frac{1}{i_1 + i_2} = \frac{1}{i_A} + \frac{1}{i_S}$$

Table 5.8. (Continued)

SR + E

$$\frac{1}{i_1} = \frac{1}{i_A} + \left[\frac{1 + \sigma\left(1 - \frac{i_1}{i_E}\right)}{i_k i_S\left(1 - \frac{i_1}{i_E}\right)}\right]^{1/2}$$

$$i_2 = 0$$

ER + S

$$\frac{1}{i_1} = \frac{1}{i_A} + \frac{1}{i_S} + \frac{i_1}{i_k i_E} \frac{\sigma^2}{2[\sigma - \ln(1 + \sigma)]}$$

$$\frac{1}{i_1 + i_2} = \frac{1}{i_A} + \frac{1}{i_S}$$

E

$$\frac{1}{i_1} = \frac{1}{i_E}$$

$$i_2 = 0$$

S + E

$$\frac{1}{i_1} = \frac{i_S}{i_S + i_E} \frac{1}{i_A} + \frac{1}{i_S + i_E}$$

$$i_2 = 0$$

S

$$\frac{1}{i_1} = \frac{1}{i_A} + \frac{1}{i_S}$$

$$i_2 = 0$$

[a]In terms of A:

at the first wave

$$\frac{d^2a}{dy^2} - \frac{i_k}{i_S}\frac{a\left\{1 - \frac{i_S}{i_E}\left[\frac{i_A}{i_S}(1-a_1)y - a + a_0\right]\right\}}{1 + \sigma\left\{1 - \frac{i_S}{i_E}\left[\frac{i_A}{i_S}(1-a_1)y - a + a_0\right]\right\}} = 0$$

with

$$\left(\frac{da}{dy}\right)_0 = 0 \qquad 1 - a_1 = \frac{i_A}{i_S}\left(\frac{da}{dy}\right)_1 \qquad i_1 = i_A(1 - a_1)$$

In terms of Q

$$\frac{d^2q}{dy^2} - \frac{i_k}{i_E}\frac{q\left\{1 - \frac{i_E}{i_S}\left[q_1 - q - \left(\frac{dq}{dy}\right)_0\left(1 + \frac{i_S}{i_A} - y\right)\right]\right\}}{1 + \sigma q}$$

with

$$q_0 = 1 \qquad \left(\frac{dq}{dy}\right)_1 = 0 \qquad i_1 = -i_E\left(\frac{dq}{dy}\right)_0$$

In terms of A:

at the second wave

$$\frac{d^2a}{dy^2} - \frac{i_k}{i_S}\frac{a\left[1 - \frac{i_A}{i_E}(1-a_1)y + \left(\frac{i_S}{i_E}\right)a\right]}{1 + \sigma\left[1 - \frac{i_A}{i_E}(1-a_1)y + \left(\frac{i_S}{i_E}\right)a\right]} = 0$$

with

$$a_0 = 0 \qquad 1 - a_1 = \frac{i_S}{i_A}\left(\frac{da}{dy}\right)_1 \qquad i_1 + i_2 = i_A(1 - a_1)$$

In terms of Q

$$\frac{d^2q}{dy^2} - \frac{i_k}{i_E}\frac{q\left\{\frac{i_A}{i_A + i_S}\left[1 + \frac{i_E}{i_S}(1-q_1)\right]y - \frac{i_E}{i_S}(1-q)\right\}}{1 + \sigma q} = 0$$

with

$$q_0 = 1 \qquad \left(\frac{dq}{dy}\right)_1 = 0 \qquad i_1 + i_2 = \frac{i_A i_S}{i_A + i_S}\left[1 + \frac{i_E}{i_S}(1 - q_1)\right]$$

Table 5.9. Preactivation Mechanism: Kinetic Control by the Preactivation Step

R1

$$\frac{1}{i_1} = \frac{1}{i_A} + \frac{1}{i_{k_1}}$$

$$\frac{1}{i_1 + i_2} = \frac{1}{i_A} + \frac{1}{i_S}$$

R1+S

$$\frac{1}{i_1} = \frac{1}{i_A} + \frac{1}{(i_k^{1} i_S)^{1/2} \tanh\left(\frac{i_k^{1}}{i_S}\right)^{1/2}}$$

$$\frac{1}{i_1 + i_2} = \frac{1}{i_A} + \frac{tanh\left(\frac{i_k^{1}}{i_S}\right)^{1/2}}{(i_k^{1} i_S)^{1/2}}$$

ER1

$$\frac{1}{i_1} = \frac{1}{i_A} + \frac{i_1}{2 i_k^{1} i_E}$$

$$\frac{1}{i_1 + i_2} = \frac{1}{i_A} + \frac{1}{i_S}$$

SR1

$$\frac{1}{i_1} = \frac{1}{i_A} = \frac{1}{(i_k^{1} i_S)^{1/2}}$$

$$i_2 = 0$$

ER1+S

$$\frac{1}{i_1} = \frac{1}{i_A} + \frac{1}{i_S} + \frac{i_1}{2 i_{k_1} i_E}$$

$$\frac{1}{i_1 + i_2} = \frac{1}{i_A} + \frac{1}{i_S}$$

E

$$\frac{1}{i_1} = \frac{1}{i_E}$$

$$i_2 = 0$$

S+E

$$\frac{1}{i_1} = \frac{i_S}{i_S + i_E} \frac{1}{i_A} + \frac{1}{i_S + i_E}$$

$$i_2 = 0$$

S

$$\frac{1}{i_1} = \frac{1}{i_A} + \frac{1}{i_S}$$

$$i_2 = 0$$

that two successive waves (one for the mediated oxidation of Fe^{2+} and another one, at a more positive potential, for the direct oxidation of Fe^{2+} that permeates through the film) were observed, each exhibiting a linear Koutecky–Levich behavior. Increasing the coating thickness resulted in an increase of the plateau current of the first wave at the expense of the second wave, as was expected for a "R + S" situation in which the catalytic reaction occurs in a sizeable portion of the film rather than at a monolayer located at the film–solution interface.

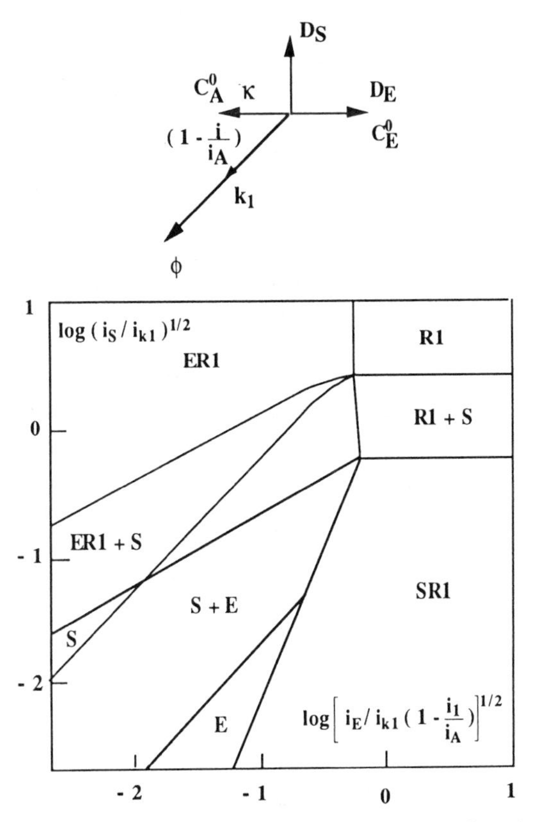

Figure 5.9. Zone diagram for the preactivation mechanism in the case where the reactivation step is rate determining.

Simultaneous analysis of the limiting current densities of the two waves by means of the two equations:

$$\frac{1}{i_1} = \frac{1}{i_A} + \frac{1}{(i_k i_S)^{1/2} \tanh(i_k/i_S)^{1/2}} \tag{5.93}$$

$$\frac{1}{i_1 + i_2} = \frac{1}{i_A} + \frac{\tanh(i_k/i_S)^{1/2}}{(i_k i_S)^{1/2}} \tag{5.94}$$

allowed the separate determination of i_k (Eq. 5.63) and i_S (Eq. 5.18). The electron-transport characteristic current density, i_E (Eq. 5.12), as measured from potential-step experiments (see Section 5.2.2), was found to be larger than both i_S and i_k thus confirming the assignment of the rate-controlling factors.

Similar results were found when the oxidation of Fe^{2+} was mediated by a poly(vinylpyridine) film to which a chloro-bisbipyridine)-Ru^{2+}/Ru^{+} complex was coordinatively attached (one ruthenium complex for four protons) (105). The data and their interpretation are summarized in Fig. 5.10. A complete analysis of the kinetics allowed the determination of the rate constant of the reaction of Fe^{2+} with the Ru(III) complex and also that of the reverse reaction ($Fe^{2+} + Ru^{II}$).

The backward rate constant could be determined from the forward rate constant and the equilibrium constant but also by the direct study of the mediation of the reduction of Fe^{3+} by the same coating. The latter process provides an example of a "case *B*" situation (Section 5.4.2.2, Table 5.5). Another one was found in the oxidation of $Ru(bpy)^{2+}$ mediated by a poly$[(bpy)_2Os(vpy)_2]^{3+/2+}$ film (93). Also interesting was the investigation of the half-wave potentials of the catalytic wave. These potentials were found to be in good quantitative agreement with the theoretical predictions developed in Ref. 78. The same was observed with the $[Ru(bpy)]^{3+/2+}$ – poly$[(bpy)_2Os(vpy)_2]^{3+/2+}$ system (93). Similarly, a "R + S" kinetic control was identified in the oxidation of ascorbate mediated by films of acrylamide–vinylpyridine gels loaded with the $[Fe(CN)_6]^{3-/4-}$ couple (109) (the locations of the experimental data points in the kinetic zone diagram are shown in Figs. 7 and 8 of Ref. 109).

Examples of "SR" behaviors were found in the oxidation of catechol and L-DOPA mediated by a film of the ternary copolymer shown below in which the $[IrCl_6]^{3-/4-}$ couple was incorporated

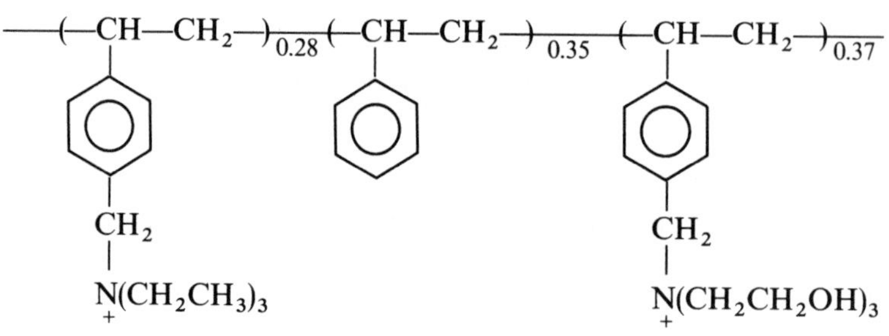

$$—(CH—CH_2)_{0.28}—(CH—CH_2)_{0.35}—(CH—CH_2)_{0.37}—$$

(first unit: phenyl–CH_2–$\overset{+}{N}(CH_2CH_3)_3$; second unit: phenyl; third unit: phenyl–CH_2–$\overset{+}{N}(CH_2CH_2OH)_3$)

(102). The term i_S was determined by using a surrogate coating containing the $[Fe(CN)_6]^{3-/4-}$ couple (not active in the potential region of interest) instead of the $[IrCl_6]^{3-/4-}$ couple. This strategy thus allowed to extract i_k from the intercept $(1/(i_S i_k)^{1/2})$ of the Koutecky–Levich plots. The ensuing values of κk and D_S allowed the determination of the thickness of the reaction layer located at the film–solution side (Eq. 5.77). It was found much larger than the size of a monolayer thus confirming the occurrence of a volume rather than of a surface reaction. The potential-step chronoamperometric determination of i_E confirmed that charge transport was too fast to be involved in the rate control.

The reduction of O_2 in aqueous acid solutions mediated by a polymer

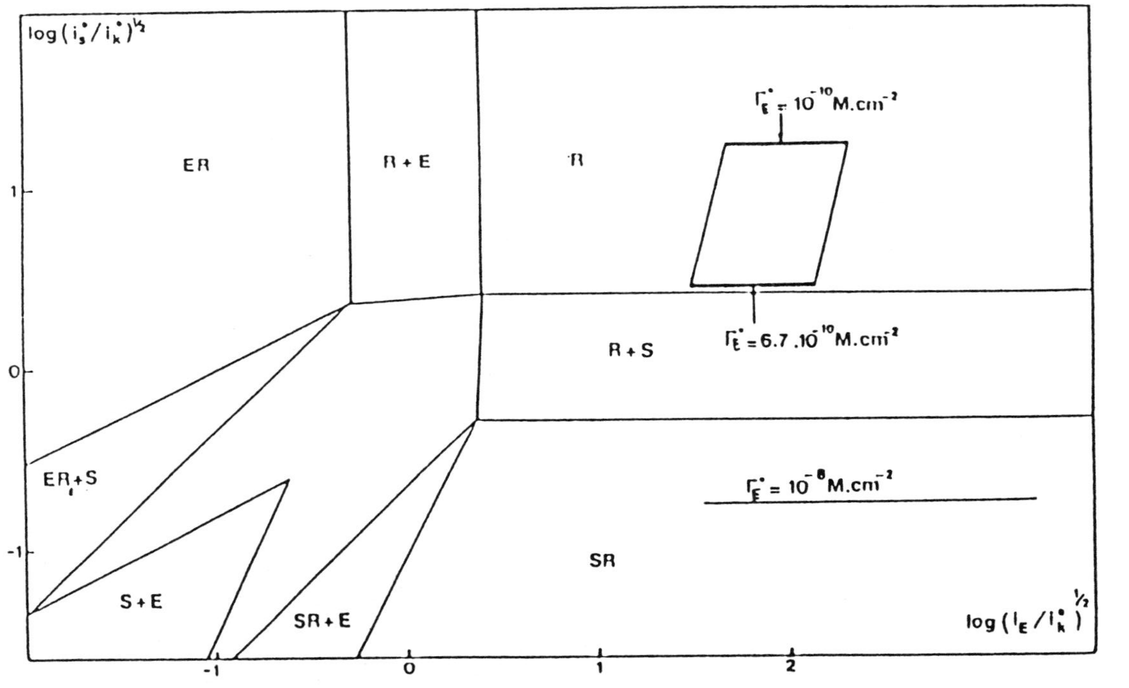

Figure 5.10. Oxidation of Fe^{2+} in 1*M* HCl mediated by a chloro-bis(bipyridine)poly(vinylpyridine) coating. Location of the rotating disk electrode data (varying the film thickness and the rotation rate) in the kinetic zone diagram (see Section 5.4.1 and Fig. 5.5). [Reprinted with permission from C. P. Andrieux, O. Hass, and J.-M. Savéant, *J. Am. Chem. Soc.*, 108, 8175, (1986). Copyright © (1986) American Chemical Society.]

coating having the structure shown below, deposited on a graphite electrode, also gives rise to an "R + S"

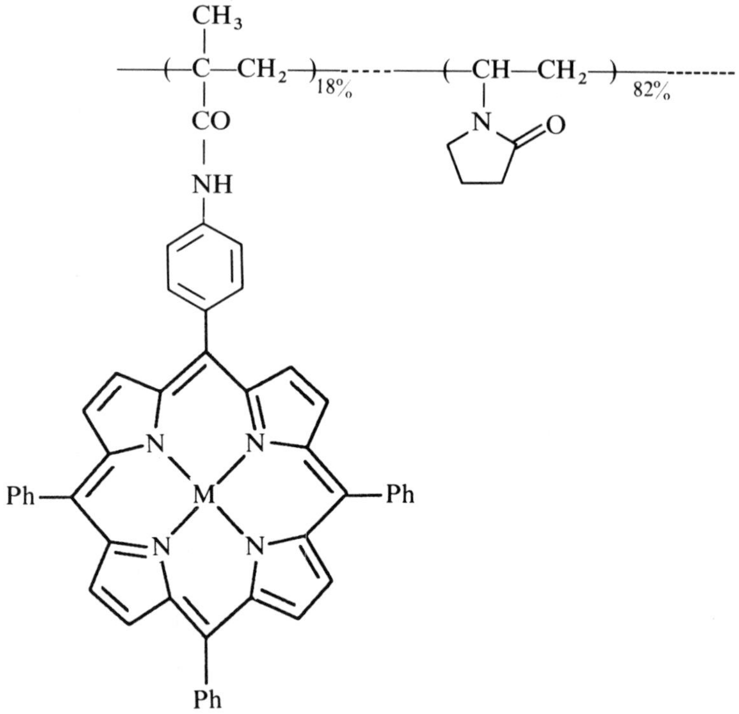

kinetic control (97). The strategy employed for the determination of i_S was different from that followed in the preceding case. The graphite electrode was plated with platinum, which substantially accelerates the reduction of O_2 at the electrode surface and makes the system pass from case *A* to case *C*. Since the equilibrium constant *K* is large, the first wave, which now corresponds to the direct reduction of O_2 at the electrode, simply obeys (see Table 5.6) the following equation:

$$\frac{1}{i_1} = \frac{1}{i_A} + \frac{1}{i_S} \tag{5.17}$$

Koutecky–Levich plots of the plateau heights then allows the determination of i_S.

An interesting reaction, in the sense that it provides an example of the interference of electron transport as rate-controlling factor, is the reduction of O_2 in dimethyl sulfoxide catalyzed by a polymer containing anthraquinone moieties (poly-[*p*-(9,10-anthraquinone-2-carbonyl)styrene]-*co*-styrene) (108). Diffusion of O_2 through such coatings is fast, whereas electron transport is slow [as also noted with other quinone containing polymers (2)]. Analysis of the rotating electrode responses as a function of the rotation rate and of the film thickness indicated the occurrence of a "R + E" control that passes from a "R" to an "ER" situation as the film thickness is increased (see Fig. 5.5).

Examples of case *C* (see Section 5.4.2.2) were reported. One particularly clear case is provided by the oxidation of $[Co(tpy)_2]^{2+}$ (tpy = 2,2′,2″-terpyridine) in aqueous solution mediated by a protonated poly(lysine) polymer containing the $[Mo(CN)_8]^{3-/4-}$ or the $[W(CN)_8]^{3-/4-}$ couple (92). The waves are observed, the first of which corresponds to the direct reduction at the electrode surface of the substrate that has diffused through the coating. At the second wave the oxidation of the substrate is mediated by the attached redox couple and is fast enough to give rise to a "SR" kinetic control (see Table 5.6) as attested by the fact that the plateau current is independent of the film thickness. The equilibrium constant of the cross-exchange reaction, *K* is large enough (~ 104) for the cosh–tanh expression of the first wave height (Table 5.6) to be approximated with an excellent precision by:

$$\frac{1}{i_1} = \frac{1}{i_S} + \frac{1}{i_A} \tag{5.17}$$

This provided a quite convenient means for determining i_S. The use of the term "catalysis" in such a case *C* situation is somewhat unwarranted since the direct oxidation of the substrate occurs at a potential less positive than that of the "catalyst." In fact, the presence of the film inhibits the oxidation of the substrate at the electrode surface and the effect of the incorporated redox couple is to restore, through redox mediation, a part of the original oxidation current.

Several systems have been described where the catalytic reaction takes place at a monolayer located at the film–solution rather than within a sizeable portion of the polymer coatings. They belong either to case *C* or to case *B*. Among the former, the oxidation of hydroquinone, of Fe^{2+}, and of $[Fe(CN)_6]^{3-}$ mediated by the $Ru(bpy)_3]^{3+/2+}$ couple incorporated in Nafion coatings (94) are remarkable examples of film–solution surface reactions, where the overall kinetics at the mediated wave are controlled jointly by the electron transport in the film (i_E) and the diffusion of the substrate from the bulk of the solution to the film boundary (i_A). The three systems belong to case *C* with a large value of the equilibrium constant of the cross-exchange reaction, *K*. Under such conditions, there is a single wave in rotating disk electrode voltammetry, the plateau current density being given by (Table 5.7):

$$\frac{1}{i_1} = \frac{1}{i_A} + \frac{1}{i_E} + \frac{1}{i_k} - \frac{i_1}{i_E i_A} \tag{5.95}$$

where i_k is the monolayer characteristic current density defined by Eq. 5.44. Seen in Fig. 5.11 is the example of hydroquinone. The characteristic current

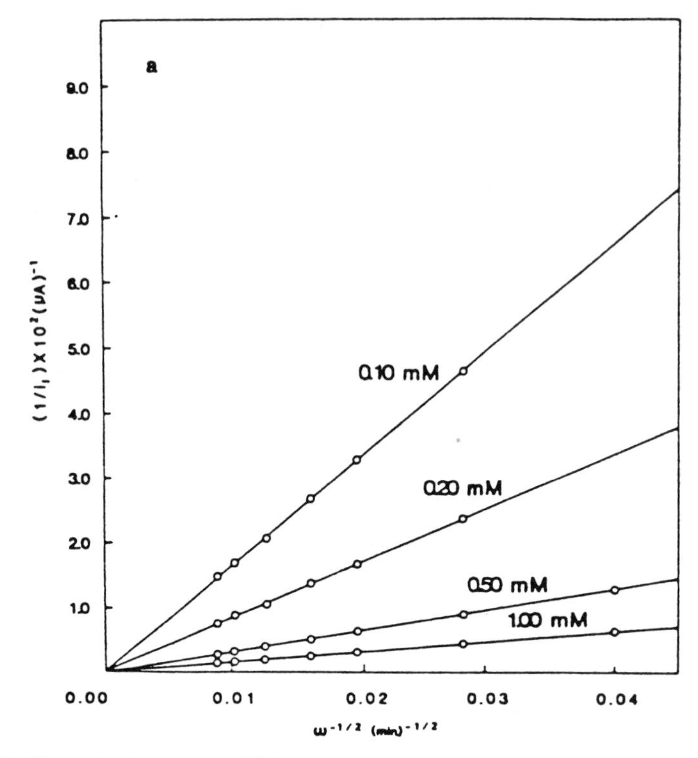

Figure 5.11. Plots of i^{-1} versus $\omega^{-1/2}$ for the oxidation of hydroquinone mediated by Nafion containing the $[Ru(bpy)_3]^{3+/2+}$ couple. [Reprinted with permission from M. Krishnan, X. Zhang, and A. J. Bard, *J. Am. Chem. Soc.*, **106**, 7371 (1984). Copyright © (1984) American Chemical Society.]

density i_k is so large, as compared to either i_E or i_A, that the experimental plots practically obey the relationship:

$$\left(\frac{1}{i_1} - \frac{1}{i_A}\right)\left(\frac{1}{i_1} = \frac{1}{i_E}\right) = 0 \tag{5.96}$$

which predicts an extremely sharp transition between the two controls.

Quite similar kinetics were observed in the oxidation of ferrocene-1,1′-disulfonate in acetonitrile mediated by poly(styrene sulfonate) films containing the electrostatically bound $[Os(bpy)_3]^{3+/2}$ couple (107) as well as for the reduction of $[Ru(NH_3)_6]^{3+}$ mediated by a polymer containing cobaltocinium moieties (103).

Kinetics belonging to case *B*, with a mediated reaction confined to a monolayer located at the film–solution boundary, were found in the oxidation of a series of iron, ruthenium, and osmium bipyridine and phenanthroline complexes in acetonitrile mediated by poly$[(bpy)_2Os(vpy)_2]^{3+/2+}$ (93). The cross-exchange electron transfer is an uphill reaction with an equilibrium constant K ranging from 10^{-2} to 10^{-8}, but the reaction is forced to completion by the fast electron transfer controlled reoxidation of the incorporated Os^{2+} complex. The expression of the plateau current density appropriate for these systems is (Table 5.7 case *B*) is, since K is small:

$$\frac{1}{i_1} = \frac{1}{i_A} + \frac{1}{i_E} + \frac{1}{i_k} + \frac{i_1}{K i_E i_A} \tag{5.97}$$

(i_k being defined by Eq. 5.44).

However, the electron transport is so fast that the $1/i_E$ terms can be neglected and the equation simplifies into:

$$\frac{1}{i_1} = \frac{1}{i_k} + \frac{1}{i_A} \tag{5.43}$$

in accordance with the fact that linear Koutecky–Levich plots were observed. Their intercepts could thus be used to obtain the values of rate constant k in the series. These values were found to correlate with the respective standard potentials in the Marcus fashion, pointing to the outer-sphere character of the electron transfer betweeen the complexes in solution and the immobile osmium complex at the film–solution interface. Catalysis by such coatings, if any, should then be of the redox type. The results just described could thus give, at first sight, the impression that redox catalysis can take place at a monolayer at variance with the discussion of this point in Section 5.2.1. However, a careful examination of the positive displacement of the half-wave potentials of the mediated wave as compared to the standard potential of the incorporated redox couple showed that they agree with the theoretical predictions (78) and that they are located at the same potential where the direct oxidation at a bare electrode takes place. Catalysis is thus negligible in accord with the prediction developed in Section 5.2.1.

More generally, most of the catalytic reactions described in this section involve outer-electron transfers, that is, are of the redox catalysis type. Under these conditions, as predicted by theory, effective catalysis requires that the reaction involve more than a monolayer of catalytic material. Chemical catalysis is clearly involved in the reduction of O_2 by means of the iron–porphyrin containing polymer that we discussed above. The reduction of O_2 by the anthraquinone polymer described earlier may also be of the chemical catalysis type.

With both redox and chemical catalytic coatings, the beneficial effect of an increase of the film thickness, predicted by theory has been demonstrated experimentally within "R + S" or "R + E" kinetic controls. It has also been observed that this effect levels off, leading to an optimum of the film thickness that corresponds to the "SR" or "ER" controls. An increase of the film thickness beyond this optimal value not only results in a waste of the catalytic material but may also have a deleterious effect when limitations by electron transport (SR + E) or substrate diffusion (ER + S) through the film come into play.

As discussed above, the theory developed in the preceding sections has been applied successfully in a number of cases. Exceptions are, the oxidation of ascorbate and of ferrocyanide mediated by $[Os(bpy)_3]^{3+/2+}$–Nafion coatings at high concentrations of the substrate (96, 110). Although, at low concentrations, the current appears to be controlled jointly by the rate of the surface reaction taking place at the film–solution interface and by the rate of electron transport through the film, the plateau current density reaches, at high concentrations, a constant value that is significantly smaller than i_E. Similar observations were also made in the oxidation of ferrocyanide mediated by a quaternized poly(pyridine) film containing the $[IrCl_6]^{3-/4-}$ couple (104). These behaviors have been rationalized assuming that the electron-transfer reaction requires the formation of precusor complexes at the surface of the film leading to saturation effects and thus to an i_k current density that ceases to increase with the substrate concentration (110). It remains to be proven that these observations could not be explained by a different mechanism and to identify the chemical nature of these complexes.

5.5.2 The Catalyst and the Electron Carrier Are Different

In the most cases the two functions, catalysis and electron transport, are carried out by the same redox couple. This is not a necessary requirement. Good catalysts may be poor electron carriers and vice versa. It may thus be advantageous to incorporate two different couples in the polymer coating performing each of the two functions separately. One example of such a system is the mediation of O_2 reduction in aqueous solutions by a Nafion coating in which electron transport is carried out by the $[Ru(NH_3)_6]^{3+/2+}$ couple and catalyst by the cobalt(III/II)-tetrakis(4-*N*-methylpyridyl)porphyrin (91). The latter is a good chemical catalyst of O_2 reduction but a quite poor electron carrier, whereas this function is satisfactorily performed by the ruthenium couple. In this system, O_2 is reduced into H_2O_2, thus requiring the uptake of two electrons and two protons as sketched below.

O_2
k_1
e^- Ru^{III} Co^{II} k_{-1} $Co^{II}O_2$ Ru^{II}
k_2
Ru^{II} Co^{III} $Co^{III}O_2^{2-}$ Ru^{III} e^-
H_2O_2 $2H^+$

The kinetics is of the "preactivation" type as defined in Section 5.4.2.3 since the concentration of the Co(II) complex can be regarded as constant. The theoretical analysis developed in Section 5.4.2.3 has been applied to the variations of the plateau current of the $[Ru(NH_3)_6]^{2+}$ oxidation wave with the various experimental parameters (100). As seen in Fig. 5.12*a*, the current density

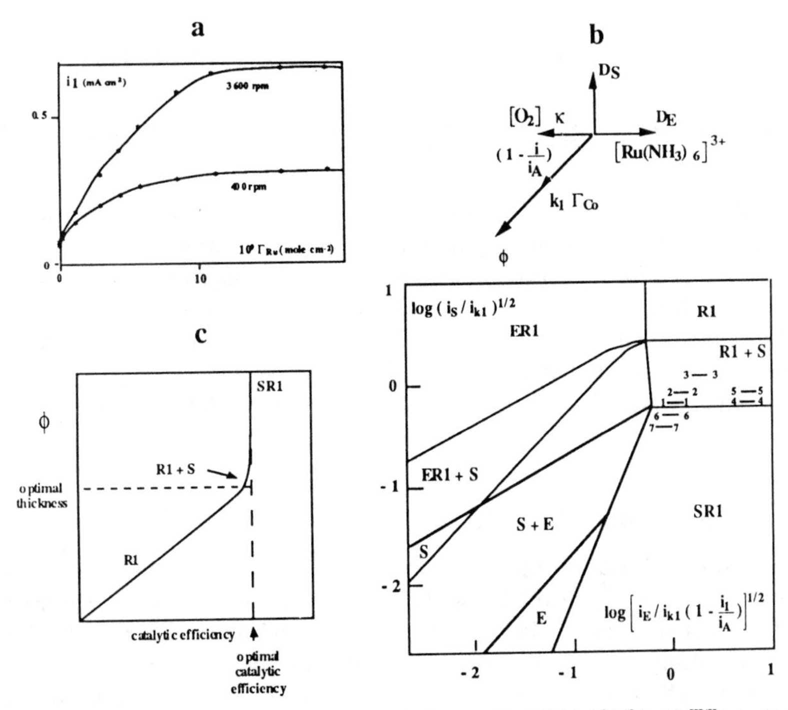

Figure 5.12. Reduction of dioxygen mediated by a $[Ru(NH_3)_6]^{3+/2+}$–$Co^{III/II}$-tetrakis(4-*N*-methylpyridyl)porphyrin–Nafion (Nf). (*a*) Limiting reduction currents as a function of the quantity of $[Ru(NH_3)_6]^{3+}$ in the coating. $\Gamma_{Co} = 3.7 \times 10^{-9}\,\text{mol cm}^{-2}$. Supporting electrolyte = 0.1*M* acetate buffer (pH 4.5) saturated with air and also containing $2 \times 10^{-6}M[Ru(NH_3)_6]^{3+}$. (*b*) Location of the data points in the kinetic zone diagram for a preactivation mechanism in which the preactivation step is rate determining (see Section 5.4.2.3 and Fig. 5.9). The experimental parameters are those listed. (*c*) Determination of the optimal thickness leading to the maximal catalytic efficiency with the minimal amount of the cobalt porphyrin catalyst.

Experiment	$10^7\Gamma_{Nf}$ (mol cm^{-2})	$10^9\Gamma_{Co}$ (mol cm^{-2})	$[O_2]$ (m*M*)
1	1.5	3.0	0.28
2	1.5	1.9	0.28
3	0.9	1.1	0.28
4	1.5	3.0	0.028
5	1.5	1.9	0.028
6	1.8	3.6	0.28
7	2.1	4.2	0.28

reaches a limiting value upon increasing the amount of $Ru(NH_3)_6$ in the coating. The kinetics is then controlled by the formation of a $Co(II)O_2$ adduct in the framework of the following reaction scheme.

$$Co^{II} + O_2 \underset{k_{-1}}{\overset{k_1}{\rightleftarrows}} Co^{II}O_2$$

$$Co^{II}O_2 + [Ru(NH_3)_6]^{2+} \xrightarrow{k_2} Co^{II}O_2^- + [Ru(NH_3)_6]^{3+}$$

$$Co^{II}O_2^- + 2H^+ + [Ru(NH_3)_6]^{2+} \xrightarrow{\text{fast}} Co^{II} + H_2O_2 + [Ru(NH_3)_6]^{3+}$$

The attainment of this limiting value corresponds to the best catalytic efficiency that can be obtained upon varying the ruthenium hexammine concentration. Under these conditions, the effect of varying the O_2 concentration, the rotation rate, and the film thickness is represented in the zone diagram of Fig. 5.12*b*. The kinetic control is of the "R1 + S" type for the conditions defined in the caption of Fig. 5.12. The effect of increasing the film thickness is shown in Fig. 5.12*c*: The optimal thickness of the coating is obtained when the system just passes into the "SR" zone.

5.6 CONCLUDING REMARKS

The essential features of the kinetics of catalytic processes in redox polymer coatings appear to be reasonably well understood at the present time as attested to by the fact that the predictions of the theoretical model have received wide experimental support. The few exceptions that have been noted seem to indicate that activity effects on the kinetics of the catalytic steps should probably be investigated more systematically in the future in a way similar to what has been started for the dynamics of electron transport. The identication of the three main rate-limiting factors, electron transport, substrate permeation, catalytic reaction, and the description of their combined involvement in the control of the overall kinetics have been illustrated by numerous experimental examples. The theoretical prediction that efficient redox catalysis requires the participation of more than one monolayer of the catalytic material has been experimentally confirmed. The volume effect that basically makes redox catalysis of electrochemical reactions by polymer coatings possible has been amply demonstrated. The increase of the catalytic current with the thickness of the coating, anticipated both with redox and chemical catalytic reactions, however, meets limitations resulting from electron transport and substrate diffusion through the film. The models developed for a theoretical understanding of the dynamics of the catalytic polymer coatings can also be used for determing optimal values of the film thickness that allow the maximal efficiency to be reached with the minimal amount of catalytic material. With some noticeable exceptions, most of the systems designed so far function in a redox catalytic manner. The development of more chemical catalytic systems would certainly be of value since chemical catalysis is superior to redox catalysis both in terms of efficiency (for the same standard potential) and chemical selectivity. In this respect the dissociation of the electron transport and catalyst roles played by the incorporated couples appears as a particularly attractive strategy. Within this framework, electron transport by means of electron hopping between the two halves of a redox couple might be advantageously replaced by ohmic conduction along conjugated polymer backbones. In many instances, the same factors that favor a persistent anchoring of catalytic polymer films onto electrode surfaces are also responsible for the slowness of the three rate-limiting factors of the catalytic process relative to corresponding phenomena in the liquid phase. Delicate compromises have thus to be sought in order to reconcile these conflicting requirements.

5.7 GLOSSARY

A/B	Substrate redox couple
P/Q	Catalyst redox couple
E^0_{AB}	Standard potential of the substrate redox couple
E^0_{PQ}	Standard potential of the catalysts redox couple
$\Delta G^{\ddagger}_{hom}$	Free energy of activation of the A + Q → B + P outer-sphere electron transfer
$\Delta G^{\ddagger}_{0,hom}$	Free-energy intrinsic barrier of A + Q → B + P outer-sphere electron transfer
$\Delta G^{\ddagger}_{el}$	Electrochemical free energy of activation
$\Delta G^{\ddagger}_{0,el}$	Intrinsic barrier free energy of the electrochemical reaction
ϕ	Thickness of the polymer film
E	Electrode potential
D_E	Diffusion coefficient for electron hopping between the oxidized and reduced halves of the redox couple in redox polymers
k_E^{pair}	Rate constant of electron hopping between two adjacent oxidized and reduced forms of the redox couple in redox polymers
k_E	Conventional bimolecular rate constant in redox polymers
D_E^{app}	Apparent diffusion coefficient for charge transport in a redox polymer
D_{pd}	Diffusion coefficient for the physical displacement of both halves of the redox couple
D	Diffusion coefficient of an unspecified species
D_A	Diffusion coefficient for the substrate in the solution
D_S	Membrane-type diffusion coefficient of the substrate in the polymer film
$(C_A)_0$	Concentration of the substrate at the electrode surface
$(C_A)_{\phi-}$	Concentration of the substrate at the film–solution interface on the film side

$(C_A)_{\phi+}$	Concentration of the substrate at the film–solution interface on the solution side
C	Concentration of an unspecified species
C_A	Space-dependent concentration of the substrate A
C_A^0	Concentration of A in the bulk of the solution
C_E^0	Total concentration of redox sites in the redox polymer
k_E^{act}	Activation controlled rate constant of electron hopping between two adjacent oxidized and reduced forms of the redox couple in redox polymers
k_D	Bimolecular diffusion limit
J_P	Flux of P
J_Q	Flux of Q
J	Flux of an unspecified species
R	Perfect gas constant
T	Absolute temperature
$\mathscr{F}$	Faraday
x	Distance from the electrode surface
z	Charge number of an unspecified species
δ_E	Average optimal electron hopping distance
δ_A	Diffusion layer thickness of A in solution
i	Current density
i_A	Levich current density
i_k	Current density characterizing the rate of the catalytic reaction
k	Rate constant of the catalytic reaction
Γ_E^0	Number of moles of redox sites per unit surface area of the electrode
ν	Kinematic viscosity
ω	Rotation rate of the rotating disk electrode
t	Time
i_1	Plateau current density at the first wave
i_2	Plateau current density at the second wave
i_E	Current density characterizing the rate of electron transport through the film
i_S	Current density characterizing the rate of diffusion of the substrate through the film
i_L	Plateau current density
C_Q	Space-dependent concentration of the active form of the catalyst
κ	Partition equilibrium constant of the substrate between solution and film
y	Distance to the electrode normalized against the film thickness
a	Concentration of substrate normalized against C_A^0
q	Concentration of the active form of the catalyst normalized against C_E^0
μ	Thickness of the reaction layer

χ_f	Rate constant for the passage of the substrate from the solution to the polymer film
χ_b	Rate constant for the passage of the substrate from the polymer film to the solution
K	Equilibrium constant of the preactivation reaction
k_1, k_{-1}	Forward and backward first order (or pseudo-first-order) rate constant of the preactivation reaction
k_2	Second-order rate constant of the electron transfer between the active form of the catalyst and the active form of the substrate in preactivation mechanisms
i_{kl}	Current density characterizing the preactivation step in preactivation mechanism
i_{CD}	Current density characterizing the constrained diffusion toward the pinholes at the surface of the coating facing the solution
k_S^{app}	Standard rate constant of electron transfer at the electrode surface uncorrected from double layer effects
k_m	Surface rate constant of the catalytic reaction at a monolayer coating
i_{k_S}	Current density characterizing the standard rate constant of electron transfer at the electrode surface
μ_{CD}	Constrained diffusion layer at the film–solution interface

5.8 REFERENCES

1. R. W. Murray, *Electroanalytical Chemistry*, A. J. Bard (Ed.), Dekker, New York, 1984. pp. 191–368.
2. C. Degrand and L. R. Miller, *J. Am. Chem. Soc.*, **102**, 5728 (1980).
3. A. Merz and A. J. Bard, *J. Am. Soc.*, **100**, 3222 (1978).
4. P. J. Peerce and A. J. Bard, *J. Electroanal. Chem.*, **112**, 97 (1980).
5. L. L. Miller and M. R. Van De Mark, *J. Am. Chem. Soc.*, **100**, 3223 (1978).
6. P. Burgmayer and R. W. Murray, *J. Electroanal. Chem.*, **135**, 335 (1982).
7. T. Saji, N. F. Pasch, S. E. Webber, and A. J. Bard, *J. Phys. Chem.*, **82**, 1101 (1978).
8. F. R. T. Fan, A. Mau, and A. J. Bard, *Chem. Physics Lett.*, **116**, 400 (1985).
9. L. Coche, A. Deronzier, and J. C. Moutet, *J. Electroanal. Chem.*, **198**, 187 (1986).
10. P. Audebert, G. Bidan, and M. Lapkowski, *J. Electroanal. Chem.*, **219**, 165 (1987).
11. F. Bedioui, A. Merino, J. Devynck, C. E. Mestres, and C. Bied Charreton, *J. Electroanal. Chem.*, **139**, 433 (1988).
12. M. Velasquez Rosanthal, T. A. Skotheim, A. Melo, and M. I. Flort, *J. Electroanal. Chem.*, **185**, 297 (1985).
13. A. Haimerl and A. Merz, *Angew. Chem. Int. Ed. Engl.*, **25**, 180 (1986).
14. H. D. Abruña, P. Denisevich, M. Umaña, T. J. Meyer, and R. W. Murray, *J. Am. Chem. Soc.*, **103**, 1 (1981).
15. H. D. Abruña, *Coord. Chem. Rev.*, **86**, 135 (1988).
16. N. Oyama and F. C. Anson, *J. Am. Chem. Soc.*, **101**, 739 (1979).
17. C. D. Ellis and T. J. Meyer, *Inorg. Chem.*, **23**, 1748 (1984).

18. S. Geraty and J. G. Vos, *J. Electroanal. Chem.*, **176**, 389 (1984).
19. I. Rubinstein and A. J. Bard, *J. Am. Chem. Soc.*, **102**, 6641 (1980).
20. D. A. Buttry and F. C. Anson, *J. Am. Chem. Soc.*, **104**, 4824 (1982).
21. M. Majda and L. R. Faulkner, *J. Electroanal. Chem.*, **169**, 77 (1984).
22. N. Oyama and F. C. Anson, *J. Electroanal. Chem.*, **127**, 249 (1980).
23. N. Oyama and F. C. Anson, *J. Electroanal. Chem.*, **127**, 247 (1980).
24. N. Oyama, T. Shimomura, K. Shigehara, and F. C. Anson, *J. Electroanal. Chem.*, **112**, 271 (1980).
25. F. C. Anson, J-M. Savéant, and K. Shigehara, *J. Am. Chem. Soc.*, **105**, 1096 (1983).
26. C. P. Andrieux, J-M. Dumas-Bouchiat, and J.-M. Savéant, *J. Electroanal. Chem.*, **87**, 39 (1978).
27. R. A. Marcus, *J. Chem. Phys.*, **39**, 1734 (1963).
28. J.-M. Savéant, *J. Am. Chem. Soc.*, **109**, 6788 (1987).
29. C. P. Andrieux, J.-M. Dumas-Bouchiat, and J.-M. Savéant, *J. Electroanal. Chem.*, **123**, 171 (1981).
30. D. Lexa, J.-M. Savéant, K. B. Su, and D. L. Wang, *J. Am. Chem. Soc.*, **109**, 6464 (1987).
31. D. Lexa, J.-M. Savéant, H. Schäfer, K. B. Su, B. Vering, and D. L. Wang, *J. Am. Chem. Soc.*, **112**, 6161 (1990).
32. M. Majda, Chapter IV in this volume.
33. E. F. Dalton, N. A. Surridje, J. C. Jermnigan, K. O. Wilbourn, J. S. Facci, and R. W. Murray, *Chem. Phys.*, **141**, 143 (1990).
34. C. P. Andrieux and J.-M. Savéant, *J. Electroanal. Chem.*, **111**, 377, (1980).
35. E. Laviron, *J. Electroanal. Chem.*, **112**, 1, (1980).
36. I. Ruff and L. Botar, *J. Chem. Phys.*, **83**, 1292 (1985).
37. J. Newman, *Electrochemical Systems*, Prentice Hall, New York, 1973.
38. R. P. Buck, *J. Electroanal. Chem.*, **46**, 1 (1973).
39. J.-M. Savéant, *J. Electroanal. Chem.*, **201**, 211 (1986).
40. J.-M. Savéant, *J. Electroanal. Chem.*, **227**, 299 (1987).
41. J.-M. Savéant, *J. Phys. Chem.*, **92**, 4526 (1988).
42. J.-M. Savéant, *J. Electroanal. Chem.*, **242**, 1 (1988).
43. C. P. Andrieux and J.-M. Savéant, *J. Phys. Chem.*, **92**, 6761 (1988).
44. F. B. Kaufman and E. M. Engler, *J. Am. Chem. Soc.*, **101**, 547 (1979).
45. M. Majda and L. R. Faulkner, *J. Electroanal. Chem.*, **137**, 149 (1982).
46. M. Majda and L. R. Faulkner, *J. Electroanal. Chem.*, **169**, 97 (1984).
47. C. Elliott and J. G. Redepenning, *J. Electroanal. Chem.*, **181**, 137 (1984).
48. J. S. Facci, R. H. Schmehl, and R. W. Murray, *J. Am. Chem. Soc.*, **104**, 4959 (1982).
49. A. Eisenberg, *Macromolecules*, **3**, 147 (1970).
50. A. Eisenberg and M. King, *Ion-Containing Polymers*; Academic: New York, 1977.
51. R. A. Komozovski and K. A. Mauritz, in *Perfluorinated Ionomer Membranes*; A. Eisenberg and H. L. Yeager (Eds.), American Chemical Society, Washington, DC, 1982. ACS Symp. Ser. No. 180, pp. 113–138.
52. J.-M. Savéant, *J. Phys. Chem.*, **92**, 1011 (1988).
53. F. C. Anson, D. N. Blauch, J.-M. Savéant, and C. F. Shu, *J. Am. Chem. Soc.*, **113**, 1922 (1991).
54. I. Fritsch-Faules and L. R. Faulkner, *J. Electroanal. Chem.*, **237**, 263 (1989).
55. P. He and X. Chen, *J. Electroanal. Chem.*, **256**, 353 (1988).
56. I. Ruff and V. J. Friedrich, *J. Phys. Chem.*, **75**, 3297 (1971).

57. D. A. Gough and J. K. Leypold, *Anal. Chem.*, **51**, 439 (1979).
58. D. A. Gough and J. K. Leypold, *Anal. Chem.*, **52**, 1126 (1980).
59. D. A. Gough and J. K. Leypold, *J. Electrochem. Soc.*, **127**, 1278 (1980).
60. A. J. Bard and L. R. Faulkner, *Electrochemical Methods. Fundamentals and Applications*, Wiley, New York, 1980.
61. C. P. Andrieux and J.-M. Savéant, "Electrochemical Reactions," *Investigation of Rates and Mechanism of Reactions*, Vol. 6, 4th ed., (Weissberger–Techniques of Chemistry), Part 2, C. F. Bernasconi (Ed.), Wiley, New York, 1986, pp. 305–390.
62. J. Leddy, A. J. Bard, J. T. Maloy, and J.-M. Savéant, *J. Electroanal. Chem.*, **187**, 205 (1985).
63. J. Leddy and A. J. Bard, *J. Electroanal. Chem.*, **153**, 223 (1983).
64. T. Ikeda, R. H. Schmehl, P. Denisevich, K. Willman, and R. W. Murray, *J. Am. Chem. Soc.*, **104**, 2683 (1982).
65. A. E. Ewing, B. J. Feldman, and R. W. Murray, *J. Phys. Chem.*, **89**, 1263 (1985).
66. F. C. Anson, T. Ohsaka, and J.-M. Savéant, *J. Phys. Chem.*, **87**, 640 (1983).
67. J.-M. Savéant, *J. Electroanal. Chem.*, **302**, 91 (1991).
68. C. Amatore, J.-M. Savéant, and D. Tessier, *J. Electroanal. Chem.*, **147**, 39 (1983).
69. W. J. Albery, A. W. Foulds, K. J. Hall, and A. R. Hillman, *J. Electrochem. Soc.*, **127**, 654 (1980).
70. W. J. Albery, W. R. Bowen, F. S. Fisher, A. W. Foulds, K. J. Hall, A. R. Hillman, R. G. Egdell, and A. F. Orchard, *J. Electroanal. Chem.*, **107**, 37 (1980).
71. N. Oyama and F. C. Anson, *Anal. Chem.*, **52**, 1192 (1980).
72. C. P. Andrieux, J.-M. Dumas-Bouchiat, and J.-M. Savéant, *J. Electroanal. Chem.*, **114**, 159 (1980).
73. R. W. Murray, *Philos. Trans. R. Soc. London A*, **302**, 253 (1981).
74. C. P. Andrieux, J.-M. Dumas-Bouchiat, and J.-M. Savéant, *J. Electroanal. Chem.*, **131**, 1 (1982).
75. C. P. Andrieux and J.-M. Savéant, *J. Electroanal. Chem.*, **134**, 163 (1982).
76. C. P. Andrieux, J.-M. Dumas-Bouchiat, and J.-M. Savéant, *J. Electroanal. Chem.*, **169**, 9 (1984).
77. W. J. Albery and A. R. Hillman, *Annu. Rep. (1981) R. Chem. Soc. London*, 1983, pp. 317–437.
78. C. P. Andrieux and J.-M. Savéant, *J. Electroanal. Chem.*, **142**, 1 (1982).
79. F. C. Anson, J.-M. Savéant, and K. Shigehara, *J. Phys. Chem.*, **47**, 214 (1983).
80. C. P. Andrieux and J.-M. Savéant, *J. Electroanal. Chem.*, **171**, 65 (1984).
81. C. P. Andrieux, A. Merz, J.-M. Savéant, and R. Tomahogh, *J. Am. Chem. Soc.*, **106**, 1957 (1984).
82. C. P. Andrieux, A. Merz, and J.-M. Savéant, unpublished results.
83. A. R. Hillman, in *Electrochemical Science and Technology of Polymers*, R. G. Lindford (Ed.), Elsevier Applied Science, New York, 1987, Chapters 5 and 6.
84. M. Kaneko and D. Wöhrle, *Adv. Polym. Sci.*, **14**, 142 (1988).
85. H. D. Abruña, in *Electrode Modification with Polymeric Reagents in Electroresponsive Molecular and Polymeric Systems*, Vol. 1, T. A. Skotheim (Ed.), Dekker, New York, 1988, pp. 98–171.
86. R. D. Rocklin and R. W. Murray, *J. Phys. Chem.*, **85**, 2104 (1981).
87. K. N. Kuo and R. W. Murray, *J. Electroanal. Chem.*, **131**, 37 (1982).
88. T. Ikeda, C. R. Leidner, and R. W. Murray, *J. Electroanal. Chem.*, **138**, 343 (1982).
89. F. C. Anson, J.-M. Savéant, and K. Shigehara, *J. Electroanal. Chem.*, **145**, 423 (1983).
90. R. H. Schmehl and R. W. Murray, *J. Electroanal. Chem.*, **152**, 97 (1983).
91. D. A. Buttry and F. C. Anson, *J. Am. Chem. Soc.*, *106*, 59 (1986).
92. F. C. Anson, T. Oksaka, and J.-M. Savéant, *J. Am. Chem. Soc.*, **105**, 4883 (1983).
93. C. R. Leidner and R. W. Murray, *J. Am. Chem. Soc.*, **106**, 1606 (1984).

94. M. Krishnan, X. Zhang, and A. J. Bard, *J. Am. Chem. Soc.*, **106**, 7371 (1984).
95. C. W. Lee, H. B. Gray, F. C. Anson, and B. G. Malmstrom, *J. Electroanal. Chem.*, **172**, 289 (1984).
96. F. C. Anson, Y. M. Tsou and J.-M. Savéant, *J. Electroanal. Chem.*, **178**, 113 (1984).
97. G. X. Wan, K. Shigehara, E. Tsushida, and F. C. Anson, *J. Electroanal. Chem.*, **179**, 239 (1984).
98. C. W. Lee and F. C. Anson, *Inorg. Chem.*, **23**, 837 (1984).
99. P. W. Geno, K. Ravichandran, and R. P. Baldwin, *J. Electroanal. Chem.*, **183**, 155 (1985).
100. F. C. Anson, C. L. Ni, and J.-M. Savéant, *J. Am. Chem. Soc.*, **107**, 3442 (1985).
101. C. Roulier, E. Waldner, and E. Laviron, *J. Electroanal. Chem.*, **191**, 59 (1985).
102. M. Sharp, D. D. Montgomery, and F. C. Anson, *J. Electroanal. Chem.*, **194**, 247 (1985).
103. R. A. Simon, T. E. Mallouk, K. A. Daube, and M. S. Wrighton, *Inorg. Chem.*, **24**, 3119 (1985).
104. B. Lindholm and M. Sharp, *J. Electroanal. Chem.*, **198**, 37 (1986).
105. C. P. Andrieux, O. Haas, and J.-M. Savéant, *J. Am. Soc.*, **108**, 8175 (1986).
106. L. J. Miller and M. Majda, *J. Electroanal. Chem.*, **207**, 49 (1986).
107. E. T. Turner Jones and L. R. Faulkner, *J. Electroanal. Chem.*, **222**, 201 (1987).
108. S. Holdcroft and B. L. Funt, *J. Electroanal. Chem.*, **225**, 177 (1987).
109. J. E. Van Koppenhagen and M. Majda, *J. Electroanal. Chem.*, **236**, 113 (1987).
110. M. Sharp, B. Lindholm, and E. L. Lind, *J. Electroanal. Chem.*, **274**, 35 (1989).

Chapter **VI**

ELECTRODES MODIFIED WITH CLAYS, ZEOLITES, AND RELATED MICROPOROUS SOLIDS

Allen J. Bard and Thomas Mallouk
Department of Chemistry, University of Texas, Austin, Texas

Molecular Design of Electrode Surfaces,
Edited by Royce W. Murray. Techniques of Chemistry Series, Vol. XXII.
ISBN 0-471-55773-0

This chapter deals with the preparation, characterization, and application of electrodes modified by forming films of inorganic microcrystalline structured materials (frequently, clays, zeolites, or related aluminosilicates) on a conductive substrate surface. Such materials are of interest for a number of reasons. They are often very stable and chemically robust materials; for example, they can withstand high temperatures and highly oxidizing conditions. Because they are usually capable of incorporating ions by an ion-exchange process, they, like polymeric ionomers, can serve as matrices for electroactive ions. Many of these materials, such as the naturally occurring clays, are very abundant and inexpensive.

Perhaps the greatest interest in these materials, however, arises because they have known and well-defined structures. Thus clays have a sheetlike structure, which after pillaring, as discussed below, leads to a well-defined interlamellar spacing. Zeolites are characterized by pores and channels of definite size. These structural factors lead to size and shape selectivity in chemical reactions carried out when these materials are used as supports for heterogeneous catalysts or reagents (1). This widespread use in catalysis can provide guidelines in the design of electrocatalysts based on these materials. The structural features in these materials can also be useful in the fabrication of analytical devices, since they have the potential for *molecular recognition* properties. Molecular recognition implies the presence of specific sites on a matrix with cavities of specific sizes, shapes, and spacings that occur naturally in a material or are formed in it during its synthesis.

The definite structures also imply that these materials can serve as templates for the assembly of more complex systems. There is considerable interest in the construction of *integrated chemical systems* (2), which are heterogeneous, multiphase systems involving several different components designed and arranged for specific functions. Several examples of such systems, involving composites of clays and zeolites with polymers, metals, and semiconductors, are described below.

6.1 INTRODUCTION TO CLAY CHEMISTRY

Clays are colloidal, layered hydrous aluminosilicates, most of which occur naturally and belong to the class of minerals called phyllosilicates (e.g., montmorillonites, kaolinites, hectorites, and nontronites) (3,4). Of special interest are the swelling (or expanding) clays known as smectite clays. The layers in these clays are assembled from sheets of SiO_4 tetrahedra (Tet) and AlO_6 octahedra (Oc) (Fig. 6.1). For example, in the widely used smectite, such as montmorillonite, each layer of the clay is composed of an Al-octahedral sheet sandwiched between two Si-tetrahedral sheets, with the oxygen atoms on the apices of the tetrahedra shared with the octahedra (Fig. 6.1). Montmorillonites are referred to as 2:1 or Tet–Oc–Tet clays. The 1:1 or Tet–Oc clays have layers that are composed of one tetrahedral and one octahedral sheet. Isomorphous (3) substitution of other cations for Si or Al causes the layers to become electrically

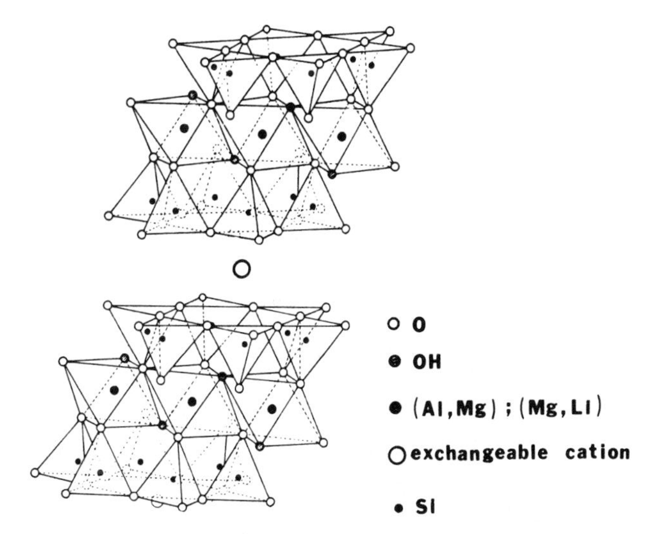

Figure 6.1. Schematic structure of 2:1 layer silicates (smectites), such as montmorillonite.

charged. For example, substitution of Mg^{2+}, Fe^{2+}, or Li^{+} for Al^{3+} results in a layer with an excess negative charge. This excess charge is compensated by exchangeable cations (e.g., Na^{+} or Ca^{2+}) adsorbed on the surface of the layers.

Most clay minerals consist of particles containing several layers stacked parallel to one another to form plates. The cohesive forces between the layers are primarily electrostatic and van der Waals forces. The interlayer spacing, called the $d(001)$, c or basal spacing and defined in Fig. 6.2, depends mainly on the extent of hydration of the exchangeable cations. For example, the spacing of air-dried Na^{+} montmorillonite is about 12.5 Å. When heated to 500°C the layers collapse to a d spacing of 9.6–10 Å. When exposed to high humidity, the clay swells with the introduction of one to four layers of water molecules within the interlayer region, with d spacings of up to 17–20 Å. With the addition of larger amounts of water, the clay can become "fully expanded" and dispersed into individual layers. Polar organic compounds (e.g., ethylene glycol or glycerol) can also be inserted between the layers of montmorillonites with corresponding increases in the basal-spacings.

The cation exchange capacity (CEC) of montmorillonites is typically 70–100 meq per 100 g, which is similar to that of a polymer like Nafion. The exchangeable cations are distributed over both the internal and external surfaces of the clay particles. The edges of the platelike clay particles are thought to resemble the surface of an alumina particle and, in acidic solutions, to carry a net positive charge. When the clay plates aggregate into larger particles, three modes of association occur: face to face, edge to face, and edge to edge. The edge-to-face and edge-to-edge associations lead to more voluminous "house-of-cards" structures (Fig. 6.3). Similar structures probably form when films of clay are deposited from colloidal suspensions onto substrate surfaces.

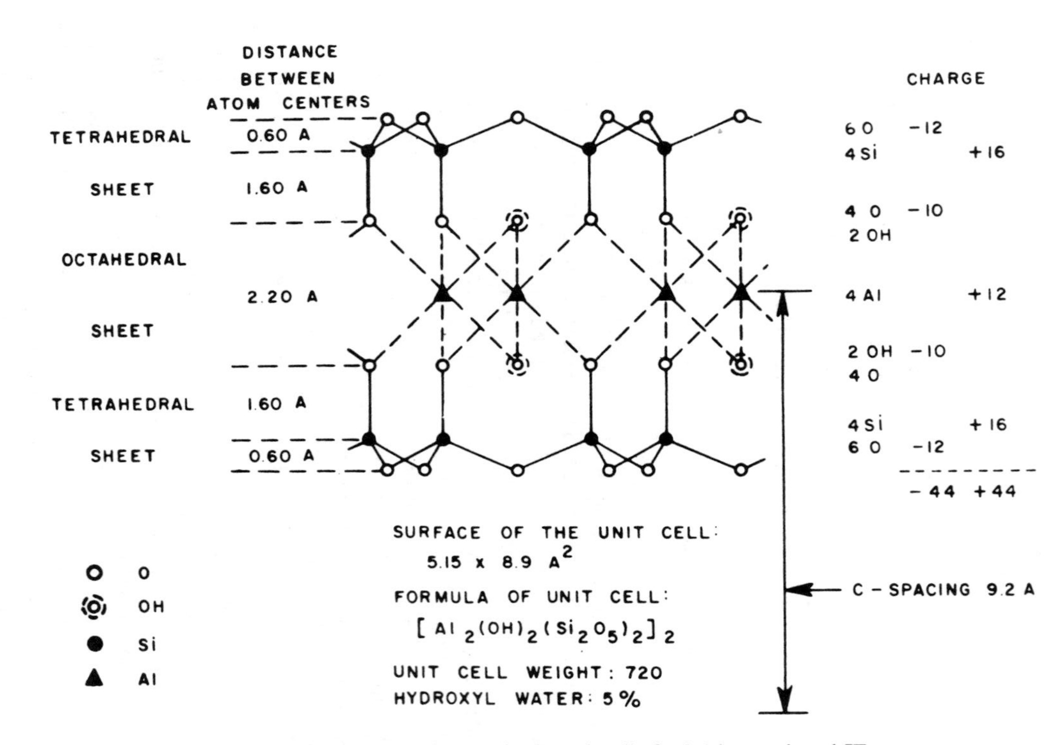

Figure 6.2. Schematic of the atomic arrangement in the unit cell of a 2:1 layer mineral [From van Olphen, H. *An Introduction to Clay Colloid Chemistry*, Wiley: New York, 1977. Copyright © (1977) John Wiley & Sons, Inc. Reprinted by permission of John Wiley & Sons, Inc.]

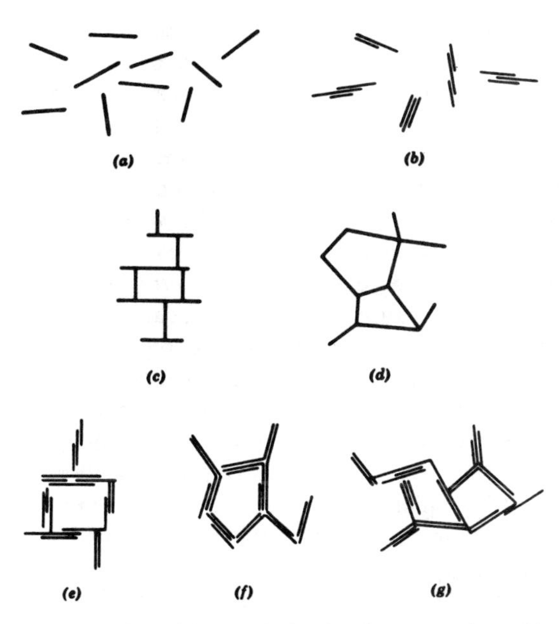

Figure 6.3. Possible modes of particle association in clay suspensions. (*a*) dispersed and deflocculated; (*b*) aggregated and deflocculated (face-to-face association of platelets); (*c*) edge-to-face flocculated; (*d*) edge-to-edge flocculated; (*e*) edge-to-face flocculated and aggregated; (*f*) edge-to-edge flocculated and aggregated; (*g*) both edge-to-face and edge-to-edge flocculated and aggregated [From van Olphen, H. *An Introduction to Clay Colloid Chemistry*, Wiley: New York, 1977. Copyright © (1977) John Wiley & Sons, Inc. Reprinted by permission of John Wiley & Sons, Inc.]

Samples of clay minerals can be obtained from the Source Clay Minerals Repository (University of Missouri, Columbia, Missouri). Most of the electrochemical studies have been carried out with Na-Wyoming montmorillonite (abbreviated Na^+SWy-1), Ca-Texas montmorillonite (STx-1), and Ca-hectorite (SHCa-1). Even these "standard" natural clays may contain large amounts of impurities that must be removed prior to use. Purification methods are discussed in the literature (3, 5). Generally, the clay is suspended in water and the larger particles are allowed to settle and are removed. The clay is converted to the desired form (e.g., the Na^+–form) by stirring with a concentrated electrolyte solution (e.g., 1*M* NaCl for 48 h). The clay particles are then separated by centrifugation and washed with water one or more times. The suspension is then dialyzed to remove excess electrolyte. Another centrifugation leads to a sediment containing clay particles less than 0.2 μm. This purified sample is sometimes subjected to freeze-drying. Treatments are also available to remove specific impurities. For example, iron is a common contaminant as iron oxide, which can be removed by treatment with dithionite and citrate (6). This treatment does not remove the structural iron in octahedral lattice positions.

Clays are most frequently characterized by X-ray diffraction (XRD) measurements (4). The *d*-spacing is readily obtained from peak positions expressed as the

Bragg angle, 2θ. Electron diffraction, scanning and transmission electron microscopy, Mössbauer spectroscopy (for iron), nuclear magnetic resonance (NMR), and electron spin resonance (ESR) (of transition metals) are also employed.

Pillared clays are prepared by exchanging the interlamellar cations by bulky, thermally stable, robust cations that can act as molecular props or pillars and provide a fixed distance between the layers (7). While alkylammonium ions, bicyclic amine cations, and metal chelate complexes can be used as pillaring agents, the most stable pillared clays probably result from treatment with polynuclear hydroxy metal cations. For example, cations like $[Zr_4(OH)_{16-x}]^{x+}$ and $Al_{13}O_4(OH)_{28}]^{3+}$ can be intercalated and then bound to the aluminosilicate layers by a heat treatment at about 550°C. This produces a fixed interlayer spacing of about 9 Å (corresponding to a *d*-spacing of ~18 Å) and surface areas of 200–500 $m^2\,g^{-1}$. Such clays show highly selective molecular sieving properties, suggesting a regular distribution of pillars and pores within the particles (7b). Other pillaring materials that have been investigated include chromium oxide, titanium oxide, nickel oxide, silicon oxide, and metal cluster cations, such as $[Nb_6Cl_{12}]^{2+}$ and $[Ta_6Cl_{12}]^{2+}$ (7a, b). Pillared clays are of interest, because they allow the preparation of materials with fixed pores and channels that are larger than those in zeolites.

While most clays are naturally occurring materials, synthetic clays have also been prepared. Most are synthesized by hydrothermal methods [reactions in aqueous media above 100°C and 1 bar (7c, 8)]. These are based on heating mixtures of the appropriate oxides or natural clays with excess water in a pressure vessel; typical reaction conditions are 300°C and 1000 bar. Of special interest are the anion-exchange materials related to the hydrotalcites. Hydrotalcite is a Mg–Al compound consisting of positively charged brucite-like layers $[Mg_6Al_2(OH)_{16}]^{2+}$ with compensating anions (e.g., CO_3^{2-}) in the interlayer regions. Synthetic hydrotalcite-like materials (also called layered double hydroxides) have been used to fabricate modified electrodes, as described below.

6.2 CHARACTERIZATION OF CLAY-MODIFIED ELECTRODES

6.2.1 Early Studies of Clay-Modified Electrodes

The first reported clay-modified electrode (CME) were prepared from mixtures of Na^+ Wyoming montmorillonite (Na^+ SWy-1) and small amounts of poly(vinyl alcohol) (PVA), with and without colloidal Pt to improve film conductivity and morphology (9). Electrodes of about 3-μm thick clay films were prepared on tin oxide on glass, Pt, and glassy carbon (GC) substrates. Species, such as $[Ru(bpy)_3]^{2+}$, $[Fe(bpy)_3]^{2+}$, methyl viologen (MV^{2+}), and $[TMAFe(Cp)_2]^+$ (TMA = trimethylammonium and Cp = cyclopentadiene), incorporated into the films by soaking the electrodes in solutions of these species, showed cyclic voltammetric (CV) responses at potentials near those found for the same species in solution (Fig. 6.4). At scan rates (v) of 5–100 $mV\,s^{-1}$, the

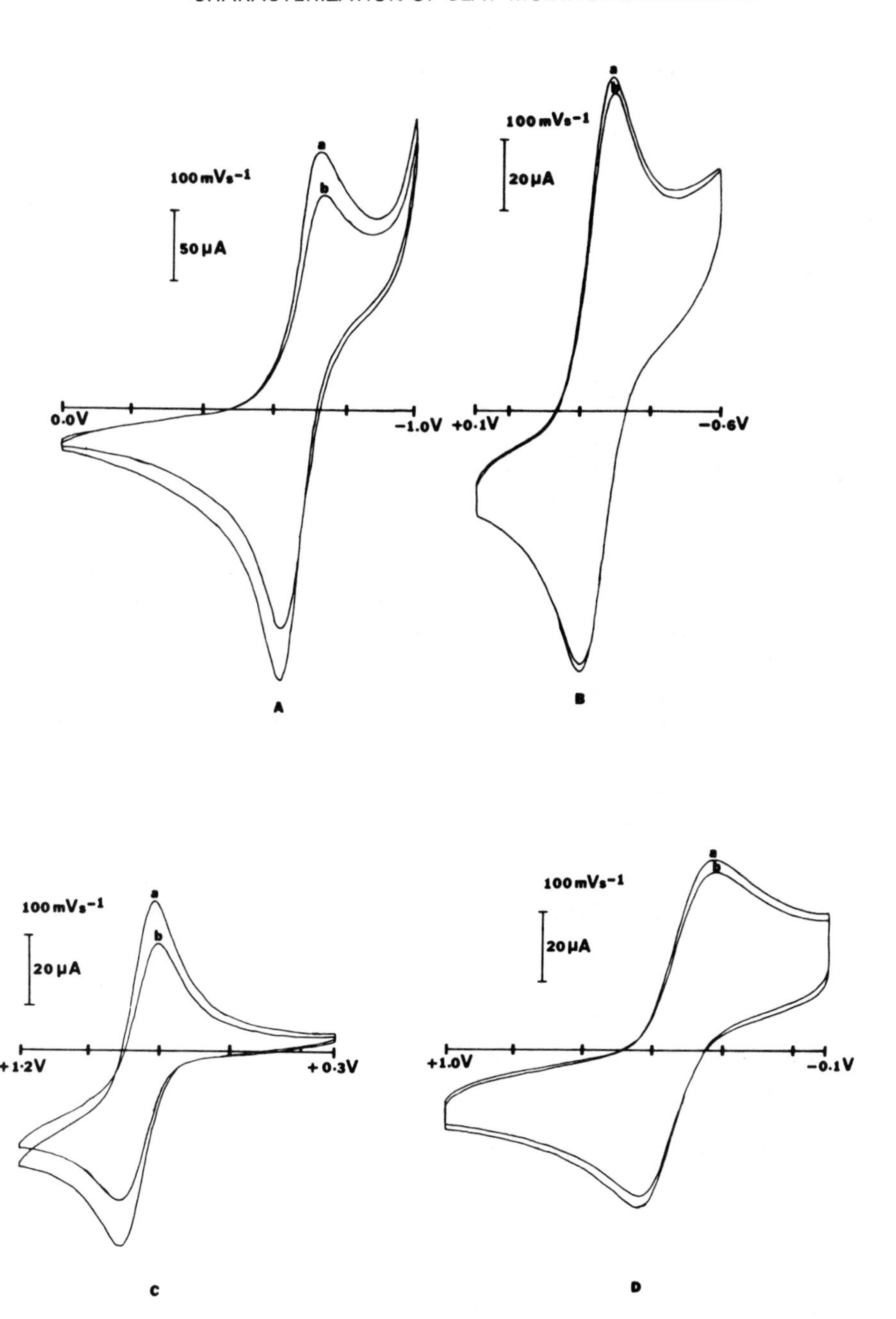

Figure 6.4. Cyclic voltammograms for several electroactive compounds incorporated into montmorillonite–PVA–Pt films on a tin oxide electrode. (*A*) MV^{2+} (a) first scan, (b) 100th scan; (*B*) $[Ru(NH_3)_6]^{3+}$ (a) first scan, (b) 105th scan; (*C*) $[Fe(bpy)_3]^{2+}$ (a) first scan, (b) 40th scan; (*D*) TMA $[FeCp_2]^+$ (a) first scan, (b) 40th scan. All voltammograms were recorded at $100\,mV\,s^{-1}$ in $0.1M\ Na_2SO_4$, pH 7 [Reprinted with permission from Ghosh, P. K.; Bard, A. J. *J. Am. Chem. Soc.* **1983**, *105*, 5691. Copyright © (1983) American Chemical Society.]

waves appeared diffusion controlled, with peak currents (i_p) proportional to $v^{1/2}$ and apparent diffusion coefficients for the incorporated species (D_{app}) of about $10^{-11}\,cm^2\,s^{-1}$. The term D_{app}, as in other modified electrodes, relates either to the actual motion of the incorporated electroactive moiety or to the rate of charge transfer between species. Only a fraction of the incorporated species (~10%) was electroactive. Several studies of CME that followed (10–13) this initial report also employed the clay–PVA formulation. Subsequent studies (13–21) showed that stable clay films on electrodes without added PVA could also be prepared and that these would incorporate electroactive species and show similar electrochemical behavior. A summary of the electrochemical studies that have been carried out with CMEs is given in Table 6.1.

6.2.2 Preparation of Clay-Modified Electrodes

The observed behavior of CME is very dependent on the method of preparation of the films, their pretreatment, and the means of incorporation of the electroactive species. Clay films are cast from colloidal suspensions prepared as described above and containing particles below 0.2 μm. While films can be cast by slow evaporation of a measured volume (typically 10–150 μL of a solution containing 2–14 g of clay L^{-1}) placed on the substrate surface, more reproducible, uniform, and thinner films can be obtained by spin coating. In this procedure the suspension, sometimes containing ethanol, is spin-coated on the cleaned and preheated substrate, which is rotated with a photoresist spinner at 3000–8000 rpm. An alternative approach is electrophoretic deposition by immersing the electrode is a suspension of the clay and applying an appropriate potential to attract the charged clay particles (24). The thickness of a clay film can be measured with a profilometer, such as the Sloan Dektak FLM profilometer. For very thin films, the thickness can sometimes be estimated from the observed interference color.

6.2.3 Peak Potentials of Incorporated Species

Cyclic voltammetric measurements of the differences in peak potentials of species incorporated in the clay film compared to those in solution can provide information about the interactions of these species with the clay layers. Several papers have described such shifts in formal potential for different redox couples. For example, the $\Delta E_p = E_{p(clay)} - E_{p(soln)}$ for $[Os(bpy)_3]^{3+/2+}$ couple in montmorillonite was +40 mV (0.2M solium acetate, pH 5) (14). Similarly, ΔE_p for $[Ru(bpy)_3]^{3+/2+}$ was +42 mV (0.2M sodium acetate, pH 5) (14) and about 100 mV (0.05M Na_2SO_4) (13). Positive ΔE_p values (40–110 mV) were also found for $[Cr(bpy)_3]^{3+/2+}$ in 0.01 and 0.1M Na_2SO_4 solutions that were functions of the concentration of redox species in solution (21). As discussed in connection with similar redox reactions in Nafion as compared to those in solution (25), the shift in peak potentials depends on the supporting electrolyte concentration in addition to any interactions or changes in equilibria induced by the clay film

(17, 21). Thus, following the model proposed for polymer electrodes (25), for the electrode reaction

$$\mathrm{Ox}^{n+}(n\mathrm{A}^-)_{\mathrm{clay}} + (\mathrm{Na}^+)_{\mathrm{soln}} + \mathrm{e}^- = \mathrm{Red}^{n-1}[(n-1)\mathrm{A}^-]_{\mathrm{clay}} + \mathrm{Na}^+(\mathrm{A}^-)_{\mathrm{clay}}$$

where (A^-) represents anionic sites on the clay and Ox^{n+} and Red^{n-1} the incorporated oxidized and reduced redox species, respectively, the apparent formal potential of the species in the clay film (i.e., uncorrected for the Donnan potential) is given by

$$(E^{\circ\prime})_{\mathrm{app,clay}} = E^{\circ\prime}_{\mathrm{clay}} + (RT/\mathscr{F})\ln([\mathrm{Na}^+]_{\mathrm{soln}}/[\mathrm{Na}^+]_{\mathrm{clay}})$$

Interactions involving the redox species in the clay are indicated by differences between $E^{\circ\prime}_{\mathrm{clay}}$ and $E^{\circ\prime}_{\mathrm{soln}}$ (the formal potentials in the clay and solution, respectively). A detailed correlation of the peak potential shifts in clays with interactions, incorporating Donnan potential effects, has not yet been reported.

The clay environment can also affect the equilibrium constant for a reaction, in addition to any interaction of the solution species with the clay matrix. For example, a study of the electrochemistry and ESR of methyl viologen (*N*,*N*′-dimethyl-4,4′-bipyridinium or MV^{2+}) in STx-1 showed that the equilibrium constant for the reaction $2\mathrm{MV}^{\cdot+} = (\mathrm{MV}^{\cdot+})_2$ was about $5 \times 10^4 M^{-1}$ in the clay environment compared to about $380 M^{-1}$ in solution (16). Similarly, a large shift in peak potential was found for tetrathiafulvalenium (TTF^+), which was ascribed to strong interactions of the species with clay as well as effects on equilibrium constants (17).

6.2.4 Electroactivity of Incorporated Species

Most of the studies of CME have shown that only a fraction (~10–30%) of the species incorporated into the clay film by soaking in a solution of the species is electroactive (13, 14, 17, 21, 22). The total amount of species incorporated into the film can be obtained from estimates based on the CEC, by actual measurements of the concentration of the species in the film by spectrophotometry, or by measurements of the decrease in the solution concentration after the film has been soaked in it. The total electroactive species in the film is measured by slow scan cyclic voltammetry under thin-layer conditions, where the integrated current is a measure of total charge passed in electrolyzing the film component. The fraction of the incorporated species that is electroactive appears to be very dependent on the film preparation conditions and the method of incorporation. For example, incorporation of the electroactive species by treating the colloidal clay suspension before casting the film on the electrode results in essentially no electroactivity of the species (19). Further soaking of this cast film in a solution of the same redox species, however, will cause the electrode to exhibit CV waves for the incorporated species (19, 21).

Although only a fraction of the incorporated species is electroactive, the CV

Table 6.1 Summary of Cyclic Voltammetry Studies on Clay-Modified Electrodes

Clay[a]	*Substrate*[b]	*Electroactive Species*[c]	D_{app} ($cm^2 s^{-1}$)	*Reference*
Clay–PVA Electrodes				
SWy–1	SnO_2, Pt, GC	$[M(bpy)_3]^{2+}$ (M = Ru, Fe)	$\sim 10^{-11}$	9
		MV^{2+}, $[TMAFe(Cp)_2]^+$, $[Ru(NH_3)_6]^{2+}$		
SWy-1	SnO_2	Thionine		10
S-M	SnO_2	$[Ru(phen)_3]^{2+}$		11
SHCa-1	SnO_2	$Ru(bpy)_2[bpy–(CO_2)_2]$	3×10^{-6}	12
STx-1	GC	H_2Q	$\sim 1 \times 10^{-7}$	13
	SnO_2	$[M(bpy)_3]^{2+}$ (M = Ru, Fe, Os)	$\sim 10^{-12}$	
SWy-1	SnO_2	PVS		
Clay (Alone)				
STx-1, SWy-1	SnO_2	$[M(bpy)_3]^{2+}$ (M = Ru, Fe, Os)	$\sim 10^{-12}$	13
		PVS		
S-M	PG	$[Os(bpy)_3]^{2+}$	3.5×10^{-11}	14
		$[Co(tpy)_2]^{2+}$	1.8×10^{-11}	
		$[Fe(CN)_6]^{3-}$		
S-M	PG	$[Ru(NH_3)_6]^{2+}$		15
STx-1	SnO_2	MV^{2+}	9×10^{-12}	16

STx-1	Pt	TTF^+	$\sim 10^{-10}$	17
STx-1	Pt, GC	MV^{2+}		18
SWy-1, LAP	PG	$[M(bpy)_3]^{2+}$ (M = Ru, Fe, Os) MV^{2+}, $[Fe(phen)_3]^{2+}$		19
S-M	GC	$[Ru(bpy)_3]^{2+}$ (various pH), $[Fe(CN)_6]^{3-}$		20
SWy-1, STx-1 SHCa-1, SAz-1	Pt	$[Cr(bpy)_3]^{3+}$	7×10^{-12}	21
P-STx-1, NON P-SHCa-1	SnO_2	$[M(bpy)_3]^{2+}$ (M = Ru, Fe, Os), $[Ru(NH_3)_6]^{2+}$	$\sim 10^{-12}$ 3.2×10^{-10}	22
Pillared Clays				
P-STx-1	SnO_2	$[M(bpy)_3]^{2+}$ (M = Fe, Os) $[Fe(CN)_6]^{3-}$	$\sim 10^{-12}$	23
P-STx-1 P-SHCa-1	SnO_2	$[M(bpy)_3]^{2+}$ (M = Ru, Fe, Os), $[Ru(NH_3)_6]^{2+}$	$\sim 10^{-11}$ 5.5×10^{-10}	22

[a]Clays: SWy-1 = Na-Wyoming montmorillonite; STx-1 = Ca-Texas montmorillonite; S-M = Na montmorillonite; SAz-1 = Arizona montmorillonite, SHCa-1 = calcium hectorite; LAP = laponite; NON = nontronite. (Most clays exchanged with Na^+ before use.)
[b]Substrates: GC = glassy carbon; PG = pyrolytic graphite; SnO_2 = conductive tin oxide on glass.
[c]Electroactive Species: bpy = 2,2′-bipyridine; tpy = 2,2′,2″-terpyridine; phen = 1,10-phenanthroline; $[bpy\text{-}(CO_2)_2]^{2-}$ = 4,4′-dicarboxy-2,2′-bipyridine; MV^{2+} = methyl viologen; PVS = propyl viologen sulfonate; Cp = cyclopentadienyl; H_2Q = 1,4-benzohydroquinone.

peak current for this species can be larger in the clay film than that of the species in the soaking solution at a bare electrode (under conditions where the CV currents are proportional to $v^{1/2}$) (13, 14). If the concentration of electroactive species in the soaking solution and clay film is given by C_s and C_f, respectively, with corresponding diffusion coefficients, D_s and D_f, this implies that in these cases $C_f D_f^{1/2} > C_s D_s^{1/2}$. The term D_f is always much smaller than D_s (see Table 6.1), so the effective electroactive concentration in the film must be significantly larger than that in solution, and an attractive interaction between the clay and the solution species must exist. This finding also implies that the electroactive species is not merely held in pores or channels within the clay layer, since any CV response from species in these sites would be much smaller than that from the species in the bulk solution. Pillaring of the clay layer decreases the CV response and the amount of incorporated electroactive $[M(bpy)_3]^{2+}$ cation compared to that in a layer of the same clay that has not been pillared (22, 23). This presumably arises from neutralization of some of the exchange sites by the pillaring agent.

6.2.5 Effective Diffusion Coefficients in the Clay

The effective diffusion coefficients of incorporated cations in clay films generally lie in the range of 10^{-12}–10^{-11} $cm^2 s^{-1}$ (Table 6.1). As with other modified electrodes, there are probably two components to the effective diffusion coefficient: the translational motion of the species to the substrate surface and electron hopping or transfer from site to site among the electroactive species. The importance of this latter component can be illustrated by several experiments with CME. For example, the larger D value found for $[Os(bpy)_3]^{2+}$ compared to $[Co(tpy)_2]^{2+}$ (Table 6.1) has been attributed to the larger contribution of electron hopping in the diffusion of the Os species compared to the Co complex (14). The self-exchange rate is known to be much larger for the Os species and similar effects have been found for these species in Nafion films (26). When $[Co(tpy)_2]^{2+}$ is incorporated in a film containing $[Os(bpy)_3]^{2+}$, the peak current found for the $[Os(bpy)_3]^{2+}$ wave is enhanced (Fig. 6.5). This enhancement is ascribed to a more rapid electron-transfer cross reaction between the Os(III) species with a larger effective diffusion coefficient and the more slowly diffusing $[Co(tpy)_2]^{2+}$ (14). A parallel effect was seen in a study of MV^{2+} in an STx-1 film (16). Reduction of the MV^{2+} leads to the formation of a dimer, $(MV^{\cdot +})_2$, that is not oxidizable in the clay film at potentials of the $MV^{2+}/MV^{\cdot +}$ wave. This fact was ascribed to very slow electron exchange between dimers in the clay, which apparently do not physically diffuse to the substrate surface. When $[Fe(CN)_6]^{4-}$, which can physically diffuse through clay films, as discussed below, was added to the solution, and $[Fe(CN)_6]^{3-}$ electrogenerated in the clay layer, the $(MV^{\cdot +})_2$ species was oxidized. In this case the $[Fe(CN)_6]^{3-}$ acts as a mediator for the oxidation of the immobile $(MV^{\cdot +})_2$. Both of these experiments demonstrate the importance of electron-transfer or self-exchange contributions to the effective diffusion coefficients of incorporated and presumably strongly bound cationic species in clay layers. The complex

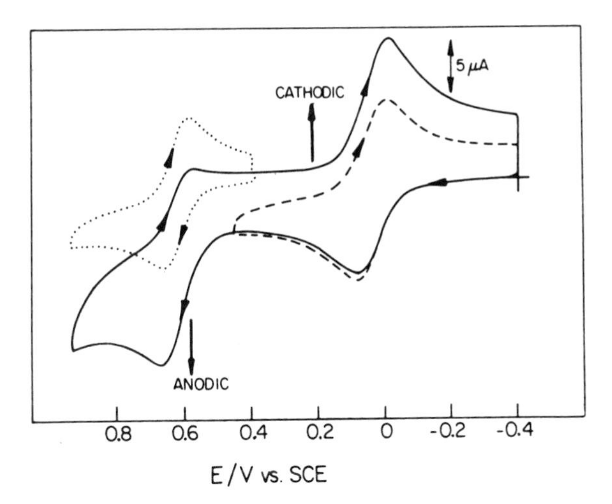

Figure 6.5. Cyclic voltammograms of mixtures of $[Os(bpy)_3]^{2+}$ and $[Co(tpy)_2]^{2+}$ in a montmorillonite film on a pyrolytic graphite electrode in 0.2*M* sodium acetate, pH 5.5 at $50\,mV\,s^{-1}$. (---) potentials where only $[Co(tpy)_2]^{2+}$ contributes to the current; (······) potentials where only $[Os(bpy)_3]^{2+}$ contributes to the current; (———) potential scanned over both waves. The larger anodic peak current for the oxidation of $[Os(bpy)_3]^{2+}$ is attributed to the cross reaction between $[Os(bpy)_3]^{3+}$ and $[Co(tpy)_2]^{2+}$ (Reprinted with permission from Ref. 14.)

$[Ru(NH_3)_6]^{3+}$ shows a much larger *D* value in clay than the corresponding bpy species and presumably physically moves more rapidly in the clay environment (22). Moreover, the *D* value is much less affected by pillaring of the clay (22) and the peak potential of the 3+/2+ couple in the clay is the same as that in solution (14). These results suggest a much smaller interaction of the Ru—NH_3 species with the clay layer compared to M═bpy species.

Uncharged species, such as hydroquinone (H_2Q) and bis(bpy) (4,4′-dicarboxy-2,2′-bipyridyl) Ru^{II}, can also penetrate the clay layer (12, 13). These show larger effective diffusion coefficients (Table 6.1), presumably because they interact less strongly with the clay film and physically move more rapidly than most cationic species. Such species are rapidly removed from the film when it is soaked in a solution that only contains supporting electrolyte. There is evidence, however, that some uncharged species can interact with, and be preferentially extracted by, the clay. For example, the peak currents for the CV of propyl viologen sulfonate (PVS) are 2–4 times larger in a clay film than at a bare electrode (both immersed in a solution containing 2 m*M* PVS) (13).

The electrochemical response (CV or chronoamperometric) in the presence of neutral species, like HQ, has been used to obtain information about the presence of pinholes or channels in the film, or the behavior of the films as membranes (13). A plot of the ratio of the current at the filmed electrode following a potential step to the diffusion-limited plateau region normalized with respect to the current at the same time at the bare electrode versus time, shows that the current ratio varies with time, as expected from both a pinhole (26–31) or membrane (28, 29) model.

Anionic species can also penetrate the clay layer. A number of studies of the electrochemical response of $[Fe(CN)_6]^{3-}$ or other anionic species have been reported (13, 14, 16, 17, 20, 23, 27). In general, the clay layer attenuates the response of these species compared to that at a bare electrode, but clear CV waves are observed. Pillaring of the clay (22, 23) or increasing the supporting electrolyte (NaCl) concentration (27) suppresses the $[Fe(CN)_6]^{3-}$ CV wave, presumably because channels within the clay film are blocked or narrowed.

6.2.6 Clay Structure in Clay-Modified Electrodes

Several papers have dealt with methods other than electrochemical for investigating the structure of CME. Scanning electron microscopy of the clay films (13, 19, 21, 46) generally shows a rough, "wafflelike" structure, with grain boundaries between the clay particles that might serve as microchannels, but no obvious large pinholes or cracks. Electron spin resonance studies of stable radicals (spin labels), can provide information about the structure and orientation of clay films (32). In a study of CME (13) the cationic radical tempamine was mixed with colloidal clay to 2–5% of the CEC before casting films (with and without PVA). Hectorite and montmorillonite films that did not contain PVA showed strongly anisotropic ESR signals, for example, spectra that depended on the orientation of the film with respect to the magnetic field, which indicated a highly ordered assembly of spin probes. In agreement with the earlier work (32), this suggested that these films consisted of ordered clay sheets parallel to the substrate surface. Films cast with montmorillonite and PVA and prepared in the same way showed ESR spectra that were independent of film orientation, suggesting a chaotic collection of microregions. This indicates that PVA disrupts the parallel plate structure. In both cases the spectra showed that the tempamine spin probe in the clay film was motionally restricted on the ESR timescale. An ESR study of $MV^{\cdot +}$ formed in a film of STx-1 showed no signal for the radical cation, and only a very weak signal for $MV^{\cdot +}$ adsorbed on a clay dispersion (16). This result was explained by the promotion of dimerization of the paramagnetic cation radical by the clay medium.

6.2.7 Electrolyte, Solvent, and pH Effects

The nature and concentration of supporting electrolyte affect the electrochemical response of CME by several mechanisms. The supporting electrolyte cation can compete with the electroactive ion for ion-exchange sites; thus high concentrations of electrolyte will tend to exchange the cation for the one incorporated in the film. The nature of the electrolyte anion can affect the extent of ion pairing within the clay film and the amount of cation that is bound in excess of the CEC (19). Electrolyte concentration can affect the clay film structure. For example, the distance between the face-to-face stacked platelets increases with dilution of the electrolyte and this can affect the permeability of the film (27). Larger structural changes with electrolyte concentration, for example, rearrangement of the platelets from a stacked to a house-of-cards configuration, might also occur. Some examples of electrolyte effects include reports that the CV response of $[M(bpy)_3]^{2+}$ (M = Fe, Os), which is readily observed in $0.1M$ Na_2SO_4, is suppressed in $0.1M$ $LiClO_4$, KCl, and tetraethylammonium chloride (TEACl) after 5-min equilibration of the film (23). The concentration of the supporting electrolyte is also important, however. Thus another report states that electroactivity was observed or $[Fe(bpy)_3]^{2+}$ in $0.1M$ solutions of electrolytes like NaCl, KCl, and TEACl, but was greatly diminished at $1M$ concentrations of the same electrolytes (22). However, the response of MV^{2+} was said to be the same in $0.05M$ NaCl and $1.0M$ tetramethylammonium chloride (16). The CV peak currents for $[Os(bpy)_3]^{2+}$ in Na_2SO_4 were 1.4–3.0 times larger than those in $NaC_2H_3O_2$ over the ionic strength range of 0.05–$0.30M$ (19). Similarly, the maximum peak currents for the reduction of $[Cr(bpy)_3]^{3+}$ in $0.01M$ Na_2SO_4 were 2.6 times those obtained in $0.02M$ NaCl (21).

Changes in pH can give rise to similar effects. For example, the CV response of $[Ru(bpy)_3]^{2+}$ in unbuffered $0.1M$ Na_2SO_4 solutions, pH 1–10, shows a uniform response for pH > 4 (20). At lower pH, the waves are greatly attenuated, although a film prepared at a higher pH and transferred to supporting electrolyte at lower pH does not show this attenuation. The authors propose that protons compete for surface sites of the Ru species at lower pH, but that preadsorbed $[Ru(bpy)_3]^{2+}$ is not readily displaced by protons when the film is immersed in the acidic solutions. The reduction of $[Fe(CN)_6]^{3-}$ is attenuated at higher pH (20). This was explained by electrostatic effects in the voids (i.e., intergranular channels) of the clay particles. Changes in the structure of the clay layer upon changing the pH, as discussed above for changing electrolyte concentrations (31), can also explain these results.

There has only been one study of CME in nonaqueous solvents [EtOH, MeOH, MeCN, *N,N*-dimethylformamide (DMF) all containing $0.1M$ $NaClO_4$ (23). The reactant, $[Fe(bpy)_3]^{2+}$, was rapidly leached from the clay layer by MeCN and DMF. The CV response was suppressed in MeOH and EtOH compared with an aqueous solution, but the response was restored when the electrode was transferred back to aqueous $0.1M$ $NaClO_4$. This suppression can be understood by the changes in basal spacing in the alcoholic media. Pillared clays showed a much smaller attenuation of the CV response, because the basal spacing was maintained by the clay pillars in MeOH and EtOH.

6.2.8 Chiral Effects

Although clays are not optically active, there have been reports over the years about chiral effects on clays (33). In connection with CME, the adsorption of enantiomeric versus racemic metal chelates have been of special interest. For smectites, like montmorillonite and hectorite, Yamagishi et al. (33–36) found that racemic mixtures of $[M(phen)_3]^{2+}$ (M = Fe, Ru, Ni) were adsorbed in larger amounts than either of the pure enantiomers (Λ and Δ). The racemate

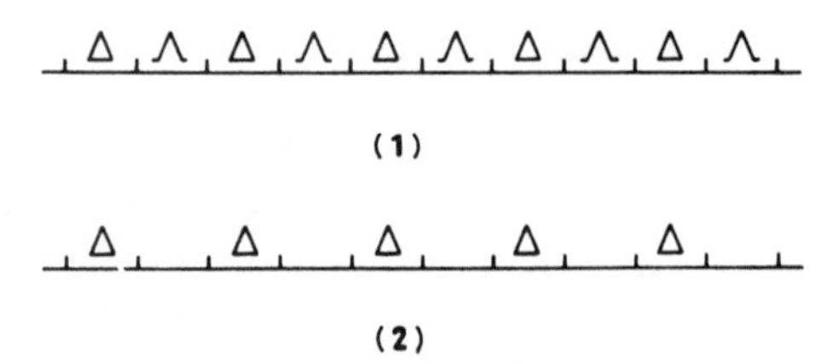

Figure 6.6. Schematic diagram suggesting packing arrangement of isomers of $[Ru(phen)_3]^{2+}$ on montmorillonite [Reprinted with permission from Ref. 11.]

adsorbed in amounts of twice the CEC, while the pure enantiomer adsorbed only up to the CEC. It was proposed that the racemic mixtures were adsorbed as cation pairs and packed more efficiently on the clay surface than the enantiomers, as suggested in Fig. 6.6. Thus for a clay surface that was saturated with the Δ isomer of $[Ru(phen)_3]^{2+}$, only the Λ isomer of the same or a closely related species could be accommodated. The oxidation of racemic $[Co(phen)_3]^{2+}$ at a CME prepared by coating SnO_2 with montmorillonite–PVA (9) that was presaturated with Δ-$[Ru(phen)_3]^{2+}$ produced Λ-$[Co(phen)_3]^{2+}$ in 7% enantiomeric excess (11). When a similarly prepared CME containing Λ-$[Ru(phen)_3]^{2+}$ was employed for the oxidation of various phenyl alkyl sulfides, enantiomeric excesses of the R isomers of the corresponding sulfoxides were produced, when the alkyl groups were bulky ones, such as cyclohexy (37). The oxidation did not apparently involve the intermediacy of Λ-$[Ru(phen)_3]^{3+}$, since the addition of the sulfide suppressed the CV wave for the Ru species, and the chirality of the reaction was ascribed to the preferential stabilization of the R isomer of the sulfoxide or an intermediate on the clay–ΛΔ $Ru(phen)_3]^{2+}$ surface.

There have been few studies that have followed up on these initial interesting observations of Yamagishi and Aramata (11, 37) concerning chiral electrode reactions on CME. In a study of $[Cr(bpy)_3]^{3+}$, larger cathodic peak heights were found at a montmorillonite-modified electrode for the racemic mixture than for the enantiomeric solutions, once the CEC had been exceeded (21). The UV–vis absorption spectra of $[RuL_3]^{2+}$ (L = bpy or phen) adsorbed on a montmorillonite suspension at low loading showed clear differences in the spectra of the enantiomers and the racemic mixture (38). These results suggest that interactions between the adsorbed optical antipodes, even at low loading, may be important in understanding the chiral effects observed. Moreover, recent experiments in our laboratory (39) show that it is important to distinguish between the total relative amounts of enantiomer and racemate adsorbed and the amounts that are electroactive. These experiments also suggest that the aggregation of the clay is affected differently by adsorption of the different forms. Overall, the results obtained so far strongly suggest the existence of chiral effects on CME. A complete understanding of the origin of the effects and details of the structures and distribution of the adsorbed species will require further investigation.

6.2.9 Effect of Lattice Iron

Considerable amounts of iron are often present in clay preparations, either as a separate impurity (e.g., Fe_2O_3) that can be removed by a suitable treatment, or as an isomorphous replacement for Si or Al in the clay sheets. There is evidence that this lattice iron can participate in redox reactions in CME. For example, reduction of tetracyanoethylene (TCNE) by lattice iron to produce the anion radical has been reported (40). This lattice iron does not exhibit direct CV electroactivity, because it is widely dispersed within the electronically insulating clay matrix and cannot move to the electrode surface (15, 22). However, when electroactive ions are incorporated into the clay film, they can interact with lattice iron and their CV behavior can be modified. Thus it is frequently noticed that the first CV oxidation wave for $[Ru(bpy)_3]^{2+}$ incorporated into SWy-1 is much larger than the cathodic wave on reversal and larger than the anodic waves on subsequent scans (13). This observation was ascribed to reaction of the electrogenerated $[Ru(bpy)_3]^{3+}$ with a lattice component of the clay, later identified to be Fe(II) (22). The effect was shown to be more pronounced with clays that contain larger amounts of lattice iron, and is especially important in nontronite, a clay with iron extensively substituted into the lattice (22). The $[Ru(NH_3)_6]^{2+/3+}$ couple, which shows a rather large D in montmorillonite films, can also mediate lattice iron redox processes very effectively (15, 22). The lattice iron was also shown to act as a catalyst in the electroreduction of H_2O_2 on a CME (15).

6.2.10 Model for Charge Transfer in Clay Films

One can propose at least five different kinds of sites or species incorporated into clay films (Fig. 6.7). The relative amounts of these different sites and their population by a given electroactive species probably depends on the mode of preparation of the film and the nature (charge, conformation and counterions) of the species. For example, a species like HQ probably does not interact strongly with the clay layer and is mainly located in pores and channels (4-sites). A strongly interacting cation like $[Ru(bpy)_3]^{2+}$ can be located at most of the sites. There is still some disagreement about which sites contribute to the observed electrochemical response. Most of the studies indicate that deeply intercalated species (1-sites) are electroinactive (19, 20, 22). Some mobile species in pores or channels (4-sites) are probably needed to carry charge from the substrate to species within the film. Based on the electroinactivity of layers that were pretreated with electroactive species before the film was cast, King et al. (19) proposed that only cations bound as ion pairs (in excess of the CEC) to the surfaces of the clay particles that border microchannels (5-sites) are electroactive. However, the observed chiral effects at CME, as discussed above, the observed strong interactions of some of the electroactive species with the clay films, and a rough correlation of the amount of electroactive species with the external surface area in the clay film, have suggested the possibility that surface adsorbed (2-sites), and perhaps edge adsorbed (3-sites) species are also electroactive (21, 22). Mediation and lattice iron redox processes seem to

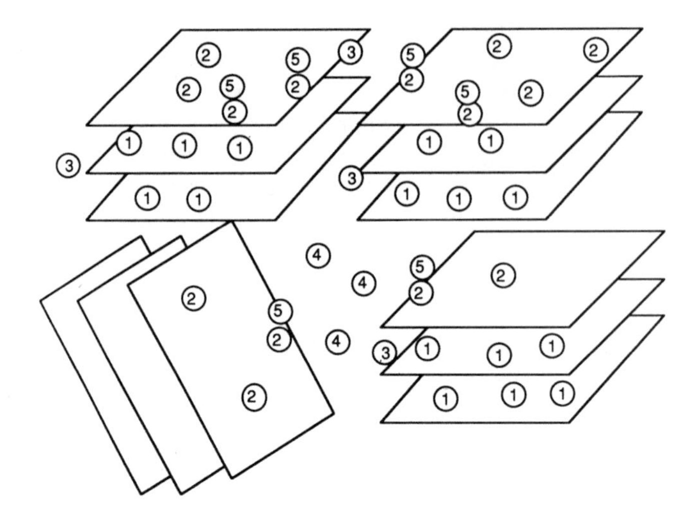

Figure 6.7. Schematic representation of different sites for an incorporated compound in a clay film on an electrode surface.

support some electroactivity of adsorbed species. Further studies and more quantitative experiments are needed, but these are difficult because of problems in preparing precisely reproducible clay films on electrodes.

6.3 APPLICATIONS OF CLAY-MODIFIED ELECTRODES

6.3.1 Catalytic Reactions

Several studies have employed CME to catalyze electrode reactions that occur sluggishly on the substrate material. The reduction of hydrogen peroxide was shown to occur more readily at a montmorillonite-coated electrode in the presence of $[Ru(NH_3)_6^{3+}$ than at the bare pyrolytic graphite substrate (15). Lattice iron, reduced to the +2 state by $[Ru(NH_3)_6]^{2+}$, was proposed as the catalytic center. Bis(bpy)(4,4′-dicarboxy-bpy) Ru(II) incorporated into a sodium hectorite–PVA film containing RuO_2 on a tin oxide substrate catalyzes the oxidation of water to oxygen (12). The uncharged Ru species is not strongly held in the film and shows a much larger diffusion coefficient that the analogous, but strongly interacting, $[Ru(bpy)_3]^{2+}$ (Table 6.1). It presumably serves the role of a mediator, just as the $[Ru(NH_3)]^{3+/2+}$ couple in the previous example, delivering charge from the substrate to the catalytic RuO_2 sites. Finally, the $PVS^{0/-1}$ couple in a montmorillonite–PVA–Pt film serves as a mediator for the reduction of protons to hydrogen (13).

6.3.2 Pillared and Synthetic Clays

Two papers have dealt with CME prepared by pillaring the clay layers cast on the substrate surface by treating with the appropriate pillaring agent, such as $[Al(OH)_2Cl]_x$ or tetraethoxysilane, followed by heating (22, 23). Pillaring was shown, by XRD, to fix the basal spacing in the range of 14.6–18.0 Å, depending on the pillaring agent and the method of treatment, and decrease the penetration of $[Fe(CN)_6]^{3-}$ through the film to the substrate. The exchange capacity of the pillared films for cations was also lower than that of the unpillared clay. Pillared clay layers are of interest because they maintain their interlayer spacings, even under swelling conditions, and may permit size selectivity (7). They may thus allow the use of CME in some nonaqueous solvents (23).

Several papers have dealt with the use of synthetic hydrotalcite-like clay (HT) (or layered double hydroxide, LDH) to prepare electrodes (41–43). These materials are of special interest because the brucite-like layers are positively charged so that they behave as anion-exchange materials. Thus species such as $[Mo(CN)_8]^{4-}$, $[IrCl_6]^{2-}$, and $[Fe(CN)_6]^{3-}$ could be incorporated into about 100-nm thick films of a HT [with layers of $[Mg_6Al_2(OH)_{16}]^{2+}$] and showed stable, diffusion-controlled CV waves with the electrodes immersed in $0.1M$ Na_2SO_4 solution (41). The observed peak potentials were near those for the same species in solution, and the estimated diffusion coefficient for $[Mo(CN)_8]^{4-}$ was $8 \times 10^{-12}\,cm^2\,s^{-1}$. Composite electrodes of the HT with $Zn_x^{2+}Al_y^{3+}(OH)_{2x+3y-z}^{z+}$ layers mixed with conductive carbon black particles in a poly(styrene) matrix were used to oxidize catechol in aqueous solutions (42). The charge-transfer kinetics at this HT composite were improved compared to that seen at a GC electrode under the same conditions. A film of the same Zn–Al HT in a poly(styrene) matrix (in the absence of added carbon) could also be cast on an electrode surface (43). As opposed to the diffusion controlled behavior observed with the Mg–Al HT, described above, $[Fe(CN)_6]^{3-}$ incorporated into this film showed thin-film or surface-confined behavior that was ascribed to the formation of a Prussian Blue-like layer on the surface of the LDH particles.

6.3.3 Composite Electrodes With Clay Films

Several reports have dealt with the application of the clay films as matrices for growing structures of other materials, such as electronically conducting polymers and semiconductors to form composite layers. For example, the introduction of electronically conductive materials into the ionically conductive clay matrix produces biconductive phases, analogous to those formed in earlier studies of the ionically conducting polymer, Nafion (44). Thus crystals of conductive $TTFBr_{0.7}$ can be formed inside STx-1 films by incorporation of

TTF^+ followed by potential cycling in 1.0*M* NaBr. These crystals were reported to promote charge transfer at a CME from the substrate to dissolved $[Fe(CN)_6]^{3-}$. Similarly, a biconductive layer could be produced by oxidizing pyrrole in an acetonitrile solution to form the electronically conductive polymer, poly(pyrrole) (PP), in a hectorite layer on a GC substrate (45). The dissolved species $[Fe(CN)_6]^{3-}$ and $[Ru(NH_3)_6]^{2+}$ could be electrolyzed at the GC–hectorite–PP electrode, although the films formed in MeCN were quite brittle. Improved films were formed from STx-1 with PP formed in aqueous solutions (18). In this case the incorporation of the polymer in the clay film enhanced the mechanical stability of the film and allowed free-standing films, separated from the electrode substrate (GC and Pt), to be prepared. The presence of MV^{2+} during the polymerization process affected the ultimate structure of the clay film. Rudzinski et al. (46) also produced clay–PP films in MeCN and aqueous media and studied their structures by scanning electron microscopy (SEM). The redox potential for the $Fe^{3+/2+}$ couple incorporated in the composite film was about 300 mV more negative than that of the couple in the clay layer before deposition of PP.

Polyaniline can be formed in a similar manner inside a montmorillonite film by the electrochemical oxidation of aniline (47). The authors noted that the basal spacing, which increased from 1.26 to 1.56 nm upon incorporation of aniline, decreased upon formation of poly(aniline) to 1.30 nm.

A composite electrode consisting of a synthetic HT and carbon in a poly(styrene) matrix was described in the preceding paragraph. Clay–carbon composites can also be fabricated by sintering a mixture of clay (mainly kaolinite) and porous carbon black in an 85:15 mixture at 800°C for 1 h (48). This material showed a very high double-layer capacitance ($\sim 10\,\mathscr{F}/cm^3$ in aq. solutions); faradaic reactions at these electrodes were not investigated. Semiconductors, such as TiO_2 and CdS, have also been incorporated into clays, but such materials were not used as electrodes (49–51).

6.3.4 Chemically Modified Electrodes Derived from Other Layered Solids

Clays and layered double hydroxides are only two of a much larger class of crystalline layered compounds that can incorporate electrochemically interesting species. Extensive studies of intercalation (52) have been carried out with van der Waals solids such as graphite, layered boron nitride, transition metal disulfides, metal phosphorus trichalcogenides, oxides, and oxyhalides. Numerous electrochemical experiments involving these materials have been performed. The impetus for this work is the considerable technological potential of insertion compounds as electrodes for high-energy density secondary batteries and for electrochromic display applications.

To date few of these materials have been examined as thin films adsorbed or chemically attached to electrode surfaces. Recently, we and others have developed techniques for making thin-film metal phosphate- and phosphonate-modified electrodes. The phosphonates are an interesting class of layered solids,

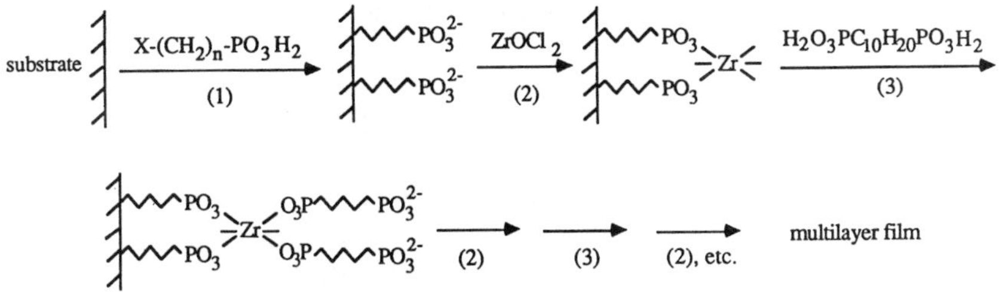

Figure 6.8. Sequential adsorption scheme for growing multilayer zirconium phosphonate films on surfaces. "X" represents a functional group that binds irreversibly to the surface, for example, a silanol for oxide surfaces or a thiol for gold surfaces [Reprinted with permission from Lee, H.; Kepley, L. J.; Hong, H.-G.; Mallouk, J. E. *J. Am. Chem. Soc.* **1988**, *110*, 618; Also Scheme I of H. Lee, L. J. Kepley, H.-G. Hong, S. Aphter, and T. E. Mallouk, *J. Phys. Chem.* **1988**, *92*, 2597. Copyright © (1988) American Chemical Society.]

which includes $Zr(HPO_4)_2H_2O$, $VO(HPO_4)0.5H_2O$, and $UO_2(HPO_4)$, with demonstrated potential for redox catalysis and molecular recognition (53). Like clays, some of these compounds can be pillared by organic groups or large cations to yield solids with molecular sieving properties (54–57).

The solid $Zr(HPO_4)_2H_2O$ can be exfoliated with primary amines such as n-$C_4H_9NH_2$ to make colloidal dispersions. This colloid can then be spin-coated onto surfaces and dried, in a manner analogous to that described above for the preparation of clay-modified electrodes. Alternatively, the solid can be suspended in a solution containing an inert binder such as poly(styrene) and then applied by dip or spin coating.

A method for depositing precisely controlled thicknesses of tetravalent metal phosphates or phosphonates involves the alternate immersion of a suitably pretreated electrode in aqueous solutions containing the appropriate metal salt and a soluble phosphate or phosphonate. In this way the solid is built up layer by layer on the electrode as shown in Fig. 6.8 (58, 59). Here a pillaring organic group is used, which leaves no space in the film for incorporation of electroactive ions or molecules. Fortunately, the technique also works with a mixture of pillaring phosphonate and inorganic phosphate, or simply pure inorganic phosphate in step 3; the result is a microporous structure containing acidic —OH groups in place of some or all of the organic pillars. The mixed films show molecular sieving properties, admitting electroactive cations smaller than the pillar height (~ 10 Å) (60), whereas the completely inorganic phosphate films admit cations of any size.

Figure 6.9 shows the cyclic voltammetry of a 10-layer $Zr(HPO_4)_2nH_2O$ film on a conductive SnO_2 electrode. The film is impervious to anions and there is no discernable cathodic or anodic current in a $[Fe(CN)_6]^{4-}$ solution. Cations such as $[Os(bpy)_3]^{2+}$ can be exchanged into the film and their presence is clearly seen by cyclic voltammetry in a solution containing only electrolyte. When the ferro–ferricyanide couple is examined at the $[Os(bpy)_3]^{2+}$ exchanged electrode, facile oxidation of $[Fe(CN)_6]^{4-}$ to $[Fe(CN)_6]^{3-}$ occurs near the $[Os(bpy)]^{3+/2+}$

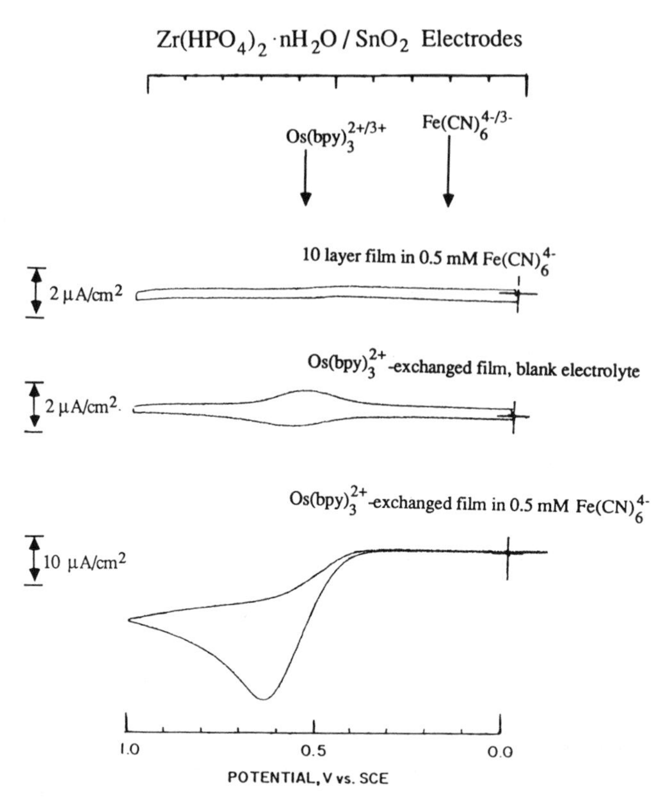

Figure 6.9. Cyclic voltammetry of a 10-layer $Zr(HPO_4)_2nH_2O$ film (top) in 0.5m*M* $[Fe(CN)_6]^{4-}$; (middle) in a blank electrolyte solution after exchange with $[Os(bpy)_3]^{2+}$ ions; (bottom) same electrode as in the middle part, in a 0.5 m*M* $[Fe(CN)_6]^{4-}$ solution (Reprinted with permission from Hong, H.-G.; Rong, D.; Mallouk, T. E., Extended Abstracts, 175th Electrochemical Society Meeting, Los Angeles, May 7–12, 1989, and Hong, H.-G.; Rong, D.; Mallouk, T. E. *Coord. Chem. Rev.*, **1990**, *97*, 237.]

potential, but reduction of $[Fe(CN)_6]^{3-}$ is not observed. The current-rectifying capability of this electrode is ascribed to mediated oxidation of $[Fe(CN)_6]^{4-}$ by $[Os(bpy)_3]^{3+}$ contained within the film. The reverse process does not occur to any measurable extent because it involves thermodynamically uphill transfer of electrons from $[Os(bpy)_3]^{2+}$ to $[Fe(CN)_6]^{3-}$.

Ingersoll and Faulkner (61) used a similar alternate immersion scheme in order to prepare films containing multiply charged anions such as isopolymolybdate, silicotungstate ($SiW_{12}O_{40}^{4-}$), and phosphotungstate ($PW_{12}O_{40}^{3-}$). The principle is the same as described above, namely, the layer by layer deposition of a highly insoluble salt via sequential adsorption of its soluble components. In this case the multiply charged anion is first adsorbed as a monolayer or submonolayer onto a glassy carbon electrode, providing an anchoring point for subsequent film growth. The electrode is dipped in an aqueous solution containing a long-chain alkylammonium salt, a dicationic transition metal complex, or a soluble cationic polymer, which binds irreversibly to the surface-confined polyanion. Repetitive alternate immersion in polyanion and cation solutions gives a multilayered film. Cyclic voltammetry, chronocoulometry, and Auger spectroscopy establish that the amount of electroactive material increases linearly with the number of immersion cycles, each complete cycle depositing approximately the same amount of polyanion that was present in the originally adsorbed monolayer.

The mechanism for film formation and film morphology are thought to depend on the nature of the cations used. With monovalent cations such as cetylpyridinium, cetyltrimethylammonium, and dodecyltrimethylammonium the film most likely resembles a Langmuir–Blodgett or bilayer membrane assembly, with polar head group–polyanion regions alternating with nonpolar tail groups. Films prepared from cationic polymers [protonated poly(vinylpyridine) or Cu^{2+}-poly(ethyleneimine)] probably are more randomly organized since anions from subsequent immersion cycles contact different cationic centers on the same polymer chain. The technique appears to be extremely versatile in that different anionic and cationic species may be incorporated in each immersion cycle, and indeed some species that do not readily form monolayer films, such as decavanadate and tungstate, may be readily immobilized by this method. This study and the metal phosphonate work described above demonstrate that very insoluble compounds, which are comprised of two individually soluble components, may be assembled layer by layer (i.e., with molecular-level control of film composition and structure) via sequential adsorption reactions.

6.4 Zeolites

Zeolites are crystalline aluminosilicates that contain openings and void spaces of approximately molecular dimensions. Like clays, zeolites occur naturally and some 40 different mineral zeolites have been identified. Beginning with R. M. Barrer's (8) pioneering work in the 1940s and 1950s, many of the mineral zeolite structures have been reproduced in laboratory syntheses, and many more new structures, for which there is no natural counterpart, have been discovered (62).

6.4.1 Composition, Structure, and Properties

The synthesis and structures of zeolites and related molecular sieves have been recently reviewed (63–65). The zeolite framework is made up of silicon and aluminum atoms tetrahedrally coordinated by oxygen. Each of the oxygen atoms is shared by two silicon or aluminum atoms, so the framework stoichiometry is MO_2. Because the AlO_2 unit has a formal -1 charge, each aluminum in the structure must be counterbalanced by a nonframework cation, typically a Group 1A (1) or IIA (2) ion, H^+, NH_4^+, or an organic cation. The Si:Al ratio of the zeolite (and hence the density of nonframework cations) is quite variable, ranging from one with zeolite A to essentially infinity with highly siliceous zeolites. The latter are zeolitic allotropes of SiO_2 produced by chemical treatment (usually with $SiCl_4$) of lower Si:Al materials.

In addition to the aluminosilicate zeolites, several new families of synthetic molecular sieves have been reported in the last 10 years (66). These include the aluminophosphate (or AlPO) family, with a framework stoichiometry $AlPO_4$, the silicoaluminophosphate (SAPO) family, and metal-containing derivatives of these (MAPOs and MAPSOs), which are doped with Li, Be, Mg, or transition elements. Many of the aluminosilicate zeolite structures are found among these newer molecular sieves; a host of other structures, some of which had been hypothesized but never prepared as aluminosilicates, have now been observed in the AlPO and SAPO families. A striking example is VPl-5, an AlPO that contains rings of 18 tetrahedrally coordinated atoms and has 13-Å diameter openings. The structure was predicted in 1984 (67) and the compound first successfully synthesized in 1988 (68).

The tetrahedrally coordinated (T) atoms in all these materials are joined by oxygen atoms into $(T—O)_n$ rings. Because of the considerable flexibility of the T—O—T bond angle, rings containing 4, 5, 6, 8, 10, and 12T atoms are commonly found. The rings join to form prisms and more complex cages, and the cages share faces to give the various frameworks shown in Fig. 6.10. In this representation each vertex denotes a T atom, and a line joining vertices indicates a T—O—T bond. Among the multitude of structures formed in this way, examples can be found of three-, two-, and one-dimensionally connected large void spaces, or cages, enclosed by the framework. For example, in the mineral faujasite and synthetic zeolites X and Y, a "supercage" of 13-Å diameter is connected via 12 rings of 7–8-Å diameter to four other supercages in a tetrahedral arrangement. In zeolite A, a cage of similar dimensions is connected to six others, again giving a three-dimensional network; here, however, the connection is through 8 rings of 3.0–4.5-Å diameter, which allow only small molecules like H_2O, O_2, CH_4, and cyclopropane to pass between cages. Zeolite L has large cages connected by 12 rings, like faujasite, X, and Y and therefore has similar molecular sieving properties; however, in the L structure each cage has only two neighbors, so the channel network consists of infinite one-dimensional tunnels. ZSM-5, a high-silica zeolite used as a catalyst for xylene isomerization and methanol-to-gasoline conversion, has 10-ring tunnels that interconnect, forming a pleated two-dimensional network.

Zeolites are synthesized hydrothermally, both in nature and in the laboratory, and the resulting structure contains water and nonframework cations in the cages. Filling of available cage volume by water accounts for as much as 50% of the weight of an apparently "dry" zeolite. Typical compositions of zeolites A and Y, for example, are $Na_{12}Al_{12}Si_{12}O_{48} \cdot 27H_2O$ and $Na_{56}Al_{56}Si_{136}O_{384} \cdot 250H_2O$. Both the water and extra framework cations are highly mobile and largely exchangeable. The water is easily removed by heating, which is the origin of the work zeolite (69) (from the Greek "zeo," boil, and "lithos," stone).

Because they are made hydrothermally, typically by crystallization of

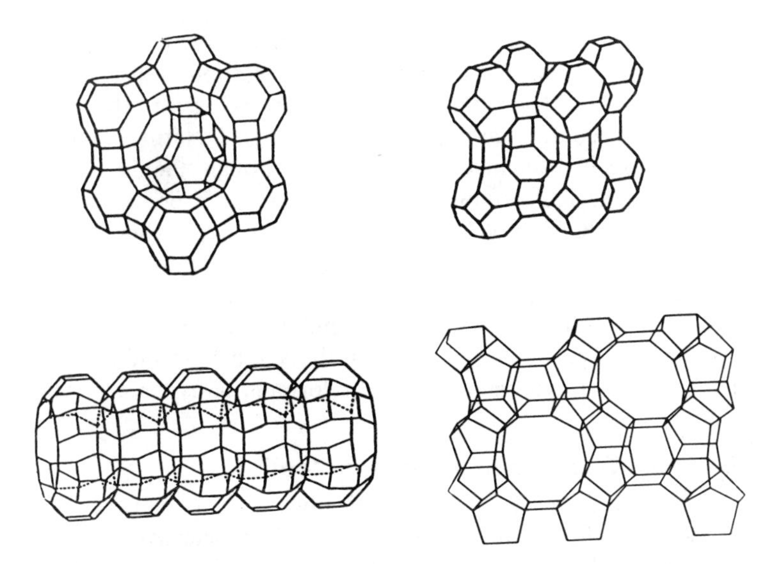

Figure 6.10. Zeolite framework structures. Clockwise from upper left: Faujasite (Zeolites X and Y), Zeolite A, ZSM-5, Zeolite L (adapted from Refs. 63 and 75).

amorphous gels, synthetic zeolites are rarely available as macroscopic single crystals. Natural crystals are sometimes of sufficiently large size that conventional single-crystal X-ray diffraction techniques are useful in solving their structure. However, for synthetic materials, single-crystal diffraction is usually impossible. The problem is particularly aggravated by the fact that zeolite syntheses often do not yield pure phases, and even the pure phases may crystallize with large unit cells in low-symmetry space groups. Data from other techniques such as X-ray and neutron powder diffraction, microcrystal diffraction with synchrotron sources, electron microscopy and diffraction, molecular sorption selectivity, and model building may be brought together to determine a single structure. This approach was recently applied successfully to the highly challenging zeolite beta structure, which consists of coherently intergrown chiral and achiral regions, both in low-symmetry space groups; it was solved by a combination of diffraction, microscopy, and model building techniques (70).

The intensive interest in zeolites on the part of both industrial and academic laboratories (some 35,000 *Chemical Abstracts* citations to date) stems largely from their utility as sorbents, ion exchangers, and catalysts (71–75). Zeolites have two special properties that make them exceptionally useful as sorbents and molecular sieves: they have tremendously high internal surface area (typically several hundred $m^2 g^{-1}$), and access to this internal space is restricted to molecules small enough to pass through their apertures, which are 3–8 Å in diameter, depending on structure type. The first commercial application of zeolites relied on this molecular sieving property, and followed R. M. Barrer's observation that O_2 and N_2 could be separated on the basis of their molecular dimensions using a zeolite of appropriate aperture size. Natural and synthetic zeolites contain freely exchangeable cations, although the highly siliceous zeolites and synthetic AlPO molecular sieves have little or no ion-exchange capacity. In the SAPO family Si is substituted for framework P and these materials are reported to act as anion exchangers (76). Before the invention of polymeric organic ion exchangers, zeolites were used commercially as water softeners. Synthetic zeolites are today of tremendous practical utility as catalysts and catalyst supports, particularly in acid-catalyzed cracking and isomerization, and in metal-catalyzed hydrogenation and reforming of hydrocarbons (77).

6.4.2 Electron-Transfer Reactions in Zeolites

Since many of the materials used in chemically modified electrodes have been chosen on the basis of their ability to adsorb analytes selectively, to exchange electroactive ions, and to function as electrocatalysts, it follows that zeolites are interesting materials with which to modify electrodes. It is surprising that such studies have begun only quite recently, especially in light of a considerable body of literature relating to the immobilization of electroactive species in zeolites and on their outer surfaces, and the use of zeolites as components of electrochemical cells. These applications have been reviewed recently by Ozin et al. (78). It is interesting to note that membranes for ion-selective electrodes incorporating zeolites (79–82) were reported as early as 1939, and that batteries in which ion-exchanged zeolites served as cathodes, anodes, and electrolytes were patented in 1965 (83).

Numerous other reports of intrazeolitic electron-transfer chemistry have appeared since these early studies. For example, catalytically important metals and metal oxides are often incorporated into zeolites (84) by ion exchange of a cationic metal complex, followed by redox reactions involving gases such as hydrogen, oxygen, or chlorine. Platinum-loaded zeolite Y, a useful hydrocracking catalyst, is prepared from a $[Pt(NH_3)_4]^{2+}$ exchanged zeolite, which is converted to Pt^{2+}—Y and finally to Pt^0—Y using oxygen and then hydrogen at 300–600°C (85). The dispersion of catalytically active Pt in zeolite L is also controlled using alternating streams of oxidizing (HCl/O_2 or Cl_2) and reducing (H_2) gases (86).

Zeolites are attractive hosts for light-induced electron-transfer reactions, because the reaction may be initiated at a photosensitizer, which is immobilized within the cage structure by ion exchange or a "ship-in-a-bottle" synthesis. Silver-, titanium-, and europium-doped zeolites have been studied as photocatalysts for water splitting. Jacobs et al. (87) detected oxygen upon photolysis of Ag(I)-exchanged zeolite Y, and Kuznicki and Eyring (88) reported hydrogen evolution from Ti(III)-exchanged zeolite A. In neither case was the process truly catalytic, and a second thermal step was required in order to regenerate the photactive zeolite. Arakawa et al. (89) studied the luminescence of Eu(III)-

exchanged zeolite Y, and observed bands assigned to Eu(II) after thermal treatment of the exchanged zeolite. Suib et al. (90) studied europium-exchanged zeolites by extended X-ray absorption fine structure (EXAFS), EPR, and Mössbauer spectroscopy. By combining these methods with time-resolved luminescence spectroscopy, they were able to determine the hydration number and coordination environment of Eu(III) in a number of different zeolite structures. Interestingly, they found that exposure of zeolite A to $Eu(OH)_2(H_2O)_x$ gives an exclusively Eu(III)-exchanged material via redox reactions like that observed for $Eu(OH)_2(H_2O)$ itself.

$$2Eu(OH)_2(H_2O) \rightarrow 2Eu(OH)_3 + H_2$$

Exposure to hydrogen or thermal treatment was shown to regenerate Eu(II)-zeolites A, X, and Y.

The $[UO_2]^{2+}$ exchanged zeolites have been found to be selective photocatalysts for oxidation of isopropanol to acetone and ethanol to acetaldehyde (91). The process involves binding of the alcohol to UO_2^{2+} and light-induced hydrogen abstraction. U(V) species are presumably generated in this process; they are reoxidized to UO_2^{2+} by traces of oxygen in the zeolite. Selective conversion of the alcohols was observed only if the zeolites had pores of intermediate or large size, UO_2^{2+}—ZSM-5 being the most effective photocatalyst. Zeolites that lost their crystallinity during the ion-exchange reaction were not active.

An interesting "ship-in-a-bottle" synthesis of $[Ru(bpy)_3]^{2+}$ in faujasitic zeolites was reported by Lunsford and his co-workers (92). The synthesis involved ion exchange of zeolite Y with the smaller $[Ru(NH_3)_6]^{3+}$ ion; the latter is small enough to enter the structure through its about 8-Å apertures. Thermal decomposition of this complex in the presence of excess bpy yields $[Ru(bpy)_3]^{2+}$ trapped within the supercages. This complex is approximately 12 Å in diameter and is therefore incapable of exiting through the 8-Å 12 rings. Faulkner et al. (93) showed that the luminescence of $[Ru(bpy)_3]^{2+}$ so entrained in zeolite X could be quenched by electron donors such as *N*,*N*,*N'*,*N'*-tetramethyl-1,4-phenylenediamine (TMPD). Size-excluded electron donors like 10-phenyl phenothiazene (10-PP) could not access the zeolite-encapsulated complex and did not quench its luminescence

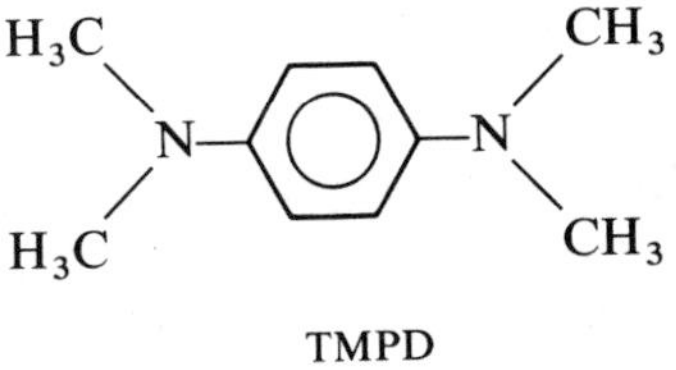

TMPD

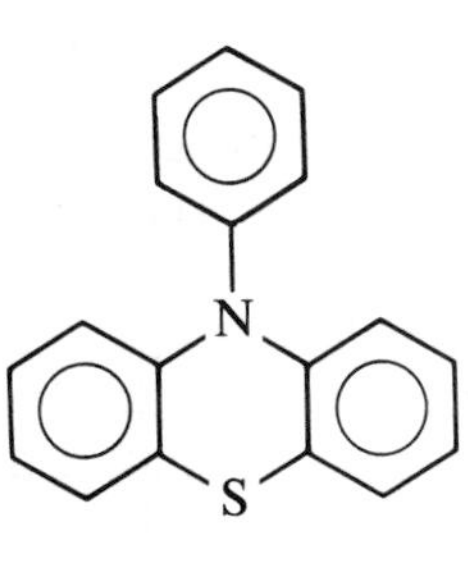

10-PP

efficiently. With TMPD, the product of the light-induced electron-transfer reaction, $TMPD^{+\cdot}$, could be detected spectroscopically. This result indicates an unusual stabilization of the energetic $\{[Ru(bpy)_3]^{3+}$ --- $TMPD^{+}\}$ product pair in zeolite X.

6.4.3 Zeolite-Modified Electrodes

The studies of intrazeolite photochemistry and redox chemistry described above predated the first reports, by Rolison and her co-workers (94), and by de Vismes et al. (95), of zeolite-containing electrode coatings. Interfacing a zeolite-based redox reaction with an electrochemical cell clearly provides an experimental handle not available in the earlier studies. Additionally, it opens the interesting possibilities of using zeolites or chemically modified zeolites in electrosynthesis and electrochemical analysis. Before these applications can be realized, several problems must be addressed. First, since zeolites and related molecular sieves are electronically insulating, ways must be found to contact them electrically. A related problem is the physical or chemical binding of the zeolite, which is ordinarily available as micron-size or larger particles, to the electrode surface. It must be determined if the locus of electron-transfer reactions is the surface of the conductive electrode or contacting material, and to what extent electron exchange occurs within the zeolite as well. Finally, mass-transport of intrazeolitic redox couples, reactants, and analytes, as well as the counterions that accompany redox reactions, must be understood.

6.4.3A Electrode Fabrication

The simplest method of preparing a zeolite-modified electrode is to use a polymeric binder to fix the solid to the electrode surface. Following techniques developed earlier for clay-modified electrodes, several groups used poly(styrene) as an electrochemically inert binder (95–99). A slurry of zeolite Y powder is made in a tetrahydrofuran (THF) or CH_2Cl_2 solution of poly(styrene), and a few drops are applied to the electrode surface. The solvent dries to leave a zeolite–polymer composite on the electrode. Scanning electron microscopy shows that on Pt electrodes, poly(styrene) adheres to the electrode surface, and aggregates of zeolite particles contact a relatively small fraction of the surface (97). With conductive SnO_2 electrodes, the polymer does not appear to wet the surface; during the drying step, most of the polymer is drawn to the outer surface of the film by capillarity and a relatively compact layer of zeolite particles is left on the surface of the electrode (98). Because this technique leaves a thick layer of insulating zeolite on the electrode, on the order of 100 μm, not all of it is accessible in short-timescale electrochemical experiments, for example, cyclic voltammetry.

An alternative to this approach is to use a redox active or electronically conducting polymer as a binder. Murray et al. (94) deposited zeolite A-containing films at a Pt rotating disk electrode by electrolyzing reducible (and possibly electropolymerizable) suspensions containing 1,4-dinitrobenzene. It was noted that rotational stirring was an important factor in promoting adhesion of particles to the surface. Thick films were obtained with dinitrobenzene and other reducible molecules [TCNQ, $[Ru(bpy)_3]^{2+}$, and $Ru(bpy)_2Cl_2$] (where TCNQ = tetracyanoquinodimethane); interestingly, no film forms with these electroactive couples if zeolite fines are not present. That the molecular sieving properties of the zeolite in the coating are maintained was shown by experiments involving molecular oxygen. With coatings prepared from zeolite A, oxygen reduction waves were seen, as were waves characteristic of the reducible species used to generate the coating. Purging the solution with N_2 or Ar caused loss of electroactive O_2 from coatings made from 4 Å, but not from 3-Å sieves. It was concluded that the persistent electroactivity observed with 3-Å sieves arose from O_2 trapped within the zeolite cavities, since O_2 adsorbed on the zeolite external surface would be displaced by nitrogen.

There have been no reports of modified electrodes in which a zeolite is contacted with an electronically conducting polymer, although Bein et al. (100) demonstrated that pyrrole and aniline can be oxidatively polymerized within the zeolite pore structure, and Dutta et al. (101) prepared poly(acetylene) on transition metal-loaded zeolites. Composite electrode coatings containing conductive carbon particles have been studied by several groups. Typically, graphite powder is ground with zeolite powder and the dry mixture is either physically pressed into a wire grid (95, 102) or added to an inert binder and applied to the electrode surface. The binder can be either a mineral oil (as is commonly used to make carbon paste electrodes) or a mixture of styrene and divinylbenzene, which are heated to make a cross-linked polymer (42). The advantage of these techniques is greatly improved electrical contact, via the carbon particles, to the electronically insulating zeolite. de Vismes et al. (95) compared the performance of zeolite Y ion-exchanged with electroactive metallotetra(*N*-methyl-4-pyridyl)porphyrins (MTMPyP) applied to electrodes with and without added graphite powder. The graphite-containing electrodes showed much higher current densities and better reversibility by cyclic voltammetry; additionally, most of the porphyrin on the electrode surface ($\sim 10^{-8}$ mol cm^{-2}) was accessible at slow scan rates or in chronoamperometric experiments. The better reproducibility attainable with carbon paste electrodes and zeolite–graphite–polymer composites make these electrodes superior for analytical applications (42, 97).

Li et al. (103) showed that a layer of zeolite particles can be covalently attached to electrode surfaces using a reactive silane (*I*). The electrode

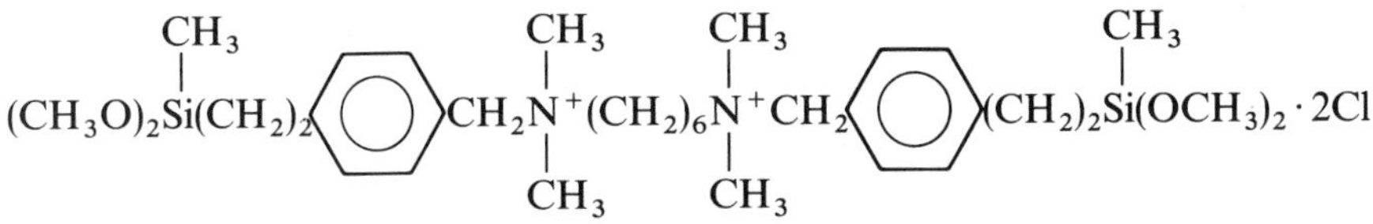

$$(CH_3O)_2\overset{\overset{\large CH_3}{|}}{Si}(CH_2)_2\text{-}C_6H_4\text{-}CH_2\underset{\underset{\large CH_3}{|}}{\overset{\overset{\large CH_3}{|}}{N^+}}(CH_2)_6\underset{\underset{\large CH_3}{|}}{\overset{\overset{\large CH_3}{|}}{N^+}}CH_2\text{-}C_6H_4\text{-}(CH_2)_2\overset{\overset{\large CH_3}{|}}{Si}(OCH_3)_2 \cdot 2Cl^-$$

I

(SnO_2 or Pt) is soaked in a solution containing structure **I**, which binds and polymerizes to give a coating a few monolayers thick. The derivatized electrode is then stirred with a suspension of zeolite Y and acetonitrile. Binding of the

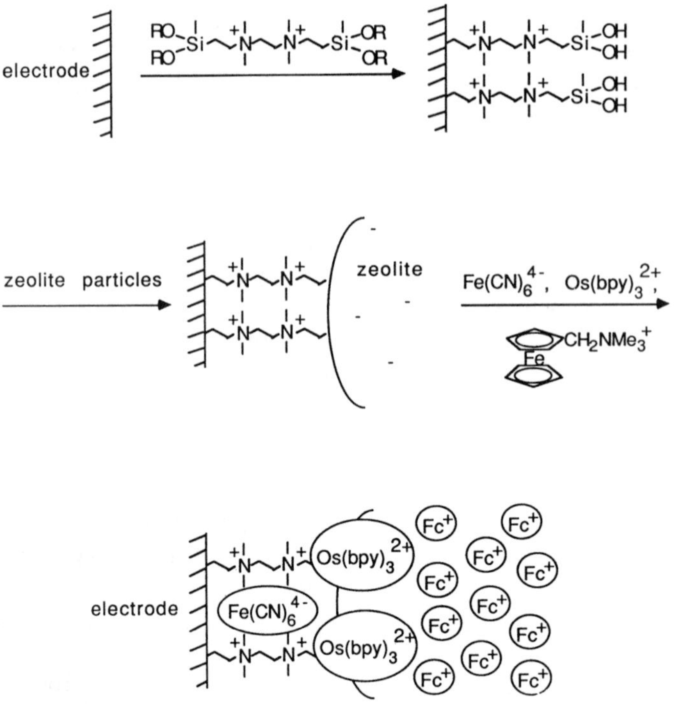

Figure 6.11. Covalent anchoring of zeolite Y particles to an electrode and self-assembly of a trimolecular redox chain. [Reprinted with permission from Li, Z.; Lai, C.; Mallouk, T. E. *Inorg. Chem.* **1989**, *28*, 178. Copyright © (1989) American Chemical Society.]

zeolite particles to the electrode is thought to involve reaction of residual silanol groups in the coating with those on the zeolite surface, forming Si—O—Si bonds. The binding process, which leaves a "monolayer" of zeolite particles on the electrode surface, is shown schematically in Fig. 6.11. Films prepared in this way are mechanically durable and can bind a variety of electroactive cations and anions. The cationic silane provides an anion exchange site that can bind $[Fe(CN)_6]^{4-}$; size-excluded cations like $[Os(bpy)_3]^{2+}$ exchange onto the external surface of the zeolite particle and smaller electroactive cations exchange into the internal pore structure. The modified electrode thus provides a template for self-assembly of a three-molecule electron-transport chain.

An alternative to the direct attachment of zeolite particles to electrode surfaces has been explored recently by Rolison et al. (104, 105). They sought to address a problem that is likely to be important in practical zeolite-based electrosyntheses, namely, the low surface area of planar chemically modified electrodes. A possible way around this problem is to deposit the "electrode" as small metallic clusters on a high surface area support, such as zeolite Y, and to effect electrochemical reactions by means of an externally applied electric field. Platinum clusters were prepared on zeolite Y at loadings of 1–10 wt.% by ion exchange and reduction of $[Pt(NH_3)_4]^{2+}$. This procedure is known to leave metallic Pt inside the supercages and on the external surface of zeolite Y. The intrazeolite metal clusters must have a diameter $\leqslant 13$ Å in order to fit inside the supercages. Those on the external surface may be larger, but were found to be smaller than the photoelectron escape depth (25–30 Å) in XPS experiments.

The zeolite-supported microelectrodes were compared to other types of microelectrodes (Al_2O_3-supported Pt and unsupported Pt microspheres) in a cylindrical cell containing two concentric feeder electrodes. Water or a benzene–water mixture containing the Pt microelectrodes was electrolyzed in this cell at applied fields of 0–300 V cm^{-1}. The current densities obtained with zeolite Y-supported Pt were one to two orders of magnitude higher than those found with the other microelectrodes. This result is especially interesting in light of the fact that only a small fraction of the platinum clusters are on the zeolite external surface, where they might be able to physically contact the feeder electrodes or clusters on the surface of other particles. Dispersion electrolyses involving ferrocene and size-excluded 1,2-dibenzylferrocene establish that only the externally sited Pt is electrochemically active. It appears therefore that Pt microelectrodes on the surface of zeolite Y are at least 1000 times as active as similar metal clusters supported on platinum.

The mechanism of the electrolysis in this system involves charging of the microelectrodes by contact with the feeder electrodes, and charge transfer via interparticle collisions. The electric field between the feeder electrodes is sufficient to induce bipolarity in the externally sited Pt particles, so that each particle can act as both an anode and cathode. It is postulated that the ionic intrazeolite environment screens internally sited Pt clusters from the applied field, so that they cannot act as bipolar electrodes. The high mobility of cations within the zeolite may contribute to the very high activity of the externally sited microelectrodes. In the absence of supporting electrolyte, the zeolite–Pt electrocatalyst acts as a mobile electrode–solid electrolyte package. For practical electrosyntheses, these zeolite-supported microelectrodes may allow reactions to be carried out at high current densities and with catalyst loadings one to two orders of magnitude lower than is possible with other catalyst supports.

6.4.3B Molecular Diffusion and Charge Transport

Zeolite particles that are contacted to electrodes are usually a few microns in diameter. Since direct electron exchange between molecules deep within these particles and the electrode surface is not possible over such long distances, the rates of molecular and charge-transport diffusion within the zeolite will be important in every application of zeolite-modified electrodes. Because of its importance in catalysis, molecular diffusion in dehydrated molecular sieves has been thoroughly studied, and excellent reviews are available (75, 106–107).

Molecules diffusing in zeolites are constantly in contact with the internal surface and the similarity of their dimensions to the pore dimension leads to "configurational" diffusion. In this size regime diffusivity is highly dependent on molecular shape and matching of molecule and pore sizes. For example, *n*-heptane and *n*-octane diffuse more slowly in dehydrated erionite, a natural zeolite with an 8 ring, three-dimensional channel system, than do shorter or longer *n*-alkanes. This anomaly is attributed to matching of the chain length to the size of the erionite cages (13 Å). Longer and shorter molecules do not fit so neatly into the cages and can diffuse more freely (108). Configurational diffusivity is also important in catalytic xylene isomerization using ZSM-5. Para-xylene can exit the structure much faster than *m*- or *o*-xylene; the latter two isomers are converted to *p*-xylene much faster than they can diffuse out of the structure (109).

To date shape-selective diffusion has not been exploited in zeolite-based electrochemical systems, although several interesting effects have been ascribed to size exclusion and are discussed later in this chapter. Electrochemical studies have provided information about charge-transport diffusion in zeolites exposed to aqueous and nonaqueous electrolyte solutions. The apparent diffusion coefficient, D_{app}, is sufficiently high under these conditions that molecules deep within a large-pore zeolite particle can be electrolyzed in cyclic voltammetry. For example, Fig. 6.12 shows the cyclic voltammogram of methylviologen (MV^{2+}) in 0.1*M* aqueous $LiClO_4$. In the upper trace one sees the two one-

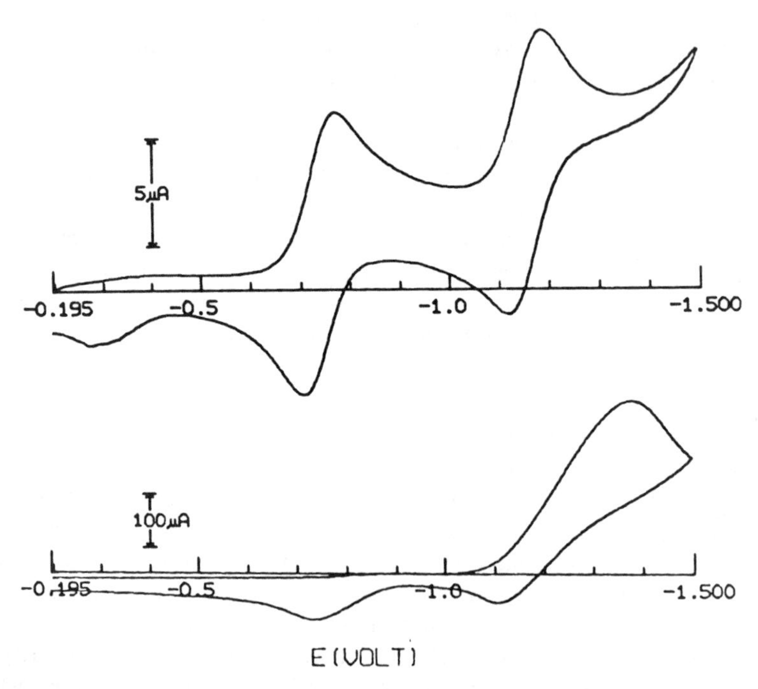

Figure 6.12. Upper trace: cyclic voltammetry of 1 m*M* $MVCl_2$ in 0.1*M* $LiClO_4$ in acetonitrile at aclean Pt electrode. Lower trace: NaY zeolite-coated electrode soaked 75 min in 1 m*M* $MVCl_2$, then transfer to 0.1*M* $LiClO_4$, second scan (from Ref. 96).

electron waves attributed to the $MV^{2+} + e^- = MV^+$ and $MV^+ + e^- = MV^0$ redox processes at a clean platinum electrode. The lower trace shows the behavior of a MV^{2+} exchanged zeolite Y-modified electrode under similar conditions. The electrochemistry is obviously altered by confining MV^{2+} to the zeolite microporous structure. The current at the modified electrode exceeds that at the bare electrode, because the zeolite concentrates the electroactive probe to a relatively high concentration C_z, and because the zeolite-bound probe exchanges electrons rapidly with the electrode. Under these conditions the current arising from electrolysis of zeolite-bound material exceeds that from the solution because $D_{app}^{1/2}C_z$ exceeds $D_s^{1/2}C_s$.

The diffusion coefficient D_{app} determined electrochemically is actually the sum of contributions from molecular diffusion and electron self-exchange between molecules. For zeolite Y ion-exchanged with small cations such as metallocenes and viologens, D_{app} is on the order of $10^{-10}\ cm^2\ s^{-1}$. This value (98) is comparable to those measured for a variety of polymer- and clay-modified electrodes and underscores the viability of incorporating zeolites into electrosynthetic or electrocatalytic systems.

The role of electrolyte cations in modulating zeolite electrochemistry is not fully understood. In order to maintain electroneutrality, each electron injected into the zeolite pore network must be accompanied by a cation. When this counterion motion is slow, or if the counterion is size-excluded, D_{app} will decrease. Several studies have been directed towards an understanding of this effect. Gemborys and Shaw (96) studied the effects of varying the cation size on the electrochemistry of the zeolite $Y-MV^{2+}$ system. When size-excluded cations such as tetrahexylammonium (THA^+) and tetrabutylammonium (TBA^+) were used (as bromide salts), essentially no electrochemistry was observed. Providing a small cation such as Li^+ allowed for charge compensation, and current densities with Li^+/THA^+ were surprisingly higher than with Li^+ alone. Interestingly, viologen ions were not leached from the films in the Li^+/THA^+ electrolyte, whereas loss of electroactive ions was fast with Li^+ alone. This suggests that THA^+ caps the zeolite pores, restricting motion of the larger MV^{2+} out of the zeolite while allowing smaller ions such as Li^+ to pass. Similar effects were observed by Li et al. (98, 99) with zeolite Y: binding of a large cation like $[Ru(bpy)_3]^{2+}$ or $ZnTMpyP^{4+}$ (where py = pyridine) to the outer surface seals up the zeolite against rapid exchange of metallocene or viologen ions contained within; however, rapid apparent diffusion was seen with these capped zeolites in aqueous solutions containing K^+, suggesting that the motion of these smaller cations is not restricted.

Two possible mechanisms exist for the reduction and oxidation of electroactive cations (E^{m+}) at a zeolite-modified electrode. The first involves direct electron transfer accompanied by motion of electrolyte cations (C^+). The second involves exchange of E^{m+} out of the zeolite and electron transfer at the electrode surface (97):

$$\text{Mechanism 1} \qquad E^{m+}(z) + ne^- + nC^+(s) = E^{(m-n)+}(z) + nC^+(z)$$

Mechanism 2
$$E^{m+}(\mathrm{z}) + mC^{+}(\mathrm{s}) = E^{m+}(\mathrm{s}) + mC^{+}(\mathrm{z})$$
$$E^{m+}(\mathrm{s}) + ne^{-} = E^{(m-n)+}(\mathrm{s})$$
$$E^{(m-n)+}(\mathrm{s}) + (m-n)C^{+}(\mathrm{z}) = E^{(m-n)+}(\mathrm{z}) + (m-n)C^{+}(\mathrm{s})$$

Experiments involving E^{m+} ions such as Cu^{2+}, Fe^{3+}, and Ag^{+} have been done with zeolite A, Y, and mordenite films in order to determine which mechanism most accurately describes the electrode process (97, 102, 110, 111). For the $Cu^{2+/+}$, $Fe^{3+/2+}$, and $Ag^{+/0}$ couples, the shape of the cyclic voltammetric waves and peak potentials are similar to those observed at unmodified electrodes with the same couples in solution, implying that Mechanism 2 is active. On the other hand, the $Cu^{+/0}$ potential is displaced about 140 mV, relative to its value at an unmodified electrode, with a mordenite–carbon paste electrode (111). Pereira-Ramos et al. (102) used electron microscopy to determine that electrochemical reduction of Ag^{+} in a mordenite–graphite film leaves Ag^{0} crystallites on the graphite as well as small Ag^{0} aggregates within the zeolite. They rationalize this behavior in terms of Mechanism 2, coupled with migration of Ag^{0} into the zeolite (see Fig. 6.13). The apparent diffusion coefficient for the $Ag^{+/0}$ couple in mordenite was found to be $4 \times 10^{-10}\,\mathrm{cm^2\,s^{-1}}$ by chronoamperometry.

6.4.4 Applications

6.4.4A Electrochemical Analysis

Zeolites, because of their microporous structure, can be used to select and preconcentrate cationic and neutral molecules for analysis. The availability of a wide variety of structure types, each with a characteristic channel size and framework charge density, should in principle allow one to choose a zeolite with maximal selectivity for a given analyte. Additionally, "ship-in-a-bottle" syntheses of analyte-specific receptors within zeolites should be possible. To date studies of zeolites as components of potentiometric and amperometric electroanalytical devices have involved predominantly simple cations, and analyte selectivity had been found to depend strongly on the choice of zeolite. Johansson et al. (112) exploited the known selectivity of mordenite for Cs^{+} (relative to other monovalent and divalent cations) to prepare cesium-specific membranes. Membranes were made by embedding synthetic mordenite crystals in an epoxy resin, and a Nernstian response was found over a concentration range of 3×10^{-5}–$0.1\,M$ Cs^{+}. Membranes prepared in a similar way from Cu^{2+} exchanged zeolite X have been shown to give a Nernstian response over five decades of Na^{+} concentration, presumably because of the high selectivity of zeolite X for monovalent ions such as sodium (113).

Wang and Martinez (114) determined silver ions in aqueous solutions by preconcentration at a zeolite A–carbon paste electrode. The electroactive Ag(I) bound to the electrode was detected by differential pulse voltammetry in a blank electrolyte ($NaNO_3$) solution. Zeolite A was chosen on the basis of its selectivity order for monovalent cations, $Ag > Tl > Na > K > NH_4 > Rb > Li > Cs$.

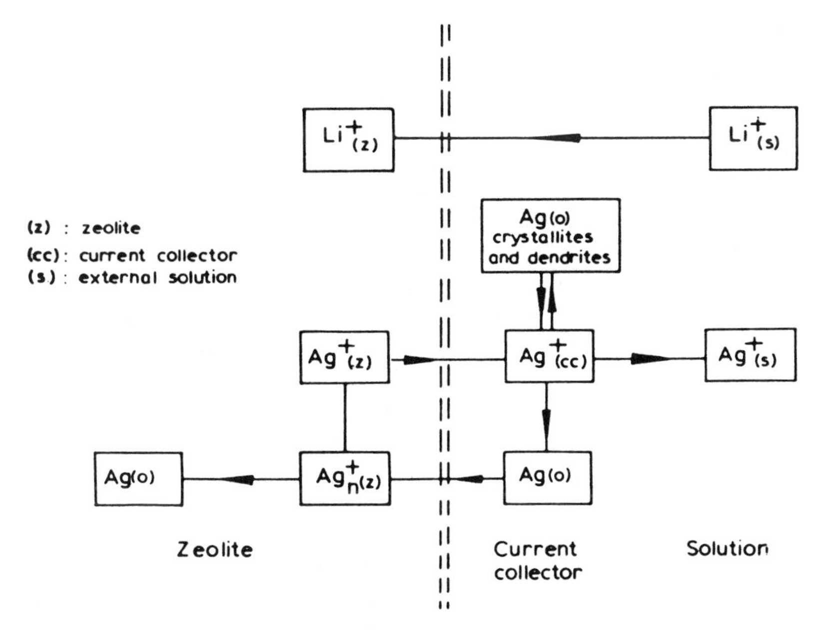

Figure 6.13. Proposed mechanism for Ag^+ electrochemical reduction at a mordenite–graphite-modified electrode (from Ref. 102).

Because of the stability of Ag-exchanged zeolite A, no interference was found with a 10-fold excess of weakly bound species like Co(II), Mn(II), Ca(II), Pb(II), Cu(II), and Ti(IV); however, more strongly bound cationic species such as Tl(I), Zn(II), Ni(II), Bi(III), and U(IV) suppressed the current attributed to the $Ag^{+/0}$ couple by competing for ion-exchange sites within the zeolite. Under favorable conditions analysis in the sub-milligram per liter range could be carried out. In these analyses the peak current depended strongly on preconcentration time; at high analyte concentrations, the current response was monotonically increasing even after relatively long times. For a given preconcentration time (1–10 min), analyses were reproducible with a standard deviation of 3.9%.

Shaw et al. (97) studied the voltammetric response of zeolite A- and Y-modified electrodes in solutions of multiply charged analytes such as viologens, Cu^{2+}, $[Ru(NH_3)_6]^{3+}$, and $[Fe(CN)_6]^{4-}$. In order to compare the response at a zeolite-modified electrode relative to that of an unmodified electrode, they defined an "enhancement factor" as the ratio of peak current density for a given analyte at the two electrodes under similar conditions. As expected, analytes that were excluded on the basis of size and/or charge (heptylviologen and ferricyanide) gave enhancement factors less than unity. Much higher enhancements (up to ~50) were found for cationic analytes such as MV^{2+} and $[Ru(NH_3)_6]^{3+}$, particularly when the latter were present in low concentration ($\leqslant 10^{-4} M$) in solution. Under these conditions, the zeolite preconcentrates the analyte in the film, and large cyclic voltammetric currents result from the high diffusion gradient from the electrode surface (e.g., $1M$ analyte) to bulk solution ($0M$ analyte). Reproducibility was generally very good with zeolite–carbon paste electrodes, compared to electrodes coated with polystyrene–zeolite suspensions. Composite electrodes prepared by polymerization of a styrene–divinylbenzene mixture containing both zeolite and carbon particles gave mechanically stable, polishable electrodes with excellent reproducibility (42).

Some of the limitations noted by Creasy and Shaw et al. (116) for analytical applications of zeolite-modified electrodes include the long equilibration time required (especially at low analyte concentration) and the difficulty of modeling or simulating the electrode process. In these heterogenous electrode coatings the analyte is concentrated to $1-2M$, many times the concentration of supporting electrolyte salts. Under these conditions the current response is governed not only by the affinity of the analyte for the zeolite and by D_{app}, but also by migration, edge, and thin-layer effects.

6.4.4B Electrocatalysis

Zeolites offer considerable potential as supporting media for both the reagents and catalysts involved in chemical reactions. Most of the reactions studied to date involve molecular oxygen, which can be concentrated effectively by zeolites A and Y. Devynck and his co-workers (115) used zeolite Y-supported $MnTMPyP^{4+}$ to effect the oxidation of 2,6-di-tertbutyl phenol.) The O_2-oxidized catalyst was regenerated either electrochemically at a Pt electrode or chemically with $NaBH_4$. The highest yield and selectivity for oxidation to the corresponding diphenoquinone was achieved electrochemically. In this experiment the $MnTMPyP^{4+}$ is bound onto the zeolite and can contact the electrode and reagents, but is inhibited from reactions with itself (e.g., formation of oxo-bridged dimers) because of its strong interaction with the negatively charged zeolite external surface.

Creasy and Shaw (116) studied the electrocatalytic reduction of oxygen at a zeolite-modified electrode in methylviologen solutions. Because heterogeneous films are difficult to model using conventional theory (117) for electrocatalysis at modified electrodes, they employed a simplex algorithm for optimization of experimental conditions. It was found that maximum catalytic currents were obtained under conditions in which the exchange of MV^{2+} for Na^+ in the zeolite was complete. Addition of zeolite A to the carbon paste–zeolite Y electrodes increased the current density, probably because of efficient oxygen preconcentration by zeolite A.

6.4.4C Multicomponent Systems

By using a combination of size excluded and smaller electroactive cations, one can fabricate zeolite-modified electrodes that act as current rectifiers. Figure 6.14 shows schematically how large and small cations will localize, respectively, closer to and farther from a metallic electrode surface when they are ion exchanged in a zeolite-containing film. Li et al. (98) studied systems in which cationic metallocenes (substituted cobaltocenes and ferrocenes) were exchanged

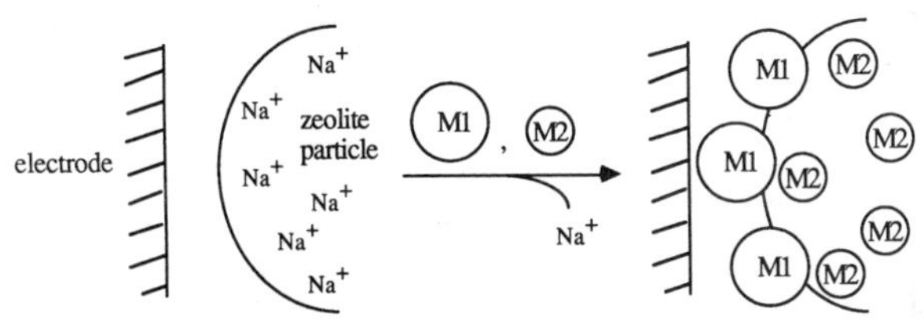

Figure 6.14. Self-assembly of a molecular bilayer at a zeolite-modified electrode. M1 and M2 represent, respectively, a size-excluded cation and a cation small enough to exchange into the bulk of zeolite Y. [Reprinted with permission from Li, Z.; Wang, C. M.; Persaud, L.; Mallouk, T. E. *J. Phys. Chem.* **1988**, *92*, 2592. Copyright © (1988) American Chemical Society.]

into the bulk of zeolite Y and large ions such as $[Ru(bpy)_3]^{2+}$ and $[Os(bpy)_3]^{2+}$ were exchanged onto the outer surface. The latter ions were present in the range $2\text{–}3 \times 10^{-6}$ mol g^{-1}, which corresponded to approximately monolayer coverage of micron-size zeolite particles. Cyclic voltammetry gave large anodic and cathodic peak currents when the formal potentials of the surface and bulk ions were within 100–200 mV of each other. When these potentials were appreciably different, one of the waves was greatly attenuated, consistent with mediated electron transfer to or from the bulk ions via the surface ions. A typical cyclic voltammogram of this type is shown in Fig. 6.15. The surface-confined ions can easily mediate reduction of bulk ions since the $[Ru(bpy)]^{2+/+}$ potential is about 500-mV negative of the $CoCp(CpCOOCH_3)]^{+/0}$ potential. However, reoxidation of the cobaltocene is slow because it requires thermodynamically unfavorable electron transfer from $[CoCp(CpCOOCH_3)]^0$ to $[Ru(bpy)_3]^{2+}$. Direct oxidation of $[CoCp(CpCOOCH_3)]^0$ does not occur, presumably because these molecules are confined to sites too far from the electrode surface.

Subsequent studies involving other pairs of small and size-excluded cations showed similar effects (99, 103). Using a surface cation that has two or more well-separated redox waves, one can prepare electrodes that show charge trapping–untrapping effects like those observed with polymer bilayer film electrodes (118). The most significant aspects of this zeolite-based study are the self-organizing nature of the electron-transport assembly and the demonstration that current rectification can be achieved with approximately one monolayer of large cations on the external surface. One can envision these molecular-scale rectifiers as active components of microelectronic devices or artificial photosynthetic systems. Recent work in our laboratories has shown that integrated chemical systems involving internally platinized zeolites with bimolecular surface "rectifiers" can be used to generate hydrogen from water photochemically (119). In this initial study, rapid charge recombination within the photogenerated reduced viologen–oxidized porphyrin pair at the zeolite surface gave resulted in a low quantum efficiency ($\sim 10^{-5}$) for the formation of chemical products. By using a slightly more elaborate donor–acceptor pair at the zeolite–solution interface (120), the quantum yield for charge separation was increased to 10^{-1}.

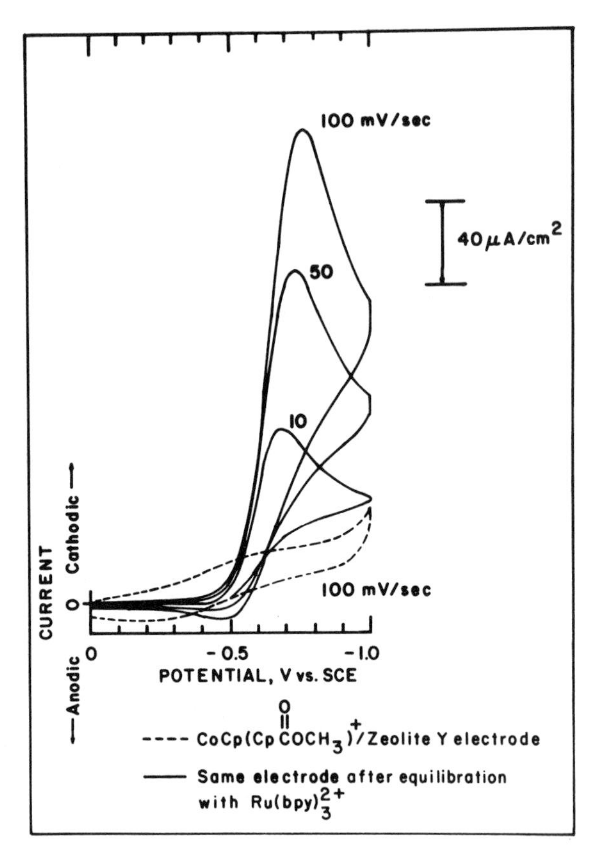

Figure 6.15. Cyclic voltammetry of a $CoCp(COOCH_3)^+$–zeolite Y electrode before and after 5-min equilibration with $[Ru(bpy)_3]^{2+}$. Electrode held 1 min at 0.0 V between scans. [Reprinted with permission from Li, Z.; Mallouk, T. E. *J. Phys. Chem.* **1987**, *91*, 643. Copyright © (1987) American Chemical Society.]

The modification of zeolites, clays, and related materials with semiconductors such as CdS and TiO_2, conducing polymers, and self-organizing molecular assemblies like those described above augurs well for future progress along the lines of photocatalysis, molecular electronics, and electrochemistry in microporous crystalline solids.

ACKNOWLEDGMENTS

We gratefully acknowledge the efforts of our co-workers, who are listed in the references. TEM acknowledges support from the Division of Chemical Sciences, Office of Basic Energy Sciences, Department of Energy, under contract DE-FG05-87ER13789, and the Camille and Henry Dreyfus Foundation for support in the form of a Teacher–Scholar Award. AJB acknowledges support from the National Science Foundation and the Robert A. Welch Foundation.

6.5 REFERENCES

1. Laszlo, P., (Ed.), *Preparative Chemistry Using Supported Reagents*, Academic: New York, 1987.
2. (a) Bard, A. J.; Fan, F-R. F.; Hope, G. A.; Keil, R. G., in *Inorganic Chemistry: Toward the 21st Century*, M. Chisholm (Ed.), American Chemical Society Symp. Ser. **1983**, No. 211, 93; (b) Krishnan, M.; White, J. R.; Fox, M. A.; Bard, A. J. *J. Am. Chem. Soc.* **1983**, *105*, 7002.
3. van Olphen, H. *An Introduction to Clay Colloid Chemistry*, Wiley: New York, 1977.
4. Brindley, G. W.; Brown, G. (Eds.), *Crystal Structures of Clay Minerals and their X-Ray Diffraction*, Mineral Soc. Monogr. 5: London, 1980.
5. (a) Tanner, C. B.; Jackson, M. L. *Soil Sci. Soc. Am. Proc.* **1947**, *12*, 60; (b) Jackson, M. L.; Wittig, L. D.; Pennington, R. P. *Sail Sci. Soc. Am. Proc.*, **1949**, *14*, 77.
6. Mehra, O. P.; Jackson, M. L. *Calys Clay Miner. Proc. Conf.* **1958**, *7*, 317.
7. (a) Barrer, R. M.; MacLeod, D. M. *Trans. Faraday Soc.*, **1955**, *51*, 1290; (b) Pinnavaia, T. J. *Science*, **1983**, *220*, 365; (c) Kellendonk, F. J. A.; Neinerman, J. J. L.; van Santen, R. A., van Olphen, H. *An Introduction to Clay Colloid Chemistry*, Wiley, New York, 1977, p. 459 and references cited therein.
8. Barrer, R. M., *Hydrothermal Synthesis of Zeolites*, Academic: New York, 1982.
9. Ghosh, P. K.; Bard, A. J. *J. Am. Chem. Soc.* **1983**, *105*, 5691.
10. Kamat, P. V. *J. Electroanal. Chem.* **1984**, *163*, 389.
11. Yamagishi, A.; Aramata, A. *J. Chem. Soc. Chem. Commun.* **1984**, 452.
12. Ghosh, P. K.; Mau, A. W-H.; Bard, A. J. *J. Electroanal. Chem.* **1984**, *169*, 315.
13. Ege, D.; Ghosh, P. K.; White, J. R.; Equey, J. F.; Bard, A. J. *J. Am. Chem. Soc.* **1985**, *107*, 5644.
14. Liu, H. Y.; Anson, F. C. *J. Electroanal. Chem.* **1985**, *184*, 411.
15. Oyama, N.; Anson, F. C. *J. Electroanal. Chem.* **1986**, *199*, 467.
16. White, J. R.; Bard, A. J. *J. Electroanal. Chem.* **1986**, *197*, 233.
17. Carter, M. T.; Bard, A. J. *J. Electroanal. Chem.* **1987**, *229*, 191.
18. Castro-Acuña, C. M.; Fan, F-R. F.; Bard, A. J. *J. Electroanal. Chem.* **1987**, *234*, 347.
19. King, R. D.; Nocera, D. G.; Pinnavaia, T. J. *J. Electroanal. Chem.* **1987**, *236*, 43.
20. Inoue, H.; Haga, S.; Iwakura, C.; Yoneyama, H. *J. Electroanal. Chem.* **1988**, *249*, 133.
21. Fitch, A.; Lavy-Feder, A.; Lee, S. A.; Kirsh, M. T. *J. Phys. Chem.* **1988**, *92*, 6665.
22. Rudzinski, W. E., Bard, A. J. *J. Electroanal. Chem.* **1986**, *199*, 323.
23. Itaya, K.; Bard, A. J. *J. Phys. Chem.* **1985**, *89*, 5565.
24. Villemeure, G., Bard, A. J., unpublished results, 1988.
25. Naegeli, R., Redepenning, J., Anson, F. C., *J. Phys. Chem.* **1986**, *90*, 6227.
26. Buttry, D. A., Anson, F. C., *J. Am. Chem. Soc.* **1983**, *105*, 685.
27. Fitch, A., Fausto, C. L., *J. Electroanal. Chem.* **1988**, *257*, 299.
28. Peerce, P.; Bard, A. J. *J. Electroanal. Chem.* **1980**, *112*, 97.
29. Leddy, J.; Bard, A. J. *J. Electroanal. Chem.* **1983**, *153*, 223.
30. Gueshi, T.; Tokuda, K; Matsuda, H. *J. Electroanal. Chem.* **1978**, *89*, 249.
31. Gueshi, T.; Tokuda, K.; Matsuda, H. *J. Electroanal. Chem.* **1979**, *101*, 29.
32. McBride, M. B. *Clays Clay Miner.* **1979**, *27*, 97.
33. Yamagishi, A. *J. Coord. Chem.* **1987**, *16*, 131, and references cited therein.
34. Yamagishi, A. J.; Soma, M. *J. Am. Chem. Soc.*, **1981**, *103*, 4640.
35. Yamagishi, A. J.; Soma, M. *J. Phys. Chem.* **1982**, *86*, 2747.
36. Yamagishi, A. *J. Phys. Chem.*, **1982**, *86*, 2472.
37. Yamagishi, A.; Aramata, A. *J. Electroanal. Chem.*, **1985**, *191*, 449.
38. Joshi, V.; Kothar, D.; Ghosh, P. K. *J. Am. Chem. Soc.*, **1986**, *108*, 4650.

39. Villemeure, G.; Bard, A. J.; *J. Electroanal. Chem.*, **1990**, *283*, 403.
40. Solomon, D. H.; Loft, B. C.; Swift, J. D. *Clay Miner.* **1968**, *7*, 399.
41. Itaya, K.; Chang, H-C.; Uchida, I. *Inorg. Chem.* **1987**, *26*, 624.
42. Shaw, B. R.; Creasy, K. E. *J. Electroanal. Chem.* **1988**, *243*, 209.
43. Shaw, B. R.; Beng, Yl; Strillacci, F. E.; Carrado, K. A.; Fessehai, M. G. *J. Electrochem. Soc.* **1990**, *137*, 3136.
44. Henning, T. P.; White, H. S.; Bard, A. J. *J. Am. Chem. Soc.* **1981**, *103*, 3937.
45. Fan, F-R. F.; Bard, A. J. *J. Electrochem. Soc.* **1986**, *133*, 301.
46. Rudzinski, W. E.; Figueroa, C.; Hoppe, C.; Kuromoto, T. Y.; Root, D. *J. Electroanal. Chem.* **1988**, *243*, 367.
47. Inoue, H.; Yoneyama, H. *J. Electroanal. Chem.* **1987**, *233*, 291.
48. Kanbara, T.; Yamamoto, T.; Tokuda, K.; Aoki, K. *Chem. Lett.* **1987**, 2173.
49. Fan, F-R. F.; Liu, H-Y.; Bard, A. J. *J. Phys. Chem.* **1985**, *89*, 4418.
50. Enea, O.; Bard, A. J. *J. Phys. Chem.* **1986**, *90*, 301.
51. Stramel, R. D.; Nakamura, T.; Thomas, J. K. *Chem. Phys. Lett.* **1986**, *130*, 423.
52. Levy, F. (Ed.), *Intercalated Layered Materials*, Reidel: Dordrecht, 1979; Whittingham, M. S.; Jacobsen, A. J. (Eds.), *Intercalation Chemistry*, Academic: New York, 1982; Schollhorn, R., in *Inclusion Compounds*, Atwood, J. L. (Ed.), Academic: London, 1984, Vol. 1, p. 249.
53. Johnson, J. W.; Jacobsen, A. J.; Butler, W. M.; Rosenthal, S. E.; Brody, J. F.; Lewandowski, J. T. *J. Am. Chem. Soc.* **1989**, *111*, 381.
54. Dines, M. B.; DiGiacomo, P. Inorg, Chem. 1981, 20, 92; Dines, M. B. Griffith, P. C. *Inorg Chem.* 1983, 22, 567; Dines, M. B.; Cooksey, R. E.; Griffith, P. C. *Inorg. Chem.* **1983**, *22*, 1003.
55. Alberti, G.; Constantino, U.; Alluli, S.; Tomassini, J. *J. Inorg. Nucl. Chem.* 1978, 40, 1113; Alberti, G.; Constantino, U.; Giovagnotti, M. L. L. *J. Chromatogr.* 1979, 180, 45; Casciola, M.; Constantino, U.; Fazzini, S.; Tosoratti, G. *Solid State Ionics* **1983**, *8*, 27.
56. Johnson, J. W.; Jacobsen, A. J.; Brody, J. F.; Lewandowski, J. T. *Inorg. Chem.* **1984**, *23*, 3844.
57. Cheng, S.; Peng, G. Z.; Clearfield, A. *Ind. Eng. Chem. Prod. Res. Dev.* **1984**, *23*, 219.
58. Lee, H.; Kepley, L. J.; Hong, H.-G.; Mallouk, T. E. *J. Am. Chem. Soc.* **1988**, *110*, 618; Lee, H.; Kepley, L. J.; Hong, H.-G.; Akhter, S.; Mallouk, T. E. *J. Phys. Chem.* **1988**, *92*, 2597.
59. Akhter, S.; Lee, H.; Hong, H.-G.; Mallouk, T. E.; White, J. M. *J. Vac. Sci. Tech. B*, **1989**, *7*, 1608.
60. Rong, D.; Kim, Y. I.; Hong, H.-G.; Kreuger, J. S.; Mayer, J. E.; Mallouk, T. E., *Coord. Chem. Rev.* **1990**, *97*, 237.
61. Ingersoll, D.; Faulkner, L. R., manuscript in preparation.
62. Szostak, R., *Molecular Sieves, Principles of Synthesis and Identification*, Van Nostrand Reinhold: New York, 1989.
63. Newsam, J. M. *Science* **1986**, *231*, 1093.
64. Meier, W. M.; Olson, D. H., *Atlas of Zeolite Structure Types*, International Zeolite Association published by Structure Commission, IZA: Zurich, 1978.
65. Breck, D. W., *Zeolite Molecular Sieves*, Wiley: New York, 1974.
66. Flanigen, E. M.; Lok, B.; Patton, R. L.; Wilson, S. T. *Pure Appl. Chem.* **1986**, *58*, 1351.
67. Smith, J. V.; Dytrych, W. J. *Nature (London)* **1984**, *309*, 607.
68. Davis, M. E.; Saldarriaga, L.; Montes, C.; Garces, J.; Crowder, C. *Nature (London)* **1988**, *331*, 698.
69. Cronstedt, A. F. *Akad. Handl. Stockholm* **1756**, *17*, 120.
70. Treacy, M. M. J.; Newsam, J. M., *Nature (London)* **1988**, *332*, 249.
71. Rabo, J. A. (Ed.), *Zeolite Chemistry and Cqtalysis*, ACS Monograph Ser., Vol. 171, American Chemical Society: Washington, DC, 1976.

71. Murakami, Y.; Iijima, A.; Ward, J. W. (Eds.), *New Developments in Zeolite Science and Technology*, Proc. 7th Int. Zeolite Conf., Tokyo, Aug. 17–22, 1986.

73. Lok, B. M.; Wilson, S. T. in Preparative Chemistry Using Supported Reagents, Laszio. P. (Ed.), Academic, New York, 1987, pp. 403–452.

74. Maxwell, I. *J. Inclus. Phenom.* **1987**, *4*, 1.

75. Dwyer, J. *Chem. Ind.* **1984**, *7*, 258; Dwyer, J.; Dyer, A. *Chem. Ind.* 237.

76. Dyer, A.; Malik, S. A.; Araya, A.; McConville, T. J., *Recent Developments in Ion Exchange*, Elsevier: Amsterdam, 1987, p. 257.

77. Bolton, A. P. (Ed.), *Zeolite Chemistry and Catalysis*, ACS Monograph Ser., Vol. 171, American Chemical Society: Washington DC, 1976, pp. 714–779.

78. Ozin, G. A.; Kuperman, A.; Stein, A., Angew, *Chem. Intl. Ed.* **1989**, *101*, 373.

79. Marshall, C. E., *J. Phys. Chem.* **1939**, *43*, 1155; Marshall, C. E.; Bergman, W. E. *J. Am. Chem. Soc.* **1941**, *63*, 1911; Marshall, C. E.; Krinbill, C. A., *J. Am. Chem. Soc.* **1942**, *64*, 1814; Marshall, C. E.; Ayers, A. D., *J. Am. Chem. Soc.* **1948**, *70*, 1297; Marshall, C. E.; Eime, L. O., *J. Am. Chem. Soc.* **1948**, *70*, 1302.

80. Wyllie, M. R. J.; Patrode, H. W. *J. Phys. Chem.* **1950**, *54*, 204.

81. Barrer, R. M.; James, S. D. *J. Phys. Chem.* **1960**, *64*, 417.

82. Thackeray, M. M.; Coetzer, *J. Solid State Ionics*, **1982**, *6*, 135.

83. Freeman, D. C., Jr. U.S. Patent 3,186,875, 1965.

84. Jacobs, P. A.; Jaeger, N. I.; Jiru, P.; Schulz-Ekloff, G. (Eds.), *Metal Microstructures in Zeolites*, Elsevier: Amsterdam, 1982.

85. Gallezot, P.; Alarcon-Diaz, A.; Dalmon, J.-A.; Renouprez, A. J.; Imelik, B. *J. Catal.* **1975**, *39*, 334; Gallezot, P. *Catal. Rev. Sci. Eng.* **1979**, *20*, 121; Felthouse, T. R.; Murphy, J. A. *J. Catal.* **1986**, *98*, 411.

86. Tauster, S. J.; Steger, J. J.; Fung, S. C.; Poeppelmeier, K. R.; Funk, W. G.; Montagna, A. A.; Cross, V. R.; Kao, J. L. U.S. Patent 4,595,668, 1983.

87. Jacobs, P. A.; Uytterhoeven, J. B.; Beyer, H. K. *J. Chem. Soc. Chem. Commun.* **1977**, 128.

88. Kuznicki, S. M.; Eyring, E. M. *J. Am. Chem. Soc.* **1978**, *100*, 6790; Kuznicki, S. M.; DeVries, K. L.; Eyring, E. M. *J. Phys. Chem.* **1980**, *84*, 535.

89. Arakawa, T.; Takata, T.; Adachi, G. Y.; Shiokawa, J. *J. Luminescence* **1979**, *20*, 325; *J. Chem. Soc. Chem. Commun.* **1979**, 453.

90. Suib, S. L.;; Zerger, R. P.; Stucky, G. D.; Morrison, T. I.; Shenoy, G. K. *J. Chem. Phys.* **1984**, *80*, 2203.

91. Suib, S. L.; Bordenianu, O. G.; McMahon, K. C.; Psaras, D., *Inorganic Reactions in Organized Media*, ACS Symposium Ser., 177, American Chemical Society: Washington, DC, 1982, pp. 225–238; Suib, S. L.; Kostapapas, A.; Psaras, D. *J. Am. Chem. Soc.* **1984**, *106*, 1614; Suib, S. L.; Tanguay, J. F.; Occelli, M. L., *J. Am. Chem. Soc.* **1986**, *108*, 6972.

92. De Wilde, W.; Peeters, G.; Lunsford, J. H. *J. Phys. Chem.* **1980**, *84*, 2306.

93. Faulkner, L. R.; Suib, S. L.; Renschler, C. L.; Green, J. M.; Bross, P. R., in *Chemistry in Energy Production*, Wymer, R. G.; Keller, O. L. (Eds.), Wiley: New York, 1982, pp. 99–114.

94. Murray, C. G.; Nowak, R. J.; Rolison, D. R. *J. Electroanal. Chem.* **1984**, *164*, 205.

95. de Vismes, B.; Bedioui, F.; Devynck, J.; Bied-Charreton, C. *J. Electroanal. Chem.* **1985**, *187*, 197.

96. Gemborys, H. A.; Shaw, B. R. *J. Electroanal. Chem.* **1986**, *208*, 95.

97. Shaw, B. R.; Creasy, K. E.; Lanczycki, C. J.; Sargeant, J. A.; Tirhado, M. *J. Electrochem. Soc.* **1988**, *135*, 869.

98. Li, Z.; Mallouk, T. E. *J. Phys. Chem.* **1987**, *91*, 643.

99. Li, Z.; Wang, C. M.; Persaud, L.; Mallouk, T. E. *J. Phys. Chem.* **1988**, *92*, 2592.

100. Bein, T.; Enzel, P. *Synth. Metals*, **1989**, *29*, E163.

101. Dutta, P. K.; Puri, M. *J.* Catal. **1988**, *111*, 453.

102. Pereira-Ramos, J.-P.; Messina, R.; Perichon, J. *J. Electroanal. Chem.* **1983**, *146*, 157.

103. Li, Z.; Lai, C.; Mallouk, T. E. *Inorg. Chem.* **1989**, *28*, 178.

104. Rolison, D. R.; Nowak, R. J.; Pns, S.; Ghoroghchian, J.; Fleischmann, M., in *Molecular Electronic Devices III*, Carter, F. L.; Siatkowski, R. E.; Wohltjen, H. (Eds.), Elsevier: Amsterdam, 1988.

105. Rolison, D. R.; Hayes, E. A.; Rudzinski, W. E., *J. Phys. Chem.*, in press.

106. Kärger, J.; Ruthven, D. M. *J. Chem. Soc. Faraday Trans. 1* 1981, 77, 1485.

107. Kärger, J. *Adv. Colloid Interface Sci.* **1985**, *23*, 129.

108. Chen, N. Y.; Lucki, S. J.; Mower, E. B. *J. Catal.* **1969**, *13*, 329.

109. Weisz, P. B. *Pure Appl. Chem.* **1980**, *52*, 2091.

110. Iwakura, C.; Miyazaki, S.; Yoneyama, H. *J. Electroanal. Chem.* **1988**, *246*, 63.

111. El Murr, N.; Kerkeni, M.; Sellami, A.; Ben Taarit, Y. *J. Electroanal. Chem.* **1988**, *246*, 461.

111. Johansson, G.; Risinger, L.; Fälth, L. *Anal. Chim. Acta* **1980**, *119*, 25.

113. Demertzis, M.; Evmiridis, N. P. *J. Chem. Soc. Faraday Trans. 1* **1986**, *82*, 3647.

114. Wang, J.; Martinez, T. *Anal. Chim. Acta* **1988**, *207*, 95.

115. de Vismes, B.; Bedioui, F.; Devynck, J.; Bied-Charreton, C.; Perrée-Fauvet, M. *Nouv. J. Chim.* **1986**, *10*, 81.

116. Creasy, K. E.; Shaw, B. R. *Electrochim. Acta* **1988**, *33*, 551.

117. Andrieux, C. P.; Dumas-Buochiat, J.-M.; Savéant, J. M. *J. Electroanal. Chem.* **1984**, *169*, 9.

118. Abruña, H. D.; Denisevich, P.; Umana, M.; Meyer, T. J.; Murray, R. W. *J. Am. Chem. Soc.* **1981**, *103*, 1; Denisevich, R.; Willman, K. W.; Murray, R. W. *J. Am. Chem. Soc.* **1981**, *103*, 4727; Willman, K. W.; Murray, R. W. *J. Electroanal. Chem.* **1982**, *133*, 211; Pickup, P. G.; Leidner, C. R.; Denisevich, P.; Murray, R. W. *J. Electroanal. Chem.* **1984**, *164*, 39; Pickup, P. G.; Kutner, W.; Leidner, C. R.; Murray, R. W. *J. Am. Chem. Soc.* **1984**, *106*, 1991; Leidner, C. R.; Murray, R. W. *J. Am. Chem. Soc.* **1985**, *107*, 551.

119. Persaud, L.; Bard, A. J.; Campion, A.; Fox, M. A.; Mallouk, T. E.; Webber, S. E.; White, J. M. *J. Am. Chem. Soc.* **1987**, *109*, 7309.

120. Krueger, J. S.; Mayer, J. E.; Mallouk, T. E. *J. Am. Chem. Soc.* **1988**, *110*, 8232.

Chapter **VII**

ELECTRON TRANSFER MEDIATION IN METAL COMPLEX POLYMER FILMS

Charles R. Leidner
Department of Chemistry, Purdue University, West Lafayette, Indiana

---

Electron-transfer mediation is fundamental to any idealized or realized application of metal complex polymer films on electrode surfaces. It is by this mechanism that the polymer communicates electrochemically with the surrounding chemical universe. Observation, quantitation, and manipulation of this process has obvious practical and intriguing fundamental appeal.

A discussion of the electron-transfer mediation reaction of metal complex polymer films on electrode surfaces is presented herein. A broad overview of polymer types, processes, and relevant theories precedes a discussion of various studies of mediation with metallopolymers. Particular attention is given to those studies that attempt to identify and quantify the many physical processes in the overall mediation processes. In addition, the more common mediation reactions involving dissolved solution reactants, those existing in structured polymer assemblies like bilayers are discussed.

Chemical microstructures on electrode surfaces (1–3) now encompass remarkably diverse classes of compounds. One of the earliest types, polymer films, remains the subject of intense study. The scope of polymer films on electrode surfaces has broadened considerably since the early work of Anson (4), Bard (5), Miller (6), Murray (7), Wrighton (8) and their co-workers. These early examples were amorphous polymer systems like poly(vinylferrocene) or metallated

---

Molecular Design of Electrode Surfaces,
Edited by Royce W. Murray. Techniques of Chemistry Series, Vol. XXII.
ISBN 0-471-55773-0

poly(vinylpyridine). More recently, organic conducting polymers (9, 10), liquid crystalline polymers (11, 12), clays (13, 14), metal oxides (15, 16), metallic bronzes (17), and metal cyanometallates (18) have blurred somewhat the distinction of "polymer films." Regardless of the scope or definition of "polymer coated electrodes," they continue as a mainstay of chemical microstructures on electrode surfaces.

Electron-transfer mediation, the process by which redox polymers transfer charge from one phase to another, is the mechanism by which the polymer communicates with its surroundings. As such, mediation is important to many aspects of polymer modified electrodes. Application of polymer coated electrodes to electrocatalysis, analysis, sensors, and devices depends largely on electron-transfer mediation. The importance to these applications and the fundamental interest in electron-transfer reactions has prompted investigations of mediation by polymers.

This chapter will provide one perspective on electron-transfer mediation with metal complex polymer films on electrodes. The remainder of this chapter is comprised of two sections, universal considerations and examples of mediation. The former describes classification of metal complex polymer films, processes within polymer modified electrodes, and identification of the locus of the electron-transfer mediation reaction. The latter section furnishes the bulk of this chapter and concentrates on the observation and quantitation of the overall mediation process. Particular attention is given to those studies that attempt to unravel the many individual physical processes constituting the overall mediation reaction. Such endeavors, although fraught with difficulties, are often rewarded with remarkably complete descriptions of electron-transfer mediation.

7.1 UNIVERSAL CONSIDERATIONS

7.1.1 Classification of Metal Complex Polymers

The diverse field of polymer coated electrodes contains numerous examples of inorganic and organic systems. Even within the inorganic systems, the discussion must be limited to metal complex polymer films to provide a more reasonable scope for this chapter. A useful classification of metallopolymer films segregates them into fixed redox site and exchange polymers.

Fixed redox site metallopolymers contain the metal complex as an integral, relative immobile, functional group chemically attached to or part of the polymer matrix. Numerous examples of such systems have been reported. One type of metallopolymer is based on metallocene functional groups like ferrocene (5, 7, 19, 20) and cobaltocene (21–23). Poly(vinylferrocene) (PVFc) homopolymers (5, 7, 20) and copolymers (19) are the most common examples. Another class of metallopolymers are those based on metal poly(pyridyl) complexes (24–34). A remarkable variety of metal centers and ligands have been reported, although those based on ruthenium bipyridine and poly(pyridine) complexes are most common. The two most common versions of metal poly(pyridyl) polymers are those that contain a metal center attached to a poly(vinylpyridine) (PVP) backbone (32–34) and those prepared by electropolymerization of metal poly(pyridyl) complexes (24–31). As seen in later sections, subtle differences exist between these two types. Metalloporphyrin polymers (35–38) comprise a third interesting class. The more limited number of such systems reflects the synthetic difficulty associated with preparing suitable metalloporphyrin precursors. While these three classes of metal complexes comprise a vast majority of the metallopolymer films, other examples have been reported (39–41).

Ion-exchange polymers provide a powerful and versatile means of preparing metallopolymer films on electrodes. Casting an ion exchange polymer onto an electrode, followed by incorporation of one or many metal complex ions generally leads to stable, electroactive films (42). The virtues of this method include the ability to prepare quite thick films, variable loading, great flexibility, and high mobility of redox ions. The liability of leaching of redox ions from the polymer is often circumvented by utilizing (additional) hydrophobic interactions, trace levels of the redox ion in the bathing solution, or short timescale experiments. The natural distinction among these types of polymers is the anion versus cation exchange properties. Typical cation exchange polymers are poly(styrene) sulfonate (PSS^-) (43–46) and the sulfonated fluorocarbon polymer Nafion (47–50); common cationic sorbents are metal poly(pyridyl) complexes. Typical anion exchange polymers are protonated (51–53) or quaternized (54) poly(vinylpyridine) and protonated poly-*l*-lysine ($PLLH^+$) (55, 56); metal cyanide and halide complexes are common sorbents for these films.

Several polymer coated electrodes do not easily fit into these two tidy classifications. These are comprised of either neutral redox molecules partitioned into ion-exchange polymers or redox molecules partitioned into neutral polymers or gels. Cobalt tetraphenylporphyrin in Nafion (57) is exemplary of the former, while neutral polymeric gels that contain neutral or ionic redox molecules (58) are examples of the latter. Conceptually, these seem more like ion-exchange polymer systems and are often treated as such.

7.1.2 Processes Within Polymer Modified Electrodes

The conceptualization, realization, and observation of the various physical processes that (can) take place within polymer modified electrodes have developed substantially over the last decade (59–70). The picture that has emerged for the possible rate-limiting steps in electron-transfer mediation reactions of metallopolymer films contacting electroactive solution species includes the processes:

1. Heterogeneous electron transfer at the polymer–electrode interface.
2. Charge transport within the polymer film.
3. Permeation of dissolved reactant within the polymer film.

4. Rate of reactant crossing at the polymer–solution interface.
5. Electron-transfer reaction between polymer sites and dissolved reactant; this process may occur at the polymer–solution interface or within the polymer film.
6. Mass transport of reactant to the polymer–solution interface.

Ancillary processes like swelling, morphological changes, and lattice motions of the polymer, differential solvation or spatial segregation of the dissolved reactant within the polymer, or ion movement within the polymer are not explicitly listed in this picture. In the attempt to categorize the primary physical processes within the polymer, the ancillary processes are normally incorporated into the primary process or even neglected. For example, the charge-transport process, at the simplest level, can be viewed as being determined by the intrinsic electron-transfer reaction between neighboring polymer sites, ion mobility, polymer lattice mobility, or mobility of electroactive ions within an ionmer film. Charge transport is usually treated as an overall process that is important, but the details of which are peripheral, to discussions of mediation.

Mediation of a dissolved reactant by a polymer coated electrode is the typical arrangement, however, mediation is also encountered with bilayer electrode assemblies. These are prepared by the sequential deposition of two discrete polymer films onto an electrode surface. Since the outer polymer (adjacent to the solution) is spatially segregated from the electrode, its redox reactions must be mediated by the inner polymer adjacent to the electrode film. The outer polymer film can communicate with the electrode only through the electron-transfer mediation reaction at the polymer–polymer interface. Charge transport must now be considered in both the inner and outer polymer films.

Theoretical developments that provide the underpinnings for discussion of these processes likewise have developed substantially over the last decade. Whether conceptualization of these processes or the theoretical treatments describing them came first is an interesting question, but both have led to a better, although admittedly incomplete, description of polymer modified electrodes. Unfortunately, a complete discussion of these various theories is beyond the scope of this chapter, Chapter V should be consulted for a more appropriate treatment. Despite the intimidating number of papers on the theory of polymer coated electrodes, we must consider several of these in minimal detail in order to discuss electron-transfer mediation.

The numerous papers by Savéant and his co-workers, a portion of which is given at the end of this chapter (59–64), have been a major force in developing an understanding of the physical processes in polymer modified electrodes. A perusal of these papers provides a chronology of this development. One particular treatment (62) has been extensively used by many workers in the field and has been revised by several workers (63, 64, 67). In the Savéant formalism the primary candidates for rate-limiting processes in the overall mediation process are given the symbols R, S, and E for processes 5, 3, and 2, respectively, listed above. Mixed control is designated by combinations (ER, R + S,

SR + E, ···) of these letters. This relatively simple terminology is the basis for an extremely powerful means of analyzing steady state (rotated disk voltammetry) results. Repeated references, at an admittedly superficial level, will be made to this treatment in ensuing sections. Interested readers are urged to consult the original papers to fully grasp the treatment.

Several alternative treatments contemporary to those of Savéant have also played an important role in unraveling the delicate puzzle of polymer modified electrodes. A seminal paper by Anson (65) utilizing a simple Marcus theory treatment of outer-sphere redox reactions between polymeric mediators and dissolved reactants set the stage for several experimental studies (70–74). The development by Ikeda et al. (70) for mediation by impermeable polymer films can now be viewed as a subset of the Savéant treatment (62, 63). This treatment of mediation has been used successfully in several instances (70–72, 74). The treatment by Albery and Hillman (67) shares many of the concepts and conclusions of Savéant, but provides an alternative method of data analysis.

All of these theoretical treatments, to one degree or another, highlight the importance of considering all conceivable rate-determining steps in the overall mediation event before extracting kinetic information from the data. For example, to extract the electron-transfer rate constant in process 5 above one must know the charge transport and permeability characteristics of the polymer film. This requires numerous, carefully performed ancillary experiments. One must know accurately the charge-transport rates of the film, preferably under both transient and steady state conditions. One must also know the extent to which the solution reactant penetrates the polymer film during the mediation process. This knowledge is necessary to eliminate permeation considerations from the treatment if the reactant does not penetrate significantly beyond the polymer–solution interface or to unravel the mixed kinetic control if the reactant does penetrate. It must be emphasized that measurement of charge transport and permeation rates should be made under conditions identical to those existing during the mediation experiments. Satisfying this requirement is often difficult for permeation since the mediation reaction interferes with the permeation measurement. Often such measurements are made with an inactive form of the polymer mediator or with analogous solution reactants. It is clear that one must take extreme care when attempting to quantify these physical processes. Utilization of these theoretical treatments and measurement of the physical processes accompanying the electron-transfer mediation step in metal complex polymer films play a major role in the studies of mediation discussed in the remainder of this chapter.

7.2 EXAMPLES OF ELECTRON-TRANSFER MEDIATION

The following sections are not meant to be exhaustive literature reviews of the considerable amount of work done on mediation with metal complex polymers, but instead are designed to provide an overview of the field. Representative examples from the many types of polymers and metal complexes are presented in an effort to illustrate the breadth of investigations in this field. Several of these examples are discussed in greater detail for they provide considerable insight into the observation and quantitation of electron-transfer mediation. The bulk of the ensuing discussion investigates mediation of dissolved reactants by metal complex polymer films and is organized by whether the metallopolymer film is fixed redox site or ion exchange. The third and much shorter section treats mediation in structured polymer assemblies like bilayer electrodes and interdigitated arrays.

7.2.1 Fixed Redox Site Polymers

A series of papers by Murray and his co-workers (70–75) document a substantial effort toward an understanding of the redox, charge transport, permeability, and electron-transfer properties of a family of metal poly(pyridyl) polymers. An exemplary paper (74) details the kinetic study of the electron-transfer reactions between poly-$[Os(bpy)_2(vpy)_2]^{3+}$ (where bpy = 2,2′-bipyridine and bpy = 4-vinylpyridine) and three series of $(ML_3)^{2+}$ complexes in acetonitrile:

$$\text{Pt/poly-}(Os)^{3+} + (ML_3)^{2+} \rightarrow \text{Pt/poly-}(Os)^{2+} + (ML_3)^{3+}$$

The authors demonstrated the well-defined electrochemistry and rapid charge-transport rates for the poly-$(Os)^{2+/3+}$ film. The low permeability of the film to bulky, cationic species like the $(ML_3)^{2+}$ complexes was demonstrated by rotated disk voltammetry. This low permeability of the cationic film to bulky cations is characteristic of the electropolymerized metal poly(pyridyl) films (75). Determination of the low permeability and high charge-transport rates of the poly-$(Os)^{2+/3+}$ film was crucial for the data analysis because this made possible the identification of the locus of the mediation reactions to be exclusively the polymer–solution interface. This situation is most easily handled using the analyses of Ikeda et al. (70) or Savéant and his co-workers (62). Limiting currents of the rotated disk voltammograms were analyzed by Koutecky–Levich plots to yield heterogeneous electron-transfer rate constants. The key feature of this and other similar studies by Murray and his co-workers (70–72, 74) is that mediation reactions with favorable free-energy changes are mass transport controlled, but those with unfavorable free-energy changes possess rates that are small enough to measure.

Electron-transfer rates were measured for the mediation of nine substituted phenanthroline and bipyridine Fe(II) complexes. Each of these reactions were thermodynamically unfavorable with ΔE° values (E°_{soln}-E°_{surf}) ranging from 0.092 to 0.338 V for the series; these correspond to equilibrium constants from 0.028 to 1.9×10^{-6}. The heterogeneous electron-transfer rate constants were converted to homogeneous rate constants (k_{12}) by considering that the mediation reactions occurred exclusively at the outermost monolayer of the polymer film. The k_{12} values decreased with decreasing driving force or equilibrium constant.

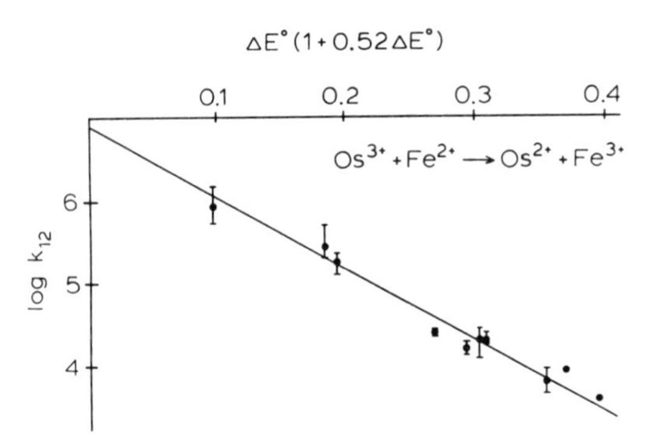

Figure 7.1. Marcus plot of electron-transfer mediation rate constants (k_{12}) as a function of driving force of reaction ($\Delta E^\circ = E^\circ_{\text{soln}} - E^\circ_{\text{surf}}$). Line is drawn with theoretical slope of $-8.47\,\text{V}^{-1}$. [Reprinted with permission from C. R. Leidner and R. W. Murray, *J. Am. Chem. Soc.* *106*, 1606 (1984). Copyright © (1984) American Chemical Society.]

The k_{12} values were fit to a rearranged form of the Marcus equation,

$$\log k_{12} = -\tfrac{1}{2} - \tfrac{1}{2}\log(k_{11}k_{22}) - 8.47\Delta E^\circ(1 + \chi\Delta E^\circ)$$

where χ is a factor accounting for the quadratic dependence of ΔG^* on ΔG°, and k_{11} and k_{22} are the self-exchange rate constants for the interfacial polymer sites and the solution reactants. The Marcus plot for these data are shown in Fig. 7.1. A least-squares fit of the data yields a slope of $-7.96\,\text{V}^{-1}$, which is in close enough agreement to justify use of the theoretical slope of $-8.47\,\text{V}^{-1}$ shown in Fig. 7.1. Extrapolation of the line in Fig. 7.1 to zero driving force provides a $k_{11}k_{22}$ value of $6.3 \times 10^{13} M^{-2}\text{s}^{-2}$, which is good agreement with that obtained from the known self-exchange rate constants for $[\text{Os(bpy)}_3]^{2+/3+}$ and $[\text{Fe(bpy)}_3]^{2+/3+}$ in CH_3CN, $8.1 \times 10^{13} M^{-2}\text{s}^{-2}$.

These mediation studies were performed on five series of polymer–solution reactions utilizing three different electropolymerized metal poly(pyridyl) films and various metal poly(pyridyl) complexes in solution (70–74). Composite Marcus plots are given in Fig. 7.2 for the five systems. The various correlations span a range of five orders of magnitude in k_{12} and 0.54 V in ΔE°. Comparison of the $k_{11}k_{22}$ values in Table 7.1 obtained by the extrapolation of the five lines to zero driving force reveals that $k_{11}k_{22}$(polymer) values for each of the polymer–solution reactions agree quite well with the known $k_{11}k_{22}$(solution) values. The correlations in Fig. 7.2 and the rate constants in Table 7.1 demonstrate the applicability of free-energy relationships developed for solution reactions to electron-transfer mediation reactions of metal complex polymer films. This was suggested previously by Anson (65) and appears to be correct for interfacial reactions of this class of fixed site metal poly(pyridyl) polymers, but should not be extended to all polymer systems.

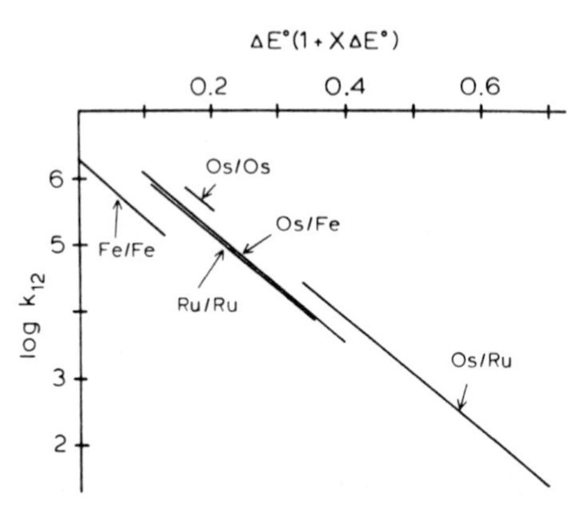

Figure 7.2. Composite Marcus plot of five series of polymer–solution mediation reactions. Figure reproduced from Ref. 71 with permission of The Electrochemical Society. This paper was originally presented at the Spring 1983 Meeting of the Electrochemical Sodiety, Inc. held in San Francisco, California.

An elegant study was recently reported by Savéant and his co-workers (76) where they used a $[Ru(bpy)_2Cl(PVP)]^+$ film on glassy carbon to mediate the redox reactions of $Fe^{2+/3+}$ in $1M$ HCl. Saveánt studied both the polymer-mediated oxidation of Fe^{2+} and reduction of Fe^{3+} and applied the Saveánt theory to extract rate constants for both electron-transfer reactions. This example is more complex than the above studies in that the solution reactant is capable of permeating into the polymer film. Savéant and his co-workers first characterized the redox and charge-transport properties of the poly-$(Ru^{II/III})$ film using cyclic voltammetry and chronoamperometry and the permeability of the film to Fe^{3+} by rotated disk voltammetry. With these preliminary data in hand they studied the mediated oxidation reaction:

$$\text{poly-}(Ru^{III}) + Fe^{2+} \xrightarrow{k_+} \text{poly-}(Ru^{II}) + Fe^{3+}$$

Table 7.1. Comparison of Self-Exchange Rate Constants from Polymer–Solution and Homogeneous Solution Reactions

System *Polymer–Solution*	$k_{11}k_{22}$(*Polymer*) ($M^{-2}s^{-2}$)	$k_{11}k_{22}$(*Solution*) ($M^{-2}s^{-2}$)
Fe–Fe	5×10^{12}	14×10^{12}
Ru–Ru	6.5×10^{13}	6.9×10^{13}
Os–Fe	6.3×10^{13}	8.1×10^{13}
Os–Ru	4×10^{14}	1.8×10^{14}
Os–Os	4×10^{14}	4.8×10^{14}

which was thermodynamically favorable by 220 mV. The rotated disk data collected at various rotation rates, polymer coverages, and Fe^{2+} concentrations conformed well to theory. At coverages less than 7×10^{-10} mol cm^{-2} the mediated oxidation was limited solely by the electron-transfer reaction; this corresponds to the "R" case in Savéant theory. Under these conditions the entire film was involved in the mediation reaction. At thicker films ($\Gamma_T = 10^{-8} \times$ mol cm^{-2}) the reaction was of the "SR" variety, meaning that both the mediation reaction and substrate permeation control the overall reaction. Under these conditions the mediation reaction occurred in a reaction layer adjacent to the polymer–solution interface; only 20% of the polymer participated in the mediation reaction.

This analysis permitted the calculation of the electron-transfer rate constant $k_+ = 4.5 \times 10^4 M^{-1}s^{-1}$. The corresponding solution reaction proceeds with a driving force of 510 mV and at a rate constant of approximately $10^6 M^{-1}s^{-1}$. Correction of this value to a driving force of 220 mV leads to a value of approximately $10^4 M^{-1}s^{-1}$. Within the limits of these approximations and slight differences in ionic environment, the rate of reaction within the polymer film is of the same order of magnitude as that in homogeneous solution. Similarly, the authors were able to measure the rate constant (k_-) for the thermodynamically unfavorable mediated reduction of Fe^{3+}. The value of $11 M^{-1}s^{-1}$ provides a measure of the equilibrium constant for the favorable reaction of 4×10^3; this corresponds well with the value obtained from the formal potentials, 6×10^3. This report is significant because it provides an example of a mediation reaction that occurs within the polymer film and still possesses a rate constant close to that of solution reactions. As mentioned above for the polymer–solution interfacial reactions, this observed similarity between polymer and solution rate constants should not be extended a priori to all systems.

Vos and his co-workers (77–79) have employed the same basic polymer system, $[Ru(bpy)_2Cl(PVP)]^+$, in mediation studies in a series of papers. These studies have a slightly different emphasis in that the authors are developing theoretical and experimental means of using nonstationary methods to study electron-transfer mediation. They have employed rapid scan rotated disk (78) and hydrodynamically modulated rotated disk (HMRDE) voltammetry (77, 79) to study the mediated oxidation of $[Fe(CN)_6]^{4-}$. A comparison (78) of a theoretical model and experimental data for rapid scan rotated disk voltammetry reveals qualitative agreement, however, extraction of rate parameters was not possible. Cassidy and Vos (79) concluded that the $[Ru(bpy)_2Cl(PVP)]^+$ polymer is quite porous, so that $[Fe(CN)_6]^{4-}$ readily penetrates the film to involve a substantial portion of the polymer film. The kinetics are further complicated by $[Fe(CN)_6]^{4-}$ migration within the polymer film.

The utilization of stationary and rotating microelectrodes has recently been reported by Wrighton and his co-workers (22) for the measurement of rapid electron-transfer mediation reactions. Although a majority of these data were obtained with viologen polymers, the mediated reduction of $[Ru(NH_3)_6]^{3+}$ was studied using a cobalticenium silane film poly-$[Co(CpR)_2]^+$:

$$\text{poly-}[Co(CpR)_2] + [Ru(NH_3)_6]^{3+} \xrightarrow{k} \text{poly-}[Co(CpR)_2]^+ + [Ru(NH_3)_6]^{2+}$$

The poly-$[Co(CpR)_2]^{+/\circ}$ film is impermeable to $[Ru(NH_3)_6]^{3+}$ (23) and possesses a moderately rapid charge-transport rate (22, 23). The $[Ru(NH_3)_6]^{3+}$ reduction reaction is thermodynamically favorable by 420 mV and is rapid as evidenced by the large mediation currents observed at the poly-$[Co(CpR_2)]^+$-coated microdisk electrodes. Figure 7.3 shows these data for a film with $\Gamma_T = 6 \times 10^{-8}\,\text{mol cm}^{-2}$ in a 8.0 mM $[Ru(NH_3)_6]^{3+}$ solution in 0.4M KCl in 4:1 H_2O/CH_3CN. The limiting currents for the mediation reaction increase rapidly with increasing rotation rate up to approximately 2000 rpm, but increase more gradually up to 6200 rpm. A Levich plot of these data is included in Fig. 7.3. A least-squares analysis of these data yields a second-order electron-transfer rate constant $k = 8 \times 10^6 M^{-1}s^{-1}$. Attempts at correlating the electron-transfer rate constant obtained from this reaction (and those from the viologen polymers) with calculated values for the corresponding solution reactions were only moderately successful. The cobalticene/$[Ru(NH_3)_6]^{3+}$ value is a factor of 20 lower than calculated for the solution reactions; the viologen systems were likewise lower than solution values. The authors attribute much of this discrepancy to the use of self-exchange rate constants for the polymer sites derived from charge-transport measurements. This is often difficult and possibly erroneous, so the authors likely have located the weak link in this calculation. Despite these quantitative discrepancies, this paper is a significant advance in the measurement of electron-transfer mediation rates. The rotating microelectrode methodology that they have developed has extended the upper limit of measureable mediation rate constants to approximately $2 \times 10^7 M^{-1}s^{-1}$.

7.2.2 Ion-Exchange Polymers

The generality of electrostatic trapping of redox ions into ionomer films leads directly to a much greater variety of metal complex polymer films based on ion-exchange polymers. This has naturally translated into many mediation studies. In addition to all of the processes discussed above for fixed redox site polymers, these systems also contain redox mediators that migrate–diffuse within the ion-exchange film. Conceptually, this is more complex than for fixed site polymers, but operationally the redox site movement is handled within the charge-transport step. The data analysis, usually based on Savéant's treatment (62), is thus identical. Although numerous examples of electron-transfer mediation with ion-exchange polymers exist (80–94), only selected studies will be discussed. Notable are those by Anson, Savéant, Sharp, Faulkner, and Bard for they address observation and quantitation of the mediation step.

One of the earliest examples of metal complex polymer films based on ion-exchange polymers and the first critical investigation of the processes that contribute to the overall mediation of a solution complex by a polymer film was

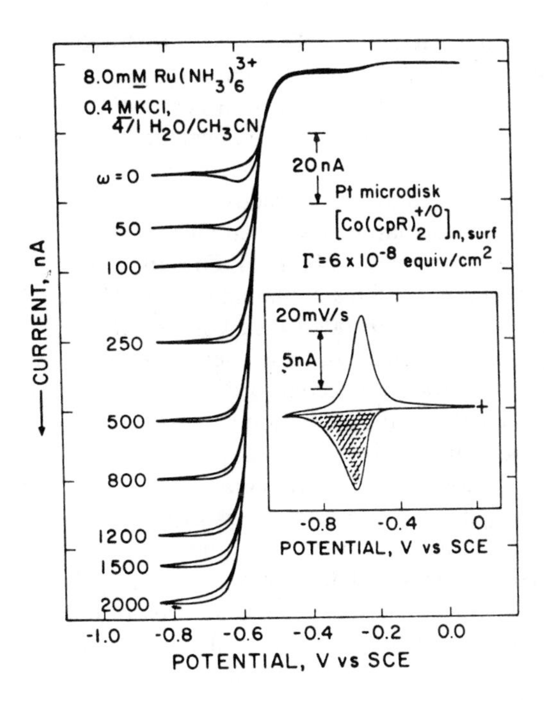

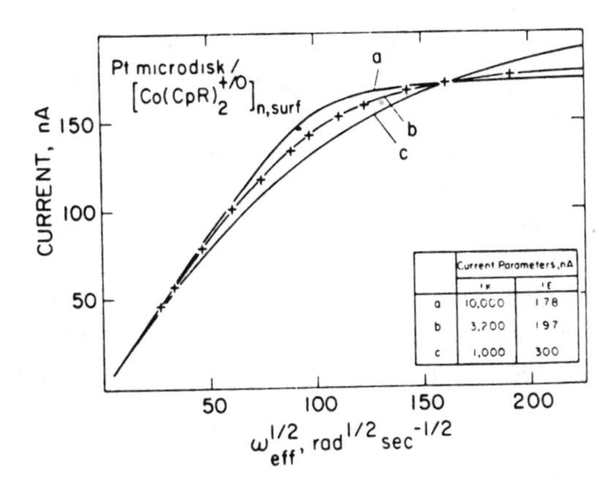

	Current Parameters, nA	
	i_K	i_E
a	10,000	178
b	3,200	197
c	1,000	300

Figure 7.3. Top: Current–voltage curves ($10\,mV\,s^{-1}$) for Pt rotating microdisk electrode, derivatized with $6 \times 10^{-8}\,mol\,cm^{-2}$ of poly-$[Co(CpR)_2]^{+/\circ}$, for the reduction of $[Ru(NH_3)_6]^{3+}$ in $0.4\,M$ KCl, 4:1 H_2O:CH_3CN with rotation rates from 0 to 2000 rpm. The inset shows the cyclic voltammogram ($20\,mV\,s^{-1}$) recorded during characterization of the poly-$[Co(CpR)_2]^{+/\circ}$ film. The shaded area is proportional to coverage. Bottom: Levich plot of the data from the above figure. Curve *b* is the best fit least-squares analysis; curves *a* and *c* are the least-squares fits with the kinetic currents raised and lowered, respectively, by a factor of about 3, and the polymer current was adjusted to give the best fit. [Reprinted with permission from T. E. Mallouk, V. Cammarata, J. A. Craston, and M. S. Wrighton, *J. Phys. Chem. 90*, 2150 (1986). Copyright © (1986) American Chemical Society.]

that by Oyama and Anson (51). This study employed protonated PVP films containing $[IrCl_6]^{2-}$ to mediate the oxidation of Fe^{2+} in solution. Oyama and Anson realized the importance of reactant permeation within the polymer film and demonstrated this interplay between mediated and nonmediated oxidation of Fe^{2+} at the $PVPH^+$–$[IrCl_6]^-$ electrode by employing a film that contained a very low $[IrCl_6]^{2-}$ loading. At lower rotation rates the Fe^{2+} was oxidized exclusively by bound mediator sites, but at higher rotation rates direct oxidation of the underlying electrode was observed. This knowledge was used to interpret the mediation reactions at films containing higher $[IrCl_6]^{2-}$ loadings. Oyama and Anson noted that the rotated disk voltammetry data fit the Koutecky–Levich equation and thus obtained effective mediation rate constants expressed as $k\Gamma$, where k is the second-order rate constant and Γ is the coverage of film that participates in the mediation reaction. Because only a portion of the polymer film participated in the mediation reaction, k values obtained by dividing $k\Gamma$ by the total $\Gamma_{[IrCl_6]^{2-}}$ were not constant. With the insight available to us today, we can see the shortcomings of the authors' analysis. But more importantly, we can also see how the authors were able to see (some of) the important physical processes that must be considered when studying mediation. This study stands as seminal work from which this author and many others have learned. This initial study has been followed by many interesting papers (81–87) from the Anson laboratory. Two of these studies are discussed here.

Anson, Savéant, and Shigehara (83) later reported on a related system but with a more sophisticated data analysis. A copolymer of 95:5 PVP–PLL was employed because this copolymer apparently forms more uniform coatings than does PVP. As in the previous work (51), the $[IrCl_6]^{2-}$-mediated oxidation of Fe^{2+} was studied using rotated disk voltammetry. Films of constant coverage of $PVPH^+$, but varying loadings (0–13%) of $[IrCl_6]^{3-}$ were employed to measure the permeation, charge transport, and mediation rates. A consistent picture emerged using the Savéant "ER" treatment of the data. A calculation of the reaction layer thickness from the data indicates that the outer (approximately) 45% of the film participates in the mediation reaction. The second-order rate constant for this reaction is approximately a factor of 2700 lower than for the same reaction in homogeneous solution. The authors attribute this disparity in rate constants to the inevitable differences between solution and polymer environments.

The report by Anson, Tsou, and Savéant (85) on the oxidation of ascorbate by $[Os(bpy)_3]^{3+}$ incorporated in Nafion illustrates the difficulty in performing the experiments necessary to analyze mediation kinetics and the occasional shortcomings of prevailing theory. Anson et al. measured the modest charge-transport rate of the Nafion-$[Os(bpy)_3]^{2+/3+}$ polymer film and noted their concerns about the ability of the film to deliver ample charge to the reaction layer during the mediation reaction. The difficulties in this study surfaced when the necessary permeation studies were attempted. Since ascorbate oxidation is inhibited at Nafion coated electrodes, an inactive form of the mediator film could not be employed in permeation experiments. The surrogate reactant

$Fe(edta)^-$ (where edta = ethylenediaminetetraacetate) yielded poorly defined limiting results at the underlying graphite, so there remained some uncertainty about the permeability of the mediator film. Mediation results at lower ascorbate concentration conformed to the "SR" case yielding a reaction layer thickness of 3 nm (the outermost monolayer or two) and homogeneous rate constant of $3 \times 10^5 M^{-1}s^{-1}$. The rate constant is considerably lower than calculated ($8 \times 10^7 M^{-1}s^{-1}$) for the analogous homogeneous solution reaction. At higher ascorbate concentrations the limiting currents could not be described using the prevailing Savéant theory. The authors attributed this failure of the Savéant theory to the microheterogeneity of Nafion films that would affect the description of the charge transport and permeability characteristics of the mediator film.

Lindholm and Sharp (88) recently reported the utilization of quaternized, cross-linked PVP containing $[IrCl_6]^{2-}$ for the mediated oxidation of Fe^{2+}. The polymer composition differs slightly from that employed by Anson and his co-workers (51, 83), and the mediation results likewise differ slightly. Lindholm and Sharp (88) describe the preparation and characterization of a series of anion exchange polymers prepared by quaternization and cross-linking of PVP with various α–ω dihalides. These film incorporate $[Fe(CN)_6]^{4-}$ and $[IrCl_6]^{3-}$ to form stable polymer modified electrodes with rapid charge-transport rates. Predictably, the metallated films are quite permeable to a small, neutral molecule like catechol, but are less permeable to cations like $[Fe(phen)_3]^{2+}$ and Fe^{2+}. Analysis of the data for the mediated oxidation of Fe^{2+} by $[IrCl_6]^{2-}$ films indicates that the system is best described as an "SR" case. The calculated reaction layer thickness of 6–7 nm indicates that only approximately 2–3 monolayers of the film participate in the mediation reaction. The electron-transfer rate constant for the reaction between $[IrCl_6]^{3-}$ and Fe^{2+} was 1.0 $(\pm 0.1) \times 10^5 M^{-1}s^{-1}$. This value is approximately 100 times that found by Anson (83) for the similar PVP–PLL system, but still approximately 30 times smaller than that for the analogous solution reaction. A dramatic difference between two seemingly similar polymer systems is not uncommon in this field.

Jones and Faulkner (45) reported a study of the mediated oxidation of ferrocene-1,1′-disulfonate (FDS^{2-}) by $[Os(bpy)_3]^{3+}$ incorporated into high molecular weight PSS^-. Despite the extreme difficulties in preparing suitable mediator films and eccentricities of the resultant PSS-$[Os(bpy)_3]^{3+/2+}$ films, they were able to perform a kinetic analysis using Savéant theory. The PSS–$[Os(bpy)_3]^{2+/3+}$ mediator film exhibited complicated redox and charge-transport responses, but supported sufficient charge-transport fluxes to be an efficacious mediator. The permeability of the FDS^{2-} within the mediator film was estimated by using a PSS–$[Ru(bpy)_3]^{2+/3+}$ surrogate film. At potentials near the $FDS^{2-/-}$ formal potential, the Ru polymer is essentially inactive as a mediator. Rotated disk experiments revealed that the PSS–$[M(bpy)_3]^{2+}$ film exhibited very limited permeability to the FDS^{2-}. Consideration of the charge transport and permeability characteristic of the polymer film and examination of the mediation results led Jones and Faulkner to conclude that the system should be described by "SR" or "SR + E" behavior, meaning that the overall mediation reaction is limited by the electron-transfer reaction with minimal permeation and possible charge-transport contributions. This example appears to be one in which the mediation reaction occurs very near the film–solution interface with only limited (if any) charge-transport limitations; this is a qualitatively similar situation to that seen for the fixed redox site metal poly(pyridyl) films (70–72, 74).

An alternative method for probing electron-transfer mediation has been reported by Bard and his co-workers (89) in which spectral sensitization of SnO_2 electrodes coated with Nafion containing $[Ru(bpy)_3]^{2+}$ was employed. Bard and his co-workers combined the more traditional voltammetric techniques with polymer films on glassy carbon with photocurrent measurements on the SnO_2–polymer assemblies. These photocurrent measurements provide an interesting and illustrative means of studying mediation, however, most of the quantitation was performed using the voltammetric methods; only the latter will be discussed here. The mediated oxidation of three solution species, hydroquinone (HQ), Fe^{2+}, and $[Fe(CN)_6]^{4-}$, was studied with both thin (0.3 μm) and thick (3 μm) $[Ru(bpy)_3]^{2+}$—containing Nafion films. Permeation measurements indicated that HQ permeates somewhat through thin films, but little through thick films. The Fe^{2+} ion permeates slightly more than HQ, while $[Fe(CN)_6]^{4-}$ is completely excluded from the film. A Savéant analysis of data revealed slight differences in the mediation process for those solutes, but all three systems exhibit charge-transport effects. The $[Fe(CN)_6]^{4-}$ system is simplified by the lack of permeation and is characterized by charge-transport control or case "E". The HQ system is similar in that charge-transport limits at high mass transport rates, except that HQ permeates slightly into the film; this corresponds to case "S + E". The Fe^{2+} system is slightly more complicated since the polymer–solution interfacial electron-transfer reaction is slower than for HQ or $[Fe(CN)_6]^{4-}$. The rate constant for this process is quite small, $80 M^{-1} s^{-1}$. These electrochemical results provided an explanation of the photochemical results by indicating that the charge-transport rates with the Nafion–$[Ru(bpy)_3]^{2+}$ films became rate limiting for thick films.

7.2.3 Structured Polymer Assemblies

The above examples of polymer films reacting with solution reactants are not the only examples of polymer electrode assemblies that exhibit electron-transfer mediation. Replacement of the solution reactant with a second polymer film should lead to an analogous mediation event. Structured polymer assemblies like bilayer electrodes (24, 94–98) containing two discrete polymer films or interdigitated arrays containing dissimilar redox polymers (99) provide another opportunity to study electron-transfer mediation.

Redox bilayer electrodes consist of two spatially segregated polymer films on an electrode. Since the outer film does not directly contact the underlying electrode, its redox reaction must be mediated by the inner film at the polymer–polymer interface. It is crucial to realize that the polymer films are redox conductors, that is they can pass charge only over limited potential ranges near their formal potentials; the mediation of the outer film can take place only at potentials near the formal potentials of the inner film. If the inner and outer film are selected to have a large difference in formal potentials, the polymer–polymer interface becomes a charge rectifying interface (24).

The challenge of measuring the polymer–polymer mediation rate is a difficult one due to the complexity of the system and the number of possible rate-limiting steps (see above). Murray and his co-workers (94–97) approached this problem in several ways. One illustrative system is the Pt/poly-$[Os(bpy)_2(vpy)_2]^{2+/3+}$/ poly-$[Ru(vbpy)_3]^{2+/3+}$ (where vbpy = 4-vinyl-4′-methyl-2,2′-bipyridine) bilayer (96). The cyclic voltammetric response of a Pt/$(Os)^{2+}$/$(Ru)^{2+}$ electrode is shown in Fig. 7.4. The sharp cathodic peak at +1.0 V is due to the favorable (405 mV) mediation reaction:

$$Pt/(Os)^{2+}/(Ru)^{3+} \rightarrow Pt/(Os)^{3+}/(Ru)^{2+}$$

Examination of this peak indicated that this favorable mediation reaction is limited by the charge-transport rate within the inner $(Os)^{2+/3+}$ film. It has been universally observed that favorable (i.e., rapid) bilayer mediation reactions are limited by film charge-transport rates and *not* the film–film interfacial electron-transfer rate (94–97, 99).

Careful examination of the voltammogram in Fig. 7.4 reveals a small, broad anodic peak at about 1.05 V. This peak is due to the unfavorable mediation reaction

$$Pt/(Os)^{3+}/(Ru)^{2+} \rightarrow Pt/(Os)^{2+}/(Ru)^{3+}$$

Variation of scan rate and film coverages reveal the expected kinetic nature of this wave, but quantitation of the unfavorable mediation reaction is not straightforward for these transient, thin-film experiments. In order to overcome the difficulties in quantitation of the cyclic voltammetric data and in the

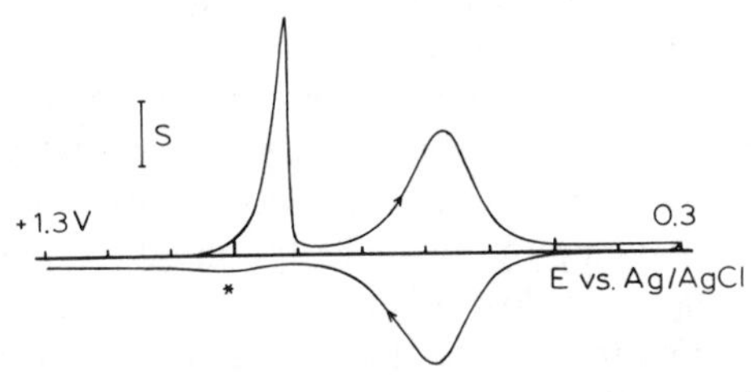

Figure 7.4. Cyclic voltammetric response of a Pt/poly-$[Os(bpy)_2(vpy)_2]^{2+}$/poly- $[Ru(vbpy)_3]^{2+}$ bilayer with $\Gamma_{inner} = 3.65 \times 10^{-9}$ mol cm^{-2} and $\Gamma_{outer} = 1.92 \times 10^{-9}$ mol cm^{-2} at 50 mV s^{-1} in 0.1 *M* Et_4NClO_4/CH_3CN. S = 73 μA cm^{-2}. The potential was held at +1.4 V versus Ag/AgCl (add 50 mV to correct to SSCE) for 20 min before initiating the negative potential scan shown. [Reprinted with permission from C. R. Leidner and R. W. Murray, *J. Am. Chem. Soc. 107*, 551 (1985).]

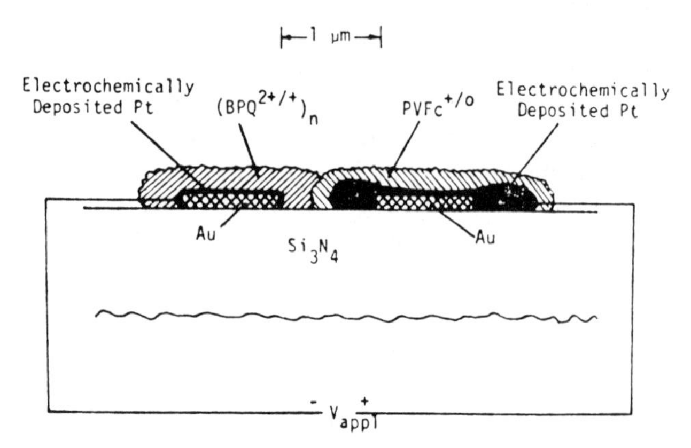

Figure 7.5. Cross-sectional view of "open-faced" bilayer electrode on interdigitated array. [Reprinted with permission from G. P. Kittleson, H. S. White, and M. S. Wrighton, *J. Am. Chem. Soc. 107*, 7373 (1985). Copyright © (1985) American Chemical Society.]

formulation of appropriate equations to describe the mediation event, $[Os(Me_2bpy)_3]^{2+}$ was employed as a solution reductant. This permitted steady state (rotated disk) measurement of the overall rate of the unfavorable mediation reaction.

An extensive data set of rotated disk voltammograms was analyzed using a theory adapted from polymer–solution studies (62, 70). Using this rather simple treatment the data indicated the polymer–polymer interfacial rate constant corresponded to a second-order homogeneous rate constant of $13–32 M^{-1}s^{-1}$. This range of values is due to a subtle, but discernible decrease in the overall rate with increasing film thicknesses; this is a characteristic of (partial) charge-transport control. The authors concluded that mixed interfacial and charge-transport control was operative and that the rate constants obtained at small coverages are the most reliable (lower limit) rate constants. The value of $32 M^{-1}s^{-1}$ is approximately 28 times slower than calculated for the analogous homogeneous reaction. This result represents the first (and only) estimate of a polymer–polymer interfacial electron-transfer rate constant. This study indicates that these polymer–polymer electron-transfer reactions are not dramatically different than corresponding polymer–solution or solution reactions.

Interdigitated arrays with dissimilar redox polymers (99) represent an "open-faced" sandwich arrangement of a bilayer between two metal electrodes (97). Wrighton's array method employs two closely spaced ($<1\ \mu m$) microelectrodes each coated with a different redox polymer; the polymer films contact between the two electrodes. Figure 7.5 illustrates this arrangement for a viologen polymer (BPQ^{2+}) and PVFc. However, the mediation reaction

$$BPQ^{+}/PVFc^{+} \rightarrow BPQ^{2+}/PVFc$$

is thermodynamically favorable by 1.0 V and is therefore very rapid. This reaction is so fast that charge-transport limitations within the polymer films become the rate-limiting steps, rather than the polymer–polymer mediation reaction. Although it was impossible to measure the interfacial rate constant in this particular example, this methodology may prove useful in other studies.

7.3 PROSPECTS FOR FUTURE STUDIES

The above overview of electron-transfer mediation in metal complex polymer films illustrates the remarkable progress of the last decade in unraveling a rather complicated situation. The theoretical developments, particularly those of Savéant, are most impressive. Complex theoretical derivations have been distilled down to quite workable equations that can be employed provided the experimentalist carefully approaches the problem. The experimental progress likewise has been impressive, however, we are not yet at the point where we can completely understand electron-transfer mediation. With the recent theoretical and experimental developments, we are approaching the juncture at which such understanding is possible. As more examples of electron-transfer mediation are reported, we should be in a much better position to address the comment made by Anson in 1980, and echoed by Faulkner and his co-workers in 1987 (45), that "each new system has appeared more particular than representative, and one is surely not yet in a position to offer reliable general recipes for preparing electrodes surfaces to achieve particular catalytic objectives" (51).

ACKNOWLEDGMENTS

I would like to sincerely thank the many individuals who provided reprints and preprints of articles on electron-transfer mediation with metal complex polymer films and other related topics. Many of these articles are cited in this chapter and all were useful in its preparation. Some have hopefully helped make this chapter interesting, but all were enlightening for me.

REFERENCES

1. R. W. Murray, *Electroanal. Chem. 13* 191 (1984).
2. R. W. Murray, *An. Rev. Mater. Sci. 14*, 145 (1984).
3. M. S. Wrighton, *Science 231*, 32 (1986).
4. N. Oyama and F. C. Anson, *J. Am. Chem. Soc. 101*, 3450 (1979).
5. A. Merz and A. J. Bard, *J. Am. Chem. Soc. 100*, 3222 (1978).
6. L. L. Miller and M. R. Van De Mark, *J. Am. Chem. Soc. 100*, 639 (1978).
7. P. Daum and R. W. Murray, *J. Electroanal. Chem. 103*, 289 (1979).
8. M. S. Wrighton, R. G. Austin, A. B. Bocarsly, J. M. Bolts, O. Haas, K. D. Legg, L. Nadjo, and M. C. Palazzotto, *J. Electroanal. Chem. 87*, 429 (1978).
9. W. J. Albery, A. W. Foulds, K. J. Hall, and A. R. Hillman, *J. Electrochem. Soc. 127*, 654 (1980).
10. A. F. Diaz, W. Y. Lee, J. A. Logan, and D. C. Green, *J. Electroanal. Chem. 108*, 377 (1980).
11. L. Polyak, D. R. Rolison, R. J. Kessler, and R. W. Howak, 163rd Electrochemical Society Meeting, San Francisco, CA, May 8–13, 1983, Vol. 83-1, No. 547.

12. J. C. Jernigan, K. O. Wilbourne, and R. W. Murray, *J. Electroanal. Chem. 222*, 193 (1987).
13. A. J. Bard and P. K. Ghosh, *J. Am. Chem. Soc. 105*, 5691 (1983).
14. R. D. King, D. G. Nocera, and T. J. Pinnavaia, *J. Electroanal. Chem. 236*, 43 (1987).
15. W. C. Dautremont-Smith, *Displays* 3 (1982).
16. M. J. Natan, T. E. Mallouk, and M. S. Wrighton, *J. Phys. Chem. 91*, 648 (1987).
17. D. Ingersoll, P. J. Kulesza, and L. R. Faulkner, 172nd Electrochemical Society Meeting, Honolulu, HI, October 18–23, 1987, No. 1402.
18. S. Sinha, B. D. Humphrey, and A. B. Bocarsly, *Inorg. Chem. 23*, 203 (1984).
19. S. Nakahama and R. W. Murray, *J. Electroanal. Chem. 158*, 303 (1983).
20. P. J. Peerce and A. J. Bard, *J. Electroanal. Chem. 112*, 97 (1980).
21. E. Laviron, L. Roullier, and F. Waldner, *J. Electroanal. Chem. 139*, 99 (1982).
22. T. E. Mallouk, V. Cammarata, J. A. Craston, and M. S. Wrighton, *J. Phys. Chem. 90*, 2150 (1986).
23. R. A. Simon, T. E. Mallouk, K. A. Daube, and M. S. Wrighton, *Inorg. Chem. 24*, 3119 (1985).
24. H. D. Abruña, P. Denisevich, M. Umana, T. J. Meyer, and R. W. Murray, *J. Am. Chem. Soc. 103*, 1 (1981).
25. P. Denisevich, H. D. Abruña, C. R. Leidner, T. J. Meyer, and R. W. Murray, *Inorg. Chem. 21*, 2153 (1982).
26. J. M. Calvert, R. H. Schmehl, B. P. Sullivan, J. S. Facci, T. J. Meyer, and R. W. Murray, *Inorg. Chem. 22*, 2151 (1983).
27. C. D. Ellis, L. D. Margerum, R. W. Murray, and T. J. Meyer, *Inorg. Chem. 22*, 1283 (1983).
28. C. R. Leidner, B. P. Sullivan, R. A. Reed, B. A. White, M. T. Crimmins, R. W. Murray, and T. J. Meyer, *Inorg. Chem. 26*, 882 (1987).
29. P. K. Ghosh and T. G. Spiro, *J. Electrochem Soc. 128*, 1281 (1981).
30. P. K. Ghosh and A. J. Bard, *J. Electroanal. Chem. 169*, 113 (1984).
31. T. F. Guarr and F. C. Anson, *J. Phys. Chem. 91*, 4037 (1987).
32. J. M. Calvert and T. J. Meyer, *Inorg. Chem. 20*, 27 (1987).
33. O. Haas and J. G. Vos, *J. Electroanal. Chem. 113*, 139 (1980).
34. O. Haas, M. Kriens, and J. G. Vos, *J. Am. Chem. Soc. 103*, 1318 (1981).
35. G. X. Wan, K. Shigehara, E. Tsuchida, and F. C. Anson, *J. Electroanal. Chem. 179*, 239 (1984).
36. K. A. Macor and T. G. Spiro, *J. Am. Chem. Soc. 105*, 5601 (1983).
37. B. A. White and R. W. Murray, *J. Electroanal. Chem. 189*, 345 (1985).
38. R. J. H. Chau and T. Kuwana, *J. Electroanal. Chem. 139*, 93 (1980).
39. K. Y. Wong and F. C. Anson, *J. Electroanal. Chem. 237*, 69 (1987).
40. J. Massaux, P. Burgmayer, E. Takeuchi, and R. W. Murray, *Inorg. Chem. 23*, 4417 (1984).
41. G. Brenard and J. Simonet, *J. Electroanal. Chem. 112*, 117 (1980).
42. N. Oyama and F. C. Anson, *J. Am. Chem. Soc. 101*, 739 (1979).
43. L. R. Faulkner, B. R. Shaw and G. P. Haight, Jr., *J. Electroanal. Chem. 140*, 147 (1982).
44. M. Majda and L. R. Faulkner, *J. Electroanal. Chem. 169*, 77 (1984).
45. E. T. T. Jones and L. R. Faulkner, *J. Electroanal. Chem. 222*, 201 (1987).
45. E. T. T. Jones and L. R. Faulkner, *J. Electroanal. Chem. 222*, 201 (1987).
46. J. R. Schneider and R. W. Murray, *Anal. Chem. 54*, 1508 (1982).
47. D. A. Buttry and F. C. Anson, *J. Am. Chem. Soc. 104*, 4824 (1982).
48. D. A. Buttry and F. C. Anson, *J. Am. Chem. Soc. 105*, 685 (1983).
49. I. Rubenstein and A. J. Bard, *J. Am. Chem. Soc. 102*, 6641 (1980).
50. C. R. Martin, I. Rubenstein, and A. J. Bard, *J. Am. Chem. Soc. 104*, 4817 (1982).
51. N. Oyama and F. C. Anson, *Anal. Chem. 52*, 1192 (1980).

52. N. Oyama, K. Shigehara, and F. C. Anson, *Inorg. Chem. 20*, 518 (1981).
53. J. S. Facci and R. W. Murray, *Anal. Chem. 54*, 772 (1982).
54. K. Doblhofer, H. Braun, and W. Storck, *J. Electrochem. Soc. 130*, 807 (1983).
55. F. C. Anson, T. Ohsaka, and J.-M. Savéant, *J. Phys. Chem. 87*, 640; 3174 (1983).
56. F. C. Anson, J.-M. Savéant, and K. Shigehara, *J. Am. Chem. Soc. 105*, 1096 (1983).
57. D. A. Buttry and F. C. Anson, *J. Am. Chem. Soc. 106*, 59 (1984).
58. J. E. Van Koppenhagen, and M. Majda, *J. Electroanal. Chem. 258*, 113 (1987).
59. C. P. Andrieux and J.-M. Savéant, *J. Electroanal. Chem. 88*, 43 (1978).
60. C. P. Andrieux, J.-M. Dumas-Bouchiat, and J.-M. Savéant, *J. Electroanal. Chem. 114*, 159 (1980).
61. C. P. Andrieux, J.-M. Dumas-Bouchiat, and J.-M. Savéant, *J. Electroanal. Chem. 123*, 171 (1981).
62. C. P. Andrieux, J. M. Dumas-Bouchiat, and J.-M. Savéant, *J. Electroanal. Chem. 131*, 1 (1982).
63. C. P. Andrieux, J. M. Dumas-Bouchiat, and J.-M. Savéant, *J. Electroanal. Chem. 169*, 9 (1984).
64. J.-M. Savéant, F. C. Anson, and K. Shigehara, *J. Phys. Chem. 87*, 214 (1983).
65. F. C. Anson, *J. Phys. Chem. 84*, 3336 (1980).
66. F. C. Anson, J.-M. Savéant, and K. Shigehara, *J. Electroanal. Chem. 145*, 423 (1983).
67. W. J. Albery and A. R. Hillman, *J. Electroanal. Chem. 170*, 27 (1984).
68. R. W. Murray, *Philos. Trans. R. Soc. London A302*, 253 (1981).
69. J. Leddy, A. J. Bard, J. T. Maloy, and J.-M. Savéant, *J. Electroanal. Chem. 187*, 205 (1985).
70. T. Ikeda, C. R. Leidner, and R. W. Murray, *J. Electroanal. Chem. 138*, 343 (1982).
71. C. R. Leidner, R. H. Schmehl, P. G. Pickup, and R. W. Murray, in *The Chemistry and Physics of Electrocatalysis*, J. D. E. McIntyre, M. J. Weaver, and E. B. Yeager (Eds.). The Electrochemical Society, Pennington, NJ 389 (1984).
72. T. Ikeda, C. R. Leidner, and R. W. Murray, *J. Am. Chem. Soc. 103*, 7422 (1981).
73. R. H. Schmehl and R. W. Murray, *J. Electroanal. Chem. 152*, 97 (1983).
74. C. R. Leidner and R. W. Murray, *J. Am. Chem. Soc. 106*, 1606 (1984).
75. T. Ikeda, R. H. Schmehl, P. Denisevich, K. W. Willman, and R. W. Murray, *J. Am. Chem. Soc. 104*, 2683 (1982).
76. C. P. Andrieux, O. Haas, and J.-M. Savéant, *J. Am. Chem. Soc. 108*, 8175 (1986).
77. J. F. Cassidy and J. G. Vos, *J. Electroanal. Chem. 218*, 341 (1987).
78. J. F. Cassidy and J. G. Vos, *J. Electroanal. Chem. 235*, 41 (1987).
79. J. F. Cassidy and J. G. Vos, *J. Electrochem. Soc. 135*, 863 (1988).
80. X. Chen, P. He and L. R. Faulkner, *J. Electroanal. Chem. 22*, 223 (1987).
81. K. Shigehara, N. Oyama, and F. C. Anson, *Inorg. Chem. 20*, 518 (1981).
82. K. Shigehara and F. C. Anson, *J. Electroanal. Chem. 132*, 107 (1982).
83. F. C. Anson, J.-M. Savéant, and K. Shigehara, *J. Electroanal. Chem. 145*, 423 (1983).
84. F. C. Anson, J.-M. Savéant, and K. Shigehara, *J. Am. Chem. Soc. 105*, 1096 (1983).
85. F. C. Anson, Y. M. Tsou, and J.-M. Savéant, *J. Electroanal. Chem. 178*, 113 (1984).
86. D. A. Buttry and F. C. Anson, *J. Am. Chem. Soc. 106*, 59 (1984).
87. F. C. Anson, C.-L. Ni, and J.-M. Savéant, *J. Am. Chem. Soc. 107*, 59 (1984).
88. B. Lindholm and M. Sharp, *J. Electroanal. Chem. 198*, 37 (1986).
89. M. Krishnan, X. Zhang, and A. J. Bard, *J. Am. Chem. Soc. 106*, 7371 (1984).
90. N. Oyama, K. Sato, and H. Matsuda, *J. Electroanal. Chem. 115*, 149 (1980).
91. P. W. Geno, K. Ravichandran, and R. P. Baldwin, *J. Electroanal. Chem. 183*, 155 (1985).
92. D. Belanger and M. S. Wrighton, *Anal. Chem. 59*, 1426 (1987).
93. K. N. Kuo and R. W. Murray, *J. Electroanal. Chem. 131*, 37 (1982).

94. P. G. Pickup, C. R. Leidner, P. Denisevich, and R. W. Murray, *J. Electroanal. Chem. 164*, 39 (1984).
95. C. R. Leidner, P. Denisevich, K. W. Willman, and R. W. Murray, *J. Electroanal. Chem. 164* 63 (1984).
96. C. R. Leidner and R. W. Murray, *J. Am. Chem. Soc. 107*, 551 (1985).
97. P. G. Pickup, W. Kutner, C. R. Leidner, and R. W. Murray, *J. Am. Chem. Soc. 106*, 1991 (1984).
98. N. Oyama, S. Yamaguchi, M. Kaneko, and A. Yamada, *J. Electroanal. Chem. 139*, 215 (1982).
99. G. P. Kittleson, H. S. White, and M. S. Wrighton, *J. Am. Chem. Soc. 107*, 7373 (1985).

Chapter **VIII**

VOLTAMMETRIC DIAGNOSIS OF CHARGE TRANSPORT ON POLYMER COATED ELECTRODES

Noboru Oyama* and Takeo Ohsaka†

**Department of Applied Chemistry, Faculty of Technology, Tokyo University of Agriculture and Technology, 2-24-16 Naka-machi, Koganei, Tokyo 184, Japan*

†Department of Electronic Chemistry, Graduate School at Nagatsuta, Tokyo Institute of Technology, 4259 Nagatsuta, Midori-ku, Yokohama 227, Japan

Molecular Design of Electrode Surfaces,
Edited by Royce W. Murray. Techniques of Chemistry Series, Vol. XXII.
ISBN 0-471-55773-0

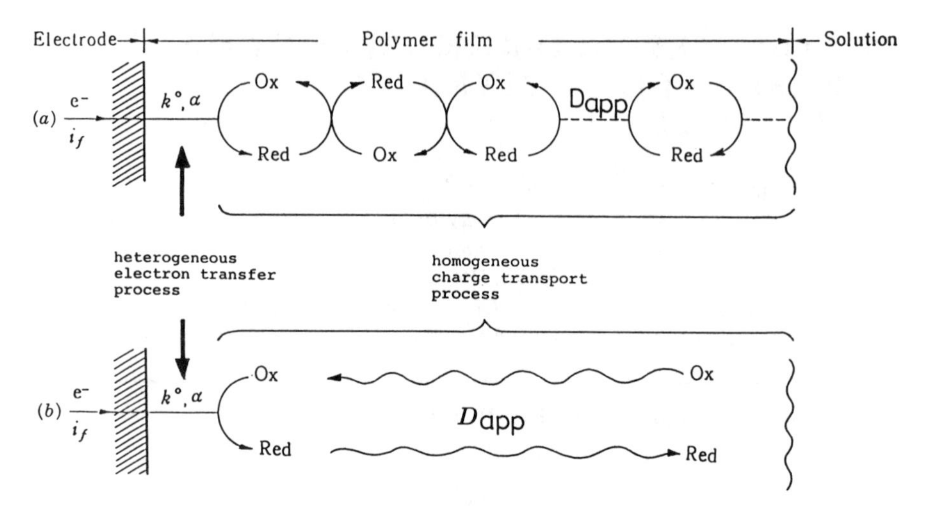

Scheme 8.1. Schematic depiction of the overall electrode reaction at polymer-coated electrodes, where electroactive species (or sites) are confined in the polymeric matrix. The terms $k°$ and α are the standard rate constant and the transfer coefficient, respectively, of the heterogeneous electron-transfer process. The term D_{app} is the apparent diffusion coefficient for the homogeneous charge-transport process within the polymer film. The faradaic current is i_f. The symbols Ox and Red represent the oxidized and reduced species (or sites), respectively.

At polymer-coated electrodes, where electroactive species (or sites) are confined in the polymeric matrix, the overall electrode reaction consists of a "heterogeneous" electron-transfer process between the electrode and the electroactive species as well as a "homogeneous" charge-transport process within the polymer film, as schematically illustrated in Scheme 8.1. The latter process is believed to occur via physical diffusion of the electroactive species itself and/or electron self-exchange (electron hopping) between adjacent pairs of oxidized (Ox) and reduced (Red) reactants. This electron exchange serves to shuttle electrons from the electrode to reactants that are located away from the electrode surface. Essentially, both processes are followed by the motions of a charge-compensating counterion, which is necessarily coupled to electron transfer for charge neutrality, the motion of solvent molecules for solvation, the segmental motion of polymer chains, and other processes. By the interplay of these processes the electrochemical response of polymer films can be greatly complicated compared with the normal diffusion process in solution for the electrode reaction at the bare electrode. As a result, the voltammetric responses obtained are much more difficult to interpret and, in fact, a great deal of effort has been focused on the unraveling of these processes (1–242). We will refer to the above-mentioned events collectively as the transport of electrochemical charge in the polymer film. The heterogeneous electron-transfer process is also complicated compared with that at the conventional bare electrode, for example, because electroactive species are confined in polymeric domains at coverages (concentrations) as high as several moles.

In this chapter, we will focus our attention on the charge-transport reaction at a polymer-coated electrode in a solution containing only a supporting electrolyte. An exception is the case where, because of the development of the discussion, we must refer to the charge-transport behaviors at polymer-coated electrodes in a solution containing redox species. These charge-transport behaviors at polymer-coated electrodes in a solution containing dissolved electroactive species have been successfully investigated experimentally and theoretically by many researchers, for example, from a viewpoint of an electrocatalysis (16, 81, 127–132, 134–138, 169–171, 243–246).

8.1 CHARGE TRANSPORT WITHIN POLYMER FILMS

In most films, especially in the case of covalent or strong coordinative attachment of the redox groups to the polymer backbone, electrochemical charge transport is thought, as proposed originally by Kaufman et al. (122, 123), to occur by an electron self-exchange reaction between neighboring oxidized and reduced sites (5, 6, 9–11, 25, 32, 33, 35, 37, 41, 42, 56, 57, 68, 71, 83, 84, 111, 112, 124, 139, 208). This electron-hopping process is mathematically representable (1, 70, 125, 126, 140) by diffusion laws in which an apparent diffusion coefficient, D_{app}, for a charge transport is introduced to measure its rate. Also, charge transport can occur via physical diffusion of electroactive reactants themselves (1–42, 44–49, 51–53, 70, 104, 123, 124, 133, 141, 143), especially in the case where the electroactive reactants incorporated in films are not part of the polymeric structure or permanently attached to functional groups present in the films. In this case the charge-transport rate is also characterized by D_{app}.

8.1.1 Methods Used to Study Charge Transport

Thus far a variety of electrochemical methods, including potential-step chronoamperometry (PSCA), potential-step chronocoulometry (PSCC), chronopotentiometry (CP), cyclic voltammetry (CV), normal pulse voltammetry (NPV), alternating current (ac) impedance method, rotating disk (and ring) electrode voltammetry (RDEV), microstructured electrode-based methods (sandwich electrode, bilayer electrode, array electrode, etc.), and other methods (e.g., spectroelectrochemistry and electrochemical luminescence measurements), have been applied to study charge-transport processes and to estimate D_{app} values (243–246).

Most of these methods are exactly the same as those used in conventional electrochemical experiments (247). Note that the thickness of polymer films at polymer-coated electrodes is usually less than about 10 μm. For this reason (a) the measurements should be confined to times sufficiently short to ensure that semiinfinite linear diffusion prevails or (b) in the analysis of the data obtained on a longtimescale the contribution of finite diffusion must be considered. Figure 8.1 shows a schematic depiction of time- and D_{app}-dependent profiles of the concentrations of fixed oxidized (C_{Ox}) and reduced (C_{Red}) sites in the films. These

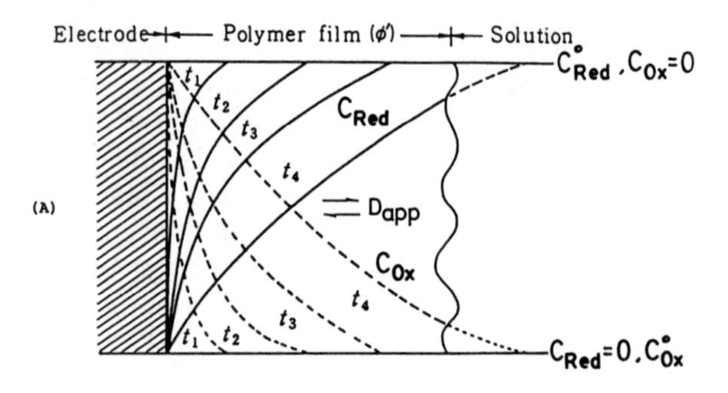

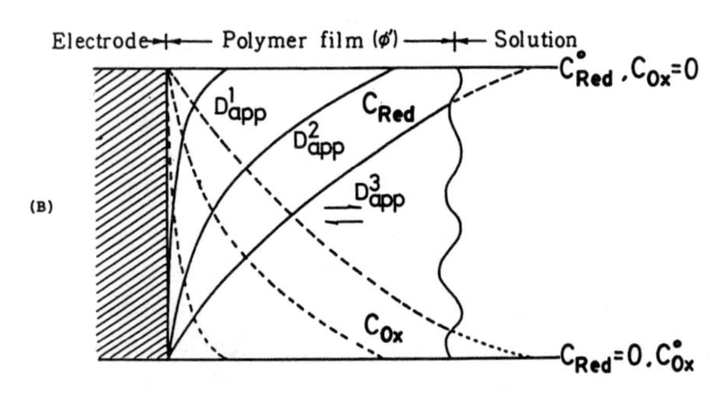

Figure 8.1. Schematic depiction of time- and D_{app}-dependent profiles of the concentrations of fixed oxidized (C_{Ox}, ---) and reduced (C_{Red}, ——) sites in the films. The electrode potential is stepped from the potential at which no oxidation of Red occurs substantially to the potential at which the oxidation of Red is diffusion controlled. (*A*) $t_1 < t_2 < t_3 < t_4$, (*B*) $D^1_{app} < D^2_{app} < D^3_{app}$. The term C^0_{Red} is the bulk concentration of Red confined initially in the film and C^0_{Ox} is the bulk concentration of the corresponding Ox.

profiles depend on the parameter $D_{app}t/\phi'^2$, where t is the experimental timescale and ϕ' is the polymer film thickness. If $D_{app}t/\phi'^2 \ll 1$, a semiinfinite electrochemical charge diffusion condition prevails. On such a timescale, the conventional electrochemical analysis procedures are applicable as they are. On the other hand, if $D_{app}t/\phi'^2 > 1$, as readily seen from Fig. 8.1, a finite diffusion must be considered, for example, as mentioned in Section 8.1.2.

8.1.2 Potential-Step Chronoamperometry, Potential-Step Chronocoulometry, and Chronopotentiometry

The following finite diffusion relationships have been derived for PSCA, PSCC, and CP, by Murray and his co-workers (70) and Oyama et al. (1, 28) for the simple electrode Reaction 8.1

$$\text{Red} \rightleftharpoons \text{Ox} + ne^- \tag{8.1}$$

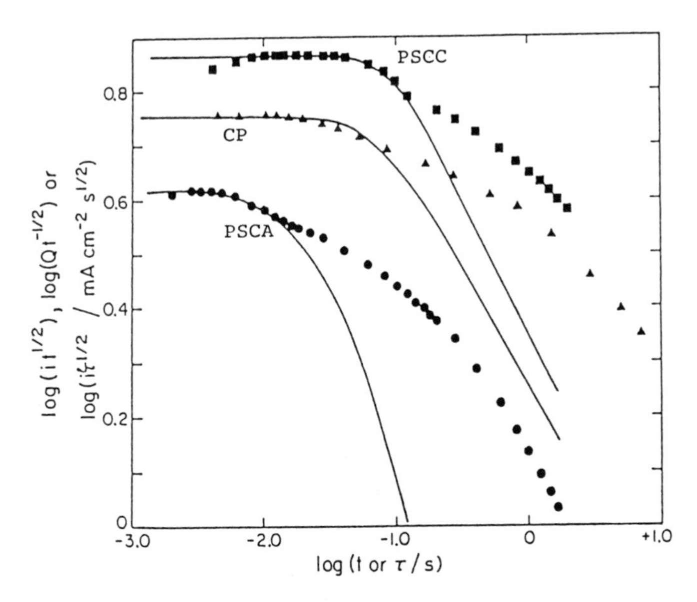

Figure 8.2. Calculated and experimental diffusional responses from $[Mo(CN)_8]^{4-}$ anions incorporated in a protonated poly(4-vinylpyridine) (PVP) coating containing 5.37×10^{-7} mol cm^{-2} of pyridine units. The solid lines are calculated from Eqs. 8.2–8.4. The symbols ●, ▲, and ■ represent the experimental points by PSCA, and PSCC, respectively. The coatings contain 1.06×10^{-7} mol cm^{-2} of $[Mo(CN)_8]^{4-}$ for PSCC and PSCA and 0.9×10^{-7} mol cm^{-2} for CP. (Reprinted with permission from Ref. 1.)

in which the reduced half (Red) of the redox couple is incorporated into a film of thickness ϕ'.

$$\text{PSCA} \qquad i = \frac{n\mathcal{F} C^{\circ}_{\text{Red}} D^{\text{a}}_{\text{app}}{}^{1/2}}{(\pi t)^{1/2}} \left(1 + 2 \sum_{m=1}^{\infty} (-1)^m \exp\left(\frac{-m^2\phi'^2}{D^{\text{a}}_{\text{app}} t}\right)\right] \tag{8.2}$$

$$\text{PSCC} \qquad Q = \frac{2n\mathcal{F} C^{\circ}_{\text{Red}} D^{\text{a}}_{\text{app}}{}^{1/2} t^{1/2}}{\pi^{1/2}} \left\{1 + 2 \sum_{m=1}^{\infty} (-1)^m \left[\exp\left(\frac{-m^2\phi'^2}{D^{\text{a}}_{\text{app}} t}\right) - \frac{m\phi'\pi^{1/2}}{(D^{\text{a}}_{\text{app}} t)^{1/2}} \operatorname{erfc}\left(\frac{m\phi'}{(D^{\text{a}}_{\text{app}} t)^{1/2}}\right)\right]\right\} \tag{8.3}$$

$$\text{CP} \qquad i' = \frac{n\mathcal{F} (\pi D^{\text{a}}_{\text{app}})^{1/2} C^{\circ}_{\text{Red}}}{2\tau^{1/2}} \left[1 + 2 \sum_{m=1}^{\infty} \left\{\exp\left(\frac{-m^2\phi'^2}{D^{\text{a}}_{\text{app}} t}\right) - \frac{m\phi'\pi^{1/2}}{(D^{\text{a}}_{\text{app}}\tau)^{1/2}} \operatorname{erfc}\left(\frac{m\phi'}{(D^{\text{a}}_{\text{app}}\tau)^{1/2}}\right)\right\}\right]^{-1} \tag{8.4}$$

In these cases i is the current density, Q is the charge per unit area, i' is the constant current density, t is the time, τ is the transition time, C_{Red} is the initial

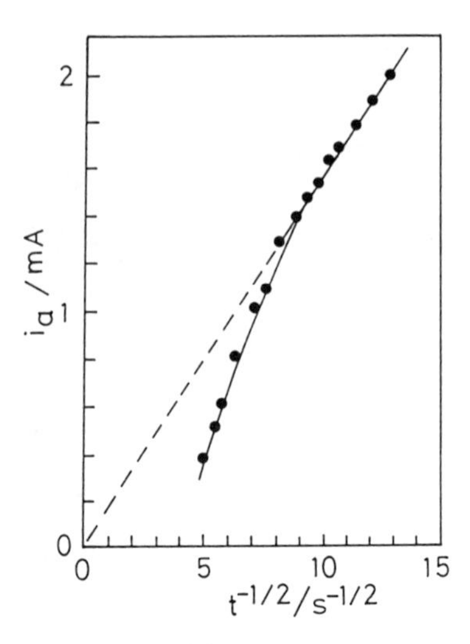

Figure 8.3. Potential-step chronoamperometry of plasma-polymerized vinylferrocene film (surface concentration of electroactive ferrocene sites: 4.9×10^{-9} mol cm^{-2}) in 1*M* $LiClO_4$ aqueous solution. Potential step: 0–0.8 V versus SSCE. The solid line is calculated from Eq. 8.2 for $(D_{app})^{1/2}$, $C = 2.24 \times 10^{-8}$ mol cm^{-2} s$^{-1/2}$, and electrode area = 0.130 cm^2. The dashed line corresponds to a semiinfinite diffusion process. The symbol ● represents the experimental points. [Reprinted with permission from P. Daum and R. Murray, *J. Phys. Chem.*, *85*, 389 (1981). Copyright © (1981) American Chemical Society.]

concentration of Red in the film, D^a_{app} is the apparent diffusion coefficient for the anodic process, and n and $\mathcal{F}$ have their usual meanings.

The leading terms on the right-hand side (rhs) of all three equations correspond to the response that is obtained when the diffusion is semifinite. The first terms in Eqs. 8.2 and 8.3 are well known as the Cottrell equation. The succeeding infinite series allow for the effects of finite diffusion within the film. Figure 8.2 shows typical plots for finite diffusional response (1). As expected, the values of $it^{1/2}$, $Qt^{-1/2}$, and $i'\tau^{1/2}$ are constants independent of t (or τ) when t (or τ) is sufficiently short because in these cases the effects of finite diffusion are negligible. At longer times, however, the finiteness of the diffusion becomes significant and these values decrease with increasing t (or τ). The experimental results follow the expected ones, except that at longer times the experimental points deviate markedly from the calculated curves. This is the result of nonuniformity in the film thickness (144). The effects of finite diffusion have been easily realized in the conventional Cottrell plots (e.g., see Fig. 8.3) (69).

Potential-step chronoamperometry and PSCC were employed to estimate D_{app} values for most films because of their experimental ease (Table 8.1). In addition, "iR" losses may be prevented by addition of an appropriate voltage to ensure that the reactant at the electrode–film interface is consumed rapidly. Under these conditions film charge propagation is rate limiting and a diffusionally limited response is observed.

8.1.3 Normal Pulse Voltammetry

Normal pulse voltammetry is also useful in the estimation of D_{app} (7–9, 22–24, 39, 50, 56–58, 111–116), because PSCA and NPV are based on the same principle. At sufficiently short times, when the depletion of the reactant has not reached the film–solution interface, the NPV Cottrell (Cott) equation is obeyed:

$$(i_d)_{Cott} = \frac{n\mathscr{F} A D_{app}^{\alpha\ 1/2} C_{Red}^{\circ}}{(\pi \tau_s)^{1/2}} \tag{8.5}$$

where $(i_d)_{Cott}$ is the limiting current, τ_s is the sampling time, A is the electrode area and C_{Red}° is the concentration of the reactant in the film. Equation 8.5 is equivalent to the conventional Cottrell plot, which is expressed by the first term on the rhs in Eq. 8.2. The $(i_d)_{Cott}$ value can be easily obtained from the plateau portion of the voltammogram, in analogy with dc (direct current) polarography for a solution reactant (247). Also, kinetic information (standard rate constant k° and transfer coefficient α) for the heterogeneous electron-transfer process at the electrode–film interface can be obtained from the rising portion of the voltammogram, as mentioned in Section 8.2.1. A typical example of such a NPV Cottrell plot is shown in Fig. 8.4 (56).

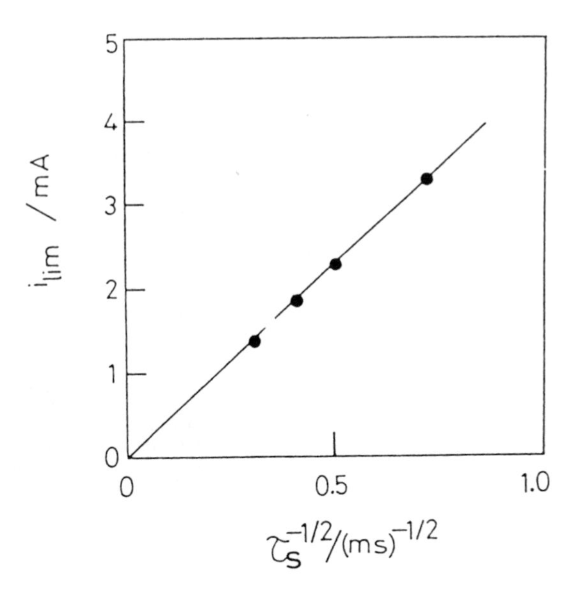

Figure 8.4. A typical normal pulse voltammetric Cottrell plot for the one-electron reduction of the viologen dication as poly(styrene-*co*-chloromethylstyrene) pendant viologen (PMV) coated on basal-plane pyrolytic graphite (BPG) electrode in a 0.2*M* KCl solution (pH 3.0). Electrode area = 0.17 cm^2. Concentration ($C_{MV^{2+}}$) of electroactive viologen sites = 2.1 × 10^{-4} mol cm^{-3}. [Reprinted with permission from N. Oyama, T. Ohsaka, H. Yamamoto, and M. Kaneko, *J. Phys. Chem.*, *90*, 3850 (1986). Copyright © (1986) American Chemical Society.]

Table 8.1 Charge-Transport Rates Within Polymer Films on Electrode Surfaces[a]

System	$D_{app}(cm^2\,s^{-1})$	Medium
PVP·H$^+$,M		
M = $[Fe(CN)_6]^{3-}$	3.0×10^{-9}	H_2O, 0.2*M* CF_3COONa,
M = $[Fe(CN)_6]^{4-}$	3.9×10^{-9}	0.01*M* CF_3COOH
M = $[W(CN)_8]^{4-}$	11.5×10^{-9}	
M = $[Mo(CN)_8]^{4-}$	$(2.5-85) \times 10^{-9}$	
M = $[Fe(CN)_6]^{3-}$	$10^{-11} \sim 3 \times 10^{-9}$	H_2O(pH 1.5), 0.2*M* CF_3COONa, $\Gamma_M/\Gamma_{PVP} = 0.0032 \sim 0.32$
M = $[IrCl_6]^{2-}$	$(0.9 \sim 11) \times 10^{-8}$	H_2O(pH 2), 0.1*M* CF_3COONa $C = 0.064 \sim 0.43M$
M = $[Fe(CN)_6]^{4-/3-}$		H_2O(pH 2), 0.1*M* CF_3COONa
M = $[Fe(CN)_6]^{3-}$	$\sim 10^{-9}$	H_2O, 0.1*M* KCl
M = $[Fe(CN)_6]^{3-}$	$(2.0 \pm 0.5) \times 10^{-9}$	H_2O, 0.1*M* KCl
M = $Fe(CN)_6]^{3-}$ for$[Fe(CN)_6]^{4-}\}$	$1 \times 10^{-9} \sim 4.2 \times 10^{-7}$	H_2O(pH 1.5), 0.2*M* CF_3COONa, CF_3COOH $\Gamma_M/\Gamma_{PVP} = 0.005 \sim 0.18$
M = $[Mo(CN)_8]^{4-}$	$2.5 \times 10^{-9} \sim 4.4 \times 10^{-8}$	H_2O(pH 1.5), 0.2*M* CF_3COONa, CF_3COOH $\Gamma_M/\Gamma_{PVP} = 0.0025 \sim 0.22$
M = $[IrCl_6]^{3-}$	$3.2 \times 10^{-9} \sim 2.5 \times 10^{-7}$	H_2O(pH 1.5), 0.2*M* CF_3COONa, CF_3COOH $\Gamma_M/\Gamma_{PVP} = 0.0039 \sim 0.11$
	$(4 \sim 10) \times 10^{-9}$	H_2O(pH 2), 0.2*M* CF_3COONa, $C = 0.22 \sim 0.57M$
M = $[W(CN)_8]^{4-}$	$(1.8 \pm 0.2) \times 10^{-8}$	H_2O(pH 1.5), 0.2*M*

Process–Species Responsible	Comments–Method	References
	Chronoamperometric, Chronocoulometric, and Chronopotentiometric data in good agreement	1
Probably physical motion at low loadings	Decrease in D_{app} as Γ_M/Γ_{PVP} increases from 0.03 to 0.15, thereafter constant	
Electrostatic cross-linking	AC impedance method D_{app} and k° decrease with increasing C $k^\circ = 10^{-4} \sim 10^{-2}\,\text{cm s}^{-1}$ $\alpha_c = 0.5$(at $C = 1.1M$), 0.41(at $C = 0.41M$)	2
Ionic movement and electron hopping	AC impedance method	3
	AC impedance method $k^\circ = (2.7 \pm 0.7) \times 10^{-4}\,\text{cm s}^{-1}$ at $C = 0.023 - 0.2M$ k° decreases from 4.1×10^{-5} to $0.8 \times 10^{-5}\,\text{cm s}^{-1}$ when C increase from 0.044 to $0.1M$	4
Diffusion–migration mechanism	RDEV	5
Physical motion Insignificance of electron hopping	PSCA $D_{app} = (3.6 \pm 1) \times 10^{-9}\,\text{cm}^2\,\text{s}^{-1}$ from film resistance' measurement	6
Electrostatic cross-linking	NPV $k^\circ = 1.6 \times 10^{-4} \sim 3.2 \times 10^{-3}\,\text{cm s}^{-1}$ and $\alpha_c = 0.5 \sim 0.3$ at $\Gamma_M/\Gamma_{PVP} = 0.005 \sim 0.18$	7
	NPV, $k^\circ = 1.7 \times 10^{-4} \sim 9.7 \times 10^{-4}\,\text{cm s}^{-1}$ and $\alpha_c = 0.5$ at $\Gamma_M/\Gamma_{PVP} = 0.0025 \sim 0.22$	8
Electrostatic cross-linking Electron hopping	NPV, $k^\circ = 2.8 \times 10^{-4} \sim 2.2 \times 10^{-3}\,\text{cm s}^{-1}$ and $\alpha_a = 0.32 \pm 0.04$ at $\Gamma_M/\Gamma_{PVP} = 0.0039 \sim 0.11$ $E_a = 31.4 \pm 2.1\,\text{kJ mol}^{-1}$ and $D_0 = 2.3 \times 10^{-1}\,\text{cm}^2\,\text{s}^{-1}$ at $\Gamma_M/\Gamma_{PVP} = 0.0071$, and $E_a = 32.6 \pm 2.1\,\text{kJ mol}^{-1}$, and $D_0 = 1.5 \times 10^{-2}\,\text{cm}^2\,\text{s}^{-1}$ at $\Gamma_M/\Gamma_{PVP} = 0.074$	9
	PSCC Cross-linked PVP	253
	$k^\circ = 7.1 \times 10^{-4}\,\text{cm s}^{-1}$ and α_a	9

Table 8.1 Continued

System	D_{app}(cm^2 s^{-1})	Medium
		CF_3COONa, CF_3COOH $\Gamma_M/\Gamma_{PVP} = 0.006 \sim 0.1$
PVP·H$^+$/(γ-methacryl-oxypropyl)trimethoxy-silane copolymer $-[-CH_2-CH(C_5H_4N^+H)-)_{0.94}-(-CH_2-C(CH_3)(C{=}O-O(CH_2)_3Si(OCH_3)_3)-)_{0.062}-]-$ 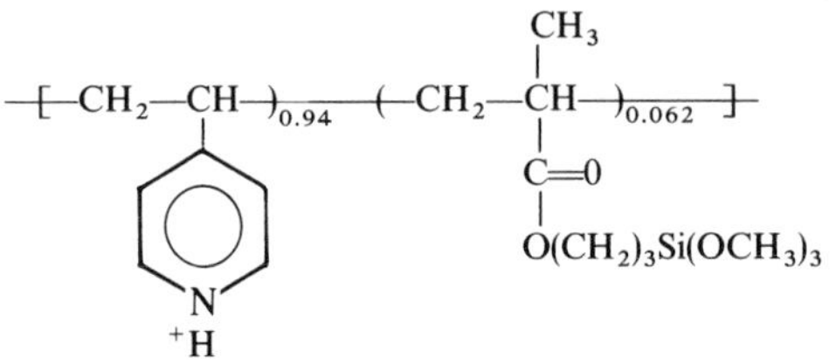		
$[Fe(CN)_6]^{3-}$	$(0.3-3.0) \times 10^{-8}$	H_2O(pH 2.8), 2*M* LiCl $C = 0.011 \sim 1.1M$
$[IrCl_6]^{3-}$	$(0.05-2.8) \times 10^{-8}$	$C = 0.011 \sim 0.7M$
Quaternized and cross-linked poly(4-vinylpyridine)$_{0.83}$-Co-(styrene)$_{0.17}$, $[IrCl_6]^{2-}$	3×10^{-9} (low permeability region) 2.3×10^{-8} (high permeability region)	H_2O, 0.1*M* KCl
Vinylpyridine–acrylamide copolymeric gels, MV^{2+}	7.8×10^{-6}	H_2O, 0.1*M* KCl/H_2O(pH 2.5), 0.1*M* KCl/H_2O(pH 2.5), 0.1*M* KCl, 1 m*M* $K_4Fe(CN)_6$
PVP impregnated microporous aluminum oxide films		
$[Fe(CN)_6]^{4-}$	$\sim 10^{-8}$	H_2O(pH 3), 0.1*M* CF_3COONa, CF_3COOH
$[Fe(CN)_6]^{4-}$	$(1.0 \pm 0.2) \times 10^{-8}$	H_2O(pH 2.7), 0.1*M* KCl, 0.05*M* $CH_2ClCOOH$
	$(1.3 \pm 0.3) \times 10^{-8}$	
Poly(*l*-lysine)·H$^+$,M M = $[Co(C_2O_4)_3]^{3-}$	2×10^{-7}	H_2O(pH 1.5), 0.2*M* CF_3COONa,
	2×10^{-6}	CF_3COOH

Process–Species Responsible	Comments–Method	References
	$=0.32\pm0.04$ at $\Gamma_M/\Gamma_{PVP}=$ $0.006\sim0.1$ $E_a=30.5\pm2.9$ kJ mol^{-1} and $D_0=2.3\times10^{-3}$ cm^2 s^{-1} at $\Gamma_M/\Gamma_{PVP}=0.034$	
	PSCA Cross-linking or coulombic repulsion lowers D_{app} at higher loadings ($>0.3M$) Rates enhanced by electron transfer in $[IrCl_6]^{3-}/[Fe(CN)_6]^{3-}$ mixtures	10, 25
Ion motion	RDEV $D_{app}=(3.5\pm1)\times10^{-9}$ cm^2 s^{-1} for film resistance measurement	11
	RDEV	12
	PSCA	13
	PSCC	14
	RDEV	
	PSCC. Co-species inside Donnan domains PSCC. Co-species outside Donnan domains	15

Table 8.1 Continued

System	D_{app}($cm^2\,s^{-1}$)	Medium
$M = Fe^{II}edta^{2-}$	7.3×10^{-6}	H_2O(pH 5.5), 0.2*M* CH_3COONa
$M = Fe^{III}edta^{-}$	2.1×10^{-6}	
$M = [Mo(CN)_8]^{4-}$	$\sim 8 \times 10^{-7}$	H_2O(pH 5.5), 0.2*M* CH_3COONa
$M = [W(CN)_8]^{4-}$	$\sim 8 \times 10^{-7}$	
$M = [Co(tpy)_2]^{2+}$	3×10^{-6}	
Random ternary copolymer (I)		H_2O(pH 4.5), 0.1*M*
PVI(MW = 7×10^4)/I, $[Fe(CN)_6]^{4-}$	1.6×10^{-6}	CH_3COONa, CH_3COOH
PVP(MW = 7.5×10^5)/I $[Fe(CN)_6]^{4-}$	1.4×10^{-6}	
PVP(MW = 7×10^4)/I, $[Fe(CN)_6]^{4-}$	1.1×10^{-6}	
PVP(MW = 1×10^4)/I, $[Fe(CN)_6]^{4-}$	1.5×10^{-6}	
ND1/I, $[Fe(CN)_6]^{4-}$	1.5×10^{-6}	
ND2/I, $[Fe(CN)_6]^{4-}$	1.9×10^{-6}	
ND3/I, $[Fe(CN)_6]^{4-}$	4.9×10^{-6}	
PEI(MW = 1.8×10^3)/I, $[Fe(CN)_6]^{4-}$	3.6×10^{-6}	
l-PEI/I, $[Fe(CN)_6]^{4-}$	2.9×10^{-6}	
PLL(MW = 1.8×10^5)/I, $[Fe(CN)_6]^{4-}$	2.1×10^{-6}	
PVI/I, $[Fe(CN)_6]^{4-}$	1.4×10^{-6}	
Block copolymer (B)		H_2O(pH 4.5), 0.1*M* CH_3COONa, CH_3COOH
B-30, $[Fe(CN)_6]^{4-}$	1.9×10^{-9}	
B-46, $[Fe(CN)_6]^{4-}$	6.6×10^{-9}	
B-80, $[Fe(CN)_6]^{4-}$	9.2×10^{-9}	
Random copolymer (R)		H_2O(pH 4.5), 0.1*M*
R-50, $[Fe(CN)_6]^{4-}$	1.5×10^{-9}	CH_3COONa, CH_3COOH
PVP/copolymer IV, $[Fe(CN)_6]^{4-}$	$(0.8-4) \times 10^{-8}$	H_2O(pH 4.5), 0.1*M* CH_3COONa, CH_3COOH
PVP/copolymer V, $[Fe(CN)_6]^{4-}$	$(2-9) \times 10^{-7}$	H_2O(pH 4.5), 0.1*M* CH_3COONa, CH_3COOH

$$\left[-N^{+}(R_1)(R_2)-C_6H_4- \right]_n,\ M$$

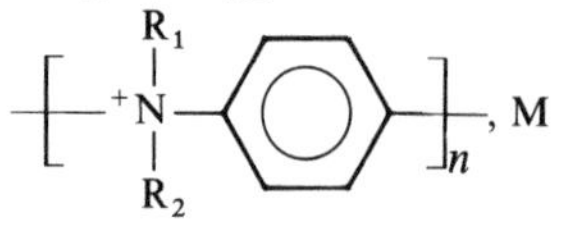

Process–Species Responsible	Comments–Method	References
	(RDEV determination of κ)	
No electron hopping, ion-hopping mechanism Two-phase model	PSCC. $D_{app} = 6.7 \times 10^{-6}$ $cm^2\,s^{-1}$ from RDEV D_{app} independence of $C (0.002 \sim 0.049 M)$	16
Diffusion outside Donnan domains	RDEV using $[Co(C_2O_4)_2(NH_3)_2]^-$ as surrogate ion	16
	PSCC. Inside Donnan domains	17
	PSCC. Inside Donnan domains	17
	RDEV	17
Ion-hopping mechanism within Donnan domains	PSCC	18, 19
	PSCC. D_{app} decreases with increasing C	20
	PSCC	20
Single-file diffusion within Donnan domains	PSCC	21
Single-file diffusion within Donnan domains	PSCC	21

Table 8.1 Continued

System	D_{app}($cm^2\,s^{-1}$)	Medium
$R_1 = R_2 =$—CH_3, $[Fe(CN)_6]^{3-}$	$(1.8 \pm 0.2) \times 10^{-9}$	H_2O(pH 1), 0.2*M* CF_3COONa, CF_3COOH $C = 1.0M$
$R_1 = R_2 =$—CH_3, $[Fe(CN)_6]^{3-}$	$(1.6 \pm 0.3) \times 10^{-9}$	$C = 0.25M$
	$(2.5 \pm 0.3) \times 10^{-9}$	
$R_1 =$—CH_3, $R_2 =$—C_2H_5, $[Fe(CN)_6]^{3-}$	$(2.8 \pm 0.3) \times 10^{-9}$	$C = 0.27M$
	$(5.8 \pm 0.4) \times 10^{-9}$	
$R_1 = R_2 =$—C_2H_5, $[Fe(CN)_6]^{3-}$	$(1.1 \pm 0.2) \times 10^{-8}$	$C = 0.23M$
	$(2.0 \pm 0.3) \times 10^{-8}$	
	$(1.7 \pm 0.2) \times 10^{-8}$	
$R_1 = R_2 =$—CH_3, $[W(CN)_8]^{4-}$	$(3.6 \pm 0.8) \times 10^{-10}$	$C = 0.42M$
	$(4.1 \pm 0.8) \times 10^{-10}$	
	$(5.3 \pm 1.1) \times 10^{-10}$	
$R_1 = R_2 =$—CH_3, $[Mo(CN)_8]^{4-}$	$(6.8 \pm 1.3) \times 10^{-10}$	$C = 0.32M$
$R_1 = R_2 =$—CH_3, $[Ru(CN)_6]^{4-}$	$(3.2 \pm 0.7) \times 10^{-10}$	$C = 0.18M$
$R_1 = R_2 =$—CH_3, $[Os(CN)_6]^{4-}$	$(2.8 \pm 0.6) \times 10^{-10}$	$C = 0.31M$
	$(5.4 \pm 1.1) \times 10^{-10}$	
$R_1 = R_2 =$—CH_3, $[Os(CN)_6]^{3-}$	$(3.0 \pm 0.6) \times 10^{-10}$	$C = 0.27M$
$R_1 = R_2 =$—CH_3, $[IrCl_6]^{2-}$	$(1.1 \pm 0.3) \times 10^{-9}$	$C = 0.50M$
	$(9.2 \pm 1.8) \times 10^{-10}$	
$R_1 = R_2 =$—CH_3, $Fe^{II}[phen($—$\phi SO_3^-)_2]_3^{4-}$	$3.1 \times 10^{-9} \sim 2.0 \times 10^{-10}$	H_2O(pH 3), 0.2*M* CF_3COONa, CF_3COOH $C = 0.021 \sim 0.23M$
Poly[3-(2-amino-ethylamino)propyl-trimethoxysilane], $[Fe(CN)_6]^{4-}$	3.4×10^{-9}	H_2O(pH 3.2), 0.1*M* KCl, 0.5*M* glycine
Poly[4-(β-tri-methoxysilyl)-ethylpyridine],	9×10^{-9}	H_2O(pH 2.5), 0.4*M* $ClCH_2COONa$, 0.5*M* KCl

Process–Species Responsible	Comments–Method	References
	NPV, PSCA, PSCC $k^{\circ}=(1.3\pm0.2)\times10^{-4}\,\mathrm{cm\,s^{-1}}$, $\alpha_c=0.31\pm0.02$	22
Single-file diffusion, electrostatic cross-linking	NPV $k^0=(8.9\pm1.8)\times10^{-5}\,\mathrm{cm\,s^{-1}}$, $\alpha_c=0.31\pm0.02$ PSCA, PSCC D_{app} decreases with increasing C	23
	NPV $k^0=(1.0\pm0.5)\times10^{-4}\,\mathrm{cm\,s^{-1}}$, $\alpha_c=0.31\pm0.05$ PSCA	23
	NVP $k^0=(2.2\pm0.5)\times10^{-4}\,\mathrm{cm\,s^{-1}}$, $\alpha_c=0.26\pm0.02$ PSCA PSCC	23
	NPV $k^0=(7.0\pm2.9)\times10^{-5}\,\mathrm{cm\,s^{-1}}$, $\alpha_c=0.56\pm0.03$ PSCC PSCA	23
	NPV $k^0=(1.8\pm0.3)\times10^{-4}\,\mathrm{cm\,s^{-1}}$ $\alpha_c=0.33$. D_{app} decreases with increasing C	23
	PSCA	23
	PSCA	23
	NPV	
	PSCA	23
	PSCA	23
	PSCC	
	NPV. $k^0=6.7\times10^{-5}\sim$ $2.4\times10^{-5}\,\mathrm{cm\,s^{-1}}$ $\alpha_a=0.71-0.74$ $\alpha_c=0.23-0.25$	24
Probably electron hopping	PSCA. D_{app} decreases at high loadings	26
	PSCA. RDEV experiment gives $D_{app}=$ $8.1\times10^{-9}\,\mathrm{cm^2\,s^{-1}}$	27

Table 8.1 Continued

System	$D_{app}(cm^2\,s^{-1})$	Medium
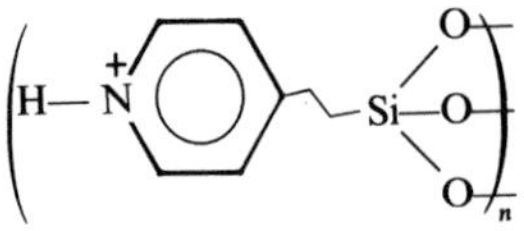 $(H-\overset{+}{N}\langle\bigcirc\rangle\text{-}CH_2CH_2\text{-}Si(O)_3)_n$		
$[Fe(CN)_6]^{4-}$		
PVP-M		
$M = Ru^{II}edta^{2-}$		H_2O(pH 3.4), 0.2M CF_3COONa
$M = [Fe^{II}(CN)_5]^{3-}$	$(0.63 \sim 1.1) \times 10^{-7}$	H_2O(pH 3), 0.2M CF_3COONa
$M = [Fe^{II}(CN)_5]^{3-}$ {or$[Fe^{III}(CN)_5]^{2-}$}	$(0.5 \sim 8) \times 10^{-8}$	$\Gamma_M/\Gamma_{PVP} = 0.005 \sim 0.075$
$M = Fe^{II/III}$	$(0.5 \sim 7.0) \times 10^{-11}$	H_2O, 5M HCl
Nafion, M		
$M = [Ru(bpy)_3]^{2+}$	2.7×10^{-9}	H_2O, 0.2M Na_2SO_4 $C = 0.5M$
	4×10^{-10}	H_2O, 0.2M Na_2SO_4
	5.8×10^{-10}	H_2O, 0.1M K_2SO_4 (or H_2SO_4, $CH_3\phi SO_3H$)
	$(1.7 \pm 0.2) \times 10^{-10}$	H_2O(pH 3.0), 0.2M CF_3COONa
	1.2×10^{-8}	H_2O(pH 3.3), 0.2M CF_3COONa
	4×10^{-10}	H_2O, 0.2M Na_2SO_4
$M = [Ru(bpy)_3]^{3+}$	1.2×10^{-9}	H_2O, 0.2M Na_2SO_4 $C = 0.5M$
	6.8×10^{-10}	H_2O, 0.1M H_2SO_4
	5.8×10^{-10}	H_2O, 0.1M K_2SO_4 (or $CH_3\phi SO_3H$)
	$(1.8 \pm 0.2) \times 10^{-10}$	H_2O(pH 3.0), 0.2M CF_3COONa
$M = [Ru(bpy)_3]^{3+/2+}$	1.5×10^{-10} 2.3×10^{-9}	H_2O, 0.45M Na_2SO_4, 0.05M H_2SO_4
$M = [Fe(bpy)_3]^{3+}$	$(1.0 \pm 0.3) \times 10^{-10}$	H_2O(pH 3.0), 0.2M CF_3COONa
$M = [Os(bpy)_3]^{2+}$	0.7×10^{-10}	H_2O, 0.2M Na_2SO_4

Process–Species Responsible	Comments–Method	References
	PSCC. $E_a \sim 19.2\,\mathrm{kJ\,mol^{-1}}$	28
	PSCA. Values increase to a plateau at high loading	29
	NPV. $k^0 = 1.5 \times 10^{-4} \sim 1.5 \times 10^{-3}\,\mathrm{cm\,s^{-1}}$ $\alpha_c = 0.52 \pm 0.05$ $\alpha_a = 0.40 \pm 0.05$	7
	PSCA. Redox film prepared by glow-discharge polymerization of vpy and $Fe(CO)_5$	30
Not counterion motion	PSCA. CV $k^0 = 2.4 \times 10^{-5}\,\mathrm{cm\,s^{-1}}$	31
Not physical motion	PSCA For physical motion $D_m = (0.2 \sim 0.3) \times 10^{-10}\,\mathrm{cm^2\,s^{-1}}$	32
Not counterion motion	PSCA. CV. No concentration dependence $k^0 = 4 \times 10^{-4}\,\mathrm{cm\,s^{-1}}$	33
	PSCC Electron hopping diffuser	34
	PSCA. PSCC No concentration dependence	35
	Convolution voltammetry $k^0 = 10^{-4}\,\mathrm{cm\,s^{-1}}$, $\alpha_a = 0.61$ at $\phi' = 0.22 \times 10^{-4}\,\mathrm{cm}$	36
Not counterion motion	PSCA	31
Not counterion motion	PSCA. CV. No concentration dependence	33
Not counterion motion	PSCA. CV. No concentration dependence	33
	PSCC Electron-hopping diffuser	34
Two parallel diffusion paths	AC impedance method	38
	PSCC	34
	PSCA	32

Table 8.1 Continued

System	$D_{app}(cm^2 s^{-1})$	Medium
	$4.5 \times 10^{-10} \sim 4.7 \times 10^{-9}$	H_2O(pH 5.5), 0.2*M* CF_3COONa, CH_3COONa, CH_3COOH $C = 0.032 \sim 0.32M$
	2.2×10^{-10}	H_2O, 0.01*M* $LiClO_4$ $\Gamma_M = 7.1 \times 10^{-9}$ mol cm^{-2}
	1.5×10^{-8}	H_2O, 0.01*M* $LiClO_4$ $\Gamma_M = 7.1 \times 10^{-9}$ mol cm^{-2}
$M = [Fe(o\text{-phen})_3]^{2+}$	$(0.4 \pm 0.1) \times 10^{-10}$	H_2O(pH 3.0), 0.2*M* CF_3COONa
$M = [Fe(o\text{-phen})_3]^{3+}$	$(0.4 \pm 0.1) \times 10^{-10}$	H_2O(pH 3.0), 0.2*M* CF_3COONa
$M = [CpFeCp'—N(CH_3)_3]^{+}$	1.7×10^{-10}	H_2O, 0.2*M* Na_2SO_4 $C = 0.007 \sim 1.2M$
	$(1.3 \pm 0.7) \times 10^{-10}$	H_2O(pH 3.0), 0.2*M* CF_3COONa
	7.5×10^{-10}	H_2O(pH 5.5), 0.2*M* CF_3COONa, CH_3COONa, CH_3COOH $C = 0.032 \sim 0.32M$
$M = [CpFeCp'—N(CN_3)_3]^{2+}$	$(3.8 \pm 0.2) \times 10^{-10}$	H_2O(pH 3.0), 0.2*M* CF_3COONa
$M = [Co(bpy)_3]^{2+}$	$(0.3 \sim 3.4) \times 10^{-11}$	H_2O, 0.5*M* Na_2SO_4 $C = 0.15$ to $1.25M$
	$(1.7 \sim 4.0) \times 10^{-12}$	H_2O, 0.5*M* Na_2SO_4 $C = 0.15$ to $1.25M$
	1.5×10^{-9}	H_2O(pH 3.3), 0.2*M* CF_3COONa
	$(2 \pm 1) \times 10^{-12}$	H_2O(pH 3.0), 0.2*M* CF_3COONa
$M = [Co(bpy)_3]^{3+}$	$(2 \pm 1) \times 10^{-11}$	H_2O(pH 3.0), 0.2*M* CF_3COONa
$M = [Ru(NH_3)_6]^{2+}$	$(3.4 \pm 0.1) \times 10^{-9}$	H_2O(pH 3.0), 0.2*M* CF_3COONa
	3.6×10^{-8}	H_2O(pH 3.3), 0.2*M* CF_3COONa
$M = [Ru(NH_3)_6]^{3+}$	$(2.3 \pm 0.1) \times 10^{-9}$	H_2O(pH 3.0), 0.2*M* CF_3COONa
	$\sim 2 \times 10^{-8}$	H_2O(pH 3.3), 0.2*M* CF_3COONa

Process–Species Responsible	Comments–Method	References
	For physical motion $D_m = (0.2 \sim 0.3) \times 10^{-10}\ cm^2\ s^{-1}$	
Electrostatic cross-linking. Single-file diffusion	PSCA. PSCC. NPV AC impedance method D_{app} decreases with increasing C. k^0 decreases from 3×10^{-4} to $7 \times 10^{-5}\ cm\ s^{-1}$ with increasing C from 0.032 to 0.32M $\alpha_a = 0.48$	39
	PSCC	242
	Steady state dual-electrode method Square-wave voltammetry	242
	PSCC	34
	PSCC	34
	PSCA No concentration dependence For physical motion $D_m = (1.6 - 1.8) \times 10^{-10}\ cm^2\ s^{-1}$	32
True ionic diffuser	PSCC	34
	PSCA, PSCC. NPV AC impedance method No concentration dependence $k^0 = \sim 10^{-4}\ cm\ s^{-1}$, $\alpha_a = 0.29$	39
True ionic diffuser	PSCC	34
	PSCC. D_{app} increases with C for reduction to Co(I)	37
	PSCC. D_{app} decreases with C for oxidation to Co(III)	37
	PSCA. PSCC	35
True ionic diffuser	PSCC. D_{app} decreases with C for oxidation to Co(III)	34
True ionic diffuser	PSCC	34
True ionic diffuser	PSCC	34
	PSCA. PSCC	35
True ionic diffuser	PSCC	34
	PSCA. PSCC	35

Table 8.1 Continued

System	$D_{app}(cm^2\,s^{-1})$	Medium
$M = [Co(NH_3)_6]^{3+}$	$\sim 2 \times 10^{-8}$	H_2O(pH 3.3), 0.2*M* CF_3COONa
M = Tetrathia-fulvalenium	8×10^{-7}	H_2O, 0.1*M* KBr
$M = (NH_3)_5RuC_5H_4NCH_2$—NHC(=O)—CpFeCp: RuLFe		H_2O(pH 4.5), 0.1*M* CF_3COONa, 0.1*M* CH_3COONa
$Ru^{III}LFe^{II} \rightarrow Ru^{II}LFe^{II}$	$(1.3 \pm 0.2) \times 10^{-9}$	
$Ru^{II}LFe^{II} \rightarrow Ru^{III}LFe^{II}$	$(2.1 \pm 0.2) \times 10^{-9}$	
$Ru^{III}LFe^{III} \rightarrow Ru^{III}LFe^{II}$	$(0.46 \pm 0.05) \times 10^{-9}$	
$Ru^{III}LFe^{II} \rightarrow Ru^{III}LFe^{III}$	$(0.57 \pm 0.05) \times 10^{-9}$	
M = dopamine (DA^+)	$(1-3) \times 10^{-9}$	H_2O(pH 7.4), phosphate buffer
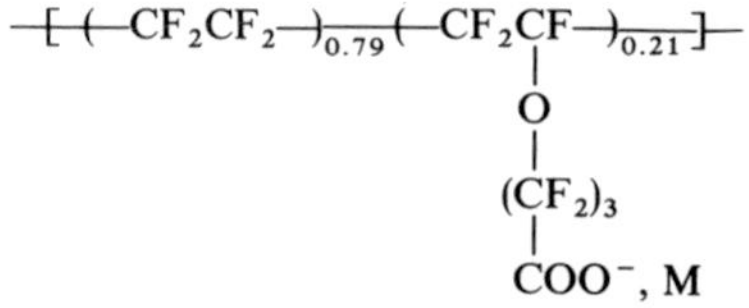 $-[(CF_2CF_2)_{0.79}(CF_2CF)_{0.21}]-$ [CF bearing —O—$(CF_2)_3$—COO^-, M]		
$M = [Os(bpy)_3]^{2+}$	$5.1 \times 10^{-11} \sim 8.3 \times 10^{-10}$	H_2O(pH 5.5), 0.2*M* CF_3COONa, 0.1*M* CH_3COONa, CH_3COOH $C = 0.032 \sim 0.32M$
$M = [CpFeCp'—N(CH_3)_3]^+$	$\sim 5 \times 10^{-11} \sim 8 \times 10^{-10}$	H_2O(pH 5.5), 0.2*M* CF_3COONa, 0.1*M* CH_3COONa, CH_3COOH $C = 0.032 \sim 0.32M$
Cationic perfluoro-polymer(CPFP), alizarin red S	$(1.5 \pm 0.1) \times 10^{-10}$	H_2O(pH 1.0), 0.2*M* CF_3COONa, CF_3COOH
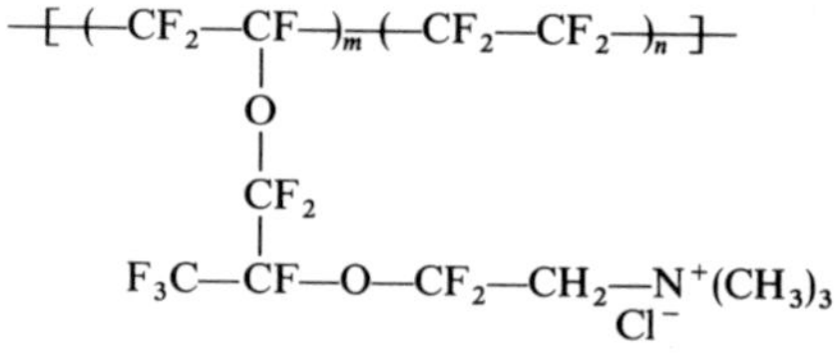 $-[(CF_2—CF)_m(CF_2—CF_2)_n]-$ [CF bearing —O—CF_2—CF(CF_3)—O—CF_2—CH_2—$N^+(CH_3)_3$ Cl^-]	$(6.1 \pm 0.1) \times 10^{-11}$	H_2O(pH 1.0), 0.2*M* CF_3COONa, CF_3COOH
Poly(styrene) functionalized with TTF carboxylate		CH_3CN, 0.1*M* TEA^+X^-

Process–Species Responsible	Comments–Method	References
	PSCA. PSCC	35
	Reduction much faster	40
Intermolecular electron self-exchange. Single-file diffusion. Electrostatic cross-linking	PSCC D_{app} decreases with increasing C	41
	PSCA	47
Electrostatic cross-linking. Single-file diffusion	PSCA, PSCC. AC impedance method. D_{app} decreases with increasing C $K^0 = 7 \times 10^{-6} \sim 2 \times 10^{-5}$ cm s^{-1}, depending on C. $\alpha_a = 0.51$ As pH is increased from 1 to 5.5, C decreases from 0.014 to 0.32M and D_{app} decreases from 1.4×10^{-9} to 2.1×10^{-10} cm^2 s^{-1}	39
Electrostatic cross-linking. Single-file diffusion. Hydrophobic interaction	PSCA, PSCC. AC impedance method. D_{app} decreases with increasing C $k^0 = 7 \times 10^{-6} \sim 2 \times 10^{-5}$ cm s^{-1}, depending on C. $\alpha_a = 0.24$	39
Electron hopping	PSCA, PSCC For reversible process	42
Molecular motion	PSCA, PSCC For irreversible process	42
	PSCC	43

Table 8.1 Continued

System	$D_{app}(cm^2\,s^{-1})$	Medium
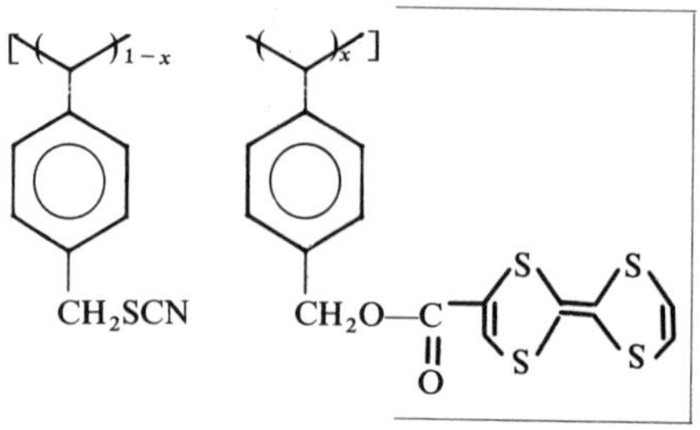 $[\ ]_{1-x}$ $[\ ]_{x}$ CH_2SCN $CH_2O-C(=O)-$ S S S S		
$x=0.15$	1.0×10^{-9}	$X^-=PF_6^-$
		$C=1.5M$
$x=0.30$	2.3×10^{-10}	$X^-=ClO_4^-$
	6×10^{-10}	$X^-=PF_6^-$ $C=2.96\pm0.13M$
	1.3×10^{-10}	$X^-=CH_3\phi SO_3^-$
Poly(styrene-sulfonate), $[Ru(bpy)_3]^{2+}$	2×10^{-9}	CH_3CN, $0.01M$ TBAT
		$C=0.4M$
		CH_3CN, $0.01M$ TBA^+X^-
	$(1.3\pm0.3)\times10^{-9}$	$X^-=BF_4^-$
	$(1.3\pm0.2)\times10^{-9}$	$X\ =ClO_4^-$
	$(0.78\pm0.2)\times10^{-9}$	$X^-=PF_6^-$
	$(1.3\pm0.3)\times10^{-9}$	$X^-=CF_3SO_3^-$
	$(0.53\pm0.2)\times10^{-9}$	$X^-=CH_3\phi SO_3^-$
	$(1.2\pm0.3)\times10^{-9}$	CH_3CN, $0.01M$ TEAT
Random copolymer of styrenesulfonate and 4-vinylpyridine, $[Os(bpy)_3]^{2+}$	2×10^{-8}	H_2O, $0.01M$ $LiClO_4$ $\Gamma_M=1.2\times10^{-9}\,mol\,cm^{-2}$
	8.5×10^{-7}	H_2O, $0.01M$ $LiClO_4$ $\Gamma_M=8.6\times10^{-10}mol\,cm^{-2}$
Nafion, $MV^{\cdot+}$	$(4\pm2)\times10^{-10}$	H_2O(pH 3.0), $0.2M$ CF_3COONa
Nafion, MV^{2+}	$(18\pm0.4)\times10^{-10}$	H_2O(pH 3.0), $0.2M$ CF_3COONa
	$(5.5\sim14.6)\times10^{-10}$	H_2O, $0.2M$ Na_2SO_4 $C=0.3\sim0.7M$
	4×10^{-9}	PC, 240°C
	8×10^{-9}	DMF, 150°C
	2×10^{-10}	H_2O, 100°C
	$5.6\times10^{-10}\sim4\times10^{-9}$	H_2O(pH 3.4), $0.2M$ $NaClO_4$ $\Gamma_M=3.2\times10^{-9}\sim3.2\times10^{-8}mol\,cm^{-2}$

Process–Species Responsible	Comments–Method	References
PF_6^-		
ClO_4^- PF_6^- $CH_3\phi SO_3^-$		
Dynamics of the motion of counterions	PSCC Diffusion coefficient corresponding to electron exchange and excitation migration $>1.0\times10^{-7}\,cm^2\,s^{-1}$ from photoluminescence experiments	44, 46
Uptake of anions	PSCC No concentration dependence at $C=0.08\sim0.38M$	45
	PSCC	242
	Steady state dual-electrode method Square-wave voltammetry	242
True ionic diffuser	PSCC	34
True ionic diffuser	PSCC	34
Physical diffusion	PSCA. D_{app} decreases with increasing C	48
	CV. $\phi'=0.013\,cm$	49
Significant contribution of migration Physical motion Electrostatic cross-linking. Electrostatic repulsions	NPV. D_{app} decreases with increasing Γ_M $k^0=6.3\times10^{-5}\sim1.8\times10^{-4}$ $cm\,s^{-1}$, depending on Γ_M $\alpha_c=0.58$	50

Table 8.1 Continued

System	$D_{app}(cm^2\ s^{-1})$	Medium
Nafion–agarose, MV^{2+}	$(0.56 \sim 1.5) \times 10^{-8}$	H_2O, 0.2*M* Na_2SO_4 $C = 0.0066 \sim 0.072M$
Montmorillonite, MV^{2+}	9×10^{-12}	H_2O, 50 m*M* NaCl
Poly(*N*-vinylbenzyl-*N*′-methyl-4,4′-bipyridinium hexafluorophosphate) 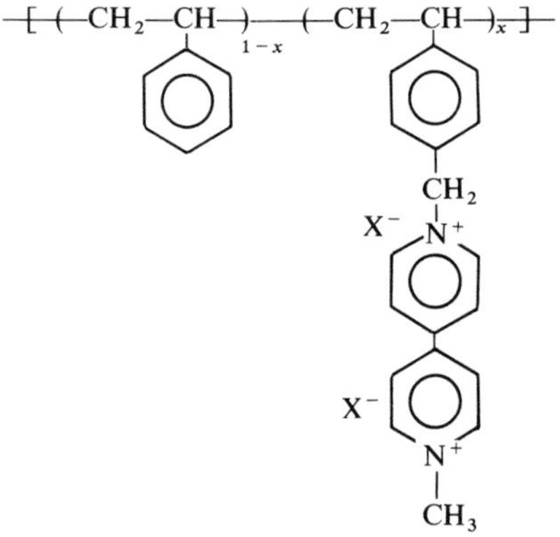 $-[-(-CH_2-CH-)_{1-x}-(-CH_2-CH-)_x-]-$		
$x = 1.0$ $(X^- = PF_6^-)$	3×10^{-12}	H_2O, 1*M* $LiClO_4$ $C = 2.6M$
$x = 0.5$ $(X^- = Cl^-, I^-)$		H_2O(pH 3.1), 0.2*M* KCl
V^{2+}	3.3×10^{-11}	
V^+	1.3×10^{-11}	
V^{2+}	1.1×10^{-11}	$[Fe(CN)_6]^{4-}$ counterion
V^+	0.37×10^{-11}	$[Fe(CN)_6]^{4-}$ counterion
$[Fe(CN)_6]^{3-}$	2.0×10^{-11}	
$[Fe(CN)_6]^{4-}$	4.6×10^{-11}	
$[IrCl_6]^{2-}$	1.4×10^{-11}	
$[IrCl_6]^{3-}$	4.2×10^{-11}	
$[Mo(CN)_8]^{3-}$	2.2×10^{-11}	
$[Mo(CN)_8]^{4-}$	6.9×10^{-11}	
Poly(*p*-styrene-sulfonate)/Poly-(*p*-xylylviologen)	$D_{app}/\phi'^2 \sim 1s$ $(D_{app} \sim 10^{-10}\ cm^2 s^{-1}$, $\phi' \sim 9 \times 10^{-6} cm)$	

Process–Species Responsible	Comments–Method	References
Heterogeneous structure. Coupling of diffusional pathways between solution and polymer phases within films. Electron exchange	PSCA. D_{app} increases with increasing C	48
Electron hopping mechanism	CV	51
	PSCA. Carbon particles embedded Extended area effects studied	52
Polymer motion and counterions	PSCA. Data are for viologen and electrostatically bound ion charge transport processes as in column 1	53
	PSCA	54

Table 8.1 Continued

System	$D_{app}(cm^2\,s^{-1})$	Medium
Poly(*p*-styrene-sulfonate)/Poly-(*m*-xylylviologen)	5.6×10^{-10}	H_2O(pH 3.4), 0.2*M* $NaClO_4$
Poly(*p*-xylyl-viologen)	$(1-2) \times 10^{-8}$	H_2O, 1*M* $NaClO_4$ $C = 1.1M$
Poly(styrene-*co*-chloromethylstyrene) pendant viologens (PMV) $-[-(-CH_2-CH-)_x-(-CH_2-CH-)_y-(-CH_2-CH-)_z-]-$ CH_2 / N^+ Cl^- / N^+ Cl^- / CH_3 ; CH_2Cl 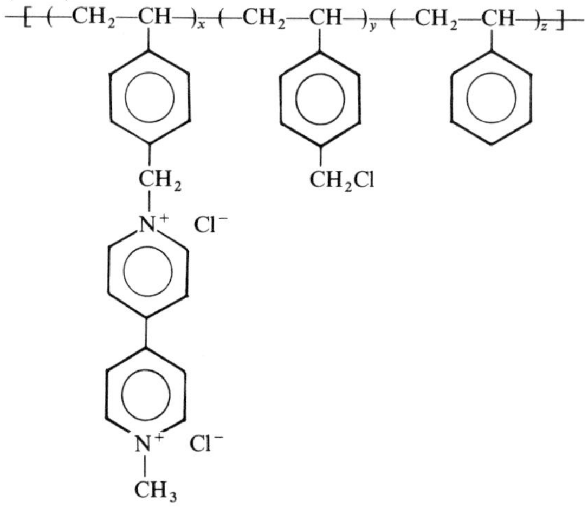	$2.5 \times 10^{-11} \sim 3.9 \times 10^{-10}$	H_2O(pH 3.0), 0.2*M* KCl $C = 0.067 \sim 0.33M$; $x = 9 \sim 34\%$
	1.2×10^{-10}	H_2O(pH 3.4), 0.2*M* $NaClO_4$, $x = 34\%$
		H_2O(pH 3.0), 0.2*M* KCl $x = 34\%$, $y = 7\%$, $z = 59\%$
		$x = 15\%$, $y = 26\%$, $z = 59\%$
		$x = 13\%$, $y = 28\%$, $z = 59\%$
		H_2O(pH 3.0), 0.2*M* electrolyte $x = 34\%$, $y = 7\%$, $z = 59\%$
	$(9.5 \pm 1.5) \times 10^{-10}$	NaCl

Process–Species Responsible	Comments–Method	References
	NPV $k^0 = (1.1 \pm 0.3) \times 10^{-4}\,\mathrm{cm\,s^{-1}}$, $\alpha_c = 0.20$	50
	PSCA	55
Electron hopping	NPV. D_{app} increases with increasing C $k^0 = 3.0 \times 10^{-5} \sim 8.9 \times 10^{-5}\,\mathrm{cm\,s^{-1}}$, $\alpha_c = 0.42 \pm 0.04$	56
	NPV $k^0 = (3.9 \pm 0.4) \times 10^{-5}\,\mathrm{cm\,s^{-1}}$, $\alpha_c = 0.56$	50
	NPV $E_a = 28.9\,\mathrm{kJ\,mol^{-1}}$, $D_0 = 4.0 \times 10^{-5}\,\mathrm{cm^2\,s^{-1}}$ $E_a = 23.0\,\mathrm{kJ\,mol^{-1}}$, $D_0 = 1.6 \times 10^{-6}\,\mathrm{cm^2\,s^{-1}}$ $E_a = 16.7\,\mathrm{kJ\,mol^{-1}}$, $D_0 = 5.0 \times 10^{-8}\,\mathrm{cm^2\,s^{-1}}$ Thermodynamic data for heterogeneous electron-transfer reaction are also estimated	57
Counterion motion Polymer motion Electron hopping	NPV. Data for PMV/Nafion and PMV/poly(*p*-styrenesulfonate) polymer complexes are also estimated $k^0 = (1.7 \pm 0.3) \times 10^{-4}\,\mathrm{cm\,s^{-1}}$, $\alpha_c = 0.40 \pm 0.02$	58

Table 8.1 Continued

System	$D_{app}(cm^2\ s^{-1})$	Medium
	$(1.3 \pm 0.2) \times 10^{-10}$	$NaClO_4$
	$(1.0 \pm 0.2) \times 10^{-9}$	$CH_3\phi SO_3Na$
	$(1.5 \pm 0.5) \times 10^{-9}$	CsCl
	$(6.8 \pm 1.3) \times 10^{-10}$	tetraphenyl-phosphonium chloride
	$(1.3 \pm 0.2) \times 10^{-10}$	poly(*p*-styrenesulfonate) sodium salt
Poly(propyl-viologensilane)	$(0.4 - 3) \times 10^{-10}$	H_2O, 0.1–4*M* LiCl [or $LiClO_4$, LiBr, Li_2SO_4, NaCl, KCl, $(CH_3)_4NCl$, $MgSO_4$]
Benzylviologen-silane polymer	$\sim 4 \times 10^{-12}$	CH_3CN, 0.1*M* TEAP
N,N'-bis(*p*-(trimethoxy-silyl)benzyl)-4,4'-bipyridinium dichloride derivative polymer (BPQ)		
$BPQ^{2+/\cdot+}$	$D_{app}C^2 = 7.3 \times 10^{-15}\ mol^2/(s\ cm^4)$	H_2O/CH_3CN(9/1), 1*M* LiCl
$BPQ^{\cdot+/0}$	$D_{app}C^2 = 4.8 \times 10^{-14}\ mol^2/(s\ cm^4)$	
$BPQ^{2+/\cdot+}$	0.58×10^{-9}	H_2O, 0.1*M* KCl
	2.2×10^{-9}	H_2O, 1*M* KCl $C = 2.2M$
	4.1×10^{-9}	H_2O, 4*M* KCl
	3.3×10^{-9}	H_2O, 4*M* KCl
$BPQ^{2+/\cdot+}$	$\sim 10^{-9}$	H_2O, 1*M* $LiClO_4$
N,N'-bis(trimethoxy-silyl)propyl-4,4'-bipyridinium dibromide derivative polymer (PQ)		
$PQ^{2+/\cdot+}$	$D_{app}C^2 = 7.2 \times 10^{-15} \times mol^2/(s\ cm^4)$	H_2O/CH_3CN(9/1), 1*M* LiCl
$PQ^{\cdot+/0}$	$D_{app}C^2 = 7.2 \times 10^{-14} \times mol^2/(s\ cm^4)$	
Siloxane-based polymer having two benzylviologen subunits flanking a benzoquinone subunit $(BV\text{-}Q\text{-}BV^{6+})_n$	2.7×10^{-10}	H_2O(pH 7.2), 1*M* LiCl, tris buffer
	3.0×10^{-10}	H_2O(pH 9.0), 1*M* KCl, tris buffer
	4.1×10^{-10}	H_2O(pH 6.0), 0.5*M* NaOTs, acetate buffer

Process–Species Responsible	Comments–Method	References
	$k^0 = (3.1 \pm 0.6) \times 10^{-4}\,\mathrm{cm\,s^{-1}}$, $\alpha_c = 0.46 \pm 0.02$ $k^0 = (6.7 \pm 1.3) \times 10^{-5}\,\mathrm{cm\,s^{-1}}$, $\alpha_c = 0.39 \pm 0.02$ $k^0 = (2.2 \pm 0.4) \times 10^{-4}\,\mathrm{cm\,s^{-1}}$, $\alpha_c = 0.44 \pm 0.02$ $k^0 = (9.5 \pm 1.5) \times 10^{-5}\,\mathrm{cm\,s^{-1}}$, $\alpha_c = 0.43 \pm 0.02$ $k^0 = (1.4 \pm 0.3) \times 10^{-5}\,\mathrm{cm\,s^{-1}}$, $\alpha_c = 0.18 \pm 0.02$	
Electron hopping	PSCA. Little variation with electrolyte at fixed ionic strength	59
	Potential-step chronoabsorbance method	60
	RDEV	61
	PSCA	62
	Two adjacent microelectrodes method	63
	RDEV	61
	Microarray electrode method. $D_{app} = 0.8 \times 10^{-10}\,\mathrm{cm^2\,s^{-1}}$ from PSCA $E_a = 49\,\mathrm{kJ\,mol^{-1}}$	64

Table 8.1 Continued

System	$D_{app}(cm^2\,s^{-1})$	Medium
	4.7×10^{-10}	H_2O(pH 7.2), 0.5*M* $CH_3\phi SO_3Na$, tris buffer
	6.3×10^{-10}	H_2O(pH 8.1), 0.5*M* $CH_3\phi SO_3Na$, tris buffer
	8.8×10^{-10}	H_2O(pH 9.0), 0.5*M* $CH_3\phi SO_3Na$, tris buffer
Benzylviologen polymer $(BV^{2+})_n$	7.28×10^{-9}	H_2O(pH 7–9), 1*M* LiCl, tris buffer
	5.25×10^{-9}	H_2O(pH 7–9), 0.5*M* $CH_3\phi SO_3Na$, tris buffer
N-methyl-*N*′-octadecyl-4,4′-bipyridinium chloride assemblies ($C_{18}MV^{2+}$-OTS-Al_2O_3)	4×10^{-8}	H_2O, 0.1*M* KCl
	2×10^{-7}	H_2O, 0.1*M* KCl, 3m*M* 1-octanol
OTS/$C_{18}MV^{2+}$ bilayer assemblies		
$C_{18}MV^{2+}$	$7.0\times10^{-8}(\pm24\%)$	H_2O, 0.1*M* KCl
	$1.3\times10^{-7}(\pm28\%)$	H_2O, 0.1*M* KCl, 1-octanol
$C_{18}MV^{+}$	$2.3\times10^{-8}(\pm43\%)$	H_2O, 0.1*M* KCl
	$4.7\times10^{-8}(\pm36\%)$	H_2O, 0.1*M* KCl, 1-octanol
OTS/octadecyl ferrocene derivative ($C_{18}Fc^{+}$)bilayer assemblies		
$C_{18}Fc^{+}$	$2.7\times10^{-8}(\pm27\%)$	H_2O, 0.1*M* KNO_3
	$7.7\times10^{-8}(\pm19\%)$	H_2O, 0.1*M* KNO_3, 1-octanol
$C_{18}Fc^{2+}$	$3.5\times10^{-8}(\pm16\%)$	H_2O, 0.1*M* KNO_3
	$8.7\times10^{-8}(\pm26\%)$	H_2O, 0.1*M* KNO_3, 1-octanol
OTS/octadecyl naphthoquinone derivative ($C_{18}Q^{+}$) bilayer assemblies		
$C_{18}Q^{+}$	$(0.25\pm0.2)\times10^{-8}$	H_2O(pH 5.0), 0.1*M* acetate buffer
	$(2.5\pm1.0)\times10^{-8}$	H_2O(pH 5.0), 0.1*M* acetate buffer 1-octanol
PVF (plasma polymerized)	9.8×10^{-12}	CH_3CN, 0.1*M* TEAT $C=3M$

Process–Species Responsible	Comments–Method	References
	$E_a = 59.4\,kJ\,mol^{-1}$	
	$E_a = 60.3\,kJ\,mol^{-1}$	
	Microarray electrode method. $D_{app} = 1.61 \times 10^{-9}\,cm^2\,s^{-1}$ from PSCA $E_a = 33.1\,kJ\,mol^{-1}$ $E_a = 51.0\,kJ\,mol^{-1}$	64
Lateral electron transport	Self-assembled at *n*-octadecyltrichlorosilane (OTS)-treated porous aluminium oxide film (Al_2O_3) PSCC	65
Lateral charge transport. Minor electron-hopping contribution. Translation diffusion of $C_{18}MV^{2+}$ molecules along OTS layer	PSCC. Porous Al_2O_3 film. Self-assembled bilayer assemblies No concentration dependence	66, 67
Lateral charge transport	PSCC. Porous Al_2O_3 film. Self-assembled bilayer	66
Lateral charge transport	PSCC. Porous Al_2O_3 film. Self-assembled bilayer assemblies	67
	$E_a = (109 \pm 8)\,kJ\,mol^{-1}$	
	$E_a = (38 \pm 4)\,kJ\,mol^{-1}$	
	CV	68

Table 8.1 Continued

System	D_{app}(cm^2 s^{-1})	Medium
	7×10^{-11}	H_2O, 1M $LiClO_4$ $C = 2.4M$
	3.5×10^{-11}	
	$10^{-13} - 10^{-12}$	C_3H_7CN, 0.1M TBAP $T = -84 \sim -50$°C
PVF (electro-deposited)	$(2-3) \times 10^{-10}$	CH_3CN, 0.1M TBAP
	4×10^{-8}	CH_3CN, 0.5M TBAT
	1×10^{-7}	
	$\sim 10^{-9}$	CH_3CN, 0.1M TBAP
	1.3×10^{-10}	H_2O, 0.5M $NaClO_4$
	0.77×10^{-9}	CH_3CN, 0.1M TBAT $\phi' = 1.13 \times 10^{-4}$cm
	1.19×10^{-9}	
PVF (silyl methacrylate copolymer) 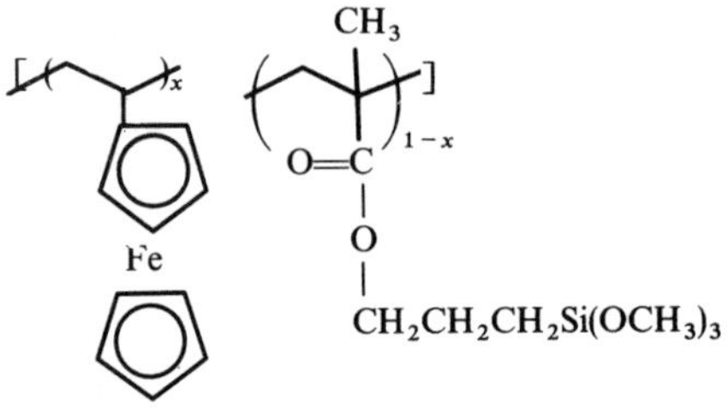 [ferrocene (Fe) vinyl unit]$_x$ – [CH_3-substituted methacrylate, $O=C$–O–$CH_2CH_2CH_2Si(OCH_3)_3$]$_{1-x}$		CH_3CN, 0.1M $LiClO_4$
$x = 0.88$	3.4×10^{-10}	$T = -44$°C
$x = 0.72$	6.2×10^{-10}	$T = -48$°C
$x = 0.59$	10×10^{-10}	$T = -36$°C
$x = 0.38$	6.2×10^{-10}	$T = -33$°C
$[M(vbpy)_3]_n$ M = Ru(III/II)	2.2×10^{-10}	CH_3CN, saturated $LiClO_4$, $C = 1.6M$
	$(1-4) \times 10^{-9}$	CH_3CN, 0.1M TEAP
	$D_{app}^{1/2}C = (2 \sim 5) \times 10^{-8}$ $\times$ mol cm^{-2}s$^{-1/2}$	CH_3CN, 0.1M TEAP
	$(0.6-1.9) \times 10^{-9}$	CH_3CN, 0.1M TEAP $C = 1.6M$

Process–Species Responsible	Comments–Method	References
	PSCA. Anodic step	69
	Cathodic step	
Cooperative chain motion	PSCA. $-85 < T < -50°C$ $D_0 \sim (0.2-1.0) \times 10^{-8}\,cm^2\,s^{-1}$ $E_a \sim 15.5\,kJ\,mol^{-1}$, $\Delta S \sim -150$ to $-130\,JK^{-1}\,mol^{-1}$	70
	CV Values for three types of interconverting sites	71
Membrane model	PSCA. For diffusion of MV^{2+} in PVF film For diffusion of benzoquinone in PVF film	72
	Two adjacent micro-electrode method	63
	PSCC	73
	Convolution voltammetry Data change with ϕ'. For oxidation step. $k° = 9.4 \times 10^{-5}\,cm\,s^{-1}$, $\alpha_a = 0.61$ For reduction step $k^0 = 2.8 \times 10^{-5}\,cm\,s^{-1}$, $\alpha_c = 0.77$	36
	PSCA. $T = -36°C$ $35 > E_a > 17\,kJ\,mol^{-1}$ Apparent dispersion removed by correction for partial electroactivity	74
Polymer lattice mobility	Sandwich device measurement. $k_{ex} = 1.7 \times 10^5\,M^{-1}\,s^{-1}$	75
	Sandwich device measurement	76
	RDEV	81
	Bilayer electrode experiments	85

Table 8.1 Continued

System	$D_{app}(cm^2\ s^{-1})$	Medium
$Ru^{II/I}$	$(1-3)\times 10^{-8}$	CH_3CN, 0.1M TEAP $C=1.6M$
$Fe^{III/II}$	9×10^{-11}	CH_3CN, 0.1M TEAP $C=1.6M$
$[M(bpy)_2(vpy)_2]_n$		
M = Os(III/II)	$(8\pm 3)\times 10^{-9}$	CH_3CN, 0.1M TEAP $C=1.5M$
$Os^{II/I}$	$(24\pm 9)\times 10^{-9}$	
$Os^{I/0}$	$(200\pm 70)\times 10^{-9}$	
$Ru^{III/II}$	7×10^{-10}	CH_3CN, 0.1M TEAP
	4×10^{-10}	
$Os^{III/II}$	5.4×10^{-9}	CH_3CN, 0.1M TEAP
	$(2.8\pm 0.6)\times 10^{-8}$	CH_3CN, 0.1M TEAP
$Ru^{III/II}$	$D_{app}^{1/2}C=1.7\times 10^{-8}$ $\times$ mol $cm^{-2}s^{-1/2}$	$AlCl_3$/N-(1-butyl)-pyridinium chloride (or $AlCl_3$/1-methyl(3-ethyl)imidazolium chloride)molten salts
	$D_{app}^{1/2}C=4.4\times 10^{-8}$ $\times$ mol $cm^{-2}s^{-1/2}$	CH_3CN, 0.1M TEAP
$Ru^{III/II}$	1.1×10^{-8}	CH_3CN, 0.1M TEAP
$Os^{III/II}$	$(2.0\pm 0.7)\times 10^{-8}$	
$[M(bpy)_2(cinn)_2]_n$		
M = Os(II)	$(1.5\pm 0.9)\times 10^{-9}$	CH_3CN, 0.1M TEAP
M = Os/Ru copolymer	$(0.2\sim 13)\times 10^{-10}$	CH_3CN, 0.1M TEAP $C=0.1\sim 1.2M$

Process–Species Responsible	Comments–Method	References
	Bilayer electrode experiments	85
	Bilayer electrode experiments	85
	Sandwich device measurement. $E_a = 24\,kJ\,mol^{-1}$ $D_0 = 2.7 \times 10^{-4}\,cm^2\,s^{-1}$ $E_a = 19\,kJ\,mol^{-1}$ $D_0 = 5.4 \times 10^{-5}\,cm^2\,s^{-1}$ $E_a = 14\,kJ\,mol^{-1}$ $D_0 = 4.5 \times 10^{-5}\,cm^2\,s^{-1}$	76
	Sandwich device measurement	76
	PSCA	76
	RDEV	77
Electron diffusion	Bilayer-electrode cyclic voltammetry	78
Electron exchange	Linear sweep voltammetry. RDEV	79
	Interdigitated array electrode	80
	Sandwich device measurement	82
Electron exchange	PSCA Ru is structurally similar diluent. $x_{Os} < 0.15$ D_{app} increasing with C $E_a = 46.8\,kJ\,mol^{-1}$ $0.15 < x_{Os} < 0.5$; D_{app} independent of C, $E_a = 26.8$ $kJ\,mol^{-1}$. $0.50 < x_{Os} < 1.0$ D_{app} increasing with C $E_a = 15.0\,kJ\,mol^{-1}$	83

Table 8.1 Continued

System	$D_{app}(cm^2 s^{-1})$	Medium
M = Os/Ru copolymer	$(3.19 \sim 20.3) \times 10^{-12}$	CH_3CN, 0.1*M* TEAP $x_{Os} = 0.1 \sim 0.36$
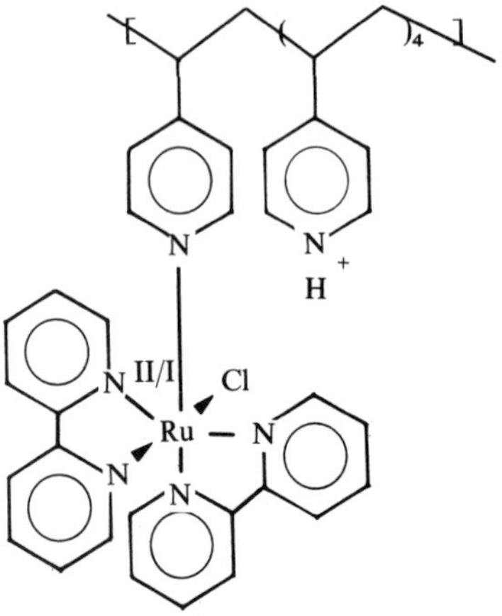	2.9×10^{-9}	H_2O, 0.1*M* HCl $C \sim 10M$
	2×10^{-7}	H_2O, 1*M* HCl
$Poly[Ru(bpz)_3]^{2+}$	2.2×10^{-9}	H_2O, 0.05*M* KNO_3 $C = 0.22M$
Cobaltocene polymer ($[CoC_pR_2]^{+/0}$) R = —CO—$C_pCoC'_p$ (50% functionalized)	2.5×10^{-9}	H_2O, 1*M* KCl
R = —CO—ϕ—CH_2— (50% functionalized)	$(0.4 - 1.6) \times 10^{-9}$	H_2O(pH 3.2), 0.2*M* $LiClO_4$
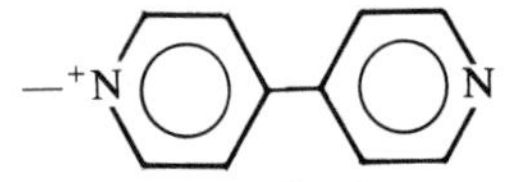	$(1.3 - 3.2) \times 10^{-12}$	H_2O(pH 1–6), 0.1*M* $LiClO_4$ $C = 1.5M$
$Fe^{II}[phen(-\phi SO_3^-)_2]_3$	$D_{app}/\phi'^2 \sim 5 s^{-1}$ $D_{app} \sim 1.6 \times 10^{-9} cm^2 s^{-1}$, $\phi' \sim 1.8 \times 10^{-5}$ cm)	H_2O, 1*M* KCl (or NaCl, LiCl)
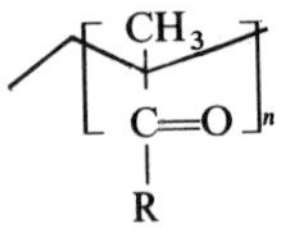 R 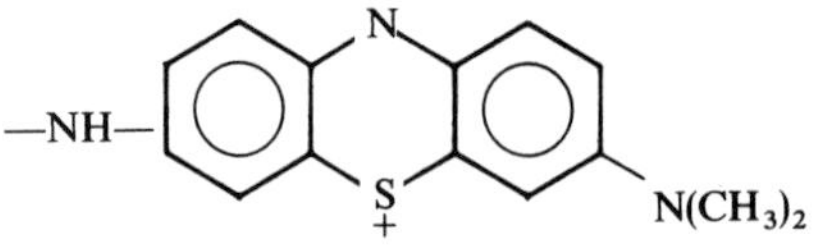 = —NH— (phenothiazinium, $N(CH_3)_2$)	1.6×10^{-9}	H_2O(pH 1–7), 0.1*M* KCl (or HCl, H_2SO_4, $HClO_4$)

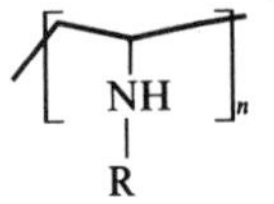

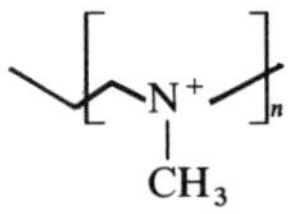

Process–Species Responsible	Comments–Method	References
Electron exchange	RDEV D_{app} increases with x_{Os}	84
	RDEV	86
	Double-coated RDEV For physical diffusion of Fe^{3+} in film	87
	CV	88
	Chronoamperometric, CV and RDEV data in good agreement	89
	PSCA. D_{app} for oxidation $\sim D_{app}$ for reduction	90
Counterion or polymer motion	CV	91
	PSCA	92
Counterion or polymer motion	PSCA	91

Table 8.1 Continued

System	$D_{app}(cm^2\,s^{-1})$	Medium
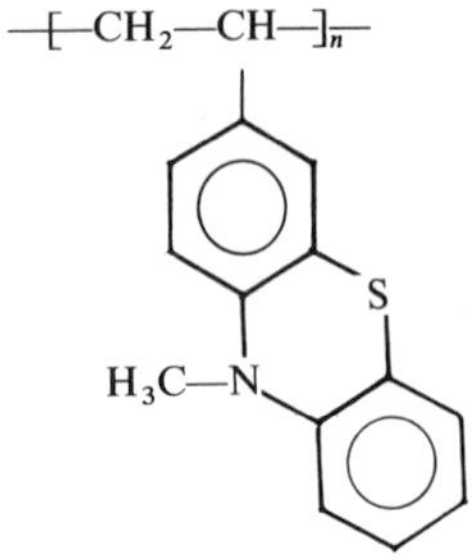 $-[-CH_2-CH-]_n-$ with pendant N-methylphenothiazine (H_3C-N, S)	1.6×10^{-9} 2.6×10^{-9}	CH_3CN, 0.4M $LiClO_4$
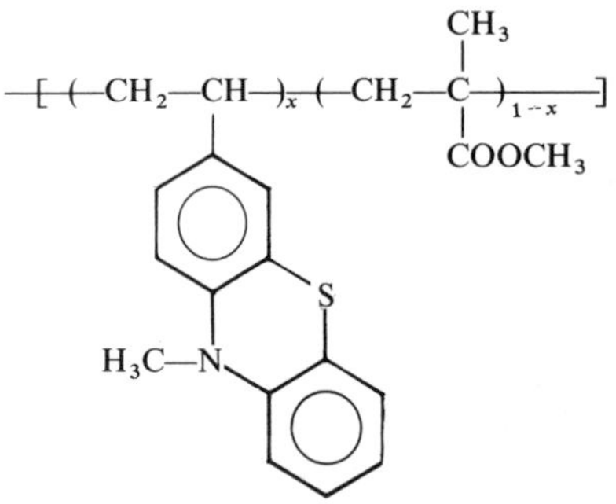 $-[-(-CH_2-CH-)_x-(-CH_2-C(CH_3)(COOCH_3)-)_{1-x}-]-$ with pendant N-methylphenothiazine (H_3C-N, S)	$\sim(0.45\sim1.6)\times10^{-9}$ $\sim(0.25\sim2.5)\times10^{-9}$	CH_3CN, 0.4M $LiClO_4$
Thick crystalline deposits		
phenazine	$D_{app}^{1/2}C=(2.6-7.9)\times10^{-8}$ $\times$ mol $cm^{-2}s^{-1/2}$	H_2O(pH 2), 0.2M KNO_3 $\Gamma_M=(6.7-10)\times10^{-8}$ mol cm^{-2}
chloranil	$D_{app}^{1/2}C=1.48\times10^{-9}$ $\times$ mol $cm^{-2}s^{-1/2}$	H_2O, 0.1M $HClO_4$ $\Gamma_M=1.46\times10^{-8}$ mol cm^{-2}
azobenzene	$D_{app}^{1/2}C=2.1\times10^{-9}$ $\times$ mol $cm^{-2}s^{-1/2}$	H_2O(pH 2), 0.2M KNO_3
TCNQ polyester	$nD_{app}^{1/2}C=(0.84\sim7.4)\times10^{-8}$ $\times$ mol $cm^{-2}s^{-1/2}$	CH_3CN, 0.1M TEAP $\Gamma_M=(2.23\sim23.8)\times10^{-9}$ $\times$ mol cm^{-2}
TCNQ–BOEB terpolymer		CH_3CN, 0.1M TEAP
TCNQ poly(urethanes)		DMA, 0.1M TEAP
$-[-TCNQ^{0/-}-MDI-]_n-$	4.0×10^{-7}	
$-[-TCNQ^{0/-}-TDI-]_n-$	7.7×10^{-8}	
$-[-TCNQ^{0/-}-HMDI-]_n-$	4×10^{-8}	
2,5-bis(2′-hydroxy-ethoxy)-7,7′,8,8′-tetra-cyanoquinodimethane	4.1×10^{-6}	

Process–Species Responsible	Comments–Method	References
	CV. PSCC For anodic process For cathodic process	241
	PSCC. For anodic process $k_{ex} = 4.0 \times 10^5\, M^{-1}\, s^{-1}$ For cathodic process $k_{ex} = 5.9 \times 10^5\, M^{-1}\, s^{-1}$	241
	PSCA Active and inert layers	121
Counterion motion	PSCC $E_a = 30\, kJ\, mol^{-1}$	93
	PSCC $E_a = 39\, kJ\, mol^{-1}$	93
Charge transport via anion/dianion oxidation states	PSCC	94

Table 8.1 Continued

System	$D_{app}(cm^2\,s^{-1})$	Medium
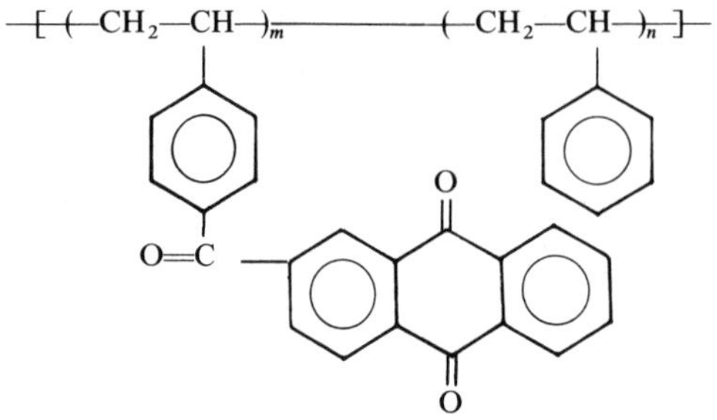 $-[(-CH_2-CH-)_m-(-CH_2-CH-)_n]-$ (O=C, O, O)	$(3.6\sim 6.8)\times 10^{-11}$	DMSO, 0.1M TEAP
Prussian Blue films		H_2O(pH 4), 0.5M electrolyte
	0.80×10^{-9}	$NaNO_3$
	$(3.9\pm 1.3)\times 10^{-9}$	KNO_3
	$(8.2\pm 2.8)\times 10^{-9}$	NH_4Cl
	$(6.6\pm 2.0)\times 10^{-9}$	RbCl, $RbNO_3$
	3.3×10^{-9}	CsCl
	$(0.3\pm 0.04)\times 10^{-9}$	KNO_3 (dry condition)
	$(2.8\pm 1.1)\times 10^{-9}$	KCl
	5×10^{-9}	H_2O, 0.9M KCl
	2.7×10^{-9}	
	$(3.1-9.8)\times 10^{-9}$	H_2O, 1M $(NH_4)_2SO_4$ (or 1M K_2SO_4)
Ruthenium Purple films	$(1.7-3.3)\times 10^{-9}$	H_2O, 1M $(NH_4)_2SO_4$ (or 1M K_2SO_4)
Nickel hexacyano-ferrate films $[Ni^{II}(NC)Fe^{II/III}(CN)_5]_n^{2-/1-}$	$(0.65-2.3)\times 10^{-11}$	H_2O, 0.2M $NaNO_3$ $\Gamma_M=(1.6-4.0)\times 10^{-8}$ mol cm^{-2}
	$(0.91-4.4)\times 10^{-11}$	
Hydrous iridium oxide films	1.5×10^{-9}	H_2O, 1M H_2SO_4 C=6.8M
	1.0×10^{-7}	
	2.3×10^{-10}	
	3.0×10^{-10}–5.1×10^{-9}	
Poly(acetylene)	6×10^{-12}	Sulfolane, 30% $LiBF_4$
	4×10^{-18}	PC, 1M $LiClO_4$

Process–Species Responsible	Comments–Method	References
	PSCA	239
Electron self-exchange Counterion motion	Interdigitated array electrode method. Data for Fe(III/II)–Fe(II/II) mixed-valent state (reduction process) $k_{ex} = 1.3 \times 10^6\ M^{-1}\ s^{-1}$	95
		80
	PSCA For Fe(III/II)–(II/II) mixed-valent state For Fe(II/II)–Fe(III/II) mixed-valent state (oxidation process)	96
Migration and diffusion of charge carriers	PSCA	97
Migration and diffusion of charge carriers	PSCA	97
	PSCC For reduction process For oxidation process Chronocoulometric and diffuse reflectance spectroscopic data in good agreement	98
	CV For negative sweep For positive sweep PSCA. Cathodic process PSCA. Anodic process Dependence on oxidation potential	99
	Galvanostatic pulse method. Dopant $= BF_4^-$	100
	Open-circuit voltage decay measurement Dopant $= ClO_4^-$	101

Table 8.1 Continued

System	$D_{app}(cm^2 s^{-1})$	Medium
Poly(pyrrole)	3.9×10^{-10}	CH_3CN, 0.1*M* $LiClO_4$
	6.2×10^{-10}	
	3×10^{-10}	PC, 1*M* $LiClO_4$
	$(0.7 \sim 2.1) \times 10^{-9}$	
	$(4.2 - 6.2) \times 10^{-9}$	CH_3CN, 0.2*M* TEAT $\phi' = 0.35 \sim 0.73\ \mu m$
Poly(aniline)	$\sim 10^{-10} - \sim 10^{-8}$	H_2O, 2*M* HCl
	$(9.4 \pm 1.8) \times 10^{-9}$	H_2O(pH 1.5), 0.2*M* CF_3COONa, CF_3COOH
	$(8.5 \pm 0.5) \times 10^{-8}$	H_2O(pH 3.0), 0.2*M* CF_3COONa, CF_3COOH
	$\phi'^2/D_{app} \sim 15$ ms $D_{app} \sim 2 \times 10^{-9}$, $\phi' \sim 5 \times 10^{-6}$ cm)	H_2O, 1*M* HCl
Poly(thiophene)		H_2O(pH 1, 3), 0.05*M* HX
	9.1×10^{-13}	$X = HSO_4^-$
	6.5×10^{-13}	$X = Cl^-$
	1.7×10^{-13}	$X = CF_3COO^-$
	0.9×10^{-13}	$X = CH_3\phi SO_3^-$
	$\sim 4 \times 10^{-12}$	H_2O(pH 5), 0.05*M* KCl
Poly(3-methyl-thiophene)/PVC composite films	$D_{app}^{1/2}C = (0.39 \sim 6.4) \times 10^{-7}$ $\times$ mol $cm^{-2}s^{-1/2}$	CH_3CN, 0.1*M* $LiClO_4$
	$D_{app}^{1/2}C = (1.4 \sim 18.2) \times 10^{-7}$ $\times$ mol $cm^{-2}s^{-1/2}$	
Poly(*N*-methyl-aniline)	$(1.2 \pm 0.1) \times 10^{-8}$	H_2O(pH 1.0), 0.2*M* $NaClO_4$, $HClO_4$
	$(1.5 \pm 0.1) \times 10^{-8}$	

Process–Species Responsible	Comments–Method	References
	PSCC For oxidation process For reduction process From potential step chrono-absorptometry, $7.7 \times 10^{-10}\ cm^2\ s^{-1}$ for oxidation process and $12 \times 10^{-10}\ cm^2\ s^{-1}$ for reduction process	102
	PSCA. For reduction process Film prepared at 0.32 V No film thickness dependence Film prepared at 0.84 V D_{app} decreases with increasing film thickness	107
	Small amplitude current-pulse method	103
	AC impedance method D_{app} increases with increased positive potential	104
	NPV $k^0 = (2.5 \pm 0.9) \times 10^{-4}\ cm\ s^{-1}$ $\alpha_c = 0.26 \pm 0.05$	105
	RDEV For diffusion of H^+ in film	106
	PSCA	108
Counterion (X^-) motion	PSCA. $E_a = 35 \pm 3\ kJ\ mol^{-1}$	109
Electron hopping	PSCA For oxidation process	110
Probably undoping process is determined by cation transport	For reduction process D_{app} decreases as PVC concentration used for film preparation is increased. Undoping process is faster than doping one	
	NPV. For oxidation process. $k^0 = (4.2 \pm 0.8) \times 10^{-4}\ cm\ s^{-1}$, $\alpha_a = 0.86 \pm 0.02$ $\alpha_c = 0.12 \pm 0.02$ For reduction process.	111

Table 8.1 Continued

System	$D_{app}(cm^2 s^{-1})$	Medium
Poly(*N*-ethylaniline)	$(4.2 \pm 2.0) \times 10^{-8}$	H_2O(pH 1.0), 0.2*M* $NaClO_4$, $HClO_4$
	$(2.3 \pm 1.1) \times 10^{-8}$	
Poly(*o*-aminophenol)	$(1 \sim 6) \times 10^{-10}$	H_2O(pH 1.0), 0.2*M* $NaClO_4$, $HClO_4$
Poly(2,3-diamino-naphthalene)	2.8×10^{-8}	H_2O(pH 1.0), 0.2*M* $NaClO_4$, $HClO_4$
Poly(1-pyrenamine)	$(1.5 \pm 0.5) \times 10^{-10}$	H_2O(pH 1.0), 0.2*M* $NaClO_4$, $HClO_4$
Poly(*o*-phenylene-diamine)	$(1.1 \pm 0.1) \times 10^{-8}$	H_2O(pH 1.0), 0.2*M* $NaClO_4$, $HClO_4$
	$(2.9 \pm 0.2) \times 10^{-8}$	
	$(2 \sim 6) \times 10^{-8}$	H_2O(pH 1.0), 0.2*M* $NaClO_4$ (or NaCl, CF_3COONa, Na_2SO_4 $CH_3\phi SO_3Na$)
	$(1.2 \sim 3.6) \times 10^{-8}$	H_2O(pH 5.0), 0.2*M* $NaClO_4$ (or NaCl, CF_3COONa, Na_2SO_4, LiCl, TEACl)
	$(0.19 \sim 0.62) \times 10^{-8}$	
	$(3.5 \pm 0.5) \times 10^{-7}$	H_2O(pH 3.0), 0.2*M* CF_3COONa, CF_3COOH
Poly(4,4′-diamino-diphenylether)	$\sim 2 \times 10^{-7}$	H_2O(pH 3.0), 0.2*M* CF_3COONa, CF_3COOH
Poly(2,6-dimethyl-1,4-phenyleneoxide)	$(1.34 \pm 0.54) \times 10^{-7}$	NM, 0.1*M* TBAP
	$(1.3 \pm 0.3) \times 10^{-7}$	H_2O(pH 2.7), 0.2*M* CF_3COONa, CF_3COOH
	$(3.5 \pm 0.4) \times 10^{-6}$	

Process–Species Responsible	Comments–Method	References
	NPV. For oxidation process. $k^0 = (5.1 \pm 2.2) \times 10^{-4}$ cm s^{-1}, $\alpha_a = 0.84 \pm 0.02$ $\alpha_c = 0.16 \pm 0.02$ For reduction process	111
	PSCA. PSCC. NPV For oxidation and reduction process $k^0 = (1.1 \pm 0.3) \times 10^{-4}$ cm s^{-1} $\alpha_a = 0.75 \pm 0.02$ $\alpha_c = 0.21 \pm 0.02$	112, 113
	PSCA. PSCC	114
	PSCA. PSCC. NPV $k^0 = (1.9 \pm 0.4) \times 10^{-5}$ cm s^{-1} $\alpha_a = 0.67 \pm 0.03$	115
	NPV. For oxidation process. $k^0 = (5.8 \pm 0.6) \times 10^{-4}$ cm s^{-1}, $\alpha_a = 0.83 \pm 0.03$ $\alpha_c = 0.23 \pm 0.03$ For reduction process	111
	PSCA. PSCC No electrolyte and film thickness dependences D_{app} for oxidation process $\sim D_{app}$ for reduction one For reduction process No electrolyte dependence	116
	For reduction process D_{app} decreases with increasing pH	
	RDEV For diffusion of H^+ in film	106
	RDEV For diffusion of H^+ in film	106
	RDEV For diffusion of ferrocene in film	117
	RDEV For diffusion of H^+ in film Film prepared electro-chemically from basic methanol solution For diffusion of H^+ in film Film prepared electro-chemically from basic acetonitrile solution	118

Table 8.1 Continued

System	$D_{app}(cm^2 s^{-1})$	Medium
	$(1.1 \pm 0.3) \times 10^{-8}$	
	$(3.4 \pm 1.0) \times 10^{-8}$	
Poly[4-(2-amino-ethyl)-1,2-phenyleneoxide]	$(1.00 \pm 0.30) \times 10^{-9}$	NM, 0.1*M* TBAP
	$(1.77 \pm 0.53) \times 10^{-9}$	H_2O, 1*M* $HClO_4$
Poly(2-hydroxy-methyl-1,4-phenyleneoxide)	$(5.00 \pm 3.00) \times 10^{-8}$	NM, 1*M* TBAP
	$(1.23 \pm 0.74) \times 10^{-8}$	H_2O, 1*M* $HClO_4$
Poly(2-cyano-1,4-phenyleneoxide)	$\sim 3.7 \times 10^{-8}$	H_2O, 1*M* $HClO_4$
Cuprophane	4.80×10^{-7}	H_2O, 1*M* NaOH
	4.18×10^{-7}	H_2O(pH 7.3), 0.01*M* sodium phosphate buffer
EBBA–polycarbonate composite membranes	$(3.4 \pm 0.9) \times 10^{-7}$	H_2O(pH 2.6), 0.2*M* Na_2SO_4
Poly(ethylene oxide)		Doped with $Li(CF_3SO_3)$ Acetonitrile vapor
(Trimethylamino)-methylferrocene	$(2-5) \times 10^{-6}$	
$[Ru(bpy)_3]^{2+}$	$(2.4-6) \times 10^{-7}$	
$[Os(bpy)_3]^{2+}$	$(3-4.5) \times 10^{-7}$	
$[Co(bpy)_3]^{2+}$	$(2-5) \times 10^{-7}$	

Process–Species Responsible	Comments–Method	References
	For diffusion of Fe^{3+} in film. Film prepared electrochemically from basic acetonitrile solution	
	For diffusion of $[Fe(CN)_6]^{3-}$ in film. Film prepared electrochemically from basic acetonitrile solution	
	RDEV	117
	For diffusion of ferrocene in film	
	For diffusion of quinhydrone in film	
	RDEV	117
	For diffusion of ferrocene in film	
	For diffusion of quinhydrone in film	
	RDEV	117
	For diffusion of quinone in film	
	RDEV	119
	For diffusion of $[Fe(CN)_6]^{3-}$ in membrane at 30°C	
	RDEV	120
	For diffusion of $[Fe(CN)_6]^{3-}$ in membrane in a nematic state at 50°C	
	Interdigitated array electrode. Solid-state linear sweep and CV at microdisk electrodes	240
	$D_{app} = (0.9 \sim 1.2) \times 10^{-5}$ cm^2 s^{-1} in acetonitrile liquid	

8.1.3A Abbreviations in Table 8.1

CV	Cyclic voltammetry
PSCA	Potential-step chronoamperometry
PSCC	Potential-step chronocoulometry
NPV	Normal pulse voltammetry
RDEV	Rotating disk electrode voltammetry
D_{app}	Apparent diffusion coefficient
D_m	Diffusion coefficient for permeation of solution-phase species in film
k^0	Standard rate constant of heterogeneous electron transfer reaction
k_{ex}	Electron self-exchange rate constant
α_c	Cathodic transfer coefficient
α_a	Anodic transfer coefficient
n	Number of electrons involved in heterogeneous electron-transfer reaction
E_a	Activation energy in Arrhenius equation $[D_{app}=D_0 \exp(-E_a/RT)]$
D_0	Preexponential factor in Arrhenius equation
κ	Partition coefficient
ϕ'	Film thickness
C	Volume concentration of electroactive species (or site) confined in film
Γ_M	Surface concentration of electroactive species (or site) confined in film
Γ_{PVP}	Surface concentration of pyridinium groups in PVP film
x_{Os}	Mole fraction of osmium sites in film
MW	Average molecular weight
TBAP	Tetrabutylammonium perchlorate $(n\text{-}C_4H_9)_4N^+ClO_4^-$
TEACl	Tetraethylammonium chloride $(C_2H_5)_4N^+Cl^-$
TEAT	Tetraethylammonium tetrafluoroborate $(C_2H_5)N^+BF_4^-$
TBAT	Tetrabutylammonium tetrafluoroborate $(n\text{-}C_4H_9)_4N^+BF_4^-$
$CH_3\phi SO_3Na$	Sodium *p*-toluenesulfonate
$CH_3\phi SO_3H$	*p*-toluenesulfonic acid
NM	Nitromethane
PC	Propylene carbonate
DMSO	Dimethyl sulfoxide
DMA	*N*,*N*-Dimethylacetamide
DMF	*N*,*N*-Dimethylformamide
bpy	2,2′-Bipyridine

cinn	*N*-(4-Pyridyl) cinnamamide
vbpy	4-vinyl-4′-methyl-2,2′-bypyridine
vpy	4-Vinylpyridine
phen	*o*-Phenanthroline or 1,10-phenanthroline
phen(—$\phi SO_3^-)_2$	Bathophenanthroline disulfonate
bpz	2,2′-Bipyrazine
tpy	2,2′,2″-Terpyridine
C_2O_4	Oxalate
edta	Ethylenediamine-*N*,*N*,*N*′,*N*′-tetraacetate or ethylenediamine tetracetic acid
ϕ	Benzene nucleus
C_p	Cyclopentadienide
C_p'	Substituted cyclopentadienyl
PVF	Poly(vinylferrocene)
PVP	Poly(4-vinylpyridine)
PVP·H^+	Protonated poly(4-vinylpyridine)
poly(*l*-lysine)·H^+	Protonated poly(*l*-lysine)
Random ternary copolymer (I)	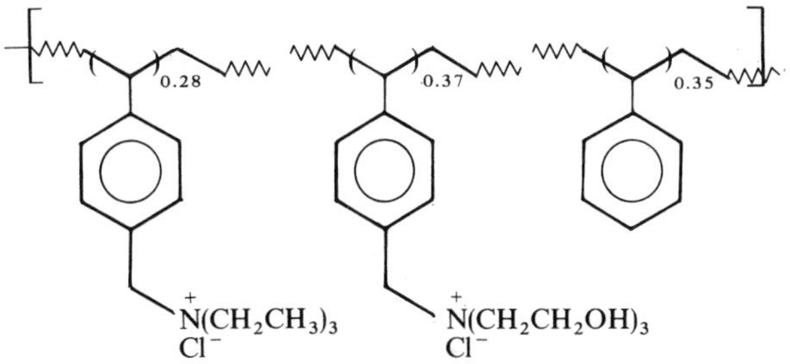 0.28, 0.37, 0.35; $\overset{+}{N}(CH_2CH_3)_3$ Cl^-; $\overset{+}{N}(CH_2CH_2OH)_3$ Cl^-
PVI	Poly(*N*-vinyl-2-methylimidazole)
ND1	Nylon derivative 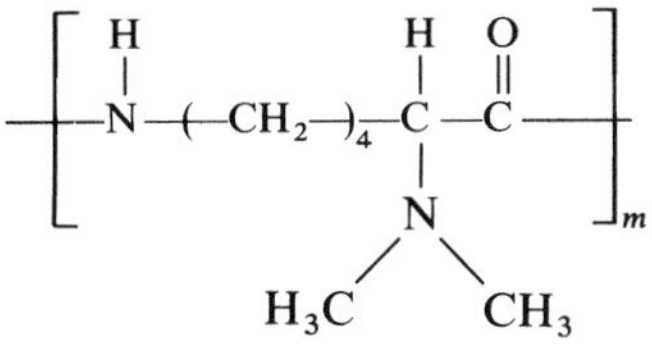 $\left[-NH-(CH_2)_4-CH(N(CH_3)_2)-C(=O)- \right]_m$
ND2	Nylon derivative; 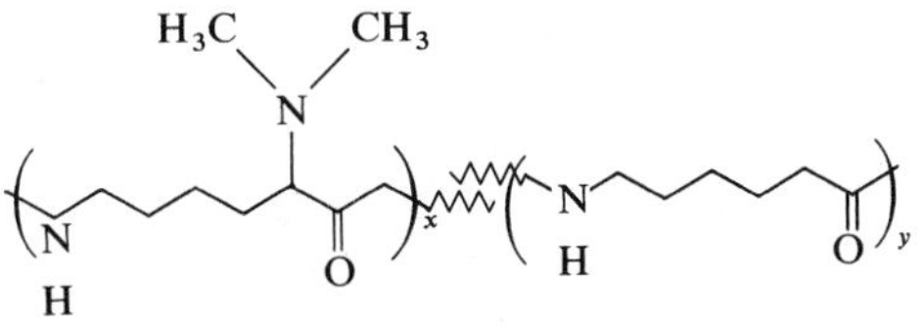 $x=0.52$, $y=0.48$
ND3	Nylon derivative of ND2 type with $x=0.33$ and $y=0.67$
PEI	Branched polyethyleneimine
l-PEI	Linear polyethyleneimine

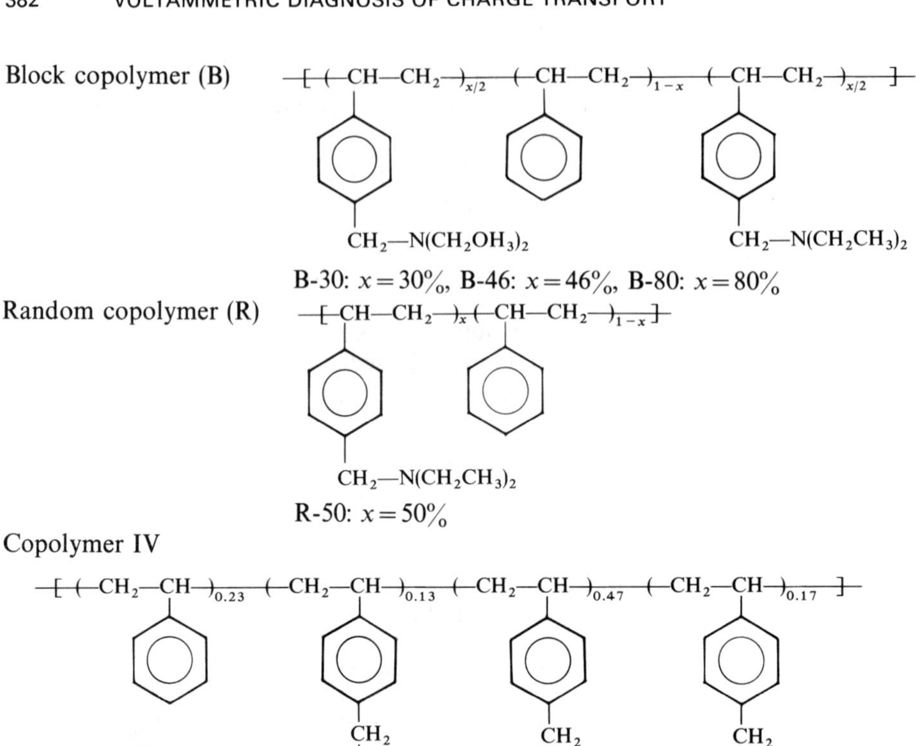

Copolymer V

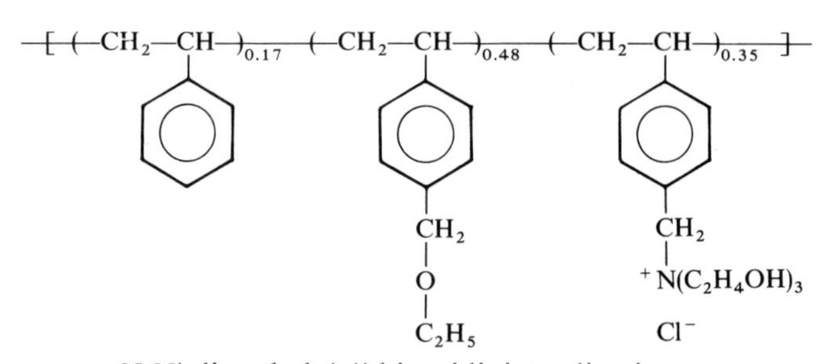

MV^{2+}	*N*,*N*′-dimethyl-4,4′-bipyridinium dication
$MV^{\cdot +}$	*N*,*N*′-dimethyl-4,4′-bipyridinium monocation radical
V^{2+}	Oxidized viologen centers of poly(viologens)
$V^{\cdot +}$	Reduced viologen centers of poly(viologens)
EBBA	*N*-(4-Ethoxybenzlidene)-4′-*n*-butylaniline

CH_3CH_2O—C_6H_4—CH=N—C_6H_4—$(CH_2)_3$—CH_3

TCNQ	Tetracyanoquinodimethane

BOEB	Poly(bis-oxyethanol benzene adipate)
MDI	Methylenebis(4-phenylene isocyanate)
TDI	1,4-Toluenediyl diisocyanate
HMDI	1,6-Hexamethylene diisocyanate
TTF	Tetrathiafulvalene

8.1.4 Other Methods

Only a few groups used the ac impedance method for study of charge-transport polymer films (2–4, 38, 39, 104). However, more widespread use may be expected in the future because of the quality of the data obtained. Alternating current impedance behavior similar to that expected for a solution–reactant was observed. The impedance analysis, such as the Cole–Cole and Randles plots, allows us to estimate D_{app}, as well as k° and α, the film resistance, double-layer capacity, etc.

Recently, "microstructured electrodes," such as sandwich, array, bilayer, and ion-gate electrodes have been developed as "devices" to study the transport of electrons and ions through the polymer films and as "pioneer structures" that may in the future lead to useful macromolecular electronic devices as studied by Murray and his co-workers (63, 76–78, 80, 82, 95, 142, 145–151, 162–164], Wrighton (63, 152–156), and others (157–159). In the experiments based on the aforementioned PSCA, PSCC, NPV, ac impedance method, and so on, the electrodes with the conventional structures and size have been used to detect simultaneous electron and ion motion over macroscopic distances. On the other hand, at ion-gate electrodes, ion transport can be measured (150, 160, 161), and sandwich, array, and bilayer electrodes are well suited for examining electron conduction in the absence of macroscopic ion motion (63, 76–78, 82, 142, 145, 148, 149, 154, 155, 162–164).

The experiments based on polymer-coated RDEV have been increasingly carried out for the study of the kinetics of electron transfers between redox-active sites in the polymer coating and redox species dissolved in the contacting solution, the electron conduction rates through electroactive polymer films (by using a nonpermeating redox mediator), the permeation rates of redox species through the polymer film, etc. The rotated disk experiment has been preferred because of its steady state plateau current responses, control of solution–reactant flux via electrode rotation rate, and a recent, remarkable development in the related theory (127–132, 134–138, 142).

In addition to the methods mentioned above, CV and a small amplitude current pulse method (103), have been used for the determination of kinetic data at polymer-coated electrodes. The data reported are summarized in Table 8.1.

8.2 HETEROGENEOUS ELECTRON TRANSFER AT ELECTRODE–FILM INTERFACES

Compared with D_{app} values, the kinetic parameters (i.e., standard rate constant k° and transfer coefficient α) for the interfacial electron-transfer reaction have not been reported as frequently (Table 8.1). Normal pulse voltammetry, the ac impedance method, CV, and convolution voltammetry were used to estimate the kinetic data (2, 4, 7–9, 22–24, 31, 33, 36, 39, 50, 56–58, 105, 111, 112, 115).

8.2.1 Normal Pulse Voltammetry

Normal pulse voltammetry has been successfully applied by Oyama and his co-workers (7–9, 22–24, 39, 50, 56–58, 105, 111, 112, 115) to the kinetic study of the heterogeneous electron-transfer process of various types of polymer-coated electrodes. This process has been found to obey the conventional Butler–Volmer equation, and thus the relevant kinetic parameters k° and α have been estimated from the analysis of normal pulse voltammograms by using the following current–potential relationship (248):

$$E = E^{\mathrm{r}}_{1/4} \pm \frac{RT}{\alpha n \mathscr{F}} \ln\left(\frac{4k^\circ\sqrt{\tau_s}}{\sqrt{3}\sqrt{D}}\right) \mp \frac{RT}{\alpha n \mathscr{F}} \ln\left\{x\left[\frac{1.75 + x^2(1+\exp(\mp\xi))^2}{1 - x(1+\exp(\mp\xi))}\right]^{1/2}\right\} \tag{8.6}$$

with

$$x = i/(i_d)_{\mathrm{Cott}} \tag{8.7}$$

$$\xi = (n\mathscr{F}/RT)(E - E^{\mathrm{r}}_{1/2}) \tag{8.8}$$

$$D = (D^{\mathrm{a}}_{\mathrm{app}})^{\alpha_a}(D^{\mathrm{c}}_{\mathrm{app}})^{\alpha_c} \tag{8.9}$$

where the top and bottom signs in the double signs ($\pm$ or $\mp$) used in Eq. 8.6 correspond to reduction and oxidation, respectively. The term E is the electrode potential, i is the normal pulse voltammetric current, $E^{\mathrm{r}}_{1/2}$ is the voltammetric reversible half-wave potential, and α denotes cathodic and anodic transfer coefficients (α_c and α_a for the cathodic and anodic reactions, respectively). The terms $D^{\mathrm{a}}_{\mathrm{app}}$ and $D^{\mathrm{c}}_{\mathrm{app}}$ are the apparent diffusion coefficients for anodic and cathodic processes, respectively, and $\mathscr{F}$, R, and T have their usual meanings. The term $(i_d)_{\mathrm{Cott}}$ is expressed by Eq. 8.5.

Equation 8.6 is applicable to the case where infinite diffusion conditions prevail. Positive feedback techniques are used to compensate for film and solution resistance. Figures 8.5 and 8.6 (56) show typical normal pulse voltammograms for the one-electron reduction of the viologen dication to the corresponding radical monocation as pendant viologen polymers coated on electrode surfaces at various τ_s values and the modified log plots for these voltammograms, respectively. The parameters k° and α can be determined from the intercept and slope of the plots of Fig. 8.6, and the D_{app} values are obtained

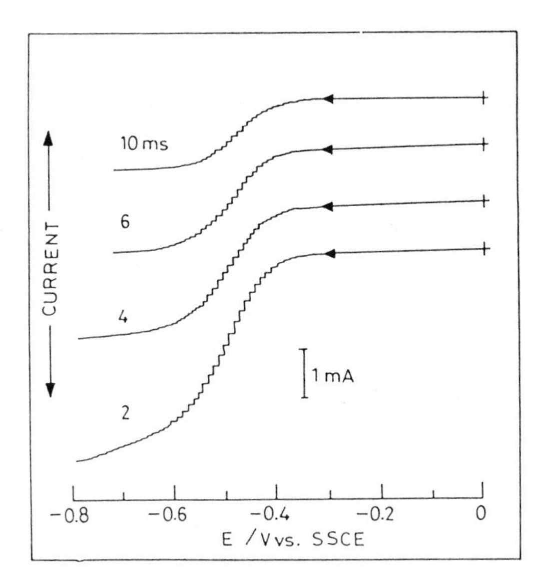

Figure 8.5. Typical normal pulse voltammograms for the one-electron reduction of the viologen dication as PMV coated on BPG electrode at various sampling times in a 0.2*M* KCl solution (pH 3.0). The term $C_{MV^{2+}} = 2.1 \times 10^{-4}\,\text{mol cm}^{-3}$. Electrode area $= 0.17\,\text{cm}^2$. Sampling times (ms) are given on each voltammogram. [Reprinted with permission from N. Oyama, T. Ohsaka, H. Yamamoto, and M. Kaneko, *J. Phys. Chem.*, *90*, 3850 (1986). Copyright © (1986) American Chemical Society.]

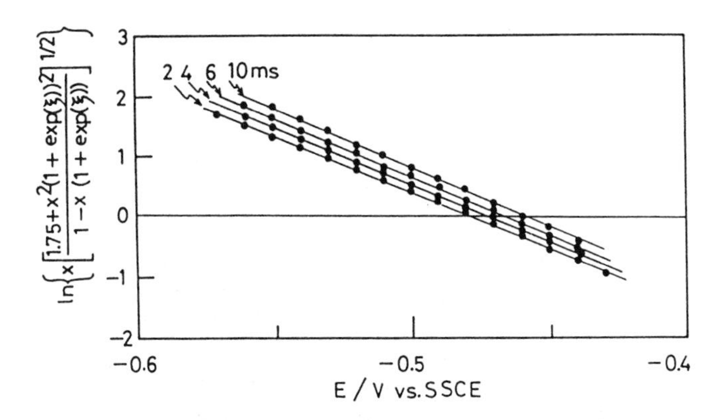

Figure 8.6. Modified log plots for the normal pulse voltammograms shown in Fig. 8.5. Sampling times (ms) are indicated on each straight line.

from the τ_s dependence of the limiting current by using Eq. 8.5 (56). The facility to measure k°, α, and D_{app} in one experiment and the apparent precision of the data indicate that this technique should be applied more widely.

8.2.2 Alternating Current Impedance Methods

The ac impedance method is also useful in kinetic studies, but thus far only a few groups (2–4, 38, 39, 104) applied this method to the study of a charge-transport reaction at polymer-coated electrodes. In some cases the Cole–Cole and Randles plots, which are essentially similar to those (247) obtained for a solution-phase reactant at an uncoated electrode, have been observed. Typical examples are shown in Fig. 8.7 (2) and the related parameters have been determined. The disagreement of the data obtained by ac impedance methods and by pulse methods such as PSCA, PSCC, and NPV remains to be resolved (2, 39).

8.2.3 Convolution Voltammetry and Cyclic Voltammetry

Leddy and Bard (36) used convolution voltammetry under semiinfinite conditions in order to evaluate the rates of heterogeneous electron transfer of poly(vinylferrocene) (Fig. 8.8) (36) and $[Ru(bpy)_3]^{2+}$ in Nafion-modified electrodes. This technique is a good method of characterizing polymer-modified electrodes, because this allows corrections for uncompensated resistance and double-layer capacitance, as well as the determination of D_{app}, k°, and α and $E_{1/2}$ from a single cyclic voltammogram.

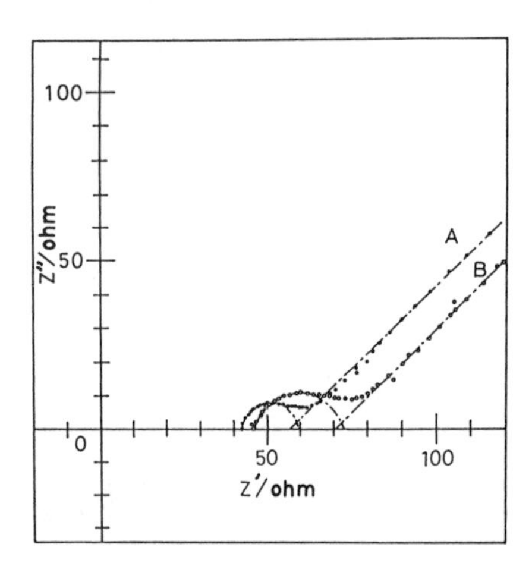

Figure 8.7. Cole–Cole plots for the redox reaction of the $[Fe(CN)_6]^{4-/3-}$ couple confined in the protonated PVP film on BPG electrodes in a 0.2*M* CF_3COONa solution (pH 1.5). The thickness of PVP films is 8.7×10^{-5} cm and the surface concentration (Γ_{PVP}) of the pyridine site of PVP is 5.6×10^{-7} mol cm^{-2}. Concentrations of $[Fe(CN)_6]^{3-}$ in PVP films are (*A*) 1.1×10^{-3} and (*B*) 4.1×10^{-4} mol cm^{-3}. The dc potentials are set at (*A*) 0.275 and (*B*) 0.265 V versus SSCE, which are formal redox potentials in cases *A* and *B*, respectively. The amplitude of an imposed ac voltage is 5.0 mV. [Reprinted with permission from Ref. 2.]

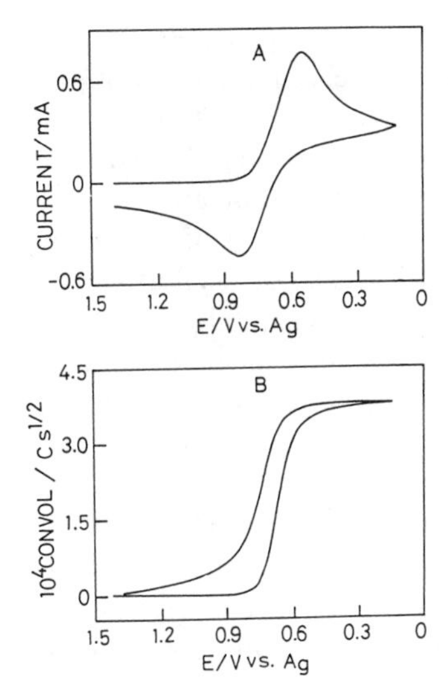

Figure 8.8. Cyclic (*A*) and convoluted (*B*) voltammograms for the reduction of a 1.13-μm thick poly(vinylferrocene) film at $1.0\,Vs^{-1}$ in $0.1M$ $TBABF_4$ in acetonitrile solution. Both peak splitting in (*A*) and the hysteresis in (*B*) are indications of quasireversible behavior. [Reprinted with permission from Ref. 36.]

Cyclic voltammetry can be used to determine k° (249–251). Sharp et al. (165, 166) attempted to use this technique to measure k° for silane- and alkylsilane-ferrocenes covalently bound to platinum. In this case, the contribution of an uncompensated resistance to the anodic and cathodic peak splitting has been suggested by Roullier and Laviron (167). Peerce and Bard (71) successfully applied digital simulation to the cyclic voltammetry of poly(vinylferrocene) on platinum by nonequivalent redox sites with interconversion between such sites, electron-transfer kinetics at electrode–film interface, and diffusion within the film.

8.3 MECHANISMS OF CHARGE TRANSPORT WITHIN POLYMER FILMS

As mentioned above, charge (electron and/or ion) transport through the solvent-swollen polymer films undergoing electrolysis [i.e., oxidation or reduction of incorporated species (or sites)] is generally believed to occur via an electron-hopping (i.e., intrinsic electron exchange) process between redox species and/or the physical diffusion (displacement) of redox species themselves (which are temporarily confined in polymer domains). The mechanism of charge transport via electron hopping has been dubbed "redox conduction" by Murray and his co-workers (76, 82, 148, 149). In this case, electron self-exchange reactions in the polymer can be driven by a concentration gradient of fixed oxidized and reduced redox sites and do not require the presence of an electrical potential gradient. That is to say, a significant electrical field gradient is thought to be present only at the electrode–polymer film interface, where electron transfers to/from the electrode occur. Both electron hopping and physical diffusion require, for charge neutrality, concurrent uptake of counterions into the polymer matrix or expulsion of co-ions initially present in the films as an ion pair. In addition, the segmental motion of the polymer chain and the motion of the solvent may also occur. Thus, distinguishing between these motions requires analysis of the functional dependence of D_{app} on variables such as the structure (morphology) and swelling of the polymer, concentration and solvation of redox sites (or species) in the film, size, charge, and concentration of supporting electrolyte ions, temperature, solvent, ionic strength, and so on, as illustrated in Table 8.1 and discussed in Sections 8.3.1 to 8.3.9. The interpretation of D_{app} in terms of the physical process is very important not only fundamentally but also practically, because successful applications of polymer-modified electrodes depend on ensuring rapid charge transport.

8.3.1 Concentration Dependence of Charge-Transport Rates

Dependence of D_{app} on the concentration (C) of electroactive species (or sites) confined in polymer domains allows us to estimate whether charge transport involves dominantly intrinsic electron exchange, physical motion of redox species, or both of these (2, 3, 7–10, 15, 16, 20, 23–26, 29, 32–35, 37, 39, 41, 45, 48, 50, 56, 58, 66, 67, 83, 84, 241). An increase in D_{app} with C has been reported for (a) the $[Co(bpy)_3]^{2+/+}$-Nafion system studied by Buttry and Anson (37), (b) for the transition metal bipyridyl redox polymer film systems studied by Murray and his co-workers (83, 84), (c) for the MV^{2+}–agarose-impregnated Nafion system studied by Moran and Majda (48), (d) for the pendant viologen polymer and its polymer complex systems studied by Oyama and Ohsaka and their co-workers (56, 58), and (e) for the 3-vinyl-10-methylphenothiazine–methylmethacrylate copolymer studied by Morishima et al. (241). The results are qualitatively understood by the use of the Dahms–Ruff electron-hopping charge-transport mechanism (252–256), which was originally developed for charge transport in solutions. On the other hand, decreases in D_{app} with increasing C and constant D_{app} values regardless of C have been also observed and appear to be more common (1–3, 7–9, 20, 23, 26, 32–35, 37, 39, 41, 45, 48, 50, 66, 67) (see Table 8.1), especially in cases where multiple-charged redox species are incorporated into polymer films by an electrostatic interaction (1–3, 7–9, 20, 23, 26, 32–35, 37, 39, 41, 45, 48, 50). In these cases, "electrostatic cross-linking" (1–42, 44–51, 53, 62, 70, 88, 141, 168, 172–181, 183, 190, 236, 237) of the films by electrostatic interactions between redox species and polymer films containing them and/or "single-file diffusion" (37, 39, 41, 182, 257), which represents the competition of diffusing species for the sites of attachment within the film, have been considered as dominant effects. An increase of the electrostatic cross-linking with increasing C causes a decrease in the diffusion rate of redox species itself, as well as decreases in the rates of the charge-compensating counterion motion, the motion of solvent and/or the segmental motion of the polymeric chain. In addition, the diffusing species, which must move between more or less fixed sites within a polymeric matrix, are considered to have their rate of motion limited by the decreasing availability of the site as C increases (single-file diffusion effect). These both result in the decrease of the overall rate of the charge transport with increasing C and thus decreased D_{app} values are observed. Even in these cases, the presence of the contribution from electron self-exchange to the overall diffusional charge-transport process is not completely excluded, because the contribution of electron self-exchange may be masked by the contributions of electrostatic cross-linking and/or single-file diffusion. In other words, (a) when the former contribution is dominant, the measured D_{app} values would increase with increasing C; (b) when the latter contributions are dominant, D_{app} values would decrease with an increase in C and; (c) when these contributions are comparable, almost constant D_{app} values would be observed irrespective of C. Typical examples of three types of C dependence of D_{app} are shown in Fig. 8.9 (9, 56, 83).

Table 8.1 contains several clear-cut examples of rate-limiting counterion diffusion. For a given redox species–polymer system, one can see correlations between counterion diffusion rates and ionic size–charge or strength of interaction with charged polymer sites. Based on comparative photoluminescence and electrochemical measurements, Majda and Faulkner (44, 46) definitely elucidated that for the $[Ru(bpy)_3]^{2+}$–poly(styrenesulfonate) film, counterion diffusion is the rate-limiting step of the oxidation–reduction process.

8.3.2 Effect of Polymeric Domain on Charge-Transport Rates

Polymer structure (or morphology) has been realized as one of the most significant factors controlling charge-transport rates. Anson and his co-workers (18–20, 133) examined the electrochemical response of $[Fe(CN)_6]^{4-}$ incorporated into protonated random and block copolymers containing various ratios of styrene and *p*-(diethylaminomethyl)styrene. With ratios of styrene to (aminoethyl)styrene groups up to about 2:1, the electrochemical responses observed for $[Fe(CN)_6]^{4-}$ anions that are incorporated are very different from the coatings prepared from random and block copolymer coatings of the same composition. The D_{app} values for incorporated $[Fe(CN)_6]^{4-}$ anions are notably larger in block than in random copolymer coatings. However, neither type of coating exhibits D_{app} values nearly as large as those $[\sim(1-5) \times 10^{-6}\,cm^2s^{-1}]$ for a ternary copolymer–homopolymer composite, which are the largest yet reported for multiply charged redox species in any polyelectrolyte. In order to explore the origin of such attractively large D_{app} values, Anson et al. (21) further examined an additional set of copolymers, both alone and in blends, with several homopolymers. These new composite coatings are especially attractive for applications in electrocatalysis. Oyama and Ohsaka and his co-workers (22–24) examined the kinetics of charge-transport processes within electropolymerized poly(*N*,*N*-dialkylsubstituted aniline) films into which multiply charged anionic metal complexes $\{[Fe(CN)_6]^{3-}$, $[Mo(CN)_8]^{4-}$, $[W(CN)_8]^{4-}$, $[Ru(CN)_6]^{4-}$, $[Os(CN)_6]^{4-}$, $[IrCl_6]^{2-}$, and so on$\}$ are incorporated electrostatically. At the same concentrations of $[Fe(CN)_6]^{3-}$ in the films D_{app} values increased in the following order: poly(*N*,*N*-dimethylaniline) < poly(*N*-methyl-*N*-ethylaniline) < poly(*N*,*N*-diethylaniline). This may be ascribed to a different extent of electrostatic cross-linking of these films, which primarily reflect the different

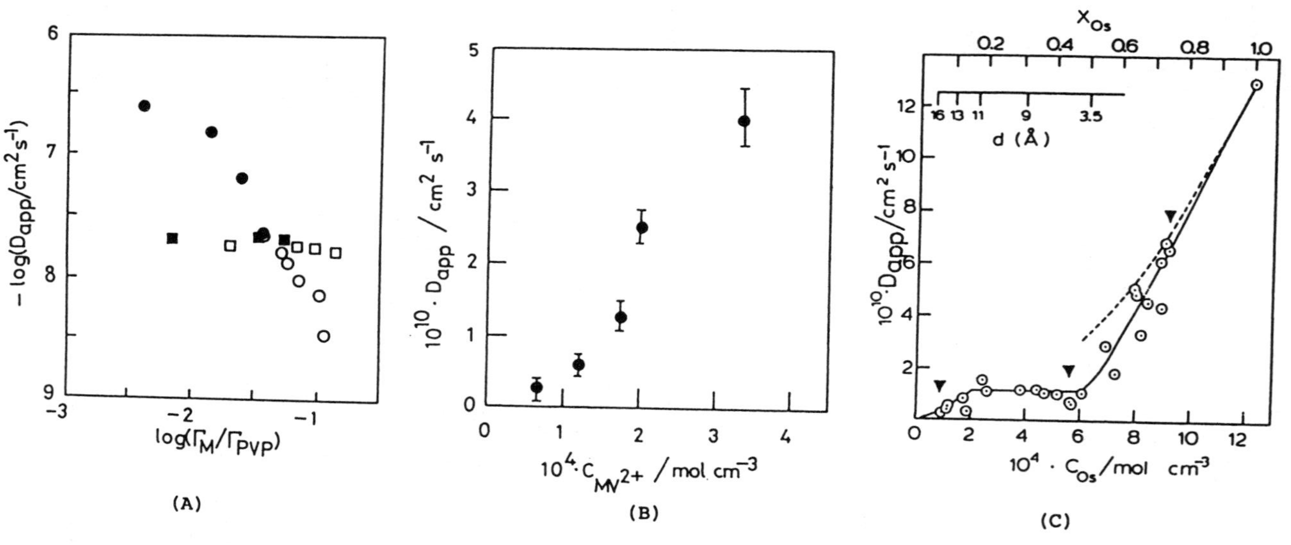

Figure 8.9. Typical examples of C dependence of D_{app}. (A) (●, ○) $[IrCl_6]^{3-}$ protonated PVP film system, (■, □) $[W(CN)_8]^{4-}$ protonated PVP film system. $\Gamma_{PVP} = 5.6 \times 10^{-7}$ mol cm^{-2} (open symbols), 1.12×10^{-6} mol cm^{-2} (solid symbols). Supporting electrolyte: 0.2M $CF_3COONa + CF_3COOH$ (pH 1.5). The term Γ_M/Γ_{PVP} represents the molar ratio of incorporated electroactive species to the protonated pyridine group of PVP. [Reprinted with permission from N. Oyama, T. Ohsaka, and T. Ushirogouchi, *J. Phys. Chem.*, *88*, 5274 (1984). Copyright © (1984) American Chemical Society.] (B) Poly(styrene-*co*-chloromethylstyrene) pendant viologen film system. Supporting electrolyte: 0.2M KCl (pH 3.0). (C) Copolymer film system of $[Os(bpy)_2(4\text{-pyNHCOCH{=}CHPh})_2]^{2+}$ and $[Ru(bpy)_2(4\text{-pyNHCOCH{=}CHPh})_2]^{2+}$. D_{app} values correspond to the $Os^{II} \rightarrow Os^{III}$ oxidation in films. The term d is the edge-to-edge separation of Os complex sites at various mol fraction (X_{Os}) of osmium sites. Activation energies measured at the pointers ($X_{Os} = 0.074, 0.45, 0.74$) are 11.2, 6.4, and 3.6 kcal mol^{-1}, respectively. [Reprinted with permission from J. S. Facci, R. H. Schmehl, and R. W. Murray, *J. Am. Chem. Soc.*, *104*, 4959 (1982). Copyright © (1982) American Chemical Society.]

steric bulkiness of methyl and ethyl groups on N atoms. Recently, Oyama et al. (39) reported the effect of polymer domain on charge-transport rates within perfluoro poly(carboxylate) and poly(sulfonate) coatings containing $[Os(bpy)_3]^{2+}$ or (trimethylammonio)ferrocene, as well as on interfacial electron-transfer rates. Many papers have been written (12, 34, 49–67) concerning the charge-transport behavior of viologen polymer films of various types. Surprisingly, the values reported for D_{app} are in the range of about 10^{-12}–10^{-7} $cm^2\,s^{-1}$ (Table 8.1), which can be considered to be due primarily to different polymer structures.

It has become increasingly apparent that most polymer films employed for modifying electrode surfaces, in general, contain an internal structure that is important to the transport of charge within them. In order to explain this wide range of D_{app} values that were reported for Nafion coatings incorporating structurally similar cations [e.g., $[Co(bpy)_3]^{3+/2+}$, $[Co(bpy)_3]^{2+/+}$, $[Co(NH_3)_6]^{3+/2+}$, $[Ru(NH_3)_6]^{3+/2+}$, and $[Ru(bpy)_3]^{3+/2+}$ couples], on the basis of the previously demonstrated two-phase structures (i.e., ionic cluster model) of Nafion (186), Buttry and Anson (37) prepared a model for the interior of Nafion coatings that includes both hydrophilic and hydrophobic phases between which incorporated reactants are partitioned. This model assumes coupling between two diffusion paths. In addition, Redepenning, Anson and their co-workers (184, 185) recently demonstrated that the effects of pH and electrolyte concentration on the formal potentials of species incorporated in Nafion coatings are due to Donnan potential effects. Rubinstein et al. (38) investigated the electrochemical processes of the $[Ru(bpy)_3]^{2+/3+}$–Nafion system by the ac impedance method, covering both semiinfinite and finite diffusion regimes, and observed a transition from semiinfinite diffusion to finite diffusion (upon decreasing the modulation frequency) with different D_{app} values. They explained their results by assuming that no coupling occurred between two parallel diffusion paths in the film, in contrast to the model of Buttry and Anson (37). On the other hand, based on a comparison of D_{app} values for the oxidized and reduced forms of various ionically diffusing redox couples, Martin and Dallard (34) considered the effect of hydrophobic interactions on the rate of ionic diffusion in Nafion films, and suggested the possibility of a much simpler model for the ionic cluster suggested by Yeager and Stech (186) and Rodmacq et al. (187). That is, it is not necessary to assume the presence of two separate ionic domain phases with associated partition equilibria between phases, if a model that provides for criss-crossing of chain material through the ionic domain is allowed. Majda and Faulkner (45) examined the electrochemical properties of $[Ru(bpy)_3]^{2+}$ confined in poly(styrenesulfonate) film and suggested that this complex is present in two distinct environments that are interconvertible, that is, in "weakly bound" and "strongly bound" forms. They also showed evidence for the nucleation, in oxidized films, of a domainlike structure in which Ru(III) centers cluster together to form tight, almost electroinactive zones. Anson et al. (15–17, 135) proposed a two-phase model under steady state conditions that reasonably explained the charge-transport behaviors of poly(*l*-lysine) films containing $Fe^{II}(edta)^{2-}$ and $Fe^{III}(edta)^{-}$ or $[Co(C_2O_4)_3]^{3-}$, which consisted of two regions, i.e., "Donnan domains," which represents the region where the counterions are confined by electrostatic forces and the remaining region of the coating that is assumed to be occupied by the supporting electrolyte solution. An analogous two-phase scheme has been also considered by Moran and Majda (48), who found that under transient conditions for MV^{2+} in Nafion–agarose matrices, D_{app} values are more than an order of magnitude higher than those in plain Nafion, and that in Nafion, D_{app} decreases with increasing MV^{2+} concentration, while in Nafion–agarose the opposite dependence is obtained. The faster rate of electron transport in Nafion–agarose films is related to the heterogeneous structure of these films and the coupling of the diffusion pathways between the solution and the polymer phases within the films.

8.3.3 Effects of Redox Cross-Reactions on Charge-Transport Rates

Enhancement of charge-transport rates by redox cross reactions between redox couples having disparate diffusion coefficients and formal potentials, which are simultaneously incorporated in an electrode coating, has been observed for some redox polymer film systems (9, 10, 25, 32, 124, 188, 189). Buttry et al. (124) demonstrated that the rate of charge transport within a Nafion coating by $[Co(tpy)_2]^{3+}$ is enhanced in the presence of $[Ru(NH_3)_6]^{2+}$ via a rapid electron-transfer cross reaction and similarly the presence of $[Cp_2FeN(CN_3)_3]^{2+}$ enhances the propagation rate of $[Co(tpy)_2]^{2+}$. Such an enhancement of charge-transport rates results because electron transfer between the more slowly diffusing complex $\{[Co(tpy)_2]^{3+}$ or $[Co(tpy)_2]^{2+}\}$, and the more rapidly diffusing $[Ru(NH_3)_6]^{2+}$ or $[Cp_2FeN(CH_3)_3]^{2+}$ complexes, allows the latter to carry charge between the electrode surface and the slower moving complex. They quantitatively confirmed their results on the basis of the theoretical calculation of the extent of the enhancement for the case of PSCA and PSCC. Qualitative evidence of mediated charge transport in an electrode coating containing two redox couples has been obtained by Facci and Murray and his co-workers (10, 25) for the $[Fe(CN)_6]^{3-}$–$[IrCl_6]^{3-}$ mixture in a polycationic copolymer film of vinylpyridine and (γ-methacryloxypropyl)trimethoxysilane. It was also obtained by Lindholm and Sharp (189) for the same redox complexes in cross-linked, protonated PVP film, by Oyama et al. (9) for the $[IrCl_6]^{3-}$–$[W(CN)_8]^{3-}$ mixture in a protonated PVP film, and by White et al. (32) for the $[Os(bpy)_3]^{2+}$–$[Ru(bpy)_3]^{2+}$ mixture in Nafion film. Miller and his co-workers (188) also observed enhanced currents at electrodes coated with a quinoid polymer when a rapidly diffusing redox mediator like bis(hydroxymethyl) ferrocene was added to the solution.

Furthermore, Tsou and Anson (41) demonstrated that in the case of the redox film incorporating a molecule that contains two separate electroactive redox groups with differing rates of electron self-exchange, the charge transport based on molecular motion and electron exchange can be differentiated depending on which redox group is employed in an electrochemical measurement of diffusion rates. For this purpose, they used the heterobinuclear metal complexes $[(NH_3)_5RuC_5H_4NCH_2NHC(=O)—CpFeCp]$ within Nafion coatings, and obtained reasonable agreement between the actually observed results and those calculated on the basis of a simple model. A similar idea has been recently applied by Ohsaka et al. (42, 182) to the Alizarine Red S dye with two separate redox centers (i.e., 3,4-dihydroxy and 9,10-dioxo groups) incorporated into a cationic perfluoropolymer film on electrodes. The oxidation of the 3,4-dihydroxy group is irreversible, while the reduction of the 9,10-dioxo group to the corresponding 9,10-diol compound and the reverse reaction are reversible. The difference in D_{app} values for these processes has been considered to be due to different contributions of the electron exchange between electroactive centers to the overall charge-transport rates.

8.3.4 Lateral Charge Transport in Organized Assemblies

Recently, Majda and his co-workers (13, 14, 65–67) prepared microporous aluminum oxide films with controlled porosity on the inner walls, which are coated with various polymers (e.g., protonated PVP) resulting in the subsequent incorporation of redox species [e.g., $[Fe(CN)_6]^{4-}$]. In addition, they employed self-organized monolayer and bilayer assemblies containing a redox-active species (e.g., viologen and ferrocene). They conducted the first direct electrochemical measurements of the "lateral charge transport" in such organized assemblies.

8.3.5 Models for Nonmediated Electrochemical Reactions of Dissolved Substances

For nonmediated electrochemical reactions of diffusing electroactive species at a film-covered electrode, the distinction between diffusion through the polymer layer itself, viewed as a homogeneous medium (membrane model), and through tortuous channels or pores of larger-than-molecular diameters in the film (pinhole model) is difficult. In pinhole and membrane models, the electrode reactions of the diffusing species occur at the exposed substrate surface and at the electrode–film interface, respectively. This problem has been dealt with by Peerce and Bard (141), Ikeda et al. (191), Gough and Leypoldt (119), and Dubois and his co-workers (117, 192).

8.3.6 Interactions between Electroactive Sites in Polymeric Domain

The presence of the interactions between electroactive sites (or species) in the film at about monolayer coverages, as well as at multilayer coverages, has been well realized. For example, this interaction can be readily seen from the wave broadening and narrowing in CV that may result from repulsive and attractive interactions, respectively. The effect has been considered first by Brown and Anson (193) and later by Laviron et al. (194–196), Matsuda and his co-workers (197–199), and others (33, 69, 71, 77, 81, 148, 149, 200–205). This is equivalent to the "activity effect." These effects originate because, in general, the redox centers are present within films at very high concentrations (e.g., several molar).

8.3.7 Temperature Dependence of Charge-Transport Rates

The temperature dependence of D_{app} (70) has helped to elucidate charge-transport processes. Arrhenius type behavior has been observed allowing the evaluation of the parameters, E_{act} (the activation energy) and $\Delta H^{\ddagger}_{diff}$, and $\Delta S^{\ddagger}_{diff}$ (the enthalpy and entropy of the diffusion process). In particular, the observation of negative $\Delta S^{\ddagger}_{diff}$ has been associated with a net increase in order around the redox sites between which charge transport occurs (9, 57, 70). Single file diffusion and electrostatic cross-linking have been observed with negative $\Delta S^{\ddagger}_{diff}$ (39). A positive $\Delta S^{\ddagger}_{diff}$ has been attributed to a hydrophobic rather than electrostatic mechanism for the incorporation of TAF^+ in Nafion (39). A novel D_{app} behavior is shown by a PVP–MMA copolymer containing pendant $[Ru(bpy)_2Cl]^+$, which displays two characteristic activation energy processes (214). At a lower temperature, an E_{act} of 40 kJ mol^{-1} with a negative $\Delta S^{\ddagger}_{diff}$ and at an elevated temperature, an E_{act} of 140 kJ mol^{-1} with a large positive $\Delta S^{\ddagger}_{diff}$. The latter has been ascribed to the substantial disorder effects caused by polymer side chain movements.

8.3.8 Migration Effect on Charge Transport

Recently, it has been recognized that the contribution of migration of redox ions (due to the electrostatic potential gradient) to charge transport in polymer coatings on electrodes associated with redox reactions can be significant (5, 6, 11, 49, 139, 140, 206–208). Yap et al. (206) and Lange and Doblhofer (5, 6, 11, 207) calculated the effect of the free moving redox ions' migration on a Cottrell-type large potential-step experiment at a polymer-coated electrode. Savéant (139, 140, 208) established the formal diffusion-migration characteristic equations appropriate for charge propagation via electron hopping between fixed redox sites. Doblhofer and his co-workers (5, 6) studied the migration contribution independent of diffusion and elucidated the diffusion–migration mechanism of charge transport across the film. They did this by measuring the redox-ion flux associated with the charging of the electrical double layer that forms at the electrode–film interface, as well as on the basis of experiments with rotated ring-disk electrode systems consisting of a polymer-coated disk electrode and an uncoated ring electrode. In their experiments $[IrCl_6]^{3-/2-}$ or $[Fe(CN)_6]^{4-/3-}$ were used as redox systems, and poly[(4-vinylpyridine)$_{0.83}$-*co*-(styrene)$_{0.17}$] and its quaternized and cross-linked derivative were employed as polymer coatings on the disk electrode, because the transference numbers of $[Fe(CN)_6]^{3-}$ or $[IrCl_6]^{2-}$ in the films could easily approach unity by choosing the appropriate experimental conditions.

Analytical solutions for diffusion–migration impedances in finite, one-dimensional charge transport of redox polymer films in bathing electrolytes were derived by Buck (209, 210, 213), who treated charge transport as "electron displacement" rather than electron hopping: The electrons behave just like ordinary moving ions, thus obeying the usual Nernst–Planck equation concerning their "diffusion" and "migration" movements. However, this is in fact not true, because the electron hops between two fixed redox sites of different oxidation states. As far as charge transport under a concentration gradient is concerned, this is equivalent to an ordinary diffusion movement, but the effect of an electric field is not strictly equivalent to the migration of an ordinary ion (140, 208). Savéant (140, 211, 212) analyzed charge transport by "electron hopping" in counterion conservative redox membranes at steady state in terms of equivalent diffusion and migration under rigorous mass conservation conditions. The treatments and results for "electron hopping" and for "electron displacement" have been compared systematically (212).

8.3.9 Charge Transport within Polynuclear Transition Metal Cyanide Films

Polynuclear transition metal cyanide films of the general formula $M'_x[M''(CN)_6]_y^{n-}$ (where M′,M″ = Fe, Ni, Os, and Ru, e.g., Prussian Blue, Ruthenium Purple, and Osmium Purple) are of interest both for their potential applications and as models for gaining an understanding of the kinetics and mechanism of electron transport of inorganic redox polymers (80, 95–98, 215–235). These inorganic polymer films have more ordered internal structures than most polymeric electroactive materials and thus are appropriate for examining electron transport in relation to the transport of charge-compensating counterions in an ordered lattice matrix. Viehbeck and DeBery (96) measured the diffusion coefficients for reduction and oxidation of the Prussion Blue–Everitt salt redox couple by PSCA. Also, Rajan and Neff (97) examined the kinetics of the reduction of Prussian Blue and Ruthenium Purple by PSCA, and they explained the kinetic response to a potential step in terms of migration and diffusion of charge carriers within the films. More recently, Feldman and Murray (95) measured the rates of redox electron conduction through Prussian Blue films by using an interdigitated array electrode. The electron diffusion coefficient in ferric ferrocyanide–ferrous ferrocyanide mixed-valent film is remarkably insensitive to the cations (K^+, NH_4^+, Rb^+, and Cs^+) that enter the lattice as a counterion to the ferrous ferrocyanide state. The implication is that the local cation motions can occur so rapidly as to not control the rate (k_{ex}) of the electron-hopping reaction (k_{ex} is estimated to be $1.3 \times 10^6 M^{-1}s^{-1}$). Bocarsly and his co-workers (98, 227–229, 231, 232) prepared $[Ni_x^{II}(M(CN)_5L_y]^{n-}$ (where M = Fe, Ru, Mn, etc. and L = CN^-, H_2O, NO, etc.) films and performed very extensive studies of this system including electrochemical characterization, diffusion reflectance measurements, ion-exchange properties, and charge-transfer kinetics. They (98) investigated the charge-transport kinetics of the $[Ni^{II}(NC)Fe^{II/III}(CN)_5]^{2-/1-}$ film by using a diffuse reflection technique and chronocoulometry, and introduced time-dependent diffusion coefficients to obtain agreement for short-time and long-time data.

8.4 CONCLUSIONS

In conclusion, there have been many interesting developments in attempting to unravel the charge-transport processes on polymer-coated electrodes in recent years. These advances have opened up whole new areas of investigation that combine modern aspects of electrochemistry, polymer chemistry, organic chemistry, inorganic chemistry, biochemistry, physics, and so on, and should prove very fruitful in future study.

ACKNOWLEDGMENTS

We wish to thank Professor R. W. Murray for his encouragement to write this chapter and for his valuable criticism. Our work was generously supported by a Grant-in-Aid for Scientific Research from the Ministry of Education, Science, and Culture, Japan.

REFERENCES

1. N. Oyama, S. Yamaguchi, Y. Nishiki, K. Tokuda, H. Matsuda, and F. C. Anson, *J. Electroanal. Chem.*, *139*, 371 (1982).
2. T. Ohsaka, T. Ushirogouchi, and N. Oyama, *Bull. Chem. Soc. Jpn.*, *58*, 3252 (1985).
3. R. D. Armstrong, B. Lindholm, and M. Sharp, *J. Electroanal. Chem.*, *202*, 69 (1986).
4. B. Lindholm, M. Sharp, and R. D. Armstrong, *J. Electroanal. Chem.*, *235*, 169 (1987).
5. K. Doblhofer, H. Braum, and R. Lange, *J. Electroanal. Chem.*, *206*, 93 (1986).
6. R. Lange and K. Doblhofer, *J. Electroanal. Chem.*, *216*, 241 (1987).
7. N. Oyama, T. Ohsaka, M. Kaneko, K. Sato, and H. Matsuda, *J. Am. Chem. Soc.*, *105*, 6003 (1983).
8. K. Sato, S. Yamaguchi, H. Matsuda, T. Ohsaka, and N. Oyama, *Bull. Chem. Soc. Jpn.*, *56*, 2004 (1983).
9. N. Oyama, T. Ohsaka, and T. Ushirogouchi, *J. Phys. Chem.*, *88*, 5274 (1984).
10. J. Facci and R. W. Murray, *J. Phys. Chem.*, *85*, 2870 (1981).
11. K. Doblhofer and R. Lange, *J. Electroanal. Chem.*, *229*, 239 (1987).
12. J. E. Van Koppenhagen and M. Majda, *J. Electroanal. Chem.*, *189*, 379 (1985).
13. C. J. Miller and M. Majda, *J. Am. Chem. Soc.*, *107*, 1419 (1985).
14. C. J. Miller and M. Majda, *J. Electroanal. Chem.*, *207*, 49 (1986).
15. F. C. Anson, T. Ohsaka, and J.-M. Savéant, *J. Phys. Chem.*, *87*, 640 (1983).
16. F. C. Anson, J.-M. Savéant, and K. Shigehara, *J. Am. Chem. Soc.*, *105*, 1096 (1983).
17. F. C. Anson, T. Ohsaka, and J.-M. Savéant, *J. Am. Chem. Soc.*, *105*, 4883 (1983).
18. D. D. Montigomery, K. Shigehara, E. Tsuchida, and F. C. Anson, *J. Am. Chem. Soc.*, *106*, 7991 (1984).
19. D. D. Montgomery and F. C. Anson, *J. Am. Chem. Soc.*, *107*, 3431 (1985).
20. K. Sumi and F. C. Anson, *J. Phys. Chem.*, *90*, 3845 (1986).
21. T. Inoue and F. C. Anson, *J. Phys. Chem.*, *91*, 1519 (1987).
22. N. Oyama, T. Ohsaka, and T. Shimizu, *Anal. Chem.*, *57*, 1526 (1985).
23. T. Ohsaka, T. Okajima, and N. Oyama, *J. Electroanal. Chem.*, *215*, 191 (1986).
24. N. Oyama, T. Ohsaka, and M. Nakanishi, *J. Macromol. Sci. Chem.*, *A24*, 375 (1987).

25. J. Facci and R. W. Murray, *J. Electroanal. Chem.*, *124*, 339 (1981).
26. K.-N. Kuo and R. W. Murray, *J. Electroanal. Chem.*, *131*, 37 (1982).
27. D. J. Harrison, K. A. Daube, and M. S. Wrighton, *J. Electroanal. Chem.*, *163*, 93 (1984).
28. N. Oyama and F. C. Anson, *J. Electrochem. Soc.*, *127*, 640 (1980).
29. K. Shigehara, N. Oyama, and F. C. Anson, *J. Am. Chem. Soc.*, *103*, 2552 (1981).
30. K. Doblhofer, W. Durr, and M. Jauch, *Electrochim. Acta*, *27*, 677 (1982).
31. I. Rubinstein and A. J. Bard, *J. Am. Chem. Soc.*, *103*, 5007 (1981).
32. H. S. White, J. Leddy, and A. J. Bard, *J. Am. Chem. Soc.*, *104*, 4811 (1982).
33. C. R. Martin, I. Rubinstein, and A. J. Bard, *J. Am. Chem. Soc.*, *104*, 4817 (1982).
34. C. R. Martin and K. A. Dollard, *J. Electroanal. Chem.*, *159*, 127 (1983).
35. D. A. Buttry and F. C. Anson, *J. Electroanal. Chem.*, *130*, 333 (1981).
36. J. Leddy and A. J. Bard, *J. Electroanal. Chem.*, *189*, 203 (1985).
37. D. A. Buttry and F. C. Anson, *J. Am. Chem. Soc.*, *105*, 685 (1983).
38. I. Rubinstein, J. Rishpon and S. Gottesfeld, *J. Electrochem. Soc.*, *133*, 729 (1986).
39. N. Oyama, T. Ohsaka, T. Ushirogouchi, S. Sanpei, and S. Nakamura, *Bull. Chem. Soc. Jpn.*, *61*, 3103 (1988).
40. T. P. Henning and A. J. Bard, *J. Electrochem. Soc.*, *130*, 613 (1983).
41. Y.-M. Tsou and F. C. Anson, *J. Phys. Chem.*, *89*, 3818 (1985).
42. T. Ohsaka, N. Oyama, Y. Takahira, and N. Nakamura, *J. Electroanal. Chem.*, *247*, 339 (1988).
43. J. Q. Chambers, F. B. Kaufman, and K. H. Nichols, *J. Electroanal. Chem.*, *142*, 277 (1983).
44. M. Majda and L. R. Faulkner, *J. Electroanal. Chem.*, *137*, 149 (1982).
45. M. Majda and L. R. Faulkner, *J. Electroanal. Chem.*, *169*, 77 (1984).
46. M. Majda and L. R. Faulkner, *J. Electroanal. Chem.*, *169*, 97 (1984).
47. G. Nagy, G. A. Gerhardt, A. F. Oke, M. E. Rice, R. N. Adams, R. B. Moore, III, M. N. Szentirmay, and C. R. Martin, *J. Electroanal. Chem.*, *188*, 85 (1985).
48. K. D. Moran and M. Majda, *J. Electroanal. Chem.*, *207*, 73 (1986).
49. C. M. Elliott and J. G. Redepenning, *J. Electroanal. Chem.*, *181*, 137 (1984).
50. T. Ohsaka, H. Yamamoto, M. Kaneko, A. Yamada, M. Nakamura, S. Nakamura, and N. Oyama, *Bull. Chem. Soc. Jpn.*, *57*, 1844 (1984).
51. J. R. White and A. J. Bard, *J. Electroanal. Chem.*, *197*, 233 (1986).
52. P. Burgmayer and R. W. Murray, *J. Electroanal. Chem.*, *135*, 335 (1982).
53. R. J. Mortimer and F. C. Anson, *J. Electroanal. Chem.*, *138*, 325 (1982).
54. H. Akahoshi, S. Toshima, and K. Itaya, *J. Phys. Chem.*, *85*, 818 (1981).
55. P. Martigny and F. C. Anson, *J. Electroanal. Chem.*, *139*, 383 (1982).
56. N. Oyama, T. Ohsaka, H. Yamamoto, and M. Kaneko, *J. Phys. Chem.*, *90*, 3850 (1986).
57. T. Ohsaka, H. Yamamoto, and N. Oyama, *J. Phys. Chem.*, *91*, 3775 (1987).
58. T. Ohsaka, N. Oyama, K. Sato, and H. Matsuda, *J. Electrochem. Soc.*, *132*, 1871 (1985).
59. D. C. Bookbinder and M. S. Wrighton, *J. Am. Chem. Soc.*, *106*, 3673 (1984).
60. K. W. Willman and R. W. Murray, *J. Electroanal. Chem.*, *133*, 211 (1982).
61. T. J. Lewis, H. S. White, and M. S. Wrighton, *J. Am. Chem. Soc.*, *106*, 6947 (1984).
62. R. N. Dominey, T. J. Lewis, and M. S. Wrighton, *J. Phys. Chem.*, *87*, 5345 (1983).
63. G. P. Kittlesen, H. S. White, and M. S. Wrighton, *J. Am. Chem. Soc.*, *107*, 7373 (1985).
64. D. K. Smith, G. A. Lane, and M. S. Wrighton, *J. Phys. Chem.*, *92*, 2616 (1988).
65. C. J. Miller and M. Majda, *J. Am. Chem. Soc.*, *108*, 3118 (1986).
66. C. J. Miller, C. A. Widrig, D. H. Charych, and M. Majda, *J. Phys. Chem.*, *92*, 1928 (1988).

67. C. A. Goss, C. J. Miller, and M. Majda, *J. Phys. Chem.*, *92*, 1937 (1988).
68. R. J. Nowak, F. A. Schultz, M. Umana, R. Lam, and R. W. Murray, *Anal. Chem.*, *52*, 315 (1980).
69. P. Daum and R. W. Murray, *J. Phys. Chem.*, *85*, 389 (1981).
70. P. Daum, J. R. Lenhard, D. Rolison, and R. W. Murray, *J. Am. Chem. Soc.*, *102*, 4649 (1980).
71. P. J. Peerce and A. J. Bard, *J. Electroanal. Chem.*, *114*, 89 (1980).
72. J. Leddy and A. J. Bard, *J. Electroanal. Chem.* *153*, 223 (1983).
73. J. Q. Chambers, *J. Electroanal. Chem.*, *130*, 381 (1981).
74. S. Nakahama and R. W. Murray, *J. Electroanal. Chem.*, *158*, 303 (1983).
75. P. Denisevich, H. D. Abruna, C. R. Leidner, T. J. Meyer, and R. W. Murray, *Inorg. Chem.*, *21*, 2153 (1982).
76. P. G. Pickup, W. Kutner, C. R. Leidner, and R. W. Murray, *J. Am. Chem. Soc.*, *106*, 1991 (1984).
77. C. R. Leidner and R. W. Murray, *J. Am. Chem. Soc.*, *106*, 1606 (1984).
78. C. R. Leidner and R. W. Murray, *J. Am. Chem. Soc.*, *107*, 551 (1985).
79. P. G. Pickup and R. A. Osteryoung, *J. Electroanal. Chem.*, *186*, 99 (1985).
80. C. E. Chidsey, B. J. Feldman, C. Lundgren, and R. W. Murray, *Anal. Chem.*, *58*, 601 (1986).
81. T. Ikeda, C. R. Leidner, and R. W. Murray, *J. Electroanal. Chem.*, *138*, 343 (1982).
82. P. G. Pickup and R. W. Murray, *J. Am. Chem. Soc.*, *105*, 4510 (1983).
83. J. S. Facci, R. H. Schmehl, and R. W. Murray, *J. Am. Chem. Soc.*, *104*, 4959 (1982).
84. R. H. Schmehl and R. W. Murray, *J. Electroanal. Chem.*, *152*, 97 (1983).
85. C. R. Leidner, P. Denisevich, K. W. Willman, and R. W. Murray, *J. Electroanal. Chem.*, *164*, 63 (1984).
86. C. P. Andrieux, O. Haas, and J. M. Savéant, *J. Am. Chem. Soc.*, *108*, 8175 (1986).
87. O. Haas and B. Sandmeier, *J. Phys. Chem.*, *91*, 5072 (1987).
88. P. K. Ghosh and A. J. Bard, *J. Electroanal. Chem.*, *169*, 113 (1984).
89. R. A. Simon, T. E. Mallouk, K. A. Daube, and M. S. Wrighton, *Inorg. Chem.*, *24*, 3119 (1985).
90. L. Roullier, E. Waldner, and E. Laviron, *J. Electroanal. Chem.*, *139*, 199 (1982).
91. K. M. O'Connell, E. Waldner, L. Roullier, and E. Laviron, *J. Electroanal. Chem.*, *162*, 77 (1984).
92. K. Itaya, H. Akahoshi, and S. Toshima, *J. Electrochem. Soc.*, *129*, 762 (1982).
93. H. Karimi and J. Q. Chambers, *J. Electroanal. Chem.*, *217*, 313 (1987).
94. C. V. Francis, P. Joo, and J. Q. Chambers, *J. Phys. Chem.*, *91*, 6315 (1987).
95. B. J. Feldman and R. W. Murray, *Inorg. Chem.*, *26*, 1702 (1987).
96. A. Viehbeck and D. W. DeBerry, *J. Electrochem. Soc.*, *132*, 1369 (1985).
97. K. P. Rajan and V. D. Neff, *J. Phys. Chem.*, *86*, 4361 (1982).
98. B. D. Humphrey, S. Sinha, and A. B. Bocarsly, *J. Phys. Chem.*, *88*, 736 (1984).
99. L. D. Burke and D. P. Whelan, *J. Electroanal. Chem.*, *162*, 121 (1984).
100. F. G. Will, *J. Electrochem. Soc.*, *132*, 743 (1985).
101. J. H. Kaufman, E. J. Mele, A. J. Heeger, R. Kaner, and A. G. MacDiarmid, *J. Electrochem. Soc.*, *130*, 571 (1985).
102. E. M. Genies, G. Bidan, and A. F. Diaz, *J. Electroanal. Chem.*, *149*, 101 (1983).
103. R. M. Penner, L. S. Van Dyke, and C. R. Martin, *J. Phys. Chem.*, *92*, 5274 (1988).
104. I. Rubinstein, E. Sabatani, and J. Rishpon, *J. Electrochem. Soc.*, *134*, 3078 (1987).
105. T. Ohsaka, K. Chiba, and N. Oyama, *Nippon Kagaku Kaishi*, 457 (1986).
106. Y. Ohnuki, H. Matsuda, T. Ohsaka and N. Oyama, *J. Electroanal. Chem.*, *158*, 55 (1983).
107. T. Ohsaka, K. Naoi, S. Ogano, and S. Nakamura, *J. Electrochem. Soc.*, *134*, 2096 (1987).
108. T. Kobayashi, H. Yoneyama, and H. Tamura, *J. Electroanal. Chem.*, *161*, 419 (1984).

109. W. G. Albery, M. G. Boutelle, P. J. Colby, and A. R. Hillman, *J. Electroanal. Chem.*, *133*, 135 (1982).
110. J. Roncali and F. Garnier, *J. Phys. Chem.*, *92*, 833 (1988).
111. K. Chiba, T. Ohsaka, and N. Oyama, *J. Electroanal. Chem.*, *217*, 239 (1987).
112. T. Ohsaka, S. Kunimura, and N. Oyama, *Electrochim. Acta*, *33*, 639 (1988).
113. S. Kunimura, T. Ohsaka, and N. Oyama, *Macromolecules*, *21*, 894 (1988).
114. N. Oyama, M. Sato, and T. Ohsaka, *Synth. Methods*, *29*, E501 (1989).
115. N. Oyama, K. Hirabayashi, and T. Ohsaka, *Bull. Chem. Soc. Jpn.*, *59*, 2071 (1986).
116. N. Oyama, T. Ohsaka, K. Chiba, and K. Takahashi, *Bull. Chem. Soc. Jpn.*, *61*, 1095 (1988).
117. P. C. Lacaze, M. C. Pham, M. Delamar, and J. E. Dubois, *J. Electroanal. Chem.*, *108*, 9 (1980).
118. T. Ohsaka, T. Hirokawa, H. Miyamoto, and N. Oyama, *Anal. Chem.*, *59*, 1758 (1987).
119. D. A. Gough and J. K. Leypoldt, *Anal. Chem.*, *51*, 439 (1979).
120. N. Oyama, T. Ohsaka, T. Okajima, T. Hirokawa, T. Maruyama, and Y. Ohnuki, *J. Electroanal. Chem.*, *187*, 79 (1985).
121. L. Roullier, E. Waldner, and E. Laviron, *J. Electrochem. Soc.*, *132*, 1121 (1985).
122. F. B. Kaufman and E. M. Engler, *J. Am. Chem. Soc.*, *101*, 547 (1979).
123. F. B. Kaufman, A. H. Schroeder, and J. Q. Chambers, *J. Am. Chem. Soc.*, *102*, 483 (1980).
124. D. A. Buttry, J.-M. Savéant, and F. C. Anson, *J. Phys. Chem.*, *88*, 3086 (1984).
125. C. P. Andrieux and J.-M. Savéant, *J. Electroanal. Chem.*, *111*, 377 (1980).
126. E. Laviron, *J. Electroanal. Chem.*, *112*, 1 (1980).
127. C. P. Andrieux and J.-M. Savéant, *J. Electroanal. Chem.*, *93*, 163 (1978).
128. R. D. Rocklin and R. W. Murray, *J. Phys. Chem.*, *85*, 2104 (1981).
129. R. W. Murray, *Philos. Trans. R. Soc. London*, *A302*, 253 (1981).
130. J. M. Dumas-Bouchart and J.-M. Savéant, *J. Electroanal. Chem.*, *114*, 159 (1980).
131. C. P. Andrieux, J. M. Dumas-Bouchiat, and J.-M. Savéant, *J. Electroanal. Chem.*, *123*, 171 (1981).
132. C. P. Andrieux, J. M. Dumas-Bouchiat, and J.-M. Savéant, *J. Electroanal. Chem.*, *131*, 1 (1982).
133. M. Sharp, D. D. Montgomery, and F. C. Anson, *J. Electroanal. Chem.*, *194*, 247 (1985).
134. N. Oyama, Y. Ohnuki, T. Ohsaka, and H. Matsuda, *Nippon Kagaku Kaishi*, 949 (1983).
135. F. C. Anson, J.-M. Savéant, and K. Shigehara, *J. Phys. Chem.*, *87*, 214 (1983).
136. C. P. Andrieux, P. Hapiot, and J.-M. Savéant, *J. Electroanal. Chem.*, *172*, 49 (1984).
137. C. P. Andrieux, J. M. Dumas-Bouchiat, and J.-M. Savéant, *J. Electroanal. Chem.*, *169*, 9 (1984).
138. C. P. Andrieux and J.-M. Savéant, *J. Electroanal. Chem.*, *134*, 163 (1982).
139. J.-M. Savéant, *J. Phys. Chem.*, *92*, 1011 (1988).
140. J.-M. Savéant, *J. Electroanal. Chem.*, *201*, 211 (1986).
141. P. J. Peerce and A. J. Bard, *J. Electroanal. Chem.*, *122*, 97 (1980).
142. H. D. Abruña, P. Denisevich, M. Umana, T. J. Meyer, and R. W. Murray, *J. Am. Chem. Soc.*, *103*, 1 (1981).
143. M. W. Espenscheid and C. R. Martin, *J. Electroanal. Chem.*, *188*, 73 (1985).
144. K. Aoki, K. Tokuda, H. Matsuda, and N. Oyama, *J. Electroanal. Chem.*, *176*, 139 (1984).
145. B. J. Feldman, P. Burgmayer, and R. W. Murray, *J. Am. Chem. Soc.*, *107*, 872 (1985).
146. J. C. Jernigan, C. E. D. Chidsey, and R. W. Murray, *J. Am. Chem. Soc.*, *107*, 2824 (1985).
147. A. G. Ewing, B. J. Feldman, and R. W. Murray, *J. Electroanal. Chem.*, *172*, 145 (1984).
148. P. Denisevich, K. W. Willman, and R. W. Murray, *J. Am. Chem. Soc.*, *103*, 4727 (1981).
149. P. G. Pickup and R. W. Murray, *J. Electrochem. Soc.*, *131*, 833 (1984).

150. P. Burgmayer and R. W. Murray, *J. Electroanal. Chem.*, *147*, 339 (1983).
151. C. E. D. Chidsey and R. W. Murray, *Science*, *231*, 25 (1986).
152. M. S. Wrighton, *Science*, *231*, 32 (1986).
153. J. W. Thackeray, H. S. White, and M. S. Wrighton, *J. Phys. Chem.*, *89*, 5133 (1985).
154. H. W. White, G. P. Kittlesen, and M. S. Wrighton, *J. Am. Chem. Soc.*, *106*, 5375 (1984).
155. G. P. Kittlesen, H. S. White, and M. S. Wrighton, *J. Am. Chem. Soc.*, *106*, 7389 (1984).
156. E. W. Paul, A. J. Ricco, and M. S. Wrighton, *J. Phys. Chem.*, *89*, 1441 (1985).
157. J. O. Howell and R. M. Wightman, *Anal. Chem.*, *56*, 524 (1984).
158. J. O. Howell and R. M. Wightman, *J. Phys. Chem.*, *88*, 3915 (1984).
159. J. A. Gerhardt, A. F. Oke, J. Nagy, B. Moghaddam, and R. N. Adams, *Brain Res.*, *290*, 390 (1983).
160. P. Burgmayer and R. W. Murray, *J. Phys. Chem.*, *88*, 2515 (1984).
161. P. Burgmayer and R. W. Murray, *J. Am. Chem. Soc.*, *104*, 6139 (1982).
162. A. G. Ewing, B. J. Feldman, and R. W. Murray, *J. Phys. Chem.*, *89*, 1263 (1985).
163. C. E. D. Chidsey, B. J. Feldman, C. Lundgren, and R. W. Murray, *Anal. Chem.*, *58*, 145 (1986).
164. B. J. Feldman and R. W. Murray, *Anal. Chem.*, *58*, 2844 (1986).
165. M. Sharp, M. Petersson, and K. Edstrom, *J. Electroanal. Chem.*, *95*, 123 (1979).
166. M. Sharp, M. Petersson, and K. Edstrom, *J. Electroanal. Chem.*, *109*, 271 (1980).
167. L. Roullier and E. Laviron, *J. Electroanal. Chem.*, *157*, 193 (1983).
168. N. Oyama and F. C. Anson, *J. Electrochem. Soc.*, *127*, 247 (1980).
169. C. P. Andrieux, J. M. Dumas-Bouchiat, and J.-M. Savéant, *J. Electroanal. Chem.*, *142*, 1 (1982).
170. C. P. Andrieux and J.-M. Savéant, *J. Electroanal. Chem.*, *171*, 65 (1984).
171. J. Leddy, A. J. Bard, J. T. Maloy, and J.-M. Savéant, *J. Electroanal. Chem.*, *187*, 205 (1985).
172. K. Shigehara, N. Oyama, and F. C. Anson, *Inorg. Chem.*, *20*, 518 (1981).
173. Y. M. Tso and F. C. Anson, *J. Electrochem. Soc.*, *131*, 595 (1984).
174. N. Oyama and F. C. Anson, *Anal. Chem.*, *52*, 1192 (1980).
175. N. Oyama, T. Ohsaka, and T. Okajima, *Anal. Chem.*, *58*, 979 (1986).
176. N. Oyama, T. Shimomura, K. Shigehara, and F. C. Anson, *J. Electroanal. Chem.*, *112*, 271 (1980).
177. I. Rubinstein and A. J. Bard, *J. Am. Chem. Soc.*, *102*, 6641 (1980).
178. D. C. Bookbinder, J. A. Bruce, R. N. Dominey, N. S. Lewis, and M. S. Wrighton, *Proc. Natl. Acad. Sci. USA*, *77*, 6280 (1980).
179. T. P. Henning, H. S. White, and A. J. Bard, *J. Am. Chem. Soc.*, *103*, 3937 (1981).
180. J. A. Bruce and M. S. Wrighton, *J. Am. Chem. Soc.*, *104*, 74 (1982).
181. C. R. Martin, T. A. Rhoades, and J. A. Ferguson, *Anal. Chem.*, *54*, 1639 (1982).
182. T. Ohsaka, Y. Takahira, O. Hatozaki, and N. Oyama, *Bull. Chem. Soc. Jpn.*, *62*, 1023 (1989).
183. H.-R. Zumbrunnen and F. C. Anson, *J. Electroanal. Chem.*, *152*, 111 (1983).
184. R. Naegeli, J. Redepenning, and F. C. Anson, *J. Phys. Chem.*, *90*, 6227 (1986).
185. J. Redepenning and F. C. Anson, *J. Phys. Chem.*, *91*, 4549 (1987).
186. H. L. Yeager and A. Steck, *J. Electrochem. Soc.*, *128*, 1880 (1981).
187. B. Rodmacq, J. M. Coey, M. Escoubes, E. Roche, R. Duplessix, A. Eisenberg, and M. Pineri, in *Water in Polymers*, S. P. Rowland (Ed.), ACS Symposium Series No. 127, American Chemical Society, Washington, Chapter 29 (1980).
188. M. Fukui, A. Kitani, C. Degrand, and L. L. Miller, *J. Am. Chem. Soc.*, *104*, 28 (1982).
189. B. Lindholm and M. Sharp, *J. Electroanal. Chem.*, *198*, 37 (1986).
190. S. Kuwabata, Y. Maida, and H. Yoneyama, *J. Electroanal. Chem.*, *242*, 143 (1988).

191. T. Ikeda, C. R. Leidner, and R. W. Murray, *J. Am. Chem. Soc.*, *103*, 7422 (1981).
192. M. Delamar, M. C. Pham, P. C. Lacaze, and J. E. Dubois, *J. Electroanal. Chem.*, *108*, 1 (1980).
193. A. P. Brown and F. C. Anson, *Anal. Chem.*, *49*, 1589 (1977).
194. E. Laviron and L. Roullier, *J. Electroanal. Chem.*, *115*, 65 (1980).
195. E. Laviron, *J. Electroanal. Chem.*, *105*, 25 (1979).
196. E. Laviron, in *Electroanalytical Chemistry*, Vol. 12, A. J. Bard (Ed.), Marcel Dekker, New York, (1982) p. 53.
197. H. Matsuda, K. Aoki, and K. Tokuda, *J. Electroanal. Chem.*, *217*, 1 (1987).
198. H. Matsuda, K. Aoki, and K. Tokuda, *J. Electroanal. Chem.*, *217*, 15 (1987).
199. K. Daifuku, K. Aoki, K. Tokuda, and H. Matsuda, *J. Electroanal. Chem.*, *183*, 1 (1985).
200. T. Kakutani and M. Senda, *Bull. Chem. Soc. Jpn.*, *53*, 1942 (1980).
201. T. Kakutani and M. Senda, *Bull. Chem. Soc. Jpn.*, *54*, 884 (1981).
202. D. F. Smith, K. Wilman, K. Kuo, and R. W. Murray, *J. Electroanal. Chem.*, *95*, 217 (1979).
203. W. G. Albery, M. G. Boutelle, P. J. Colby, and A. R. Hillman, *J. Electroanal. Chem.*, *133*, 135 (1982).
204. K. W. Willman, R. D. Rocklin, R. Nowak, K.-N. Kuo, F. A. Schultz, and R. W. Murray, *J. Am. Chem. Soc.*, *102*, 7629 (1980).
205. T. P. Henning, H. S. White, and A. J. Bard, *J. Am. Chem. Soc.*, *104*, 5862 (1982).
206. W. T. Yap, R. A. Durst, E. A. Blubaugh, and D. D. Blubaugh, *J. Electroanal. Chem.*, *144*, 69 (1983).
207. R. Lange and K. Doblhofer, *J. Electroanal. Chem.*, *237*, 13 (1987).
208. J.-M. Savéant, *J. Electroanal. Chem.*, *227*, 299 (1987).
209. R. P. Buck, *J. Electroanal. Chem.*, *210*, 1 (1986).
210. R. P. Buck, *J. Electroanal. Chem.*, *219*, 23 (1987).
211. J.-M. Savéant, *J. Electroanal. Chem.*, *238*, 1 (1987).
212. J.-M. Savéant, *J. Electroanal. Chem.*, *242*, 1 (1988).
213. R. P. Buck, *J. Electroanal. Chem.*, *243*, 279 (1988).
214. M. E. G. Lyons, H. G. Fay, J. G. Vos, and A. J. Kelly, *J. Electroanal. Chem.*, *250*, 207 (1988).
215. V. D. Neff, *J. Electrochem. Soc.*, *125*, 886 (1978).
216. D. Ellis, M. Eckhoff, and V. D. Neff, *J. Phys. Chem.*, *85*, 1225 (1981).
217. V. D. Neff, *J. Electrochem. Soc.*, *131*, 1382 (1985).
218. K. Itaya, H. Akahoshi, and S. Toshima, *J. Electrochem. Soc.*, *129*, 1498 (1982).
219. K. Itaya, H. Ataka, and S. Toshima, *J. Am. Chem. Soc.*, *104*, 4767 (1982).
220. K. Itaya, I. Uchida, and S. Toshima, *J. Phys. Chem.*, *87*, 105 (1983).
221. K. Itaya, I. Uchida, S. Toshima, and R. M. De La Rue, *J. Electrochem. Soc.*, *131*, 2086 (1984).
222. K. Itaya, I. Uchida, and V. D. Neff, *Acc. Chem. Res.*, *19*, 162 (1986).
223. R. J. Mortimer and D. R. Rosseinsky, *J. Electroanal. Chem.*, *151*, 133 (1983).
224. R. J. Mortimer and D. R. Rosseinsky, *J. Chem. Soc. Dalton Trans.*, 2059 (1984).
225. A. L. Crumbliss, P. S. Lugg, D. L. Patel, and N. Morosoff, *Inorg. Chem.*, *22*, 3541 (1983).
226. A. L. Crumbliss, P. S. Lugg, and N. Morosoff, *Inorg. Chem.*, *23*, 4701 (1984).
227. S. Sinha, B. D. Humphrey, E. Fu, and A. B. Bocarsly, *J. Electroanal. Chem.*, *162*, 351 (1984).
228. S. Sinha, B. D. Humphrey, and A. B. Bocarsly, *Inorg. Chem.*, *23*, 203 (1984).
229. H. Rubin, B. D. Humphrey, and A. B. Bocarsly, *Nature* (*London*), *308*, 5957 (1984).
230. L. F. Schneemeyer, S. E. Spengler, and D. W. Murphy, *Inorg. Chem.*, *24*, 3044 (1985).
231. A. B. Bocarsly and S. Sinha, *J. Electroanal. Chem.*, *137*, 157 (1982).
232. A. B. Bocarsly and S. Sinha, *J. Electroanal. Chem.*, *140*, 167 (1982).

233. L. M. Siperko and T. Kuwana, *J. Electrochem. Soc.*, *130*, 396 (1983).
234. T. Ozeki, I. Watanabe, and S. Ikeda, *J. Electroanal. Chem.*, *236*, 209 (1987).
235. D. Shaojun and L. Fengbin, *J. Electroanal. Chem.*, *217*, 49 (1987).
236. M. Sharp, *J. Electroanal. Chem.*, *230*, 109 (1987).
237. E. T. T. Jones and L. R. Faulkner, *J. Electroanal. Chem.*, *222*, 201 (1987).
238. B. Londholm and M. Sharp, *J. Electroanal. Chem.*, *198*, 37 (1986).
239. P. M. Hoang, S. Holdcroft, and B. L. Hunt, *J. Electrochem. Soc.*, *132*, 2129 (1985).
240. L. Geng, R. A. Reed, M. Longmire, and R. W. Murray, *J. Phys. Chem.*, *91*, 2908 (1987).
241. Y. Morishima, I. Akihara, H. S. Lim, and S. Nozakura, *Macromolecules*, *20*, 978 (1987).
242. X. Chen, P. He and L. R. Faulkner, *J. Electroanal. Chem.*, *222*, 223 (1987).
243. R. W. Murray, in *Electroanalytical Chemistry*, Vol. 13, A. J. Bard (Ed.), Marcel Dekker, New York, p. 191 (1984).
244. A. R. Hillman, in *Electrochemical Science and Technology*, Vol. 1, R. G. Linford (Ed.), Elsevier Applied Science, Netherlands, (1987) p. 103.
245. L. R. Faulkner, *Chem. Eng. News*, Feb. 27, 28 (1984).
246. H. D. Abruña, in *Electroresponsive Molecular and Polymeric Systems*, Vol. 1, T. A. Skotheim (Ed.), Marcel Dekker, New York, (1988) p. 97.
247. For example, A. J. Bard and L. R. Faulkner, in *Electrochemical Methods, Fundamentals and Applications*, Wiley, New York (1980).
248. H. Matsuda, *Bull. Chem. Soc. Jpn.*, *53*, 3439 (1980).
249. R. S. Nicholson, *Anal. Chem.*, *37*, 1351 (1965).
250. E. Laviron, *J. Electroanal. Chem.*, *101*, 19 (1979).
251. H. Angerstein-Kozlowska and B. E. Conway, *J. Electroanal. Chem.*, *95*, 1 (1979).
252. I. Ruff, *Electrochim. Acta*, *15*, 1059 (1970).
253. I. Ruff and V. Friedrich, *J. Phys. Chem.*, *75*, 3297 (1971).
254. I. Ruff, V. Friedrich, and K. Csillag, *J. Phys. Chem.*, *76*, 162 (1972).
255. I. Ruff and V. Friedrich, *J. Phys. Chem.*, *76*, 2954 (1972).
256. H. Dahms, *J. Phys. Chem.*, *72*, 362 (1968).
257. K. Heckmann, in *Biomembranes*, Vol. 3, Manson, A. L. (Ed.), Plenum Press, New York, p. 127 (1972).

Chapter **IX**

MASS AND CHARGE TRANSPORT IN ELECTRONICALLY CONDUCTIVE POLYMERS

Charles R. Martin and Leon S. Van Dyke
Department of Chemistry, Colorado State University,
Fort Collins, Colorado

9.1 INTRODUCTION

Electronically conductive polymers are an exciting new class of materials with unique electronic, electrochemical, and optical properties. Because of these unusual and useful properties, electronically conductive polymers are the focus of a massive international research effort. The players in this research effort include physicists, physical chemists, materials scientists, synthetic organic chemists, engineers, and electrochemists. Indeed, in the preface of his recent monograph on conductive polymers, Skotheim states that "this is unarguably one of the most interdisciplinary fields of science today" (1). As a result of this large and interdisciplinary research effort, the scientific literature of conductive polymers is massive and diverse. Fortunately, a number of informative review articles and monographs have recently appeared (1–6).

One of the most interesting and potentially useful aspects of these polymers is that they can be reversibly "switched" between electronically insulating and electronically conductive states. This switching reaction involves either oxidation or reduction of a nonionic and electronically insulating parent polymer to

Molecular Design of Electrode Surfaces,
Edited by Royce W. Murray. Techniques of Chemistry Series, Vol. XXII.
ISBN 0-471-55773-0

form a conductive polycationic or polyanionic daughter polymer. The electrochemical oxidation process might be represented by

$$-\!\!(\mathrm{M})_y\!\!- + n\mathrm{X}_s^- \rightarrow -\!\!(\mathrm{M}^+\mathrm{X}^-)_n\!(\mathrm{M}_{(y-n)})\!\!- + ne^- \tag{9.1}$$

where M represents a monomer unit is the nonconductive form, M^+ is the corresponding oxidized unit in the conductive form, and X^- is an anion initially present in a contacting solution phase. Equation 9.1 shows that the electrochemical switching reaction involves a charge-transport process in which oxidized monomer sites and charge compensating counterions diffuse through the polymer film.

The oxidative switching reaction shown in Eq. 9.1 could also be accomplished using a chemical oxidizing agent (Ox).

$$-\!\!(\mathrm{M})_y\!\!- + n\mathrm{Ox} \rightarrow -\!\!(\mathrm{M}^+\mathrm{Ox}^-)_n\!(\mathrm{M}_{(y-n)})\!\!- \tag{9.2}$$

Equation 9.2 shows that the chemical switching reaction also incorporates a diffuse component but, in this case, it is the chemical oxidant that diffuses through the film.

The oxidation–reduction reactions (Eqs. 9.1 and 9.2) play an integral role in nearly all of the proposed technological applications of electronically conductive polymers. For example, one of the most widely publicized applications of these polymers is as electrode materials in secondary batteries (7). In this case, the forward direction in Eq. 9.1 would correspond to the charge reaction and the reverse direction in Eq. 9.1 the discharge reaction for the battery. Other proposed applications include use as electrochromic devices (8), where the reaction in Eq. 9.1 is responsible for the color change, and as transistors (9), where this reaction is responsible for turning the device on and off.

The rates of these redox reactions are of particular importance to the proposed technological applications of conductive polymers. For example, the rate of the redox reaction will determine the current density achieved by a conductive polymer battery and will determine the switching rate of a polymer-based electrochromic or electronic device. Therefore, one of the primary objectives of the conductive polymer research effort has been to quantitatively evaluate the rates of oxidation and reduction of these polymers. In general, the rate of charge transport in an electroactive polymer can depend on a number of factors including electron self-exchange between redox sites in the polymer film, diffusion and/or migration of charge-balancing ions in the film, extent of solvent intrusion into the film and, of course, polymer film morphology (10).

In spite of the importance of the redox rate, there has, as far as we know, been no comprehensive review of this particular aspect of electronically conductive polymer research. The objectives of this chapter are to provide a critical review of the methods used to investigate the rates of redox reactions of electronically conductive polymers and to assess the current status of this important aspect of conductive polymer research. As we shall see, both chemical and electrochemical methods have been used to investigate redox reaction rates in these polymers; both of these classes of methods will be reviewed here.

This chapter is organized as follows: First, we briefly review the synthesis of electronically conductive polymers. Synthesis is important because the synthetic procedure often determines the methods used to evaluate the redox reaction rate. We then present a critical review of *electrochemical* methods for evaluating the redox reaction rates of electrochemically conductive polymers. We expose methods that are unreliable or produce questionable results and identify procedures that should yield meaningful rate data.

Following the section on electrochemical techniques, we discuss *chemical* methods for assessing redox rates in electronically conductive polymers. This section includes methods that follow the doping and undoping of the polymer and also methods that measure diffusion coefficients at fixed polymer redox potentials. Again, we attempt to identify the methods that will yield reliable results.

We then give our assessment of what is currently known and what remains to be learned about the rates of redox reactions in conductive polymers. Section 9.3 also discusses various ancillary methods that do not directly provide the redox rate but yield information that is relevant to determination of the redox rate. Finally, in Section 9.6 we discuss methods for enhancing the rates of oxidation and reduction of electronically conductive polymers.

9.2 SYNTHESIS OF ELECTRONICALLY CONDUCTIVE POLYMERS

Electronically conductive polymers can be synthesized either electrochemically or chemically. Synthesis is an important issue because, from a historical perspective, chemical methods for evaluation of redox rates have been used for chemically synthesized electronically conductive polymers; in contrast, electrochemical methods have been used almost exclusively for the electrochemically synthesized polymers.

The most common electrochemically synthesized conductive polymers include poly(pyrrole) and its analogues, poly(thiophene) and its analogues, and poly(aniline) (11, 12). These polymers are usually synthesized from a solution of the corresponding monomer via electropolymerization (11); the polymer typically precipitates as a thin film onto the electrode surface. Because the synthetic procedure yields a polymer film-coated electrode, these polymers are ideally suited for subsequent investigations using electrochemical methods.

Poly(acetylene) is the most common chemically synthesized electronically conductive polymer. Poly(acetylene) is also the first and prototypical organic conductive polymer. Poly(acetylene) is usually synthesized via the direct polymerization of acetylene over a Ziegler catalyst (13). When synthesized at room temperature, a mixture of both *cis*- and *trans*-poly(acetylene) is formed (13). The polymer can be converted to the all-trans form (the thermodynamically stable form) by heating at 200°C.

Electron microscopy shows that poly(acetylene) has a fibrous morphology. As a result, the bulk density of poly(acetylene) film is about 0.4 $g\,cm^{-3}$, whereas the density obtained by flotation techniques is about 1.2 $g\,cm^{-3}$ (14). Thus, the polymer fibrils (fibrils are small fibers) fill only about 33% of the total film volume. As we shall see, this fibrous morphology greatly complicates evaluation of rates of oxidation or reduction of this polymer. Because poly(acetylene) is synthesized chemically, the majority of the redox rate data in the literature, for this polymer, have been generated using chemical methods.

Poly(acetylene) can also be synthesized via the so-called "Durham" method: this method involves the synthesis of a precursor polymer which, when heated, undergoes elimination to yield poly(acetylene) (15). In contrast to poly(acetylene) synthesized via the conventional Shirakawa technique (16), the Durham route yields a dense, amorphous, nonfibrous form of the polymer. The transport properties of the Durham and Shirakawa poly(acetylenes) are quite different (17). Poly(acetylene) can also be synthesized by the ring-opening metathesis polymerization method (18).

Finally, it is worth noting that poly(pyrrole), poly(thiophene), poly(aniline), and others of the electrochemically synthesized polymers can also be synthesized chemically (12, 19). The chemical synthesis entails replacing the electrode with a chemical oxidizing agent. In general, chemical synthesis yields powdery deposits rather than coherent films; thus electrochemical polymerization is usually the preferred synthetic route. We have recently shown, however, that coherent conductive polyheterocyclic fibers can be obtained by chemical synthesis of the polymer in a microporous host membrane (20).

9.3 ELECTROCHEMICAL METHODS FOR EVALUATING THE RATES OF REDOX REACTIONS OF CONDUCTIVE POLYMERS

As noted above, the electrochemically synthesized electronically conductive polymers usually precipitate as thin films that coat the electrode surface. Thus, the electrosynthesis produces a "polymer-modified electrode" (10, 21), a fact that has had tremendous (and unfortunate) impact on the way in which redox reaction rates for these polymers have been investigated.

Polymer-modified electrodes have, during the last decade, been of considerable interest to electrochemists (10, 21). Most of the research effort in this area has focused on redox polymers (10, 21); these polymers contain electroactive groups and can transport charge via electron self-exchange between these groups (10, 21). The relevance of this work to electronically conductive polymers is twofold. First, the self-exchange events are initiated by using the substrate electrode to oxidize or reduce the polymer; the net reaction is identical to the reaction shown in Eq. 9.1. Second, a number of relatively simple electrochemical methods have been devised to evaluate the rate of oxidation and reduction of redox polymers (21–26).

Large amplitude potential-step experiments are the most commonly used class of methods for evaluating the redox reaction rates of redox polymers (21–26). The easiest way to explain, in qualitative terms, large amplitude methods is via reference to an example—the large amplitude potential step method (27). In a large amplitude potential step method the potential of the working electrode is stepped from some value E_1, where no redox reaction occurs, to some value E_2, where the rate of the redox reaction is question in diffusion controlled. The term "large amplitude" derives from the fact that the magnitude of the potential step usually must be at least several hundred millivolts (mV). Large amplitude methods usually assume that linear diffusion to/from the electrode surface is extant. In general, one must know the area of the electrode surface and the concentration of the redox active species. Further details can be found in Ref. 27.

When large amplitude methods are applied to redox polymer films, it is usually assumed that charge is transported by diffusion of electrons or holes from site to site through the film (21–26). This diffusion process is assumed to be linear to, or from, the substrate electrode surface. The rate of this diffusional charge-transport process is expressed in terms of an "apparent diffusion coefficient," D_{app}. The D_{app} values associated with redox reactions in redox polymers have varied from as low as about $10^{-13}\,cm^2\,s^{-1}$ (28) to as high as $10^{-5}\,cm^2\,s^{-1}$ (29).

Because large amplitude electrochemical methods proved useful for evaluations of oxidation–reduction rates in redox polymers, it seemed obvious that these methods could be applied to electronically conductive polymers (30–45). For example, large amplitude potential-step experiments were used to obtain D_{app} values (or a derivative thereof) for poly(pyrrole) (30–32), for substituted poly(pyrroles) (33–35), for poly(thiophene) (36), for poly(3-methylthiophene) (37), for poly(aniline) (38–40), for composites of poly(pyrrole) with other polymers (41–43), and for various other conductive polymers (44, 45). Indeed, these large amplitude methods provided the majority of the experimental redox-rate data available in the literature to date.

It is now well established (although perhaps less well known) that the D_{app} values obtained from large amplitude potential-step experiments at conductive polymer film-coated electrodes are essentially meaningless (46, 47). Since this is a rather brazen statement, the corroborating evidence, both theoretical and experimental, is reviewed in the following paragraphs. We then discuss electrochemical methods which, in our opinion, do yield reliable D_{app} values for electronically conductive polymers.

Fritz Will (48–50) was the first to point out a major problem associated with electrochemical (and chemical) determinations of apparent diffusion coefficients in electronically conductive polymers. With regard to electrochemical methods, this problem can be stated in terms of a question: What is the active electrode area (48–50)? This is an important question because, as noted above, nearly all electrochemical methods for determination of D_{app} are dependent on an accurate value for the electrode area (51).

Most researchers who use large amplitude methods to evaluate D_{app} for values conductive polymers assume that the active electrode area is just the area of the substrate electrode (30–45). However, because the conductive state of the polymer resembles a porous metal (52, 53), this simplifying assumption is clearly not valid (46). Thus, as pointed out by Will, the active electrode area is, in most cases, unknown, and accurate determinations of D_{app} are impossible (48–50).

Murray and his co-workers pointed out a second problem associated with the application of conventional electrochemical methods to the determination of apparent diffusion coefficients in electronically conductive polymers (54). The total current in an electrochemical experiment is composed of the faradaic current and the capacitive (double-layer charging) current. The faradaic component must be isolated from the total current if D_{app} is to be determined. Unfortunately, because these polymers behave like porous electrodes, the capacitivelike (52, 53) component of the current often overwhelms the faradaic component (53, 54).

The magnitude of the capacitivelike (52, 53) current would not be a problem if a reliable means for discriminating against or substracting this component could be devised. We have pointed out, however, that when a large amplitude perturbation is applied to a conductive polymer, it is usually impossible to separate out the capacitive component (46). For example, assume that the initial potential in a potential-step experiment is such that the polymer is in its nonconductive form. The final potential will then be to the region where the polymer is in its conductive form. Thus, the polymer is converted from an insulator to a conductor during the course of the experiment. Because the polymer is converted from an insulator to a conductor, at some time during the experiment, a large capacitivelike current must flow to charge the double layer of the newly created "porous metallike" film (53). However, the time course of delivery of this capacitivelike current is unknown. Therefore, the traditional methods based on temporal discrimination (51, 55) against the capacitive component will not work with conductive polymers (46).

We have identified several other problems associated with the use of large amplitude methods for determination of D_{app} values in conductive polymers (46). First, these methods usually assume that the rate of the redox reaction is diffusion controlled, yet the experimental data (see below) suggest otherwise. Second, as noted above, these methods assume that diffusion is linear to the substrate electrode; however, when the film is in its conductive state, there is no reason to assume that linear diffusion is obtained (47, 56).

Finally, there is the "apples-to-oranges" problem (46, 54). The reduced form of the conductive polymer is usually a nonionic, hydrophobic, insulating organic polymer. In contrast, the oxidized form is usually a polycationic, hydrophilic, electronically conductive polymer. These are very different materials, and it would be highly unlikely that they would show the same charge and mass-transport characteristics. Nevertheless, a large amplitude method converts one form into the other during the course of the experiment (i.e., converts an apple to an orange) and attempts to ascribe a single rate parameter (D_{app}) to these very different materials.

We have presented a variety of arguments suggesting that large amplitude methods cannot be used to obtain reliable and meaningful apparent diffusion coefficients for redox reactions in electronically conductive polymers. These arguments would, however, be specious if the experimental data in the literature conformed to the predictions of the relevant theoretical model. In fact, the experimental data are almost always at odds with results predicted by the relevant electrochemical model.

For example, Genies et al. (31) show plots of charge versus the square root of time associated with chronocoulometric experiments at poly(pyrrole) film-coated electrodes. If the rate of the polymer redox reaction is diffusion controlled, these plots should be linear with zero or positive intercepts (55). In fact, the experimental plots have negative intercepts (31). Furthermore, the slopes of these plots (and therefore the value of D_{app} obtained) increased with the positive limit of the applied potential step (31). The negative intercept and the potential dependent slope observed by Genies et al. (31) suggest that the charge-transport reaction is activation rather than diffusion controlled (55).

Kaneto et al. (36) conducted analogous experiments on poly(thiophene). They, again, observed potential dependent D_{app} values, something that should not happen if the redox reaction is diffusion controlled. Kaneto et al. (36) suggest that either migration or activation effects cause the D_{app} values to vary with potential. The fact that poly(pyrrole) and poly(thiophene) contain large concentrations of electrolyte suggests that migration should not be a problem for these films (46, 57). Pickup and Osteryoung (47, 56) conclusively showed that migration occurs in the oxidized form of the polymer. In any event, it is difficult to ascribe a physical meaning to D_{app} values which, in direct contradiction to simple chronocoulometric or chronoamperometric theory (27), are potential dependent.

Reynolds et al. (35) recently reported some interesting results from potential-step experiments on poly(pyrrole) and a sulfonated derivative of poly(pyrrole). Instead of plotting chronocoulometric data in the conventional fashion [charge (Q) vs. $t^{1/2}$], Reynolds et al. (35) plotted these data as log Q versus log t; these plots were roughly linear but the slopes were rarely the theoretically predicted 0.5. Indeed, the slopes of the log–log plots were both potential and ion dependent. These data, again, provide compelling evidence that a simple linear diffusion model and simple chronoamperometric theory (27), are not applicable when large amplitude potential steps are applied to conductive polymers.

So far we have discussed methods that will not yield meaningful data for redox rates for conductive polymers. As indicated above, we chose to discuss these methods first because they are the most commonly used methods. The question now becomes: What methods will yield reliable redox rate data for conductive polymers?

It might first be useful to discuss, in general terms, the characteristics that a method for evaluating redox rates of conductive polymers should possess. In our opinion, the first prerequisite is that the method involve a small amplitude electrochemical perturbation (46). As noted above, large amplitude methods by nature convert the polymer from an insulator to a conductor, and these are very different materials. Thus, we believe that small amplitude methods, which cause only minor perturbations in polymer structure and that avoid the potential region where the polymer is switched, should be employed.

The second prerequisite is that a reliable theoretical model which accurately reflects the nature of the transport process and which accounts for or obviates the effects of slow heterogeneous electron transfer must be available. With regard to transport, if the polymer is in its reduced form, a linear diffusion model is probably appropriate (see below and Ref. 46). However, if the polymer is in its conductive state an alternative model will usually be required (47, 56).

Finally, the ideal method for evaluating the rates of redox reactions for conductive polymers would be easy to use and would be applicable to a variety of polymers. The latter point is important in that, as it stands now, it is difficult to make meaningful comparisons of redox rates for different conductive polymers. These data are essential if the best polymer for a particular technological application is to be identified.

Fritz Will described a current–pulse method, which satisfies most of these conditions (48–50). This current–pulse method was applied to poly(acetylene) and was initiated in a potential region where the poly(acetylene) was in its conductive state. The polymer remained in the conductive state throughout the duration of the experiment (48–50). It is of interest to note, however, that this method is based on a linear diffusion model. Remember that because conductive polymers behave as porous metals, a simple linear diffusion model is usually not applicable (see above). This model worked for the conductive form of poly(acetylene) because Will chose a solvent that did not solvate the polymer (48–50). As a result, only the external surface of the polymer was wetted; that is, the internal surfaces, generated via the inherent porosity of the material, were not wetted. Thus diffusion was linear to or from the external surface (48–50).

The above discussion indicates that Will's method, which is based on pioneering work by Paul Delahay and his co-workers (58), solves the electrode area, diffusion model, and apples-to-oranges problems. In addition, the rate of the diffusive component of the redox reaction (as described by D_{app}) can be separated from the rate of heterogeneous electron transfer (as described by the exchange current density). Indeed, Will was the first to present reliable kinetic data for redox reactions of electronically conductive polymers (48–50).

Unfortunately, it seems to us that it is difficult to apply Will's method to most electronically conductive polymers. For example, if the conductive form of a polymer is to be analyzed [as was the case in Will's analysis of poly(acetylene)] a solvent that does not solvate the polymer must be identified. [Again, this is necessary to ensure that only the external surface is wetted (see above).] Conventional electrochemical solvents (e.g., acetonitrile, water, *N*,*N*′-dimethylformamide, etc.) solvate most conductive polymers quite strongly and both internal and external surfaces are wetted. Thus, a more exotic (and at this time unknown) solvent must be used.

Furthermore, it seems unlikely that Will's method could be applied to the reduced (i.e., insulating) form of a conductive polymer because relatively large quantities of charge are injected into the polymer during this current-step experiment. Because it usually takes only minute quantities of charge to convert the reduced form of the polymer to an electronic conductor (59, 60), it seems likely that, if Will's method was applied to the reduced form, the apples-to-oranges (see above) problem would be encountered. Furthermore, while diffusion would be linear to the substrate electrode when the polymer is an insulator (very short times), diffusion would not be linear when the polymer is converted to a conductor (long times).

What reliable yet widely applicable methods are available for evaluating D_{app} values for electronically conductive polymers? In our opinion, the best methods currently available are ac impedance methods (61–64). These are inherently low amplitude methods so that only minor perturbations of the polymer are required (61–64). Reasonable theoretical models are available that allow for a relatively straightforward evaluation of D_{app} (65). Furthermore, the effects of heterogeneous electron transfer can be isolated from the diffusive component of the charge-transport process and both heterogeneous kinetic and D_{app} data can be obtained (66). Finally, in addition to D_{app} and kinetic data, a wealth of information about both the insulating and conducting forms of these polymers can be obtained from ac impedance methods (52, 64, 67–69).

Perhaps the most negative comment that can be made about the ac impedance methods is that they assume that the electrical response of the conducting polymer film-coated electrode is identical to that of a particular electrical circuit (the "equivalent circuit"). The reliability of the data obtained is determined by the degree to which the electrical response of the polymer system actually agrees with that of the equivalent circuit. While reasonable agreement has been observed in ac-based D_{app} determinations, some discrepancies have been noted (62, 63).

We have used an ac impedance method to evaluate D_{app} values for the reduced form of poly(pyrrole) (61). Rubinstein et al. (62) conducted analogous experiments on poly(aniline). Jow and Shacklette (63) used an analogous ac impedance method to determine D_{app} values in poly(acetylene). In addition, ac impedance methods made significant contributions toward an understanding of the fundamental electrochemical properties of electronically conductive polymers (52, 61–69).

In addition to the ac impedance methods, we recently described a low amplitude current–pulse method for determination of D_{app} values for electronically conductive polymers (46). Again, the low amplitude character of this method is important; the experiment is initiated with the polymer in the reduced (nonconductive) state and the polymer remains in this state throughout the duration of the experiment. Because the polymer is nonconductive, a finite linear diffusion model, based on heat transfer in a slab of finite thickness (70), is applicable (46). The D_{app} values are obtained by matching experimental and simulated E versus time transients (46).

The down side of this new current–pulse method is that it is experimentally tedious and time consuming. The tedium arises because of the need to experimentally define the relationship between the open circuit potential and the extent of oxidation of the polymer and because background capacitive contributions must be evaluated. [Note that because the polymer remains in the insulating state throughout the duration of this experiment, capacitance associated with the porous metal (conductive) form of the polymer is avoided.] Because of these experimental difficulties we believe that the ac impedance experiment (61–63) is currently the method of choice.

The above methods are valid for determining the diffusion coefficients in the neutral insulating form of electronically conductive polymers. However, as we have pointed out earlier, the conductive form of the polymer is a very different material and therefore a different model for the determination of diffusion coefficients will be necessary.

Osteryoung and Pickup and their co-workers developed such a model, which treats the polymer film as a porous electrode (47, 56). Potential-step chronoamperometry was used to investigate charge-transport in poly(pyrrole). However, instead of modeling the data by traditional Cottrell theory (51), which we have already shown will not work, they use a porous electrode model based on the pioneering work of Posey and Morozumi (71). In this model all of the charge is assumed to be capacitive and all of the current is carried by migration. This allows the polymer to be characterized in terms of a film ionic resistance and capacitance. If one knows (assumes) the number of ionic sites in the polymer a diffusion coefficient can be calculated from the ionic film resistance (47, 72).

It should be pointed out that this model only works if the polymer remains in its oxidized conductive state during the entire experiment (i.e., the potential is stepped from a potential where the polymer is conductive to a second potential where the polymer remains conductive). Thus this too meets our requirement of being a small amplitude technique. If a large amplitude potential step is used such that the polymer is switched from its conductive form to its nonconductive form the capacitive model does not accurately fit the data. Thus, this model does not allow accurate determination of the switching rate for electronically conductive polymer. However, since a considerable amount of the charge that these polymers store is capacitive in nature, the porous electrode model should prove very valuable. It also allows for calculation of ionic diffusion coefficients for the oxidized form of the polymer (47).

One disadvantage of the chronoamperometric technique is that it requires the independent determination of uncompensated solution resistance. We have recently developed a small amplitude current step experiment (73) in which the data are treated using the porous electrode model of Posey and Morozumi (71). The polymer is again characterized in terms of capacitance and ionic film resistance. One advantage of this technique over the chronoamperometric technique is that it does not require the independent determination of uncompensated solution resistance. We have used this technique to study thin poly(pyrrole) films in their oxidized conductive state. The porous electrode model fits the data extremely well when the polymer remains in its conductive state (73).

9.4 CHEMICAL METHODS FOR EVALUATION OF THE RATES OF REDOX REACTIONS OF CONDUCTIVE POLYMERS

The chemical methods for evaluating the rates of redox reactions in conductive polymers are based on following the rate of the chemical "doping" reaction shown in Eq. 9.2. These methods were used almost exclusively with poly(acetylene) which, as noted earlier, is a chemically synthesized conductive polymer. Chien (74), in his definitive text on poly(acetylene), reviews some of the early attempts to evaluate the rate of the doping reaction in this polymer. These methods usually involved investigations of the rates of diffusion of oxidants such as I_2 into poly(acetylene) films (75–78).

As discussed in detail by Chien (74), the fibrillar morphology of poly(acetylene) complicates analyses of these diffusion data. Because of this morphology, two diffusional processes must be considered, the relatively fast diffusion of the dopant into the interfibrillar space and the slower diffusion of the dopant into the fiber itself. Furthermore, the interfiber diffusion is complicated because diffusion is accompanied by chemical reaction of the dopant with the polymer chain and by the crystallinity present within the polymer fibers (74).

Because of these complications, Chien maintains that diffusion of dopant into poly(acetylene) can never be analyzed in terms of a single Fickian process (74). More importantly, Chien states that "one must come to the conclusion that determination of the diffusion constant for the chemical doping of poly(acetylene) is a futile exercise. The results are almost always without significance."

Many groups studied the chemical doping of poly(acetylene) since Chien's review (79–84). Some recent studies have taken into account the morphological complications pointed out by Chien (83, 84). The chemical doping of Durham poly(acetylene), which does not involve the morphological complications of Shirakawa poly(acetylene) has also been studied (17). However, diffusion was complicated by reaction of the dopant with the polymer and by the physical properties of the polymer including morphology, dopant–solvent swelling, and crystallinity (17). Furthermore, the diffusion coefficients measured for initial doping, undoping, and redoping often vary widely (17). We believe therefore that Chien's conclusions are essentially correct. Furthermore, Scrosatti and his co-workers (85) analyzed some of the models used for chemical doping studies and showed them to be based on invalid assumptions.

We believe, therefore, that a new model is necessary for the proper interpretation of chemical doping of conducting polymers. Reiss and his co-workers (86) recently developed such a model for the gas-phase doping of poly(thiophene) by iodine. The model is based on thermodynamically reversible trapping of the dopant and can account for the disparity in the time required to dope and undope conductive polymers using gas-phase dopants.

Reiss et al. (86) found that the gas-phase iodine doping of poly(thiophene) is thermodynamically reversible. They show that the disparity in diffusion rates between doping and undoping can be explained by a reversible trapping of the diffusants. In this model an individual dopant molecule in the polymer can be considered to be either "free" or "trapped". The effective diffusion coefficient (D_e) for the doping process will be equal to the diffusion coefficient of the free dopant D (the maximal value) when the concentration of free dopant greatly exceeds the equilibrium constant for trapping (α). However, if α is much greater than the concentration of free dopant, then $D_e = \alpha D$ (the minimal value).

During the doping process, the concentration of free dopant is high and therefore $D_e = D$. However, during the undoping process the concentration of free dopant at the polymer surface is low, producing a region at the polymer surface where $D_e = \alpha D$. This region at the film surface, where the rate of diffusion is low, limits the rate of the undoping process, which is often orders of magnitude slower than the doping process. This model has recently been improved and confirmed by a more rigorous theoretical treatment (87). This type of theoretical model is an important step toward developing a better understanding of the doping–undoping process in electronically conductive polymers. Reiss's extremely important work deserves an additional comment. It is of interest to note that the peak current for the electrochemical reduction of poly(pyrrole) is always lower than the peak current for the oxidation of the polymer. This may result from trapping as described by Reiss and his co-workers (86, 87).

The chemical methods for the determination of diffusion coefficients discussed above rely on the diffusing species to "dope" the electronically conductive polymer. As indicated earlier this complicates the situation because the dopant both diffuses and reacts. Several methods have been developed in which the oxidation state of the polymer is held constant; thus, the reaction step is eliminated. These methods are discussed below.

The first of these methods is based on ion self-exchange. Schlenoff and Chien (88) used this method to measure diffusion coefficients in p-doped poly(acetylene) and poly(pyrrole). For both polymers, the ion self-exchange kinetics can be modeled using finite, planar Fickian diffusion. As noted earlier, however, electronically conducting polymer films often consist of two phases: the polymer and solvent filled pores. For Fickian diffusion to hold, the polymer–solvent system must act as an "effective medium." That is to say, the diffusing ionic species must make an infinite number of transitions between the polymer phase and the solvent phase during the course of the experiment. Thus, the measured diffusion coefficient is a combination of slow diffusion through the polymer phase and rapid diffusion through the solution phase. Schlenoff and Chien (88) showed that the effective medium model is applicable for poly(pyrrole) and poly(acetylene).

Anion exchange in poly(pyrrole) was also studied by Reynolds and his co-workers (89). Poly(pyrrole) was electropolymerized in an acetonitrile solution containing the tosylate anion. The poly(pyrrole) film was exposed to aqueous solutions containing various electrolytes. The leaching out (anion exchange) of the tosylate anion into the contacting solution was measured spectrophotometrically. The data were treated using a model similar to that of Schlenoff and Chien. In general, the diffusion coefficients measured by Reynolds and his co-workers (89) were about three orders of magnitude lower than those measured by the self-exchange method of Schlenoff and Chien (88). This may be a result of the different polymerization conditions and ions used in the two experiments (88, 89).

Another chemical method for the determination of diffusion in electronically conductive polymers was developed by Burgmayer and Murray (54, 72). They studied the ionic permeability of poly(pyrrole) as a function of the oxidation state of the polymer (54, 72). In this method a poly(pyrrole) film was electrochemically deposited until it completely filled the holes of a gold minigrid electrode. This poly(pyrrole) film was used as a separator membrane in a two compartment cell. The fluxes of various ions across the membrane were determined by monitoring their solution concentrations (54). This experiment allowed for an assessment of the effect of polymer oxidation state on the rate of ion transport.

9.5 WHERE THE FIELD STANDS NOW AND RECOMMENDATIONS FOR FUTURE RESEARCH

Most of the quantitative evaluations of the redox rates of the electrochemically synthesized electronically conductive polymers were made using large amplitude electrochemical methods (30–45); this is unfortunate because, as discussed in detail in the preceding section, it is not clear that these data have any real significance. However, previous investigations illuminated a number of interesting features about the redox reactions of electronically conductive polymers; we review some of these features and make recommendations about directions of future research in the following paragraphs.

First, it has been clear from the very beginning that ion transport plays a significant role in determining the rate of redox reactions in electronically conductive polymers (90, 91). Unfortunately, the ion-transport issue is not as simple as the reaction shown in Eq. 9.1 would suggest. We and others show that the reduced form of poly(pyrrole) contains large concentrations of supporting electrolyte (46, 57). Therefore, when the polymer is oxidized, charge compensation can occur either by incorporation of an anion, expulsion of a cation, or both. We now discuss two techniques that can be used to resolve this issue.

The first technique is a radiotracer method in which the adsorption of radiolabeled ions into the polymer film is monitored as a function of oxidation state of the polymer (92, 93). Inzelt and Horyani (92) used this technique to study the adsorption of labeled SO_4^{2-} and Cl^- as a function of potential for a poly(pyrrole) coated electrode. Their results suggest that there is an excess of supporting electrolyte in both the oxidized and reduced forms of the polymer. They also found that approximately only one electrolyte anion was incorporated for every four electrons removed from the film. This suggests that cation as well as anion movement is important in maintaining charge neutrality in conductive polymer films (92). Similar results were found for poly(aniline) (93).

A second, very powerful technique, which has been used to clarify the "anion in versus cation out" issue, is the quartz crystal microbalance (QCM) technique (35, 94, 95). Kaufman et al. (94) were the first to apply the QCM to the study of electronically conductive polymers. They studied poly(pyrrole) films in a propylene carbonate-containing $LiClO_4$ electrolyte. They found that both Li^+ and ClO_4^- played a significant role in the charge-transport process (94).

Orata and Buttry (95) used the QCM to study ion transport in poly(aniline). They found that the charge was carried almost exclusively by the anion. Reynolds et al. (35) used the QCM to study a sulfonated derivative of poly(pyrrole), and found that the charge was carried almost exclusively by the cations. This result is to be expected since the film contained a high number of covalently attached anionic sulfonate sites. While the QCM is a powerful technique, a general caveat should be applied to its use with polymeric systems—the technique assumes that a film on top of the QCM is a rigid solid, which oscillates at the frequency of the quartz crystal. If the film is viscoelastic, this assumption is clearly not valid. Furthermore, if the film is dendritic or fibrillar, the QCM will only register a fraction of the weight of this film.

In addition to studying ion transport, the QCM has been used to study the polymerization of poly(aniline) (95) and poly(pyrrole) (96). Other electrogravimetric techniques have also been used to study the ion transport (97, 98) and surface wetting (99) of electronically conductive polymers. Indeed, electrogravimetric techniques seem uniquely qualified for determining which of the ions of the electrolyte is the charge carrying species during redox reactions in electronically conductive polymers. Thus, a carefully planned study involving the use of the QCM to determine which ion is moving and a reliable method for evaluation of D_{app} (e.g., ac impedance) to determine the rate of diffusion for this ion would quantitatively define the role of the electrolyte in the charge-transport process.

The study suggested above would involve a number of electrolytes so that the conditions under which cation versus anion transport predominates could be evaluated. It is well known, however, that the morphology of a conductive polymer film is dependent on the salt used during the synthesis of the polymer (11). Therefore, the best approach for investigating ion effects on transport might be to synthesized all polymers in a common salt and then use an ion-exchange technique to convert the polymer to the desired salt form (100). These electrolyte-induced morphology changes are, however, interesting and should be further explored (11).

A second question that needs to be answered before accurate models for charge transport in electronically conductive polymers can be developed is: What is the geometry of the redox process? That is, does the oxidation–reduction proceed from the electrode–polymer interface, the polymer–electrolyte interface, or occur uniformly throughout the polymer? Murray (54) previously discussed some of the factors that might affect this geometry. Knowledge of the geometry of this process is important for accurate modeling of diffusional processes in these films. Ellipsometric studies showed that the oxidation of poly(aniline) occurs uniformly throughout the polymer film (60,

101). Ellipsometric studies of poly(pyrrole), however, showed that the redox conversion of poly(pyrrole) proceeds from the polymer–electrolyte interface (102). Further ellipsometric studies are necessary to better understand the geometry of both the oxidation and reduction processes in electronically conductive polymers.

9.6 ENHANCING TRANSPORT IN ELECTRONICALLY CONDUCTIVE POLYMERS

As mentioned earlier, Section 9.6 of this chapter will deal with enhancing ion transport in electronically conductive polymers. In this section we discuss composites of electronically conductive polymers with anionic polyelectrolytes, "self-doped" conducting polymers, the use of molten salt electrolytes, and optimization of the supermolecular structure of conductive polymers for charge transport.

One of the earliest attempts to enhance ion transport in electronically conductive polymers was to form composites of electronically conductive polymers with anionic polyelectrolytes (41, 42, 103–110). The anionic sites of the polyelectrolyte act as the charge balancing counterions for the positively charged electronically conductive polymer (Eq. 9.1). However, in contrast with small anions that are mobile, the anionic sites on the polyelectrolyte are immobilized within the polymer composite; this forces charge within the composite to be carried by cations during the oxidation and reduction of the electronically conductive polymer (41, 42). Unfortunately, all attempts to measure the effect of changing from anion transport to cation transport have been made by unreliable large amplitude techniques (41, 42).

The concentrations of the electronically conductive polymer and anionic polyelectrolytes within the composite can be varied to produce anion or cation exchangers (107). These composites have also found use as water deionizers (108) and as electrodes in unique polymer batteries (109). Furthermore, these composites have superior mechanical properties when compared with traditional dopants of electronically conductive polymers traditional dopants (40, 41, 103–110).

A second modification of ion transport in electronically conductive polymers was the synthesis of the "self-doped" polymers (110). While the term "self-doped" is a misnomer, these polymers are interesting because the covalently attached anionic groups apparently insure that charge is carried exclusively by cations during both oxidation and reduction of the polymer (35). Furthermore, if the size of the cation is small, the rate of charge transport is enhanced (111). Further quantitative evaluations of charge-transport rates in these interesting polymers are in order.

Another interesting electronically conductive polymer containing fixed ionic sites was recently synthesized by Mas and Pickup (112, 113). As opposed to the "self-doped" polymers described above this polymer has a high concentration of positively charged quaternary ammonium sites. In water, the reduced form of this polymer has ionic permeabilities a thousand times greater than reduced poly(pyrrole) (112).

Ion transport in electronically conductive polymers is enhanced by the use of molten salt electrolytes (114–117). Osteryoung and his co-workers investigated the electrochemistry of electronically conductive polymers in a room temperature molten salt. The salt is a mixture of aluminum chloride and 1-methyl-3-ethyl imidazolium chloride. Because of the high ionic strength of the molten salt, charge transport is much faster than in conventional aqueous and nonaqueous solvents. These media may be useful in electronically conductive polymer batteries.

Finally, ion transport in electronically conductive polymers can be enhanced by controlling the supermolecular structure of the polymer. Osaka and his co-workers (118–120) developed a method for producing highly porous poly(pyrrole) films that have higher transport rates than conventional poly(pyrrole) films. This method is based on coating an electrode surface with an insulating nitrile butadiene rubber (NBR). The electrolyte for polymerization ($LiClO_4$ in acetonitrile) etches channels through the NBR; pyrrole is then polymerized in these channels. After polymerization the NBR is extracted away with ethyl methyl ketone to leave a free-standing highly porous poly(pyrrole) film. Poly(pyrrole) films produced in this manner show clearly enhanced charge-transport characteristics when compared to conventional poly(pyrrole) films. It was shown that these films had superior performance in lithium–poly(pyrrole) batteries (118). The batteries prepared via this process had a higher charge capacity and supported a higher current density than batteries prepared in the conventional manner.

We believe that the ideal polymer supermolecular structure for fast charge transport is one in which small diameter polymer fibrils are surrounded by solution filled pores. Polymer films with this supermolecular structure have several advantages over conventional polymer films. First, the solution filled pores become fast ion conducting channels into the film. Second, while counterions ultimately must enter or exit the polymer phase (Eq. 9.1), these ions only have to traverse the narrow radius of the fiber. Finally, transport into the polymer phase is changed from a linear diffusion process, in conventional polymer films, to a cylindrical process for the fibrillar film.

We developed a procedure for producing conductive polymers with this ideal fibrillar supermolecular structure (121–124). This procedure involves synthesis of the polymer into a host membrane that has linear cylindrical pores; these pores serve as templates for the nascent polymer. After the polymerization is completed, the host membrane is extracted away leaving isolated polymer fibers.

Figure 9.1 shows an SEM of poly(pyrrole) fibers prepared by this method (123, 124). It should be noted that the density of poly(pyrrole) fibers is very high. A high density of fibers is desirable for practical applications of conductive polymers (e.g., batteries), where it is desirable to have a maximum amount of polymer in a minimal amount of volume.

A large amplitude potential-step method was used to compare the rates of

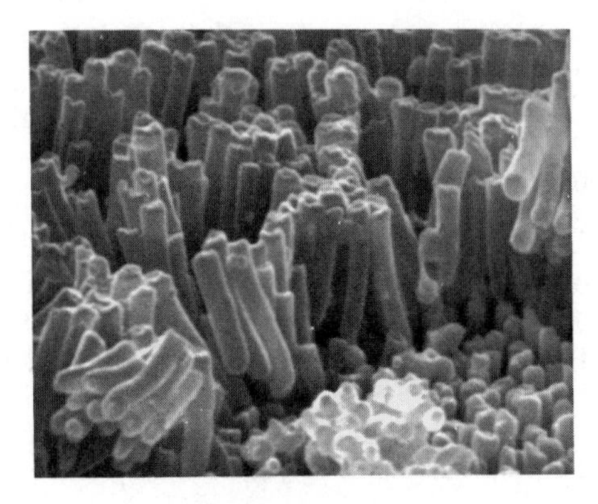

Figure 9.1. Scanning electron micrograph of 0.2-μm diameter poly(pyrrole) fibers. [Reprinted with permission from Ref. 124.]

reduction of the fibrillar and conventional poly(pyrrole) films. In this method the poly(pyrrole) film was first equilibrated at a potential where the film was quantitatively oxidized; then the film was stepped to a potential where the film is quantitatively reduced. The charge–time transient associated with the reduction of the polymer film was recorded. Because the reduction of the film is driven to completion, the charge–time transient ultimately reaches a plateau value (e.g., Fig. 9.2). The time required to achieve 95% of this plateau charge (t_{95}) was used as the qualitative measure of the rate of the reduction process. Comparisons of

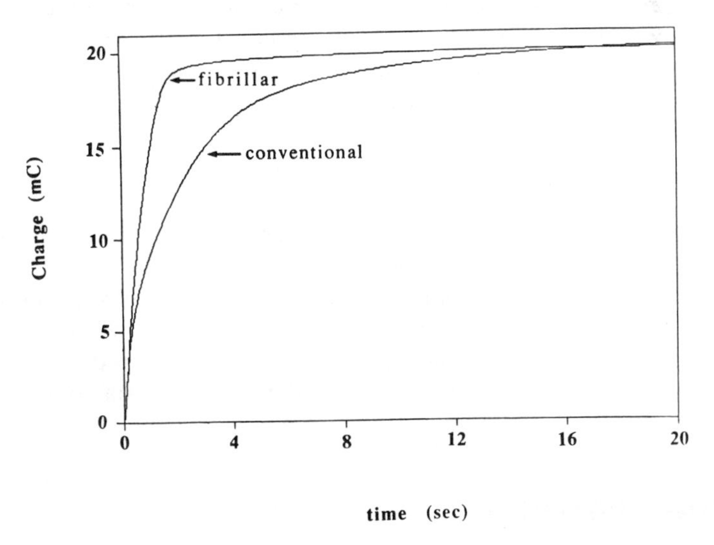

Figure 9.2. Charge versus time transients for the reduction of a conventional 1.0-μm thick conventional poly(pyrrole) film and a fibrillar poly(pyrrole) film. [Reprinted with permission from Ref. 124.]

Table 9.1. The t_{95} Values Associated with the Reduction of Various Poly(pyrrole) Films

Quantity of Poly(pyrrole) (μmol *of pyrrole*)	t_{95} (s)[a]	
	Conventional	*Fibrillar*
0.44	4.0 ± 1.2	1.7 ± 0.4
0.88	6.5 ± 0.4	2.7 ± 1.6
1.31	7.2 ± 0.8	3.7 ± 0.5
1.75	9.3 ± 0.8	5.9 ± 2.1

[a]Time required to reduce the film to 95% of maximum charge value.

t_{95} values will be valid only if the films contain the same amount of polymer (123, 124).

Figure 9.2 shows the charge–time transients for a conventional poly(pyrrole) film, and a fibrillar poly(pyrrole) film, which contains the same amount of polymer. The rate of reduction for the fibrillar film is significantly higher than for the conventional film (Fig. 9.2). This point is reinforced by the t_{95} data shown in Table 9.1. These data clearly indicate that the rate of reductive charge transport is faster in the fibrillar films than in the conventional films. It is also worth noting that in addition to having higher charge-transport rates, we have recently shown that extremely narrow conductive polymer fibers have much higher electronic conductivities than the corresponding conventional polymer films (20). We also showed that hollow tubules of electronically conductive polymers can be prepared using our microporous host membrane method (125). Such tiny tubules should have a myriad of technological applications (126).

9.7 CONCLUSIONS

The journal *Science* recently published an article entitled "Conductive Polymers Recharged" (127). This article discusses the many potential applications of conductive polymers and indicates that commercialization of conductive polymers has been achieved. The prospect for additional commercial applications combined with the inherently fascinating properties of these materials insures that conductive polymers will be an important research field in the 21st century.

ACKNOWLEDGMENTS

This work was supported by the Air Force Office of Scientific Research and the Office of Naval Research.

REFERENCES

1. T. A. Skotheim (Ed.), *Handbook of Conducting Polymers*, Marcel Dekker, New York, 1986.
2. G. K. Chandler and D. Pletcher, *Electrochemistry Volume 10*, The Royal Society of Chemistry, London, 1985.
3. A. F. Diaz, J. F. Rubinson, and H. B. Mark, Jr., in *Advances in Polymer Science*, Springer-Verlag, Berlin, 1988.
4. J. O'M. Bockris and D. Miller, in *Conducting Polymers*, L. Alacer (Ed.), Riedel, Dortrecht, 1987.
5. J. I. Kroschwitz (Ed.), *Electrical and Electronic Properties of Polymers: A A State-of-the-Art Compendium*, Wiley, New York, 1988.
6. J. R. Reynolds, *Chemtech.* 440 (1988).
7. A. G. MacDiarmid and M. R. Maxfield, in *Electrochemical Science and Technology of Polymers*, Vol. 1, R. G. Linford (Ed.), Elsevier, London, 1987.
8. M. Gazard, in *Handbook of Conducting Polymers*, T. A. Skotheim (Ed.), Marcel Dekker, New York, 1986.
9. G. P. Kittlesen, H. S. White, and M. S. Wrighton, *J. Am. Chem. Soc.*, **106**, 7389 (1984).
10. R. W. Murray, *Annu. Rev. Mater. Sci.*, **14**, 145 (1984).
11. A. F. Diaz and J. Bargon, in *Handbook of Conducting Polymers*, T. A. Skotheim (Ed.), Marcel Dekker, New York, 1986.
12. G. Tourillion, in *Handbook of Conducting Polymers*, T. A. Skotheim (Ed.), Marcel Dekker, New York, 1986.
13. J. C. W. Chien, *Polyacetylene, Chemistry, Physics, and Material Science*, Academic, Orlando, FL, 1984.
14. A. G. Macdiarmid and A. J. Heeger, *Synth. Metals*, **1**, 101 (1979).
15. J. H. Edwards, J. Feast, and D. C. Bott, *Polymer*, **25**, 395 (1984).
16. T. Ito, H. Shirakawa, and S. Ikeda, *J. Polym. Chem. Ed.*, **12**, 11 (1974).
17. P. J. S. Foot, F. Mohammed, P. D. Calvert, and N. C. Billingham, *J. Phys. D Appl. Phys.*, **20**, 1354 (1987).
18. T. M. Swager and R. H. Grubbs, *J. Am. Chem. Soc.*, **111**, 4413 (1989).
19. G. B. Street, in *Handbook of Conducting Polymers*, T. A. Skotheim (Ed.), Marcel Dekker, New York, 1986.
20. Z. Cai and C. R. Martin, *J. Am. Chem. Soc.*, **111**, 4138 (1989).
21. R. W. Murray, in *Electroanalytical Chemistry Volume 13*, A. J. Bard (Ed.), Marcel Dekker, New York, 1984.
22. P. J. Peerce and A. J. Bard, *J. Electroanal. Chem.*, **114**, 89 (1980).
23. C. R. Martin, I. Rubenstein, and A. J. Bard, *J. Am. Chem. Soc.*, **104**, 4817 (1982).
24. C. R. Martin and K. A. Dollard, *J. Electroanal. Chem.*, **159**, 127 (1983).
25. N. Oyama and F. C. Anson, *J. Electrochem. Soc.*, **127**, 247 (1980).
26. N. Oyama, T. Shimomura, K. Shigehara, and F. C. Anson, *J. Am. Chem. Soc.*, **103**, 2552 (1981).
27. A. J. Bard and L. R. Faulkner, *Electrochemical Methods Theory and Applications*, Wiley, New York, Chapter 5 (1980).
28. P. Daum, J. R. Lenhard, D. Rolison, and R. W. Murray, *J. Am. Chem. Soc.*, **102**, 4649 (1980).
29. K. Wilbourne and R. W. Murray, *J. Phys. Chem.*, **92**, 3642 (1988).
30. P. Mirebeau, *J. Physique, Supp. C3*, **6**, 579 (1983).
31. E. M. Genies, B. Bidan, and A. F. Diaz, *J. Electroanal. Chem.*, **149**, 101 (1983).
32. E. M. Genies and J. M. Pernault, *Synth. Metals*, **10**, 117 (1984/1985).
33. E. M. Genies, A. A. Syed, and M. Salmon, *Synth. Metals*, **11**, 353 (1985).

34. N. S. Sunderaresan, S. Basak, M. Pomerantz, and J. R. Reynolds, *J. Chem. Soc. Chem. Commun.*, 621 (1987).
35. J. R. Reynolds, N. S. Sudaresan, M. Pomerantz, S. Basak, and C. K. Baker, *J. Electroanal. Chem.*, **250**, 355 (1988).
36. K. Kaneto, H. Agawa, and K. Yoshino, *J. Appl. Phys.*, **61**, 1197 (1987).
37. P. Marque, J. Roncali, and F. Garnier, *J. Electroanal. Chem.*, **218**, 107 (1987).
38. N. Oyama, Y. Ohnuki, K. Chiba, and T. Ohsaka, *Chem. Lett.*, 1759 (1983).
39. J.-C. Lacroix and A. F. Diaz, *Macromol. Chem. Macromol. Symp.*, **8**, 17 (1987).
40. J.-C. Lacroix and A. F. Diaz, *J. Electrochem. Soc.*, **135**, 1361 (1988).
41. G. Nagasubramanian, S. Di Stefano, and J. Moacanin, *J. Phys. Chem.*, **90**, 4447 (1986).
42. H. Yoneyama, T. Hirai, S. Kuwabata, and O. Ikeda, *Chem. Lett.*, 1243 (1986).
43. P. Novak, O. Inganas, and R. Bjorklund, *J. Electrochem. Soc.*, **134**, 1341 (1987).
44. H. Yashima, M. Kobayashi, K.-B. Lee, D. Chung, A. J. Heeger, and F. Wudl, *J. Electrochem. Soc.*, **134**, 46 (1987).
45. N. Oyama, K. Hirabayashi, and T. Ohsaka, *Bull. Chem. Soc. Jpn.*, **59**, 2071 (1986).
46. R. M. Penner, L. S. Van Dyke, and C. R. Martin, *J. Phys. Chem.*, **92**, 5274 (1988).
47. C. D. Paulse and P. G. Pickup, *J. Phys. Chem.*, **92**, 7002 (1988).
48. F. G. Will, *J. Electrochem. Soc.*, **132**, 743 (1985).
49. F. G. Will, *J. Electrochem. Soc.*, **132**, 2093 (1985).
50. F. G. Will, *J. Electrochem. Soc.*, **132**, 2351 (1985).
51. A. J. Bard and L. R. Faulkner, *Electrochemical Methods: Theory and Applications*, Wiley, New York, p. 143 (1980).
52. R. A. Bull, F.-R. Fan, and A. J. Bard, *J. Electrochem. Soc.*, **129**, 1009 (1982).
53. S. W. Feldburg, *J. Am. Chem. Soc.*, **106**, 4671 (1984).
54. P. Burgmayer and R. W. Murray, in *Handbook of Conducting Polymers*, T. A. Skotheim (Ed.), Marcel Dekker, New York, 1986.
55. A. J. Bard and L. R. Faulkner, *Electrochemical Methods: Theory and Applications*, Wiley, New York, p. 200 (1980).
56. P. G. Pickup and R. A. Osteryoung, *J. Electroanal. Chem.*, **195**, 271 (1985).
57. Q.-Z. Zho, C. J. Kolaskie, and L. L. Miller, *J. Electroanal. Chem.*, **223**, 283 (1987).
58. T. Berzins and P. Delahay, *J. Am. Chem. Soc.*, **77**, 6448 (1955).
59. B. J. Feldman, P. Burgmayer, and R. W. Murray, *J. Am. Chem. Soc.*, **107**, 872 (1985).
60. S. Gottesfeld, A. Redondo, and S. W. Feldberg, *J. Electrochem. Soc.*, **134**, 271 (1987).
61. R. M. Penner and C. R. Martin, *J. Phys. Chem.*, **93**, 984 (1989).
62. I. Rubenstein, E. Sabantani, and J. Rishpon, *J. Electrochem. Soc.*, **134**, 3078 (1987).
63. T. R. Jow and L. W. Shacklette, *J. Electrochem. Soc.*, **135**, 541, (1988).
64. S. H. Glarum and J. H. Marshall, *J. Electrochem. Soc.*, **134**, 142 (1987).
65. C. Ho, I. D. Raistick, and R. A. Huggins, *J. Electrochem. Soc.*, **134**, 142 (1987).
66. M. J. van der Sluijs, A. F. Underhill, and B. N. Zaba, *J. Phys. D Appl. Phys.*, **20**, 1411 (1987).
67. T. Osaka, K. Naoi, S. Ogana, and S. Nakamura, *J. Electrochem. Soc.*, **134**, 2096 (1987).
68. J. Tanguy, N. Mermilliod, and M. Hoclet, *J. Electrochem. Soc.*, **133**, 1073 (1986).
69. J. Tanguy, N. Mermilliod, and M. Hoclet, *J. Electrochem. Soc.*, **134**, 795 (1987).
70. J. S. Carslaw and J. C. Jager, *Conduction of Heat in Solids*, Oxford University, London, 1959.
71. F. A. Posey and T. Morozumi, *J. Electrochem. Soc.*, **113**, 176 (1966).
72. P. Burgmayer and R. W. Murray, *J. Phys. Chem.*, **88**, 2515 (1984).
73. Z. Cai and C. R. Martin, in preparation.

74. J. C. W. Chien, *Polyacetylene, Chemistry, Physics, and Material Science*, Academic, Orlando, Fl, pp. 349–355, 1984.
75. H. Kiess, W. Meyer, D. Baeriswyl, and G. Harbeke, *J. Electron. Mater.*, **9**, 763 (1980).
76. P. Bernier, F. Schue, J. Sledz, M. Rolland, and L. Biral, *Chem. Scr.*, **17**, 151 (1981).
77. C. Benoit, M. Rolland, M. Aldissi, A. Rossi, M. Cadene, and P. Benier, *Phys. Status Solidi*, **68**, 209 (1981).
78. J. P. Louboutin and F. Beniere, *J. Phys. Chem. Solids*, **43**, 233 (1982).
79. T. Danno, K. Miyasaka, and K. Ishikawa, *J. Polym. Sci. Polym. Phys. Ed.*, **21**, 1527 (1983).
80. S. Pekker, M. Bellec, X. Le Cleac'h, and F. Beniere, *Synth. Metals*, **9**, 475 (1984).
81. S.-A. Chen and W.-C. Chan, *Mol. Cryst. Liq. Cryst.*, **117**, 117 (1985).
82. C. Riekel, H. W. Hasslin, K. Menke, and S. Roth, *Mol. Cryst. Liq. Cryst*, **117**, 117 (1985).
83. F. Radchi, P. Bernier, E. Faulqes, S. Lefrant, nd F. Schue, *J. Chem. Phys.*, **80**, 6285 (1984).
84. B. Francois, C. Mathis, and R. Nuffer, *Synthetic Metals*, **20**, 311 (1987).
85. D. D. Perlmutter and B. Scrosatti, *Solid State Ionics*, **27**, 115 (1988).
86. D. U. Kim, H. Reiss, and H. M. Raboeny, *J. Phys. Chem.*, **92**, 2673 (1988).
87. W. D. Murphy, H. M. Rabeony, and H. Reiss, *J. Phys. Chem.*, **92**, 7007 (1988).
88. J. B. Schlenoff and J. C. W. Chien, *J. Am. Chem. Soc.*, **109**, 6269 (1987).
89. E. W. Tsai, T. Pajkossy, K. Rajeshwar, and J. R. Reynolds, *J. Phys. Chem.*, **92**, 3560 (1988).
90. A. F. Diaz, J. I. Castillo, J. A. Logan, and W.-Y. Lee, *J. Electroanal. Chem.*, **129**, 115 (1981).
91. M. Salmon, A. F. Diaz, A. J. Logan, M. Krounbi, and J. Bargon, *Mol. Cryst. Liq. Cryst.*, **83**, 265 (1982).
92. G. Inzelt and G. Horanyi, *J. Electroanal. Chem.*, **230**, 257 (1987).
93. G. Inzelt and G. Horanyi, *Electrochim. Acta*, **33**, 947 (1988).
94. J. H. Kaufman, K. K. Kanazawa, and G. B. Street, *Phys. Rev. Lett.*, **53**, 2461 (1984).
95. D. Orata and D. A. Buttry, *J. Am. Chem. Soc.*, **109**, 3574 (1987).
96. C. K. Baker and J. R. Reynolds, *J. Electroanal. Chem.*, **251**, 307 (1988).
97. K. Okabayashi, F. Goto, K. Abe, and T. Yoshida, *Synthetic Metals*, **18**, 365 (1987).
98. K. Okabayashi, F. Goto, K. Abe, and T. Yoshida, *J. Electrochem. Soc.*, **136**, 1986 (1989).
99. M. A. Habib, *Langmuir*, **4**, 1302 (1988).
100. L. S. Curtin, G. C. Komplin, and W. J. Pietro, *J. Phys. Chem.*, **92**, 12 (1988).
101. A. Redondo, E. A. Ticianelli, and S. Gottesfeld, *Mol. Cryst. Liq. Cryst.*, **160**, 185 (1988).
102. C. Lee, J. Kwak, and A. J. Bard, *J. Electrochem. Soc.*, **136**, 3720 (1989).
103. N. Bates, M. Cross, R. Lines, and D. Walton, *J. Chem. Soc. Chem. Commun.*, 871 (1985).
104. R. M. Penner and C. R. Martin, *J. Electrochem. Soc.*, **133**, 310 (1986).
105. F.-R. Fan and A. J. Bard, *J. Electrochem. Soc.*, **133**, 301 (1986).
106. P. Aldebert, P. Audebert, M. Armand, G. Bidan, and M. Pineri, *J. Chem. Soc. Chem. Commun.*, 1636 (1986).
107. T. Iyoda, A. Ohtani, T. Shimidzu, and K. Honda, *Chem. Lett.*, 687 (1986).
108. T. Shimidzu, A. Ohtani, and K. Honda, *J. Electroanal. Chem.*, **251**, 323 (1988).
109. T. Shimidzu, A. Ohtani, T. Iyoda, and K. Honda, *J. Chem. Soc. Chem. Commun.* 327 (1987).
110. G. Bidan, B. Ehui, and M. Lapkowski, *J. Phys. D. Appl. Phys.*, **21**, 1043 (1988).
111. Y. Ikenoue, J. Chiang, A. O. Patil, F. Wudl, and A. J. Heeger, *J. Am. Chem. Soc.*, **110**, 2983 (1988).
112. H. Mao and P. G. Pickup, *J. Phys. Chem.*, **93**, 6480 (1989).
113. H. Mao and P. G. Pickup, *J. Electroanal. Chem.*, **265**, 127 (1989).
114. P. G. Pickup and R. A. Osteryoung, *J. Am. Chem. Soc.*, **106**, 2294 (1984).

115. J. F. Oudard, R. D. Allendoerfer, and R. A. Osteryoung, *J. Electroanal. Chem.*, **241**, 231 (1988).
116. T. A. Zawodzinski, Jr., L. Janiszewska, and R. A. Osteryoung, *J. Electroanal. Chem.*, **225**, 111 (1988).
117. L. Janiszewska and R. A. Osteryoung, *J. Electrochem. Soc.*, **134**, 2787 (1987).
118. K. Naoi, A. Ishijima, and T. Osaka, *J. Electroanal. Chem.*, **217**, 203 (1987).
119. K. Naoi and T. Osaka, *J. Electrochem. Soc.*, **134**, 2479 (1987).
120. T. Osaka, K. Naoi, M. Maeda, and S. Nakamura, *J. Electrochem. Soc.*, **136**, 1385 (1989).
121. R. M. Penner and C. R. Martin, *J. Electrochem. Soc.*, **133**, 2206 (1986).
122. C. R. Martin, R. M. Penner, and L. S. Van Dyke, in *Functional Polymers*, D. E. Bergbreiter and C. R. Martin (Eds.), Plenum, New York, 1989.
123. L. S. Van Dyke and C. R. Martin, *Langmuir*, **6**, 1118 (1990).
124. L. S. Van Dyke and C. R. Martin, *Synth. Metals*, **36**, 275 (1990).
125. C. R. Martin, L. S. Van Dyke, Z. Cai, and W. Liang, *J. Amer. Chem. Soc.*, **112**, 8976 (1990).
126. R. Pool, *Science*, **247**, 1410 (1990).
127. J. Alder, *Science*, **246**, 208 (1989).

INDEX